KU-332-798

WITHDRAWN

WITHDRAWN

WITHDRAWN

Radio Network Planning and Optimisation for UMTS

Radio Network Planning and Optimisation for UMTS

Second Edition

Edited by

Jaana Laiho and Achim Wacker
Both of Nokia Networks, Nokia Group, Finland

Tomáš Novosad
Nokia Networks, Nokia Group, USA

JOHN WILEY & SONS, LTD

Copyright © 2006 John Wiley & Sons Ltd, The Atrium, Southern Gate, Chichester,
West Sussex PO19 8SQ, England

Telephone (+44) 1243 779777

Email (for orders and customer service enquiries): cs-books@wiley.co.uk
Visit our Home Page on www.wiley.com

All Rights Reserved. No part of this publication may be reproduced, stored in a retrieval
system or transmitted in any form or by any means, electronic, mechanical, photocopying,
recording, scanning or otherwise, except under the terms of the Copyright, Designs and
Patents Act 1988 or under the terms of a licence issued by the Copyright Licensing Agency
Ltd, 90 Tottenham Court Road, London W1T 4LP, UK, without the permission in writing of
the Publisher. Requests to the Publisher should be addressed to the Permissions Department,
John Wiley & Sons Ltd, The Atrium, Southern Gate, Chichester, West Sussex PO19 8SQ,
England, or emailed to permreq@wiley.co.uk, or faxed to (+44) 1243 770620

Designations used by companies to distinguish their products are often claimed as trademarks.
All brand names and product names used in this book are trade names, trademarks or registered
trademarks of their respective owners. The Publisher is not associated with any product or vendor
mentioned in this book.

This publication is designed to provide accurate and authoritative information in regard to
the subject matter covered. It is sold on the understanding that the Publisher is not engaged
in rendering professional services. If professional advice or other expert assistance is
required, the services of a competent professional should be sought.

Other Wiley Editorial Offices

John Wiley & Sons, Inc., 111 River Street, Hoboken, NJ 07030, USA

Jossey-Bass, 989 Market Street, San Francisco, CA 94103-1741, USA

Wiley-VCH Verlag GmbH, Boschstr. 12, D-69469 Weinheim, Germany

John Wiley & Sons Australia Ltd, 42 McDougall Street, Milton, Queensland 4064, Australia

John Wiley & Sons (Asia) Pte Ltd, 2 Clementi Loop #02-01, Jin Xing Distripark, Singapore 129809

John Wiley & Sons Canada Ltd, 22 Worcester Road, Etobicoke, Ontario, Canada M9W 1L1

Wiley also publishes its books in a variety of electronic formats. Some content that appears
in print may not be available in electronic books.

Library of Congress Cataloging-in-Publication Data

Radio network planning and optimisation for UMTS / edited by
  Jaana Laiho, Achim Wacker, Tomáš Novosad
    p. cm.
  ISBN-13: 978-0-470-01575-9 (alk. paper)
  ISBN-10: 0-470-01575-6 (alk. paper)
  1. Global system for mobile communications. 2. Radio – Transmitters
and transmission. 3. Code division multiple access. I. Laiho, Jaana.
II. Wacker, Achim. III. Novosad, Tomáš. IV. Title: Radio network
planning and optimization for UMTS.
TK5103.483.R34 2005
621.384 – dc22   2005018248

British Library Cataloguing in Publication Data

A catalogue record for this book is available from the British Library

ISBN-13 978-0-470-01575-9 (HB)
ISBN-10 0-470-01575-6 (HB)

Project management by Originator, Gt Yarmouth, Norfolk (typeset in 10/12pt Times).
Printed and bound in Great Britain by Antony Rowe Ltd, Chippenham, Wiltshire.
This book is printed on acid-free paper responsibly manufactured from sustainable forestry
in which at least two trees are planted for each one used for paper production.

BAR CODE No.   5371791
X
CLASS No.
621·384 LAi
BIB
CHECK   — 1 JUN 2006   PROO
CHECK
ELW   FINAL
OS   SYSTEM NO.
LOAN CATEGORY NL

Contents

Preface

Second-generation (2G) mobile communication systems have enabled voice traffic to go wireless. More important, however, have been the accompanying standardisation, compatibility and international transparency that were simply not available to telecommunications equipment of the previous analogue generation. These features have helped 2G systems to spread rapidly around the world, with very high cellular phone penetration rates in many countries. Cellular networks have enabled certain types of communication to take place on a massive scale that previously were not possible or were at least severely limited. In the field of network building and expansion the main advances have been in planning the radio and transmission part of the network and in optimising the processes and activities necessary to run existing operational networks.

The third-generation (3G) system known as the Universal Mobile Telecommunications System (UMTS) introduces very variable data rates on the air interface, as well as the independence of the radio access infrastructure and the service platform. For users this makes available a wide spectrum of circuit-switched or packet data services through the newly developed high bit rate radio technology named Wideband Code Division Multiple Access (WCDMA). The variable bit rate and variety of traffic on the air interface have presented completely new possibilities for both operators and users, but also new challenges to network planning and optimisation.

This book gives detailed descriptions of the radio network planning and optimisation of UMTS networks based on Frequency Division Duplex (FDD) WCDMA technology up to Release 5 of the 3GPP standardisation work – i.e., one main enhancement of this second edition is the inclusion of High-speed Downlink Packet Access (HSDPA). One chapter is dedicated to the General Packet Radio System (GPRS) and Time Division Duplex (TDD) access mode of WCDMA. The optimisation and Quality of Service (QoS) aspects have, however, a wider scope, than in (W)CDMA radio technology only.

Chapter 1 introduces the history of cellular telecommunication and the changes in planning and operation of such networks. The challenges of network planning, optimisation and operation the operators and the wireless industry are facing on the way to 3G systems are introduced together with an outlook on future developments in the area towards fourth-generation (4G) systems.

Chapter 2 is in three sections. The first introduces the general background of Spread Spectrum Systems. This is followed by a section related to the Third Generation Partnership Project (3GPP), giving a panoramic view of the UMTS architecture, interfaces and functions that impact directly upon radio network planning. HSDPA physical layer properties are added as a new content in 3GPP Release 5. The third section discusses WCDMA-specific link performance indicators relevant for radio network dimensioning and planning.

Chapter 3 treats WCDMA radio network planning as a wider process that includes network dimensioning with a special section for HSDPA, detailed planning, requirements for planning tools, algorithms used for calculations in WCDMA and optimisation of the radio network plan. The relationship between network dimensioning, detailed network planning and dynamic network simulation is also discussed. The chapter closes with a discussion on cell deployment strategies with respect to the number of frequencies and the network structure. This topic is presented as a case study.

Chapter 4 covers Radio Resource Management (RRM) from the point of view of radio resource utilisation, including power control, handover control, congestion control (admission control, load control and packet scheduling), resource management and certain impacts of those functions upon network performance. A separate new section is devoted to RRM for HSPDA.

In Chapter 5, first the background noise measurement along with measured results are introduced. This part is followed by co-planning issues involving WCDMA and the Global System for Mobile communication (GSM), eventually other technologies. The third part of the chapter describes the effects of intersystem interference, together with dynamic mobile station receiver properties on network performance. The application of these methods and results is not, however, limited to the GSM–WCDMA scenario.

Chapter 6 treats various coverage and capacity enhancement techniques (beamforming, higher order receive diversity, transmit diversity, MIMO technology, mast head amplifiers, repeaters, rollout optimised configuration, sectorisation, etc.). The chapter is based on an extensive set of case studies and contains practical examples and conclusions.

Chapter 7 introduces the concept of statistical optimisation and discusses 3GPP Release 5 contributions in the management area including configuration and performance management issues. Furthermore, the TeleManagement Forum enhanced Telecom Operations Map (TMF eTOM) model is briefly introduced. A 3GPP management model for the multi-vendor environment is addressed. The management system's role in optimisation is presented and examples of management level products and their capabilities are provided.

Chapter 8 focuses on UMTS QoS mechanisms according to 3GPP Release 5 and examples of practical realisations of the QoS capabilities in network elements are introduced. Furthermore, QoS as a differentiation enabler for operators is demonstrated and differentiation possibilities with the QoS concept are presented. The optimisation loop expansion from the network layer to the service layer is described.

Chapter 9 is devoted to advanced analysis methods and automated optimisation. Several new analysis methods for network performance analysis are introduced. In the area of automated optimisation examples of optimisation logic are provided for the mobility management area, admission decision optimisation and capacity optimisation in UMTS networks.

Finally, Chapter 10 deals with two technologies that are different from the FDD mode of WCDMA. The first is the GPRS branch in GSM technology. This has brought variable rate packet data traffic into the air interface of originally circuit switched and single data rate service-oriented technology. The second, the Time Division Duplex (TDD) mode of WCDMA, represents an interesting technology for high data rate indoor users. Therefore, the radio performance properties of TDD mode are introduced.

On the *CD accompanying the first edition* of this book we included a static radio network simulator implemented in Matlab® together with detailed descriptions of the algorithms used. Most of the simulated scenarios are added, but not all the values presented can be reproduced exactly, since simulations have been done partly by using earlier versions of the tool, which used slightly different strategies. The tool is delivered in its current version and state, and the authors do not give any warranty concerning the correctness of the code. In addition, some coloured figures – in PDF format – are included. The simulator, its description and the figures can now be found at *www.wiley.com/go/laiho*.

The book is targeted at wireless operators, network and terminal manufacturers, university students, frequency regulation bodies and all those interested in radio network planning and optimisation, especially network systems RF engineering professionals. This book represents the views and opinions of the authors, which are not necessarily those of their employers.

Acknowledgements

The editors would like to acknowledge the effort and time invested by colleagues, both from Nokia and outside, who have contributed to this book. Apart from the editors, the contributors were Pauli Aikio, Simon Browne, Ted Buot, Markus Djupsund, Pauliina Erätuuli, Josef Fuhl, Jochen Grandell, Kari Heiska, Jyri Hämäläinen, Albert Höglund, Ann-Louise Johansson, Chris Johnson, Petri Jolma, Tero Kola, Anneli Korteniemi, Martin Kristensson, Mikko Kylväjä, Mats Larsson, Nilmini Lokuge, Peter Muszynski, Roman Pichna, Terhi Rautiainen, Jussi Reunanen, Vilho Räisänen, Kari Sipilä, Jussi Sipola, David Soldani, Kimmo Terävä, Mikko Toivonen, Kimmo Valkealahti, Pekko Vehviläinen and Juha Ylitalo.

The editors would like to thank Tero Ojanperä and Peter Muszynski for initial review of the first edition. During the development of the second edition many of our colleagues from various Nokia sites offered support and help in suggesting improvements, finding errors or providing figures or editorial advice. The editors would like to express their gratitude especially to Kati Ahvonen, Erkka Ala-Tauriala, Renaud Cuny, Outi Hiironniemi, Zhi-Chun Honkasalo, Salla Huttunen, Christian Joergensen, Janne Keränen, Mika Kiikkilä, Outi Keski-Oja, Pekka Kohonen, Thomas Lammert, Jani Lakkakorpi, Joni Lehtinen, Klaus Rasmussen, Mikko Rinne, Juha Räsänen, Anna Sillanpää, Kristian Skinne, Antti Toskala and Werner Trapp.

The publishing team at John Wiley & Sons, Ltd led by Mark Hammond, has done an outstanding job in the production of this book. We are especially grateful to Sarah Hinton for her patience, guidance and assistance.

We would like to express special thanks to our employer, Nokia Networks, for general permission, support and encouragement, and for providing some of the illustrations.

We also wish to acknowledge the effort of our colleagues from the Optimizer and Network System Research teams as well as from planning services, for their practical work in 3G planning studies conducted in a number of cities and environments around the world and for their valuable input from the field.

Last, but not least, we would like to say a big thank-you to our families and friends, as well as those of all the authors and reviewers, for their patience and support throughout this project.

The editors and authors welcome any comments and suggestions for improvement or changes that could be implemented in possible future editions.

Jaana Laiho, Achim Wacker and Tomáš Novosad
Espoo, Finland and Dallas, Texas

Abbreviations

16QAM	16 State Quadrature Amplitude Modulation
2G	2nd Generation
2.5G	2.5th Generation
3G	3rd Generation
3GPP	3rd Generation Partnership Project
3GPP2	3rd Generation Partnership Project 2
4G	4th Generation
8-PSK	8 Phase Shift Keying
AAL2	ATM Adaptation Layer type 2
Abis	GSM Interface BTS–BSC
AC	Admission Control
ACI	Adjacent Channel Interference
ACIR	Adjacent Channel Interference power Ratio
ACK	ACKnowledgement
ACLR	Adjacent Channel Leakage power Ratio
ACP	Adjacent Channel Protection
ACS	Adjacent Channel Selectivity
AGCH	Access Grant CHannel
AI	Acquisition Indicator
AICH	Acquisition Indicator CHannel
ALCAP	Access Link Control Application Part
AM	Acknowledged Mode
AMC	Adaptive Modulation and Coding
AMPS	Advance Mobile Phone Service
AMR	Adaptive Multi Rate
AP	Access Point; Access Preamble
AP-AICH	Access Preamble Acquisition Indicator CHannel
API	Application Programming Interface
APN	Access Point Name
APP	APPlication specific functions
ARP	Allocation Retention Priority
ARQ	Automatic Repeat reQuest
AS	Access Slot; Access Stratum
ASC	Access Service Class
AST	Active Session Throughput

ASU	Active Set Update
ATM	Asynchronous Transfer Mode
AVI	Actual Value Interface
AWGN	Additive White Gaussian Noise
AXC	ATM Cross Connect
B(T)S	Base (Transceiver) Station
BA	BCCH Allocation
BB	BaseBand
BCC	Base station Colour Code
BCCH	Broadcast Control CHannel
BCH	Broadcast CHannel
BCS	Binary Coded Signalling
BEP	Bit Error Probability
BER	Bit Error Rate
BFN	Node B Frame Number
BLER	BLock Error Rate
BM	Business Management
BMC	Broadcast/Multicast Control
BMU	Best Matching Unit
BPSK	Binary Phase Shift Keying
BSC	Base Station Controller
BSIC	Base Station Identity Code
BSS	Base Station Subsystem
BSSMAP	Base Station System Management Application Part
BTFD	Blind Transport Format Detection
BYE	Session termination
C/I	Carrier-to-Interference ratio
C_ID	Cell IDentification
C450	Analogue second-generation system in Germany
CAPEX	CAPital EXpenditure
CART	Classification And Regression Tree
CB	Cell Broadcast
CBR	Call Block Ratio
CC	Call Control; Convolutional Coding; Cumulative Counter
CCCH	Common Control CHannel
CCH	Control CHannel
CCPCH	Common Control Physical CHannel
CCTrCH	Coded Composite Transport CHannel
CD	Collision Detection
CD/CA-ICH	Collision Detection/Channel Assignment Indicator CHannel
CD-DSMA	Collision Detection-Digital Sense Multiple Access
CDF	Cumulative Distribution/Density Function
CDMA	Code Division Multiple Access
CFN	Connection Frame Number
CGI	Cell Global Identification
CI	Cell Identity

CIO	Cell Individual Offset
CM	Compressed Mode; Configuration Management
CMIP	Common Management Information Protocol
CN	Core Network
CNAME	Canonical NAME
CORBA	Common Object Request Broker Architecture
COST	European COoperation in the field of Scientific and Technical research
C-plane	Control plane
CPCH	Common Packet CHannel
CPICH	Common PIlot CHannel
CQI	Channel Quality Indicator
CRC	Cyclic Redundancy Check
CRMS	Common Resource Management Server
CRNC	Controlling RNC
CRRR	Capacity Request Rejection Ratio
CRS	Cell Resource Server
CS	Coding Scheme; Circuit Switched
CSI	Channel State Information
CSICH	CPCH Status Indicator CHannel
CSSR	Call Setup Success Ratio
CSW	Circuit SWitched (GPRS terminology)
CTCH	Common Traffic CHannel
CWND	Congestion WiNDow
D-AMPS	Digital AMPS
DCA	Dynamic Channel Allocation
DCCH	Dedicated Control CHannel
DCH	Dedicated CHannel
DCN	Data Communication Network
DCR	Drop Call Ratio
DCS1800	Digital Cellular System (GSM) at 1800 MHz band
DER	Discrete Event Registration
DGPS	Differential GPS
DHCP	Dynamic Host Client Protocol
DHO	Diversity HandOver
DiffServ	Differentiated Services
DL	DownLink
DN	Distinguished Name
DNS	Domain Name Server
DoA	Direction of Arrival
DOFF	Default OFFset
DPCCH	Dedicated Physical Control CHannel
DPCH	Dedicated Physical CHannel
DPDCH	Dedicated Physical Data CHannel
DQPSK	Differential QPSK
DRNC	Drifting RNC

DRX	Discontinuous Reception
DS	Direct Sequence
DSCH	Downlink Shared CHannel
DSCP	DiffServ Code Point
DSL	Digital Subscriber Line
DSMA-CD	Digital Sense Multiple Access–Collision Detection
DSTTD-SGRC	Double STTD with Sub-Group Rate Control
DTCH	Dedicated Traffic CHannel
DTX	Discontinuous Transmission
D-TxAA	Double Transmit Antenna Array
DVB	Digital Video Broadcasting
E1	Standard 2 Mbps transmission line
E3G	Enhanced 3G
EDGE	Enhanced Data rates for GSM Evolution
EFR	Enhanced Full Rate
EGPRS	Enhanced GPRS
EIA	Electronic Industry Alliance
EIRP	Equivalent Isotropic Radiated Power
EM	Element Manager
ERC	European Radiocommunications Committee
ES	Enterprise Systems
eTOM	Enhanced TOM
ETSI	European Telecommunications Standards Institute
FACH	Forward Access CHannel
FAUSCH	FAst Uplink Signalling CHannel
FBI	FeedBack Information
FCC	Federal Communications Commission
FCS	Frame Check Sequence
FDD	Frequency Division Duplex
FDMA	Frequency Division Multiple Access
FEC	Forward Error Correction; Forwarding Equivalence Class
FER	Frame Erasure Rate
FH	Frequency Hopping
FIFO	First In First Out
FM	Fault Management
FN	Frame Number
FP	Frame Protocol
FTP	File Transfer Protocol
FW	Firmware
G	Geometry factor
GAUGE	(Dynamic variable), used when data being measured can vary up or down during the period of measurement
GB	Guaranteed Bit rate
Gbps	Giga bits per seconds
GERAN	GSM EDGE RAN
GGSN	Gateway GPRS Support Node

GIS	Geographical Information System
GMM	GPRS MM
GMSK	Gaussian Minimum Shift Keying
GP	Guard Period
GPIB	General Purpose Interface Bus
GPRS	General Packet Radio Service
GPS	Global Positioning System
GRX	GPRS Roaming Exchange
GSM	Global System for Mobile communication
GSM1900	GSM at 1900 MHz band
GTP	GPRS Tunnel Protocol
GUI	Graphical User Interface
GW	GateWay
H-ARQ	Hybrid ARQ
HC	Handover Control
HCS	Hierarchical Cell Structure
HD	Harmonic Distortion
HDTV	High Definition TeleVision
HHO	Hard HO
HLR	Home Location Register
HLS	Higher Layer Scheduling
HO	HandOver
HSCSD	High-speed Circuit Switched Data
HSDPA	High-speed Downlink Packet Access
HS-DPCCH	High-speed Dedicated Physical Control CHannel (UL)
HS-DSCH	High-speed DSCH
HS-PDSCH	High-speed Physical DSCH
HS-SCCH	High-speed Shared Control CHannel (DL)
HSUPA	High-speed Uplink Packet Access
HTML	Hyper Text Markup Language
HTTP	Hyper Text Transfer Protocol
HW	HardWare
ID	IDentifier
IE	Information Element
IEE	The Institution of Electrical Engineers
IEEE	The Institute of Electrical and Electronics Engineers
IETF	Internet Engineering Task Force
IF-HO	Inter-Frequency HO
IIP	Input Intercept Point
IM	Information Management
IMAP	Internet Message Access Protocol
IMD	Inter-Modulation Distortion
IMEI	International Mobile station Equipment Identity
IMS	IP Multimedia Sub-system
IMSI	International Mobile Subscriber Identity
IMT	International Mobile Telecommunications

IntServ	Integrated Services
IOC	Information Object Class
IP	Internet Protocol
IPv4	IP version 4
IPv6	IP version 6
IR	Incremental Redundancy
IRP	Interface Reference Point
IS	Interim Standard (US)
IS-136	North American TDMA
IS-54	North American TDMA Digital Cellular
IS-95	North American Version of the CDMA Standard
ISCP	Interference Signal Code Power
ISDN	Integrated Services Digital Network
IS-HO	Inter-system HO
ISM	Industrial, Science, Medical (free RF band, at 2.4 GHz)
ISO	International Organisation for Standardisation
ISP	Internet Service Provider
IT	Information Technology
Itf-N	Interface-N
ITU-T	International Telecommunication Union, Telecommunication Standardisation Sector
Iu	Interconnection point between an RNC and a core network
Iub	Interface between an RNC and a Node B
Iur	Logical interface between two RNCs
JTACS	Japan TACS
kbps	Kilo bits per second
KDD	Knowledge Discovery in Database
KPI	Key Performance Indicator
KQI	Key Quality Indicator
ksps	Kilo symbols per second
L1	OSI Layer 1: Physical Layer
L2	OSI Layer 2: Radio Data Link Layer
L3	OSI Layer 3: Radio Network Layer
LA	Link Adaptation; Location Area
LAC	Location Area Code
LAN	Local Area Network
LC	Load Control
LCS	LoCation-based Services
LDAP	Lightweight Directory Access Protocol
LF	Load Factor
LLC	Logical Link Control
LLOS	Link LOSs
LNA	Low-Noise Amplifier
LoCH	Logical CHannel
LOS	Line Of Sight
LSA	Localised Service Area

MAB	Minimum Allowed Bitrate
MAC	Medium Access Control
Mbps	Mega bits per second
MCC	Mobile Country Code
MCL	Minimum Coupling Loss
Mcps	Mega chips per second
MCS	Modulation and Coding Scheme
MDC	Macro Diversity Combining
ME	Managed Element
MEHO	Mobile-Evaluated HO
MHA	Mast Head Amplifier
MIB	Management Information Base
MIM	Management Information Model
MIMO	Multiple Input Multiple Output
MISO	Multiple Input Single Output
MM	Mobility Management
MMS	Multimedia Message Service
MMUSIC	Multiparty MUltimedia SessIon Control
MNC	Mobile Network Code
MOC	Managed Object Class
MOI	Managed Object Instance
MPLS	Multi-Protocol Label Switching
MRC	Maximal Ratio Combining
MS	Mobile Station
MSC	Mobile Switching Centre
MT	Mobile Terminal
MTU	Maximum Transfer Unit
MVNO	Mobile Virtual Network Operator
N/A	Not Available; Not Applicable
N_PDU	Network Level PDU
NACK	Negative ACK
NAS	Non-Access Stratum
NB	NarrowBand
NBAP	Node B Application Part
NCC	Network Colour Code
NCx	Network Control (NC0, NC1, NC2)
NE	Network Element
NEHO	Network Evaluated HO
NF	Noise Figure
NGB	Non-GB
NLOS	Non-LOS
NM	Network Management; Network Manager
NMS	Network Management System
NMT	Nordic Mobile Telephone
Node B	WCDMA BS
NP	Network Performance; Non-Prioritised

NR	Noise Rise; Network Resource
NRM	Network Resource Model
NRT	Non-Real Time
NSS	Networking Sub-System
NW	NetWork
OCNS	Other Cell Noise Source
OFDM	Orthogonal Frequency Division Multiplexing
OH	Okumura–Hata
OMC	Operations and Maintenance Centre
OMG	Object Management Group
OPEX	OPerating EXpenditure
OS	Operations System
OSF	Operations System Functions
OSI	Open Systems Interconnection
OSS	Operations Support System
OVSF	Orthogonal Variable Spreading Factor
PACCH	Packet Associate Control CHannel
PAGCH	Packet Access Grant CHannel
PARC	Per-Antenna Rate Control
PBCCH	Packet Broadcast Control CHannel
PC	Power Control
PCCCH	Packet Common Control CHannel
PCCH	Paging Control CHannel
P-CCPCH	Primary CCPCH
PCH	Paging CHannel
PCMCIA	PC Modular Computer Interface Adapter card
PCPCH	Physical CPCH
P-CPICH	Primary CPICH
PCS	Personal Communications Systems
P-CSCF	Proxy Call State Control Function
PCU	Packet Control Unit
PDC	Pacific Digital Cellular
PDCH	Packet Data Channel
PDCP	Packet Data Convergence Protocol
PDF	Policy Decision Function; Probability Density Function
PDP	Packet Data Protocol
PDSCH	Physical DSCH
PDTCH	Packet Data Traffic CHannel
PDU	Protocol Data Unit
PG	Processing Gain
PHB	Per-Hop Behaviour
PHY	PHYsical layer
PI	Paging Indicator; Performance Indicator
PICH	Paging Indicator CHannel
PLMN	Public Land Mobile Network
PM	Performance Management

PN	PseudoNoise
PO	Power Offset
PoC	Push (to talk) over Cellular
PPCH	Packet Paging CHannel
PPP	Point-to-Point Protocol
PQ	Packet Queuing
PRACH	Physical RACH
PS	Packet Switched; Packet Scheduler
PSC	Primary Synchronisation Code
P-SCH	Primary Synchronisation CHannel
PSK	Phase Shift Keying
PSTN	Public Switched Telephone Network
PTT	Push To Talk
PU	Payload Units
PU^{2}RC	Per-User Unitary Rate Control
QM	Quality Manager; Quality Management
QoE	Quality of end-user Experience
QoS	Quality of Service
QPSK	Quadrature/Quaternary Phase Shift Keying
R	Refresh timer
RA	Routing Area
RAB	Radio Access Bearer
RAC	Routing Area Code
RACH	Random Access CHannel
RAI	Routing Area Identifier
RAKE	special receiver type used in CDMA
RAM	Radio Access Mode
RAN	Radio Access Network
RANAP	Radio Access Network Application Part
RAT	Radio Access Technique
RAU	Routing Area Update
RB	Radio Bearer
RC-MPD	Rate Control Multi-Path Diversity
RDN	Relative Distinguished Name
RF	Radio Frequency
RFC (IETF)	Request For Comments
RFN	RNC Frame Number
RL	Radio Link
RLC	Radio Link Control
RLCP	Radio Link Control Protocol
RM	Resource Manager
RMSS	Receiver Maximum Segment Size
RN	Radio Network
RNAS	RAN Access Server
RNC	Radio Network Controller
RNP	Radio Network Planning

RNS	Radio Network Subsystem
RNSAP	Radio Network Subsystem Application Part
RNTI	Radio Network Temporary Identity
ROC	Rollout Optimised Configuration
RR	Radio Resource; Receiver Report; Resource Request
RRC	Radio Resource Control; Route Raised Cosine
RRI	Radio Resource Indicator
RRM	Radio Resource Management
RRP	Radio Resource Priority
RRU	Radio Resource Utilisation
RSCP	Received Signal Code Power
RSSI	Received Signal Strength Indicator
RSVP	Resource ReSerVation Protocol
RT	Real-Time
RTCP	Real-Time Control Protocol
RTO	Roundtrip Time Out
RTP	Real-time Transport Protocol
RTT	Round-Trip Time
RTTVAR	Round-Trip Time VARiation
RTVS	Real Time Video Streaming
RU	Resource Unit
RWND	Receiver WiNDow
Rx	Receive
RxD	Receive Diversity
SA	Spectrum Analyser; Service Area
SACK	Selective ACK
SAI	Service Area Identifier
SAP	Service Access Point
SBLP	Service-Based Local Policy
S-CCPCH	Secondary CCPCH
SCH	Synchronisation CHannel
S-CPICH	Secondary CPICH
SCTP	Stream Control Transmission Protocol
SDCCH	Standalone Dedicated Control CHannel
SDES	Sender DEScription items
SDH	Synchronous Digital Hierarchy
SDP	Session Description Protocol
SDU	Service Data Unit
SF	Spreading Factor
SFN	System Frame Number
SGSN	Serving GPRS Support Node
SHO	Soft HO
SI	Status Inspection
SIGTRAN	SIGnalling TRANsport
SIM	Subscriber Identity Module
SIMO	Single Input Multiple Output

SINR	Signal-to-Interference and Noise Ratio
SIP	Session Initiation Protocol
SIR	Signal to Interference Ratio
SLA	Service Level Agreement
SLP	Service Logic Program
SM	Session Management; Service Management
SMG	Special Mobile Group
SMS	Short Message Services
SMSS	Sender Maximum Segment Size
SMTP	Simple Message Transfer Protocol
SNDCP	Subnetwork Dependent Convergence Protocol
SNR	Signal-to-Noise Ratio
SOM	Self-Organising Map
S-PARC	Selective PARC
SQM	Service Quality Manager
SR	Sender Report
SRB	Signalling RB
SRNC	Serving RNC
SRNS	Serving RNS
SRTT	Smoothed Round-Trip Time
SS	Spread Spectrum; Supplementary Services
SSC	Secondary Synchronisation Code
S-SCH	Secondary SCH
SSDT	Site Selection Diversity Technique
SSRC	Synchronisation SouRCe (identifier)
SSTRESH	Slow Start ThRESHold
STm-1	Synchronous Transport Module-1: An ITU-T-defined SDH physical interface for digital transmission in ATM at the rate of 155.52 Mbps
STTD	Space Time Transmit Diversity
SW	SoftWare
SWIS	See What I See
T1	1.544 Mbps Transmission Link
TACS	Total Access Communication System
TB	Transport Block
TBF	Temporary Block Flow
TBS	Transport Block Set
TC	Transmission Convergence
TCH	Traffic CHannel
TCP	Transmission Control Protocol
TDD	Time Division Duplex
TDM	Time Division Multiplex
TDMA	Time Division Multiple Access
TE	Terminal Equipment
TF	Transport Format
TFC	Combination

TFCI	Transport Format Combination Indicator
TFCS	Transport Format Combination Set
TFI	Transport Format Indicator
TFS	Transport Format Set
TFT	Traffic Flow Template
TGCFN	Transmission Gap Connection Frame Number
TGD	Transmission Gap start Distance
TGL	Transmission Gap Length
TGPL	Transmission Gap Pattern Length
TGPRC	Transmission Gap Pattern Repetition Count
TGSN	Transmission Gap starting Slot Number
THP	Traffic Handling Priority
TIA	Telecommunications Industry Association
TM	Transparent Mode; Telecom Management
TMF	TeleManagement Forum
TMN	Telecommunications Management Network
TMSI	Temporary Mobile Subscriber Identity
TN	Termination Node
TOM	Telecom Operations Map
TPC	Transmit Power Control
TPRC for CD-SIC	Tx Power Ratio Control for Code Domain Successive Interference Cancellation
TR	Technical Recommendation
TrCH	Transport CHannel
TRHO	Traffic Reason HO
TRX	Transmit and Receive Unit; Transceiver
TS	Technical Specification
TSG	Technical Specification Group
TSL	Time SLot
TSTD	Time Switched Transmit Diversity
TTI	Transmission Time Interval
TU3	Typical Urban 3 km/h (standard channel type specified in GSM)
Tx	Transmit
TxIMD	Transmission Inter-Modulation Distortion
UARF(C)N	UTRA Absolute Radio Frequency (Channel) Number
UDP	User Datagram Protocol
UE	User Equipment
UEP	Unequal Error Protection
UHF	Ultra High Frequency
UL	UpLink
UM	Unacknowledged Mode
UMA	Unlicensed Mobile Access
UML	Unified Modelling Language
UMTS	Universal Mobile Telecommunications System
U-NII	Unlicensed National Information Infrastructure (RF band at 5 GHz)

U-plane	User plane
URA	UTRAN Registration Area
USF	Uplink State Flag(s)
USIM	UMTS SIM
UTRA(N)	Universal Terrestrial Radio Access (Network)
Uu	Radio interface between UTRAN and UE
V	Vertical polarisation
VA	Voice Activity
VBR	Variable Bit Rate
VHF	Very High Frequency
VLR	Visitor Location Register
VoIP	Voice over IP
VV	Space diversity with vertical polarised antennas
WAP	Wireless Application Protocol
WB	WideBand
WCDMA	Wideband Code Division Multiple Access
WDP	Wireless Data Protocol
WGS-84	World Geodetic System 84
WI	Walfisch–Ikegami
WiMAX	Worldwide interoperability for Microwave Access (forum to facilitate the IEEE 802.16 standard)
WLAN	Wireless Local Area Network
WWW	World Wide Web

1

Introduction

Jaana Laiho, Achim Wacker, Tomáš Novosad, Peter Muszynski, Petri Jolma and Roman Pichna

1.1 A Brief Look at Cellular History

The history of mobile communications started with the work of the first pioneers in the area. The experiments of Hertz in the late 18th century inspired Marconi to search markets for the new commodity (to be). The communication needs in the First and Second World Wars were also supporting and accelerating the start of cellular radio, especially in terms of utilisation of ever higher frequencies. The first commercial systems were simplex, and the operator was required to place the call. In the case of mobile-originated calls the customer had to search for an idle channel manually [1]. Bell Laboratories first introduced the cellular concept as known today. In December 1971 they demonstrated how the cellular system could be designed [2].

The first cellular system in the world became operational in Tokyo, Japan, in 1979. The network was operated by NTT, known also as a strong driver for cellular systems based on Wideband Code Division Multiple Access (WCDMA). The system utilised 600 duplex channels in the 800 MHz band with a channel separation of 25 kHz. Another analogue system in Japan was the Japanese Total Access Communication System (JTACS). During the 1980s it was realised that, from the user's point of view, a single air interface was required to provide roaming capabilities. A development study was initiated in 1989 by the Japanese government, and a new digital system, Pacific Digital Cellular (PDC), was introduced in 1991.

In 1981, 2 years later than in Japan, the cellular era also reached Europe. Nordic Mobile Telephone started operations in the 450 MHz band (the NMT450 system) in Scandinavia. The Total Access Communication System (TACS) was launched in the United Kingdom in 1982 and Extended TACS was deployed in 1985. Subsequently in Germany the C450 cellular system was introduced in September 1985. Thus, at the end of the 1980s Europe was equipped with several different cellular systems that were unable to inter-operate. By then it was clear that first-generation cellular systems were becoming obsolete, since integrated circuit technology had made digital communications not only practical but also more economical than analogue technology. In the early 1990s second-generation (2G) (digital) cellular systems began to be deployed

Radio Network Planning and Optimisation for UMTS Second Edition
Edited by J. Laiho, A. Wacker and T. Novosad © 2006 John Wiley & Sons, Ltd

throughout the world. Europe led the way by introducing the Global System for Mobile communications (GSM). The purpose of GSM was to provide a single, unified standard in Europe. This would enable seamless speech services throughout Europe in terms of international roaming.

The situation in the United States was somewhat different than in Europe. Analogue first-generation systems were supported by the Advanced Mobile Phone System (AMPS) standard, available for public use since 1983. There were three main lines of development of digital cellular systems in the US. The first digital system, launched in 1991, was the IS-54 (North American TDMA Digital Cellular), of which a new version supporting additional services (IS-136) was introduced in 1996. Meanwhile, IS-95 (cdmaOne) was deployed in 1993. Both of these standards operate in the same band as AMPS. At the same time, the US Federal Communications Commission (FCC) auctioned a new block of spectrum in the 1900 MHz band. This allowed GSM1900 (PCS) to enter the US market. An interesting overview of the GSM and its evolution towards 3G can be found in [3].

During the past decade the world of telecommunications changed drastically for various technical and political reasons. The widespread use of digital technology has brought radical changes in services and networks. Furthermore, as time has passed, the world has become smaller: roaming in Japan, in Europe or in the US alone is no longer enough. Globalisation has its impact also in the cellular world. In addition, a strong drive towards wireless Internet access through mobile terminals has generated a need for a universal standard, which became known as the Universal Mobile Telecommunication System (UMTS) ([4]–[6]). These new third-generation (3G) networks are being developed by integrating the features of telecommunications- and Internet Protocol (IP)-based networks. Networks based on IP, initially designed to support data communication, have begun to carry streaming traffic like voice/sound, though with limited voice quality and delays that are hard to control. Commentaries and predictions regarding wireless broadband communications and wireless Internet access are cultivating visions of unlimited services and applications that will be available to consumers 'anywhere, anytime'. They expect to surf the Web, check their emails, download files, make real time videoconference calls and perform a variety of other tasks through wireless communication links. They expect a uniform user interface that will provide access to wireless links whether they are shopping at the mall, waiting at the airport, walking around the town, working at the office or driving on the highway. The new generation of mobile communications is revolutionary not only in terms of radio access technology, and equally the drive for new technical solutions is not the only motivation for UMTS. Requirements also come from expanded customer demands, new business visions and new priorities in life.

1.2   Evolution of Radio Network Planning

There is very little published on the Radio Network Planning (RNP) process itself. An integral approach is proposed in [7], but this is more related to the functionalities of an RNP tool than to the overall planning process. This paper challenged the existing practises in RNP by listing the following weaknesses:

- planning was based on hexagonal network layout;
- traffic density was assumed to be uniform;
- radio wave propagation was considered independent from the environment;
- base station locations were chosen arbitrarily, while in practice fixed sites were used;
- traffic region boundaries usually were not taken into account.

The discussion continued in [8], which for the first time accounted for the impacts of quality requirements in radio network planning. This paper starts to have a process approach, and capacity enhancement with base station sectorisation to support network evolution is investigated. The challenges of non-uniform traffic conditions are identified, and cell splitting as one solution is proposed.

It can be noted that radio network planning and its development through time can be easily mapped to the development of the access technologies and the requirements set by those. The first analog networks were planned based on low capacity requirements. Radio network planning was purely designed to provide coverage. Omnidirectional antennas were used and positioned high in order to keep the site density low. The Okumura–Hata model was and still is widely used for coverage calculation in macro-cell network planning ([9] and [10]). Certain enhancements and tailoring by the COST231 project have finally resulted in the still widely used COST231 Hata model, also applicable to third-generation radio networks [11]. The latest COST developments of this area can be found in [12]. Walfisch–Ikegami is another model often referred to. This model is based on the assumption that the transmitted wave propagates over the rooftops by a process of multiple diffractions ([15] and [16]). Although the Walfisch–Ikegami model is considered to be a micro-cell model, it should be used very carefully when the antenna of the transmitter is below the rooftops of the surrounding buildings. More about propagation models can be found in Section 3.2.2.2 and conclusions about their applicability in [17] and [18].

During the course of time, together with the evolution of 2G systems, the site density got higher due to increasing capacity requirements. Furthermore, the initial assumption that cellular customers would mostly be vehicular turned out to be incorrect. Thus the maximum transmit power levels of the user equipment were reduced by at least a factor of 10, causing a need to rebalance the radio link budgets. All this forced the cellular networks to omit the omnidirectional site structure and lead to the introduction of cell splitting – i.e., one site consisted of typically three sectors instead of just one ([8] and [13]). Owing to increased spectral efficiency requirements, the interference control mechanism became more important. In addition to the sectorisation antenna tilting was also introduced as a mechanism for co-channel interference reduction [14]. Furthermore, the macro-cellular propagation model was no longer accurate enough; new models were needed to support micro-cellular planning. Sectorisation, antenna tilting and link budgets are discussed later in Chapter 3.

Higher site densities also necessitated a more careful management of the scarce frequency resources. As the frequency planning and allocation methods were widely based on predicted propagation data, the propagation models had to undergo a further refinement. Examples of such more accurate models are the ones based on ray tracing. Some ray-tracing models can be found, for example, in [19] and [20].

In addition to the propagation model development it was noticed that the increasing capacity demands could only be met with more accurate frequency planning. The frequency assignment together with neighbour cell list (for handover purposes) planning and optimisation were the main issues when planning GSM networks. In the case of GSM, frequency hopping was introduced to further improve the spectrum efficiency. Advanced frequency allocation methods can be found in the literature, one example based on simulated annealing is in [23]. In [24] a method for automatic frequency planning for D-AMPS is studied. In [25] advanced features for Frequency Division Multiple Access/Time Division Multiple Access (FDMA/TDMA) systems are introduced. These features include improving the frequency reuse by applying:

- frequency hopping;
- adaptive antennas;
- fractional loading;
- hierarchical cell structures.

It can be concluded based on several papers (for example, [26]–[28]) that the prediction of propagation is of limited accuracy due to the fact that the propagation environment is very difficult to model and thus generating a generic model, which is applicable in multiple cells, is by nature accuracy-limited. This is especially applicable when the fading characteristics (both fast and slow) need to be considered. The latest radio network control activities concentrate on the closed-loop optimisation of the plan. The initial planned configuration is (semi-) automatically tuned based on statistics collected from the live network. Proposals for handover performance improvement in terms of correct neighbour cell lists can be found in [26] and [27]. The important aspect with this method is that neighbour relations that are initially based on propagation prediction are autotuned based on real measurements. Thus inaccuracies can be compensated in the optimisation phase. A similar measurement-based concept can be utilised also for WCDMA intra- and inter-system neighbour relations; this is discussed in more detail in Chapter 7.

Recently, methods for GSM frequency planning based on mobile station measurement reports have been introduced and implemented, see [21] and [22]. The possibilities offered by these reports in GSM and WCDMA should be more utilised in the network control process (planning, optimisation and integration of those two).

Another new trend in radio network planning research is plan synthesis, meaning automatic generation of base station site locations depending on a cost function output. This is briefly discussed in Section 7.3.1.1.

In cellular networks network utilisation control requires such functionality that can utilise the measured feedback information from the network and react correctly based on that. Therefore, it is crucial that the planning phase is tightly integrated into other network control functions and the network management system. This is especially important in the case of WCDMA, owing to the fact that there will be a multitude of services; that is, customer differentiation will set a multi-dimensional matrix of Quality of Service (QoS) requirements. Planning such a network very accurately is not feasible due to limited accuracy of the input data (propagation, traffic amount, traffic distribution etc.). An example of the integration of a network management system and planning for 2G systems can be found in [29].

Integrating the network management system and advanced analysis and optimisation methods for effective configuration parameter provisioning and 'pre-launch network performance' estimation are the next challenges in the radio network development and optimisation area. An example of the effective integration of the planning tools' functionalities into the Network Management System (NMS) is, for example, visualisation of statistical performance data on cell dominance areas. Furthermore, the adjacency relations can be directly generated in the NMS based on the base station coordinates and simple distance-based rules. These initial lists can be later autotuned based on statistics collected from the live network. Also WCDMA scrambling code allocation can be done in the NMS without interfacing to external planning tools by utilising the mobile station measurement reports required by both the GSM and WCDMA standards. These reports contain information that can be used to complement the information generated traditionally by the planning tool (propagation, traffic density, etc.). When mobile station positioning methods are fully in use, another huge new dimension to optimisation tasks will be opened.

Together with the introduction of the all-IP mobile world, QoS provisioning becomes very important for the operators. This directs the network control increasingly away from radio access network control to service control. In practice this means an increased abstraction level for the operator and a new era for network management.

This book concentrates on the challenges with WCDMA networks. Furthermore, one of the main motivations is to move away from the 'analytical' control of the network, and enhance the modelling and tools to give a picture as realistic as possible of the actual network performance. Network functionalities can no longer be considered as individual entities, but their interactions must be accounted for. In the analytical, ideal world this has no relevance, but in the true cellular world the understanding of these interactions and network element algorithms and their limitations is essential.

It can be stated that radio access evolution towards third generation is the first big evolutionary step after the birth of cellular systems. The large step in radio access development, the great interest in applications and services also forces the radio network planning and optimisation process to improve to fully support the offered possibilities.

1.3   Introduction to Radio Network Planning and Optimisation for UMTS

The mobile communications industry throughout the world is currently shifting its focus from 2G to third-generation (3G) UMTS technology; that is, it is investing in the design and manufacturing of advanced mobile Internet/multimedia-capable wireless networks based on the Wideband Code Division Multiple Access (WCDMA) radio access platform. While current 2G wireless networks, in particular the extremely

successful and widespread global GSM-based cellular systems, will continue to evolve
and to bring such facilities as new Internet packet data services onto the market, more
and more radio network planners and other wireless communication professionals are
becoming familiar with WCDMA radio technology and are preparing to build and
launch high-quality 3G networks. This book has been written in particular for those
RF (Radio Frequency) engineering professionals who need to thoroughly understand
the key principles in planning and optimising WCDMA radio networks, though it
should also prove useful to others in the industry.

Radio network planners, particularly, face a number of new challenges when moving
from the familiar 2G to the new 3G networks, many of them related to the design and
the planning of true multi-service radio networks, and some to particular aspects of the
underlying WCDMA radio access method. In this introductory chapter we provide a
brief outline of these challenges, which will then be discussed in much greater detail in
the following chapters of this book.

Before considering in detail what actually will be new (and different) in WCDMA
compared with GSM, for example, we summarise here some of the defining characteristics of 3G multi-service radio networks in an abstract setting, regardless of the particular
incarnation of the underlying 3G radio access protocol, such as WCDMA or EDGE.
Hence, one could attempt to characterise 3G radio access with the following attributes:

- Highly sophisticated radio interface, aiming at great *flexibility* in carrying and multiplexing a large set of voice and, in particular, data services with constant as well as
  variable throughput ranging from low to very high data rates, ultimately up to
  2 Mbps.
- Efficient support for carrying traffic under IP.
- Cell coverage and service design for *multiple services* with largely different bit rates
  and QoS requirements. Due to the great differences in the resulting radio link
  budgets, uniform coverage and capacity designs – as implemented in today's voice-
  only radio networks – could no longer be obtained economically for high bit rate
  services. Consequently, traffic requirements and QoS targets will have to distinguish
  between the different services.
- A large set of sophisticated features and well-designed radio link layer 'modes' to
  ensure very *high spectral efficiency* in a wide range of operating environments, from
  large macro-cells to small pico-cells or indoor cells. Examples of such features are
  various radio link coding/throughput adaptation schemes; support for advanced
  performance-enhancing antenna concepts, such as BS transmission diversity for the
  downlink; and the enabling of interference cancellation schemes.
- Efficient *interference-averaging mechanisms* and robustness to enable operation in a
  strongly interference limited environment in order to support very *low frequency
  reuse* schemes, with the goal of achieving high spectral efficiency. This will require
  good *dominance* of, and striving for maximum *isolation* between the cells through the
  proper choice of site locations, antenna beamwidths, tilts, orientation and so on.
  Tight frequency reuse in conjunction with interference limited operation, on the
  other hand, means that *cell breathing* effects will necessarily occur.
- Extensive use of 'best effort' provision of packet data capacity. Temporarily unused
  radio resource capacity will be made available to the packet data connections in a
  flexible and fair manner so as to improve the commonly perceived QoS. This will

result in networks operating at a *higher spectral loading* compared with today's voice-dominated networks. This higher load on the RF spectrum will result in higher interference levels, thus requiring ever-better RF planning to achieve high throughput. This trend is amplified by the significant spectrum-licensing costs which some service providers, especially in Europe, are having to bear for providing their 3G services.

- IP packet services, with their possibly 'unlimited' demand for radio capacity together with network-based best effort packet data allocation and strong interference limitations, will place a higher than ever burden on *pre- and post-deployment optimisation* of the cell sites, if satisfactory cell throughput and QoS targets are to be met. As a consequence, the effort and cost of the radio network optimisation phase will exceed that of today's 2G networks, in which the primary burden is on initial frequency planning. Furthermore, the current practice of using the ample available 2G spectrum to circumvent interference problems by appropriate frequency planning will no longer be viable for the high-throughput services relying on high spectral efficiency and tight reuse of the available spectrum.

- In order to provide ultimately high radio capacity, 3G networks must offer efficient means for multi-layered network operation, supporting micro- and pico-cellular layers, for example, and ways of moving the traffic efficiently between these layers as appropriate. This will require efficient *inter-layer handover* mechanisms, together with appropriate dimensioning and RF planning of the cell layers.

- Introduction and rollout of 3G networks will be costly and will happen within a very competitive environment, with mature 2G (e.g., GSM) networks guiding end-users' expectations of service availability and quality. Therefore, service providers will utilise their existing GSM networks to the fullest possible extent. The most obvious way to do this is to use the comprehensive GSM footprint as a coverage extension of 3G, providing 3G services initially only in limited, typically urban, areas, thus relying on *3G to GSM inter-system handover* to provide coverage continuity for basic services. Therefore, it will be important for 3G service providers to implement 3G handover to GSM cells, in order to accelerate 3G rollout and minimise upfront deployment costs. This will require RF planning methods that allow for joint 2G–3G coverage and capacity planning – i.e., some degree of integration of the tools and practices used.

- Another very important aspect is the possibility of *co-siting 3G sites with existing 2G sites*, reducing costs and overheads during site acquisition and maintenance. However, such co-siting raises a number of issues for the radio network planner to consider. Should shared antenna solutions be used? Would the RF quality of the underlying 2G network meet acceptable standards for the 3G quality targets, or would there have to be a prior optimisation phase for the 2G sites? Might there exist other constraints on site reuse, such as shelter space? Are there any potential interference-related problems in co-siting? And so forth. Again, an integrated approach, recognising the operation of 3G jointly with 2G from a *multi-radio* perspective with the goal of achieving a good cost/performance ratio for *both* systems operating concurrently, will be required.

Any generic radio access method (TDMA, FDMA, CDMA, OFDM, etc.), designed for high spectral efficiency operation and to meet the above 3G service requirements, would face the listed issues, which suggests that most of the challenges faced by network planners in moving towards 3G actually stem from dealing with an integrated multi-service, multi-data-rate system providing end-users with capacity and bit rates on demand, rather than predominantly from the radio protocol, WCDMA.

What, therefore, are the radio network planning challenges *specific to WCDMA*? Obviously there are many differences in detail between WCDMA and GSM – in the radio network parameters, for example – but let's look at the more fundamental differences:

- Planning of soft(er) handover overhead. Soft handover is a feature specific to CDMA systems, such as IS-95-based systems, or as in our case WCDMA. However, a closer look reveals that minimising soft handover overhead is closely correlated with establishing proper *cell dominance*, which we have already identified as generically desirable for maturing 2G systems as well. Thus, planning for low soft handover overheads does not require any new skills and tools, but rather adherence of good radio network planning practices known from today's systems.
- Cell dominance and isolation. These will become relatively more important in WCDMA than in 2G, due to the frequency reuse in WCDMA being 1 and the resulting *closer coupling* of mutually interfering, nearby cells. WCDMA will 'see' different and, of course, more sites/cells than GSM does. This is particularly relevant when 2G–3G co-siting and antenna sharing is attempted.
- Vulnerability to 'external' interference – e.g., interference leaking from adjacent carriers used in other systems or similar interference between different WCDMA cell layers. Again, this issue is not so much specific to WCDMA, but its importance has been dramatically increased: with an operating bandwidth of 5 MHz, a single WCDMA carrier can consume as much as 25–50% of the service provider's available spectrum. Any residual interference leaking into a WCDMA carrier and desensitising the receivers will have a much more dramatic impact on service quality than for today's 2G narrowband systems.

In this section we have taken a 'bird's eye' view of WCDMA. Summarising, we can see some new challenges and certainly much new detail for the designer to consider when planning WCDMA networks, yet in a way there is very little new about planning WCDMA: it merely requires good planning practices from today's wireless systems to be recognised and implemented in a consequent and disciplined fashion.

But exactly where do radio network planning and optimisation fit into the whole UMTS mobile network business concept? In terms of technological expertise, mobile networks represent a heavy investment in human resources. This will be even more true for 3G networks. However, not only are mobile networks technologically advanced, but the technology has to be fine-tuned to meet demanding coverage, quality, traffic and economic requirements. Operators naturally expect to maximise the economic returns from their investment in the network infrastructure – i.e., from Capital Expenditure (CAPEX). Here we should note two important aspects of network performance – planning and optimisation. Any network needs to be both planned and optimised. To what degree depends on the overall economic climate, but network optimisation is much easier and much more efficient if the network is already well-planned initially. A poorly laid out network will prove difficult to optimise to meet long-term business or technical expectations. Optimisation is a continuous process that is part of the operating costs of the network – i.e., its Operational Expenditure (OPEX). However, the concept of *autotuning* (see Section 9.3) offers new opportunities for performing the optimisation process quickly and efficiently, with minimal contribution from OPEX, in order to maximise network revenues.

Operators face the following challenges in the planning of 3G networks:

- Planning means not only meeting current standards and demands, but also complying with future requirements in the sense of an acceptable development path.
- There is much uncertainty about future traffic growth and the expected proportions of different kinds of traffic and different data rates.
- New and demanding high bit rate services require knowledge of coverage and capacity enhancement methods and advanced site solutions.
- Network planning faces real constraints. Operators with existing networks may have to co-locate future sites for either economic, technical or planning reasons. Greenfield operators are subject to more and more environmental and land use considerations in acquiring and developing new sites.
- In general, all 3G systems show a certain relation between capacity and coverage, so the network planning process itself depends not only on propagation but also on cell load. Thus, the results of network planning are sensitive to capacity requirements, which makes the process less straightforward. Ideally, sites should be selected based on network analysis with the planned load and traffic/service portfolio. This requires more analysis with the planning tools and immediate feedback from the operating network. The 3G revolution forces operators to abandon the 'coverage first, capacity later' philosophy. Furthermore, because of the potential for mutual interference, sites need to be selected in groups. This fact should be considered in planning and optimisation.

1.4   Future Trends

Even though 3G telecommunication systems are still under development and not yet in widespread operation throughout the world, it is not possible to stand still. In today's fast-moving information society, continuous improvement is essential. This applies equally to 3G systems themselves, which have already evolved a long way from the first such systems in terms of services and capacities. Although the detailed steps in their evolution since Third Generation Partnership Project (3GPP) Release '99 are still somewhat unclear, some long-term trends are already visible. One major change will be separating more or less completely the user plane from the control plane, changing more and more from circuit switched to packet switched connections, thus making the whole network ready to be based completely on IP technology. The technology for accessing a network and transporting the information will become less important, but greater emphasis will be put on the services and the quality thereof. Users will no longer even know which access technology they are using – they will just request a service and the network will decide at the time the optimum technology (GSM/EDGE, cdma2000, WCDMA, WLAN, DVB, etc.) to provide it.

1.4.1   Towards a Service-driven Network Management

After the deployment of 3G networks new challenges are ahead: even higher bit rates shall be supported, with some possible average around 2 Mbps, some peaks at 20 Mbps and in the extreme up to 200 Mbps. As a first step towards this achievement can be mentioned the introduction of the High-speed Downlink Packet Access (HSDPA) in Release 5 of the 3GPP standardisation and its counterpart the High-speed Uplink Packet Access (HSUPA), which is one of the main topics of the coming Release 6. HSDPA is also one of the major additions to this edition and will be described – e.g., in Chapters 2 and 4 – where applicable.

In general the need for higher data rates will lead to even smaller cells, self-planning dynamic topologies, full integration of IP, more flexible use of spectrum and other resources and utilisation of precise user position. However, if the radio network control and management processes are carefully designed to support 3G, the step to a wider variety of cell types and new set of services will be smooth and less revolutionary than what we face now when moving from 2G speech-oriented networks to the 3G applications- and service-driven cellular world.

Until recent years, the development of the telecommunication industry has been very technology-driven. The technical solution and its implementation paved the way, and service-handling capabilities were evaluated in a later phase. The right types of services were created in the last phase. Today, the approach is different. Services and content are the driving force, the same service type can in certain cases be delivered by several different access technologies. With Voice over IP (VoIP) the voice services will become a commodity; the mobile services and mobile content are the business drivers. The evolution of network services is depicted in Figure 1.1. The transition from voice

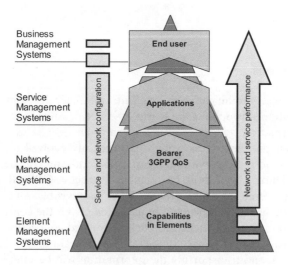

Figure 1.1   Service layers from the network point of view. Service provisioning and deployment is an evolutionary process – from best effort support to full utilisation of 3GPP Quality of Service mechanisms. The requirement for guaranteeing quality necessitates good integration between network configuration, service configuration and service assurance.

service only to QoS differentiation and Guaranteed Bit rate (GB) and non-GB separation/requirements provides possibilities, but at the same time sets new challenges for vendors and operators.

The forecast is that the Operations Support System (OSS) and especially the value adding components of the NMS are the areas to expand the business in IT as well as in telecommunications. This new trend evolves partly because the vision of the new service-driven future is becoming clearer. Further, the convergence of different technical solutions will set requirements also to OSS business.

Moreover, it is important to understand that 3G is not about technologies, but it is about services. The same service can be offered in several access technologies (GPRS, EDGE, WCDMA and the emerging WLANs, for example). Therefore, the evolution of the OSS in the service management area ought to be technology transparent and provide service management support independent of the access technology. Further-more, the level of abstraction of service management should allow the operator to focus on service configuration rather than network element configuration. This is achieved by adding intelligence to perform automatic transitions from network layer to service layer in network management systems.

1.4.2 Wireless Local Area Networks (WLANs)

Wireless Local Area Networks (WLANs) are a strongly emerging complement to cellular communication. WLAN is a short-range, high bit rate and easy-to-use radio access technology. It is, as the name says, a wireless LAN, and is used to replace wired LAN when freedom from wires is desired. Practically, a synonym to WLAN is the standard IEEE 802.11 (this is the 'full' standard name, sometimes the 'IEEE' is dropped), although there are some other technologies that could serve the same purpose. The original 802.11 was developed during 1991–1997. The first release of the standard, '802.11' – without any specifying letters – was accepted in June 1997 and consisted of a specification for the Medium Access Control (MAC) layer and three different physical layers (PHYs), frequency hopping, Direct Sequence Spread Spectrum (DS-SS) and infrared. Radio PHYs operate in the 2.4 GHz unlicensed ISM (Industrial, Scientific, Medical) band, which is available globally, and give up to 2 Mbps bit rate. Over time the DS-SS variant gained the most popularity. The MAC layer is robust and elegant, is easy to take into use and works reliably even in interfered environments. No frequency planning or other radio-related parameters are necessary to be set before use, although larger areas with a number of access points operate more efficiently when some engineering effort is spent. These characteristics are achieved at the expense of spectrum efficiency but, because spectrum is free, it is of no big concern.

Overview of the 'WLAN Standard's Alphabet'
Since wired LANs developed meanwhile from 10 to 100 Mbps, there was a need to also increase the WLAN bit rate. Therefore, enhancements to the physical layers were developed in the late 1990s, leading to a family of 802.11 standards, distinguished by an appended letter. Table 1.1 presents a summary of the 'WLAN standard's alphabet' and their main enhancements. The 802.11b, delivering up to 11 Mbps, was introduced

Table 1.1 Overview of the 802.11 standard family (for current status the reader is referred to *http://grouper.ieee.org/groups/802/11/*).

802.11 Working Group	Brief description
802.11-1997, 802.11-1999	Base standard
802.11a-1999	5 GHz extension, up to 54 Mbps
802.11b-1999	Data rate extension in 2.4 GHz IMS band up to 11 Mbps
802.11c	ID bridge addition
802.11d-2001	Regulatory domains
802.11e	QoS enhancements
802.11f-2003	Recommended practices for interaccess point communication
802.11g-2003	11a-like high-speed extension for 2.4 GHz IMS band
802.11h-2003	Dynamic channel allocation and power control extensions for European requirements
802.11i-2004	Security enhancements to MAC
802.11j-2004	Support for 4.9 GHz bands for Japan
802.11k	Radio resource measurements
802.11m	Standard maintenance, technical and editorial corrections
802.11n	High throughput >100 Mbps extension
802.11p	Wireless access in vehicular environments
802.11r	Fast BSS transition – i.e., fast handover
802.11s	Mesh networks
802.11t	Wireless performance prediction
802.11u	Inter-working with external networks
802.11v	Wireless network management

Note: If no year is indicated with the standard, then it is ongoing in 802.11 standardisation at the time of writing.

in September 1999, and 802.11a, up to 54 Mbps, in December 1999. The bit rate figures mentioned here are the 'instantaneous' PHY rates, the actual user data throughputs are $1/3 \ldots 2/3$ of those values. The 11a operates at a higher RF band, at the 5 GHz U-NII (Unlicensed National Information Infrastructure) band, which makes the equipment slightly more expensive. The exploding popularity of WLAN began with the 11b standard around 2000–2001, while the 11a has seen quite poor success in the market-place. Although 11a is technically superior – at least regarding bit rates – the 11b was a better fit to the market, was 'good enough', had compatibility with previous 802.11 DS-SS installations and was a bit cheaper. Later on (November 2001), the higher speed 11a modulations were added to the 2.4 GHz band, producing the 11g variant, which runs up to 54 Mbps. Today the 11g is the most common in the marketplace. The volume of tens of millions of 11b devices causes considerable inertia to the market, and hence most 11g equipment also includes the 11b radio interface.

802.11e MAC Enhancement

The above-described enhancements deal with the PHY layer, the original 802.11 MAC has been untouched. Practically all the 802.11 equipment has the standard Distributed Control Function (DCF) MAC, which operates in a non-connection-oriented best effort way. This supports no QoS guarantees. In the original 1997 standard there is a Point Coordination Function (PCF) option that could facilitate QoS in theory, but it has not been implemented due to some difficulties with it. So in late 1999, Task Group 11e started to reconsider the QoS issue, and finally in October 2004 the task group achieved a positive vote in the sponsor ballot. The long process is an indication of how difficult the introduction of QoS is if it is not there in the beginning. So possibly we can soon add a new 'clump into the alphabet soup'. The role of 11e is somewhat different in different environments. There are three main application environments of WLAN – enterprise, home and public – which can also be seen as different markets. The role of 11e will be slightly different at these environments. Company premises are controlled environments, so there already have been internal wireless telephone systems utilising the 802.11 WLAN. Homes are relatively low load environments so it can be expected that QoS applications might run there quite satisfactorily even without 11e, although successful application of 11e would enhance reliability. The public environment is the most difficult; until now service providers have been reluctant to sell any applications requiring QoS over WLAN in this environment. At the time of writing it remains to be seen if 11e will change this. Possibly the most obvious effect of the absence of QoS has been experienced at conference premises where a large crowd of WLAN users may collect in a single space. The best effort MAC has a bell-shaped throughput curve as a function of load – meaning that when there are too many contenders for air time, nobody no longer gets anything, the total throughput of the access point goes to a very low value. This situation may be the first to see the benefits of 11e when it gets implemented.

802.11g and the Tug of War of Market Share
The Shannon law states that low path loss – i.e., small range (typical of the WLAN usage situation) – makes very high bit rates possible, whereas the opposite, high path loss – i.e., long range (typical of the cellular usage situation) – restricts the bit rate to lower values. Today the 11g WLAN reaches 54 Mbps, with some proprietary implementation even more than 100 Mbps, whereas the fastest cellular-based radio systems reach a few 100 kbps; newer technologies are expected to operate up to a few Mbps. Considering also the low price of WLAN access points compared with a cellular base station, this obviously makes the cost of delivering data over WLAN a lot cheaper per megabyte than over cellular radio. Given that the number of public access WLAN hotspots increase rapidly, this creates an obvious tug of war between the radio technologies about the share of business of 'wireless data' for each. Both technologies have their obvious benefits in their 'home base', where they don't compete; cellular has higher coverage and more QoS mechanisms, at least for the time being, while WLAN has advantages in the indoor environment and a higher bit rate. The distinction of the technologies is deep-rooted; by natural laws, by quite separate equipment vendor companies and by different companies implementing the physical access networks. The only thing that we can conclude is that the outcome is very difficult to predict. The 'chaotic boundary area' of the technologies will potentially see dynamic behaviour in the years to come. The mechanism – be it some automatic one or the human user – that chooses the radio system over which the applications run will attract interest and be an arena of competition.

Integration of WLAN with 3G

Although many analysts and marketing departments have positioned WLAN as complementary to cellular data access both technologies obviously compete for wireless data services and revenues while having application areas where 3G is strong, and where WLAN is strong. Realising that, the cellular community has launched an effort to integrate WLAN access into cellular systems. The results are WLAN-cellular inter-working standards produced by 3GPP and 3GPP2. Inter-working was spearheaded by 3GPP in Release 6, which envisioned loose inter-working with no inter-working at the Radio Access Network (RAN) level but rather at the Core Network (CN) level. Inter-working has been introduced in so-called scenarios, which are functionally built on top of each other starting from the lowest one – no other inter-working but common customer care and billing, using authentication based on, e.g., username and password. In this scenario a user accesses the Internet over the WLAN access network and bypasses the cellular CN completely. The next inter-working scenario uses Subscriber Identity Module/UMTS SIM (SIM/USIM)-based access control and charging functionalities from the cellular CN. This is then augmented with another scenario, specifying a standardised way of diverting user traffic to the CN. Here, the user may consume all the spectrum of offered cellular services, including Internet access. There are higher scenarios being considered for the Release 7 of 3GPP standards leading up to the support of full mobility and QoS between the WLAN and cellular accesses.

Unlicensed Mobile Access (UMA) – Integration with 2G

Coverage limitation in some areas and the possibility to complement them with wireless access to 2G cellular services over home WLAN and residential broadband connection was a strong motivation to integrate the WLAN and legacy 2G services provided by cellular operators. Similarly, fixed–mobile convergence made it very attractive to offer cellular-like services from fixed network and Mobile Virtual Network Operator (MVNO) services over wide area network for fixed operators with wireless ambitions. Shared interest resulted in the Unlicensed Mobile Access (UMA) standard specifying the transport of 2G circuit switched voice and data bearers over generic IP access to cellular terminals. Although being access-technology-agnostic, UMA names WLAN and Bluetooth™ as access technologies.

802.16 and WiMAX (Worldwide Inter-operability for Microwave Access)

As the popularity of WLAN rose, the coverage limitations of the technology were soon recognised. The attention of the wireless IT community has turned to a new fledgling IT standard, IEEE 802.16, which is specified for broadband wireless access. A lot of effort is being spent on specifying mobility extensions in 802.16 enabling its use as Mobile Broadband Wireless Access. Going further, the 802.16 certification body, WiMAX, is starting efforts to specify system architecture and thus one can expect that a new standardised vendor-inter-operable wide-area wireless packet switched network will be available in the near future. Still in its inception stage it is already being recognised as yet another technology competing with the cellular technologies for the wireless data access market and it is very likely that the cellular community will try to coopt this technology by integrating a WiMAX inter-working standard into its system.

802.20

The perceived demand for high-speed wide-area wireless data networks for highly mobile users has led to the creation of IEEE 802.20. The standardisation is at a virtual standstill and may not produce the expected results but it is widely referenced.

Looking to the Future

In homes the 802.11 WLAN easily carries over the air what a broadband Internet connection delivers today and probably will do so as long as homes are connected by copper-based wires. Within the home it is able to carry several video streams or one HDTV (High Definition Television) stream. So it can be expected that the current 802.11 WLAN may serve the home environment quite well in the foreseeable future. At company premises the useful bit rate is clearly less than what wired LANs have, but the highest rates of wired LAN are not needed by the average user, except for special cases like server input/output or when the Information Management (IM) service wants to make a backup of a disk. So we don't even expect the WLAN to be needed at those points. As access connection to most users the 802.11 can be expected to serve well for quite some time. A very interesting factor for the future is the performance of 11e in the public environment. If it fulfils the expectations then we can expect the 802.11 WLAN to serve our needs for a number of years.

1.4.3   Next-generation Mobile Communication

The current 3G and 3.5G mobile communication and the variants thereof will surely not be the end of the development, even though with HSDPA and HSUPA the 3GPP radio access will be highly competitive for quite some years. The drivers behind the development can be seen basically in the WLAN and Digital Subscriber Line (DSL) technologies, both of which are capable even now to deliver high bit rates in a very cost-efficient manner. Next-generation systems denoted as fourth generation (4G) are around the corner and will ensure competitiveness even in the longer run.

The main additional requirements compared with the current 3G systems can be seen in particular in that these new systems will be developed for an optimised, pure packet switched data access with a much more distributed radio resource and network management (fully IP-based) and a multi-carrier radio access (allowing more flexible carrier bandwidths than the current 5 MHz). Furthermore, being fully IP-based, mobility management could be replaced by IP mobility as known today, improving coverage and capacity at lower costs for the operator.

As an early version of such systems can be considered an approach initiated from NTT DoCoMo, which is an adaptation of the technologies originally studied for 4G deployments onto 3G bands, internally called Enhanced 3G (E3G). With carrier bandwidths up to 20 MHz air interface peak bit rates are expected to raise up to the order of 100 Mbps, allowing practical bit rates of around 20 Mbps. For real 4G systems, carrier bandwidths are envisaged to be around 100 MHz, pushing the theoretical air interface bit rate towards the 1 Gbps border and having a spectral efficiency of roughly double that of current Release 6 plans.

However, much is still speculative and the requirements are only being formulated at the moment, but clear trends as described above can be seen.

References

[1] Jakes, W.C., *Microwave Mobile Communications*, IEEE Press, 1974, reprint 1994, p. 1.

[2] Lee, W.C.Y., *Mobile Cellular Telecommunications System*, McGraw-Hill, 1990, p. 5.

[3] Hillebrand, F. (ed.), *GSM and 3G: The Creation of Global Mobile Communication*, John Wiley & Sons, 2001.

[4] Mohr, W. and Becher R., Mobile communications beyond third generation. *Proc. VTC 2000 Spring Conf., Tokyo, Japan, May 2000*, pp. 654–661.

[5] Chandran, N. and Valenti, M.C., Three generations of cellular wireless systems. *IEEE Potentials*, **20**(1), February/March 2001, pp. 32–35.

[6] Padgett, J.E., Gunther, C.G. and Hattori, T., Overview of wireless personal communications. *IEEE Communications Magazine*, **33**(1), January 1995, pp. 28–41.

[7] Gamst, A., Beck, R., Simon, R. and Zinn, E.G., An integrated approach to cellular network planning. *Proc. VTC 1985 Conf., Boulder, Colorado*, pp. 21–25.

[8] Gamst, A., Remarks on radio network planning. *Proc. VTC 1987 Conf., June 1987*, pp. 160–165.

[9] Okumura, Y., Ohmori, E., Kawano, T. and Fukuda, K., Field strength and its variability in the VHF and UHF land mobile service. *Review Electronic Communication Laboratories*, **16**(9/10), 1968, pp. 825–873.

[10] Hata, M., Empirical formula for propagation loss in land mobile radio services, *IEEE Transactions on Vehicular Technology*, **VT-29**(3), August 1980, pp. 317–325.

[11] *Urban Transmission Loss Models for Mobile Radio in the 900 and 1800 MHz Bands*, COST 231, TD(91)73, September 1991.

[12] Correia, M. (ed.), *Wireless Flexible Personalised Communications, Cost 259: European Co-operation in Mobile Radio Research*, John Wiley & Sons, 2001.

[13] Lee, W.C.Y., *Mobile Cellular Telecommunications System*, McGraw-Hill, 1990, p. 61.

[14] Lee, W.C.Y., *Mobile Cellular Telecommunications System*, McGraw-Hill, 1990, pp. 194–199.

[15] Walfisch, J. and Bertoni H.L., A theoretical model of UHF propagation in urban environments. *IEEE Transactions on Antennas and Propagation*, **AP-36**(12), December 1988, pp. 1788–1796.

[16] Ikegami, F., Yoshida, S., Takeuchi T. and M. Umehira, Propagation factors controlling mean field strength on urban streets. *IEEE Transactions on Antennas and Propagation*, **AP-32**(8), August 1984, pp. 822–829.

[17] Cheung, J.C.S, Beach, M.A. and Chard, S.G., Propagation measurements to support third generation mobile radio network planning. *Proc. VTC 1993 Conf., Secausus, New Jersey, May 1993*, pp. 61–64.

[18] Willard, C., Rochefolle, T., Baden, C.C.E., Cheung, J.C.S, Chard, S.G., Beach, M.A., Constantinou, P. and Cupido, L., Planning tools for mobile networks. *Electronics & Communication Engineering*, **5**(5), October 1993, pp. 309–314.

[19] Wei, Q.X., Gong, K. and Gao, B.X., Ray-tracing models and techniques for coverage prediction in urban environments. *Proc. APMC 1999 Conf., Singapore, November/December 1999*, pp. 614–617.

[20] Rajala, J., Sipila, K. and Heiska, K., Predicting in-building coverage for micro-cells and small macro-cells. *Proc. VTC 1999 Conf., Houston, Texas, May 1999*, pp. 180–184.

[21] Wille, V. and King, A., Micro-cellular planning based on information from the radio network. *IEE Colloquium on Antennas and Propagation for Future Mobile Communications* (Ref. No. 1998/219), 1998, pp. 8/1–8/8.

[22] Barco, R., Canete, F.J., Diez, L., Ferrer, R. and Wille, V., Analysis of mobile measurement-based interference matrices in GSM networks. *Proc. VTC 2001 Fall Conf., Atlantic City, New Jersey, October 2001*, pp. 1412–1416.

[23] Duque-Anton, M., Kunz, D. and Ruber, B., Channel assignment for cellular radio using simulated annealing. *IEEE Transactions on Vehicular Technology*, **VT-42**(1), February 1993, pp. 14–21.

[24] Almgren, M., Frodigh, M., Magnusson, S. and Wallstedt, K., Slow adaptive channel allocation for automatic frequency planning. *Proc. ICUPC 1996 Conf., Cambridge, Massachusetts, September/October 1996*, pp. 260–264.

[25] Frullone, M., Riva, G., Grazioso, P. and Falciasecca, G., Advanced planning criteria for cellular systems. *IEEE Personal Communications*, **3**(6), December 1996, pp. 10–15.

[26] Magnusson, S. and Olofsson, H., Dynamic neighbour cell list planning in a micro-cellular network. *Proc. ICUPC 1997 Conf., San Diego, California, October 1997*, pp. 223–227.

[27] Olofsson, H., Magnusson, S. and Almgren, M., A concept for dynamic neighbour cell list planning in a cellular system. *Proc. PIMRC 1996 Intl. Symposium, Taipei, Taiwan, October 1996*, pp. 138–142.

[28] Walton, R., Wallace, M. and Howard, S., CDMA downlink performance issues. *Proc. PIMRC 1998 International Symposium, Boston, Massachusetts, September 1998*, pp. 308–312.

[29] Mende, W., Oppermann, E. and Heitzer, L., Mobile radio network management supported by a planning tool. *Proc. NOMS 1998 Symposium, New Orleans, Louisiana, February 1998*, pp. 483–492.

2

Introduction to WCDMA for UMTS

Tomáš Novosad, David Soldani, Kari Sipilä, Tero Kola and Achim Wacker

2.1 Mathematical Background of Spread Spectrum CDMA Systems

This section describes the general properties of Direct Sequence Code Division Multiple Access (DS-CDMA), along with its physical origins. Since this subject has been of interest and studied for about 70 years, it has become one of the classic topics in the theory of communication, which is covered by many books. Details may be seen, for example, in [27] or [30]. A classic reference for spread spectrum systems is [31], which includes a CDMA history section on the evolution since 1930.

2.1.1 Multiple Access

A cell in a cellular radio network could be seen as a multi-user communication system, in which a large number of users share a common physical resource to transmit and receive information. The resource in the cell is a frequency band in the radio spectrum. There are several different access techniques in which multiple users could send the information through the common channel to the receiver – see Figure 2.1.

The users may subdivide the available spectrum into a number, N, of non-overlapping, or slightly overlapping, sub-channels. This method is called Frequency Division Multiple Access (FDMA). Another method for creating multiple sub-channels is to divide the duration of a time period T_f into a number of non-overlapping sub-intervals, each of duration T_f/N. This method is called Time Division Multiple Access (TDMA).

In FDMA and TDMA the common channel is partitioned into orthogonal single-user sub-channels. A problem arises if the data from the users accessing the network is bursty in nature. A single user who has reserved a channel may transmit data irregularly so that silent periods are even longer than transmission periods. For example, a speech signal may contain long pauses. In such cases TDMA or FDMA tend to be inefficient because a certain portion of the frequency – or of the timeslots – allocated to the user

Radio Network Planning and Optimisation for UMTS Second Edition
Edited by J. Laiho, A. Wacker and T. Novosad   © 2006 John Wiley & Sons, Ltd

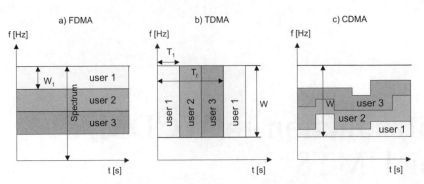

Figure 2.1   Multiple access schemes.

carries no information. An inefficiently designed multiple access system limits the
number of simultaneous users of the common communication channel. One way of
overcoming this problem is to allow more than one user to share the channel or sub-
channel by the use of spread spectrum signals. In this method each user is assigned a
unique code sequence or signature sequence that allows the user's signals to be spread
on the common channel. Upon reception, the various users' signals are separated by
crosscorrelating each received signal with each of the possible user signature sequences.
By designing these code sequences with relatively little crosscorrelation, the crosstalk
inherent in the demodulation of the signals received from multiple transmitters is
minimised. This multiple access method is called Code Division Multiple Access
(CDMA) [27].

2.1.2   Spread Spectrum Modulation

The general concept of spread spectrum modulation is presented in Figure 2.2.

Formally the operation of both transmitter and receiver can be partitioned into two
steps. At the transmitter site, the first step is modulation in which the narrowband
signal S_n, which occupies frequency band W_i, is formed. In the modulation process,
bit sequences of length n are mapped to 2^n different narrowband symbols constituting
the narrowband signal S_n. In the second step the signal spreading is carried out, in
which the narrowband signal S_n is spread in a large frequency band W_c. The spread
signal is denoted S_w, and the spreading function is expressed as $\varepsilon(\ )$.

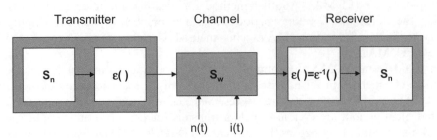

Figure 2.2   Spread spectrum system concept.

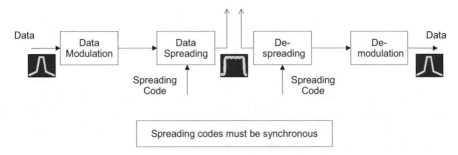

Figure 2.3 Direct Sequence Spread Spectrum concept with indicative bandwidth.

At the receiver site the first step is despreading, which can be formally presented by the inversion function $\varepsilon^{-1}(\ ) = \varepsilon(\ )$. In despreading, the wideband signal S_w is converted back to a narrowband signal S_n, which can then be demodulated using standard digital demodulation schemes. Note that the nature of spreading and despreading operation is the same and could be performed by modulation of user data bits by spreading sequence bits. Such a basic concept is depicted in Figure 2.3.

The primary reason for going to the process of spreading and despreading is to enable the CDMA multiple access method but, due to the signal spreading and the resulting enlarged bandwidth, spread spectrum signals have many other interesting properties that differ from those of narrowband signals. The most important ones are discussed in the following sections.

2.1.3 Tolerance of Narrowband Interference

A spread spectrum system is tolerant of narrowband interference, as shown in Figure 2.4.

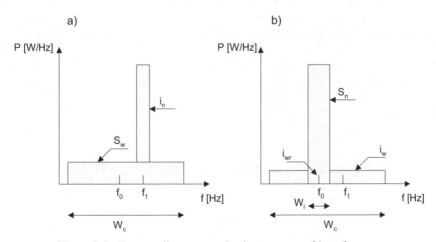

Figure 2.4 Despreading process in the presence of interference.

Let's assume that a wideband signal S_w is received in the presence of a narrowband interference signal i_n: see Figures 2.4(a) and (b). The despreading process can be presented as follows:

$$\varepsilon^{-1}(S_w + i_n) = \varepsilon^{-1}[\varepsilon(S_n)] + \varepsilon^{-1}(i_n) = S_n + i_w \tag{2.1}$$

The despreading operation converts the input signal into a sum of the useful narrowband signal and an interfering wideband signal. After the despreading operation narrowband filtering (operation $F(\ )$) is applied with a bandpass filter of bandwidth B_n equal to the bandwidth W_i of S_n. This results in:

$$F(S_n + i_w) = S_n + F(i_w) = S_n + i_{wr} \tag{2.2}$$

Only a small proportion of the interfering signal energy passes the filter and remains as residual interference, because the bandwidth W_c of i_w is much larger than W_i – see Figure 2.4(b). The ratio between the transmitted modulation bandwidth and the information signal bandwidth is called the processing gain, G_p:

$$G_p = \frac{W_c}{W_i} \tag{2.3}$$

To prevent any filter- or modulation-specific properties, from this point we equate W_c and W_i to the chip rate and user data rate, respectively.

Consider the system without error correction coding overhead, etc. In this case, the gain defined by Equation (2.3) is given by just the spectrum-spreading operation (i.e., it is in linear scale proportional to the number of times the spectrum has been expanded). Such a gain has strong narrowband interference suppression properties, as shown above. It is important to note that the term 'processing gain' as used by the Third Generation Partnership Project (3GPP) could also be defined according to Equation (2.3), but owing to the inclusion of additional signal manipulation processing (error control coding, overhead, etc.), the resulting processing gain is composed of the spreading part and the coding part.

In Figure 2.4 the effect of processing gain can be clearly seen, since the more processing gain the system has, the more the power of uncorrelated interfering signals is suppressed in the despreading process. Thus, processing gain can be seen as an improvement factor in the Signal to Interference Ratio (SIR) of the signal after despreading.

There is a certain tradeoff in the value of the transmission bandwidth W_c. For a large processing gain to give greater interference suppression, a broad transmission bandwidth is needed. In the WCDMA system the value of W_c is 3.84 Mcps which, owing to spectral sidelobes, results in a 5 MHz carrier raster.

2.2 Direct Sequence Spread Spectrum System

There are a number of techniques to spread the information-bearing signal by use of code signals. Examples are direct sequence, frequency hopping and time hopping spread spectrum techniques. It is also possible to combine these techniques in what are referred to as hybrid methods. The most common technique used in cellular radio networks is the Direct Sequence Spread Spectrum (DS-SS) technique. This is used, for example, in WCDMA technology and in the IS-95 standard.

In DS-SS systems, signal spreading is achieved by modulating the data-modulated signal a second time by a wideband spreading signal. The signal has to approximate closely to a random signal with uniform distribution of the symbols. Typical representatives of such signals in digital form are Pseudonoise (PN) sequences over a finite alphabet. Since a WCDMA system has to maximise system capacity during the spreading, the operation is done in two phases – see Section 2.4.7.1. The user signal is first spread by the channelisation code, which is a so-called Orthogonal Variable Spreading Factor (OVSF) code, its construction being based on the Hadamard matrix. The code has the property that two different codes from the family are perfectly orthogonal if in phase. Thus, its use guarantees maximum capacity, measured by the number of active users. Now all the spread users' signals are scrambled by the cell-specific scrambling sequence, which has the statistical properties of a random sequence. Thus, the system can accommodate the maximum number of users, and the output signal has quite a flat spectrum with no dominant spectral peaks, etc. Readers interested in a deeper knowledge of DS-SS techniques are referred to [31], while further information on WCDMA systems is found in [13] and [19].

2.2.1 Modulation Example

The simplest form of DS-SS employs Binary Phase Shift Keying (BPSK) as the spreading modulation. Mathematically this can be represented as a multiplication of the carrier by a function $c(t)$ that takes the value $+1$ or -1. Let's consider a data-modulated carrier with power P, frequency ω_0 and data phase modulation $\theta_d(t)$, see Equation (2.4):

$$S_n(t) = \sqrt{2 \cdot P} \cdot \cos[\omega_0 \cdot t + \theta_d(t)] \qquad (2.4)$$

The BPSK spreading is accomplished by multiplying $S_n(t)$ by $c(t)$, representing the spreading signal (Figure 2.5).

Multiplication results in the wideband signal:

$$S_w(t) = \sqrt{2 \cdot P} \cdot c(t) \cdot \cos[\omega_0 \cdot t + \theta_d(t)] \qquad (2.5)$$

Data modulation does not also have to be BPSK; there are no restrictions placed on the form of $\theta_d(t)$. But it is common to use the same type of digital phase modulation for

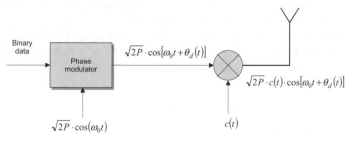

Figure 2.5 BPSK DS-SS transmitter.

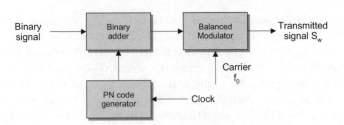

Figure 2.6 Direct sequence transmitter with BPSK data and spreading operation.

the data and the spreading code. When BPSK is used for both, the double-modulation process is replaced by a single-modulation one by the modulo-2 sum of the data and spreading code as illustrated in Figure 2.6.

The wideband signal is transmitted through a channel having a delay T_d. The signal is received together with interference and Additive White Gaussian Noise (AWGN). Despreading is accomplished by the correlation operation of the wideband signal with the appropriately delayed spreading code, as shown in Figure 2.7.

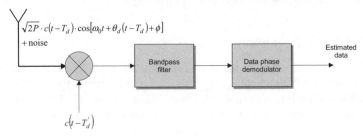

Figure 2.7 BPSK direct sequence receiver.

This demodulation or correlation of the received signal with the delayed spreading code is a critical function in all spread spectrum systems. The signal component of the output of the despreading mixer is described as:

$$S'_n(t) = \sqrt{2 \cdot P} \cdot c(t - T_d) \cdot c(t - T'_d) \cdot \cos[\omega_0 \cdot t + \theta_d(t - T_d) + \phi] \qquad (2.6)$$

T'_d is the receiver's best estimate of the transmission delay. Since $c(t)$ equals $+1$ or -1, the product $c(t - T_d) \cdot c(t - T'_d)$ will be unity if $T_d = T'_d$; that is, if the spreading code at the receiver is synchronised with the spreading code at the transmitter. When correctly synchronised, the signal component is despreaded and S_n can be demodulated using conventional coherent phase modulation [28].

Figure 2.8 illustrates both the data and the spreading and despreading operations applied to it. The processing gain is given by the ratio of chip rate to the user data rate – i.e., directly by Equation (2.3).

2.2.2 Tolerance of Wideband Interference

Tolerance of wideband interference is less straightforward than tolerance of narrowband interference as derived in Section 2.1.3, but the principle is the same –

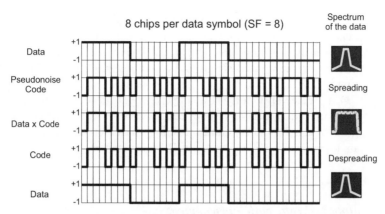

Figure 2.8 Combining of data and spreading sequences.

use of the correlation receiver. This is based on the fact that the tolerance depends on the signal structure; this can be seen from Figure 2.9.

The basic idea is that the receiver works as a correlation receiver, which means that it correlates a known (reference) code signal with an incoming signal that is composed of several different CDMA signals (from different users or channels), of general interference (from other RF systems), and of noise (of thermal nature). The output from the receiver is in the form of the autocorrelation function of the wanted signal. As far as noise and interference are concerned, there is crosscorrelation with all other signals. Consequently, a spreading sequence should have good correlation properties to facilitate separation of the wanted signal from all other signals, i.e.:

- one sharp and dominant peak of the autocorrelation function for zero phase shift;
- as small as possible values of the autocorrelation function for all out-of-phase shifts;
- as small as possible values of the crosscorrelation function for all phase shifts.

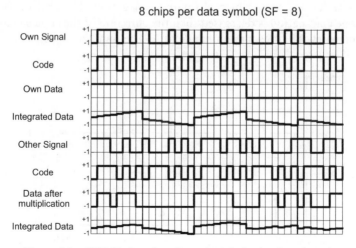

Figure 2.9 CDMA functionality principle in the time domain.

The WCDMA system works with complex spreading codes that result from combining spreading and scrambling codes (see Figure 4.20). This results in the signal having the nature of random signals that have the above properties. The despreading process for the wanted signal and a signal spread with a different code is illustrated in Figure 2.9.

Usually, a CDMA receiver is not just a single correlation receiver but a device that combines several correlation receivers under common control. Its purpose is to receive signals from paths of different delays and reflection characteristics and to combine the resulting signals together. This is known as a RAKE receiver.

2.2.3   Operation in Multi-path Environment

A radio channel can be fully characterised by its time-variant impulse response, $h(t)$. In a mobile radio channel, the impulse response consists of several time-delayed components, as shown in Figure 2.10. This kind of channel is referred to as a multi-path channel. The delayed peaks are due to reflections from surrounding objects, and time dependence is caused by the movement of the User Equipment (UE) and the other objects in the environment. Therefore, a signal transmitted in a multi-path channel will be received as several replicas of the original transmitted signal with different amplitudes and delays, both of which vary in time.

Reception of a DS-SS signal allows the use of a more efficient reception technique – namely, reception using a RAKE receiver [29], as described above. This type of receiver can exploit the energy of the multi-path channel with a resolution of one chip period. It allocates one correlator finger for each multi-path component and performs Maximal Ratio Combining (MRC) of the decorrelated narrowband signals before the symbol decision is made in the demodulator. A RAKE receiver can resolve all multi-path components whose delay difference to other multi-path components is more than one chip period. The shorter the duration of one chip period, the better the RAKE receiver can combat interference caused by multi-path propagation. The tradeoff is that if the duration of one chip period is decreased then the chip rate of the code signal increases, which also increases the transmission bandwidth. The number of fingers in the RAKE

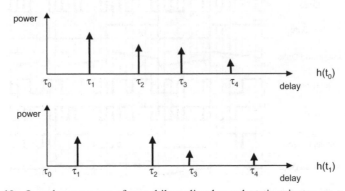

Figure 2.10   Impulse response of a mobile radio channel at time instances t_0 and t_1.

receiver sets a maximum to the number of exploitable paths, as a technology or processing limitation.

2.3 CDMA in Cellular Radio Networks

2.3.1 *Universal Frequency Reuse*

In CDMA all users in the same cell share the same frequency spectrum simultaneously. In a CDMA-based cellular network this is also true for users in different cells. In spread spectrum transmission the interference tolerance enables universal frequency reuse. This underlies all other network-level functions. For example, it enables new functions such as soft handover, but also causes strict requirements for power control.

2.3.2 *Soft Handover*

Because of universal frequency reuse, the connection of a Mobile Station (MS, or generally UE in WCDMA) to the cellular network can include several radio links. When the UE is connected to more than one cell, it is said to be in *soft handover*. If, in particular, the UE has more than one radio link to multiple cells on the same site, it is in *softer handover*. Soft handover is a form of diversity, increasing the signal-to-noise ratio when the transmission power is constant. At network level, soft handover smoothes the movement of a UE from one cell to another. It helps to minimise the transmission power needed in both uplink and downlink. It has two basic characteristics:

- *Soft handover gain* (ca. 1 to 2 dB applicable in the power budget – see Section 2.5.3.3) due to the proper combination of two or more signal branches;
- *Soft handover overhead* due to the fact that the UEs in the handover area are connected to more than one cell. The overall scenario should be clear from Figure 2.11. This overhead should be kept within reasonable limits to save the downlink traffic capacity of the cell. The usual reasonable or maximum acceptable value in CDMA networks (already applying to IS-95 and expected also for WCDMA) is 20–30% – i.e., 1.2 to 1.3 radio links per user connection.

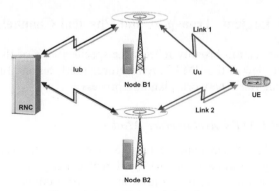

Figure 2.11   UE in soft handover.

2.3.3 Power Control

Power control ensures that each user in the network receives and transmits just enough energy to convey information while causing minimal interference to other users. This is crucial for network capacity. A secondary reason for power control is to minimise battery consumption. For the WCDMA standard, power control is applied in both the uplink and downlink.

When the UE initiates a call, it adjusts its transmission power based on the received common pilot signal power. The common pilot signal is a cell-specific signal broadcast in every cell with constant power. This provides a rough measure of the propagation path loss between the UE and the Node B. The stronger the received common pilot signal power, the less initial transmission power is needed. This type of initial power adjustment is arranged by uplink *open-loop power control*. The process has to be supported by *a priori* information which the UE receives on the cell's Broadcast Channel (BCH). To set up a connection, further control channels need to be used. An analogous procedure is used in the downlink for calculating the initial radio link power based on the E_c/N_0 measured on the common pilot symbols and reported by the terminal to the network.

Variations in the multi-path channel may mean that a fixed target value of the SIR cannot always guarantee a satisfactory quality target. Therefore the target SIR must be controlled based on the achieved bit error rate or block error rate. If the error rate is too high, the target SIR is increased until the desired error rate is met. Increasing the target SIR at the receiver end causes the *closed-loop power control* to increase the transmission power at the transmitter end until the new target SIR is reached. Control of the target SIR is named *outer-loop power control*.

The closed-loop PC function is also used for compensating the fast fading caused by the multi-path channel. In closed-loop power control, the transmission power of the UE is adjusted based on the received power measured at the Node B. The Node B compares the ratio of received energy per symbol to interference energy, SIR, with a target value and commands the UE to increase or decrease its transmission power accordingly. The same process is performed in the downlink.

WCDMA power control is described in more detail in Section 4.2.

2.4 WCDMA Logical, Transport and Physical Channels

The content of this section broadly reflects the specifications of the 3GPP. It aims at giving a panoramic view of the UMTS architecture, interfaces and functions that have a direct impact on the radio network planning process.

2.4.1 High-level UMTS Architecture Model

The Universal Mobile Telecommunications System (UMTS) has been designed to support a wide range of applications with different Quality of Service (QoS) profiles. The system is intended for long-term duration, and the modular approach adopted in 3GPP provides the necessary flexibility for its evolution [1].

In 3GPP the UMTS architecture is described in terms of its entities – UE, UMTS Terrestrial Radio Access Network (UTRAN) and Core Network (CN). The radio interface (Uu) and the CN–UTRAN interface (Iu) are the reference points between the sub-systems. The protocols over the Uu and Iu interfaces are divided into two structures: User-plane (U-plane) protocols – i.e., the protocols implementing the actual Radio Access Bearer (RAB) service; and Control-plane (C-plane) protocols – i.e., the protocols for controlling RABs and the connection between the UE and the CN. Both Uu and Iu protocols provide transparent transfer of Non-Access Stratum (NAS) messages [3].

2.4.1.1 Access Stratum and Non-Access Stratum

The high-level functional grouping into Access Stratum (AS) and Non-Access Stratum (NAS) defined in [2] is depicted in Figure 2.12.

The AS is the functional grouping of protocols specific to the access technique. It includes protocols for supporting transfer of radio-related information, for coordinating the use of radio resources between UE and access network, and for supporting the access from the serving network to the resources provided by the access network. The AS offers services through Service Access Points (SAPs) to the NAS (CN-related signalling and services) – i.e., provides the access link between the UE and CN – which consists of one or more independent and simultaneous UE–CN RAB services, and only one signalling connection between the upper-layer entities of the UE and CN. As shown in Figure 2.12, the signalling connection consists of two parts – the Radio Resource Control (RRC) connection and the Iu connection, which expands the RRC signalling connection towards the CN.

The NAS is the functional grouping of protocols aimed at Call Control (CC) for circuit switched voice and data, at Session Management (SM) for packet switched data,

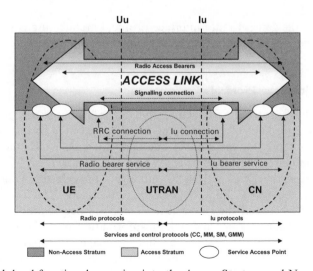

Figure 2.12 High-level functional grouping into the Access Stratum and Non-Access Stratum.

at Mobility Management (MM, GMM) and at Short Message Services (SMSs) for packet and circuit switched domains; at Supplementary Services (SS) and at RAB management for re-establishment of RAB(s) which still have active PDP (Packet Data Protocol) contexts.

The RAB is a service provided by the Access Stratum to the Non-Access Stratum in order to transfer user data between the UE and CN. A bearer is described by a set of parameters (attributes), which define that particular traffic aspect or QoS profile of that particular application or service. The QoS concept and architecture used in UMTS – i.e., the list of attributes applicable to the UMTS Bearer Service and the RAB Service – are discussed in Chapter 8.

As illustrated in Figure 2.12, a RAB Service breaks down into a Radio Bearer (RB) Service and an Iu-Bearer Service on the Uu and Iu interfaces, respectively. The RB Service covers all aspects of radio interface transport; this bearer service uses the UTRA FDD (Frequency Division Duplex) and it is brought about by the RLC-U (Radio Link Control protocol, User plane) layer between the Radio Network Controller (RNC) and UE. The Iu-Bearer Service, together with the Physical Bearer Service, provides the transport between UTRAN and CN. In the packet domain there is a one-to-one relationship between PDP context and RAB as well as between RAB and RB.

2.4.2 Radio Interface Protocol Architecture and Logical Channels

The radio interface protocols are needed to set up, reconfigure and release the Radio Bearer Services. The radio interface consists of three protocol layers – the physical layer (L1), the data link layer (L2) and the network layer (L3). L2 contains the following sub-layers: Medium Access Control (MAC), Radio Link Control (RLC), Packet Data Convergence Protocol (PDCP) and Broadcast/Multicast Control (BMC). RLC is divided into C-plane and U-plane, while PDCP and BMC exist only in the U-plane. L3 consists of one protocol, denoted as Radio Resource Control (RRC), which belongs to the C-plane [2].

In this section the general radio interface protocol architecture is presented and for each protocol the logical architecture and its functionality are pointed out.

2.4.2.1   Radio Interface Protocols

The radio interface protocol architecture and the connections between protocols are shown in Figure 2.13. Each block represents an instance of the corresponding protocol. The dashed lines represent the control interfaces through which the RRC protocol controls and configures the lower layers. The SAPs between the MAC and physical layers and between the RLC and MAC sub-layers provide the Transport Channels (TrCHs) and the Logical Channels (LoCHs), respectively. The TrCHs are specified for data transport between the physical layer and L2 peer entities, whereas LoCHs just define the transfer of a specific type of information over the radio interface [2].

As illustrated in Figure 2.13, the signalling messages transported over the radio interface consist of RRC-generated signalling messages and NAS messages generated by higher layers and mapped onto RRC messages [2].

UNIVERSITY OF CENTRAL
LANCASHIRE LIBRARY

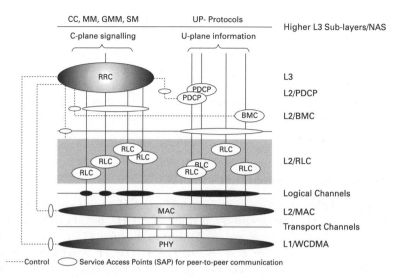

Figure 2.13   UTRA Frequency Division Duplex radio interface protocol architecture.

Examples of the termination of the air interface protocols are shown in Figure 2.14. The figure shows Release '99 data only. The special case of High-speed Downlink Packet Access (HSDPA) is discussed separately in Section 2.4.5. The protocol stacks also include other protocols of the radio network layer, which run on top of the transport network layer. They are the Frame Protocol (FP) and the Radio Access Network Application Protocol (RANAP). The RANAP belongs to the Iu C-plane and in this context expands the RRC signalling connection towards the CN [3].

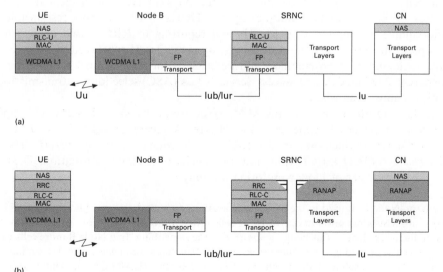

Figure 2.14   Informative examples based on [3] Release '99 data only: (a) termination of air interface U-plane protocols; (b) termination of air interface C-plane protocols.

The FP layer is an Iub (interface between an RNC and a Node B[1]) and Iur (logical interface between two RNCs) U-plane protocol on top of Asynchronous Transfer Mode (ATM) Adaptation Layer type 2 (AAL2), which is used to transfer user data, in addition to the necessary related control information, between the Serving RNC (SRNC) and a Node B. The UTRAN architecture is defined in [3]. The FP is further used to support some simple Iub/Iur procedures such as, for example, timing adjustment for UTRAN synchronisation and outer-loop power control.

2.4.2.2   Medium Access Control (MAC) Protocol

MAC is responsible for mapping *logical channels* (LoCHs) onto appropriate *transport channels* (TrCHs). MAC provides an efficient use of TrCHs; based on the instantaneous source rate(s), it selects the appropriate Transport Format (TF) within an assigned Transport Format Set (TFS) for each active TrCH. The TF is selected based on the Transport Format Combination Set (TFCS), which is assigned by the RRC protocol and produced by the admission control in the RNC when an RAB is set up or modified [5].

The functionality of the MAC layer includes priority handling between data flows of one connection (selection of a TFC for which high-priority data is mapped onto L1 with a 'high bit rate' TF); priority handling between UEs by means of dynamic scheduling and identification of UEs on common transport channels (in-band identification). MAC provides multiplexing and demultiplexing functions of RLC Packet Data Units (PDUs) into and from TBs delivered to and from the physical layer on common channels (services or better LoCHs multiplexing for CCHs), and multiplexing and demultiplexing functions of RLC PDUs into and from Transport Block Sets (TBSs) delivered to and from the physical layer on dedicated channels (services or better LoCHs multiplexing for DCHs). MAC is also responsible for traffic volume measurement on LoCHs and reporting to RRC, based on which the RRC performs TrCH switching decisions; dynamic TrCH-type switching (execution of the switching between common and dedicated transport channels); ciphering (for transparent RLC mode); and Access Service Class (ASC) selection for transmission of uplink common channels.

The data transfer services of the MAC layer are provided on LoCHs. The type of information transferred defines each LoCH type. A general classification of LoCHs is presented in [2], in which they are divided into two groups: *control channels* (CCHs) and *traffic channels* (TCHs). CCHs are used for transfer of C-plane information; TCHs are used for the transfer of U-plane information only.

---

[1] The term 'Node B' is used in the 3GPP specifications to indicate a logical node responsible for radio transmission/reception in one or more cells to/from the UE. The more generic term 'Base Station' used elsewhere in this book means exactly the same thing. In a Node B more cells can be set up; a 'cell' is defined by a Cell Identification (C-ID), configuration generation ID, timing delay (T_Cell), UTRA Absolute Radio Frequency Channel Number (UARFCN), maximum transmission power, closed-loop timing adjustment mode and primary scrambling code [17].

The meaning of the CCHs can be summarised as follows:

- *Broadcast Control Channel (BCCH)*, for broadcasting system control information in the downlink.
- *Paging Control Channel (PCCH)*, for transferring paging information in the downlink (used when the network does not know the cell location of the UE, or when the UE is in cell-connected state).
- *Common Control Channel (CCCH)*, for transmitting control information between the network and UEs in both directions (commonly used by UEs having no RRC connection with the network and by UEs using common transport channels when accessing a new cell after cell reselection).
- *Dedicated Control Channel (DCCH)*, a point-to-point bidirectional channel for transmitting dedicated control information between the network and a UE (established through the RRC connection setup procedure).

The TCHs can be described as:

- *Dedicated Traffic Channel (DTCH)*, a point-to-point channel dedicated to one UE for transfer of user information (a DTCH can exist in both uplink and downlink directions).
- *Common Traffic Channel (CTCH)*, a point-to-multi-point unidirectional channel for transfer of dedicated user information for all or a group of specified UEs.

The mapping between logical and transport channels is depicted in Figure 2.15.

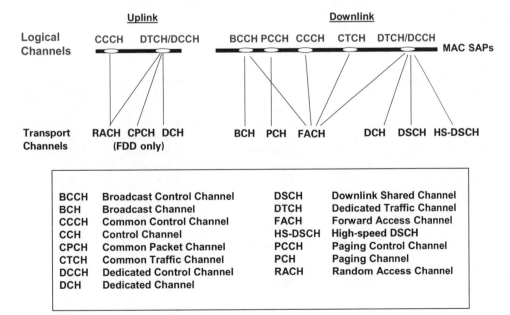

BCCH	Broadcast Control Channel	DSCH	Downlink Shared Channel
BCH	Broadcast Channel	DTCH	Dedicated Traffic Channel
CCCH	Common Control Channel	FACH	Forward Access Channel
CCH	Control Channel	HS-DSCH	High-speed DSCH
CPCH	Common Packet Channel	PCCH	Paging Control Channel
CTCH	Common Traffic Channel	PCH	Paging Channel
DCCH	Dedicated Control Channel	RACH	Random Access Channel
DCH	Dedicated Channel		

Figure 2.15 Mapping between logical channels and transport channels in uplink and downlink directions (for UTRA FDD only – i.e., without TDD channels).

2.4.2.3　Radio Link Control (RLC) Protocol

The RLC protocol provides segmentation/reassemble (Payloads Units, PU) and retransmission services for both user (RB) and control data (Signalling RB) [6].

Each RLC instance is configured by RRC to operate in one of three modes. These are Transparent Mode (TM), where no protocol overhead is added to higher layer data; Unacknowledged Mode (UM), where no retransmission protocol is in use and data delivery is not guaranteed; and Acknowledged Mode (AM), where the Automatic Repeat reQuest (ARQ) mechanism is used for error correction. For all RLC modes, Cyclic Redundancy Check (CRC) error detection is performed at the physical layer and the result of the CRC is delivered to the RLC together with the actual data.

Some of the most important functions of the RLC protocol are segmentation and reassembly of variable length higher layer PDUs into/from smaller RLC PUs; error correction, by means of retransmission in the acknowledged data transfer mode; in-sequence delivery of upper layer PDUs; flow control – i.e., rate control at which the peer RLC transmitting entity may send information; protocol error detection and recovery; Service Data Unit (SDU) discard, polling, ciphering and maintenance of the QoS as defined by upper layers.

As shown in Table 2.1, the RLC transfer mode indicates the data transfer mode supported by the RLC entity configured for that particular RB. The transfer mode for a RB is the same in both uplink and downlink directions, and is determined by the admission control in the SRNC from the RAB attributes and CN domain information.

The RLC transfer mode affects the configuration parameters of the outer-loop power control in the RNC and the user bit rate. The quality target is not affected if TM or UM RLC is used, while the number of retransmissions should be taken into account during

Table 2.1　RLC transfer modes for UMTS Quality of Service classes.

UMTS QoS class[a]	Domain	Source statistics descriptor	Service type	RLC transfer mode
Conversational	CS	Speech	CS speech	TM
		Unknown	CS T data	TM
	PS	Speech	PS speech	UM
		Unknown	PS RT data	UM
Streaming	CS	Speech	CS speech	N/A
		Unknown	CS NRT data	TM
	PS	Speech	PS speech	N/A
		Unknown	PS RT data	AM or UM[b]
Interactive	CS	N/A	—	N/A
	PS	N/A	PS NRT data	AM
Background	CS	N/A	—	N/A
	PS	N/A	PS NRT data	AM

[a] Type of application for which the UMTS bearer service is optimised [10].
[b] Transfer mode depends on the value of RAB attribute *Transfer delay*.

radio network planning if AM RLC is employed. The user bit rate is affected by the transfer mode of the RLC protocol, since the length of the L2 headers is 16 bits for AM, 8 bits for UM and 0 bits for TM. Hence, the user bit rate for radio network dimensioning is given by the L1 bit rate reduced by the L2 header bit rate.

2.4.2.4   Packet Data Convergence Protocol

This protocol exists only in the U-plane and only for services from the Packet Switched (PS) domain. The main PDCP functions are compression of redundant protocol control information (e.g., TCP/IP and RTP/UDP/IP headers) at the transmitting entity and decompression at the receiving entity; transfer of user data – i.e., receiving a PDCP_SDU from NAS and forwarding it to the appropriate RLC entity and vice versa; and multiplexing RBs into one RLC entity [7].

2.4.2.5   Broadcast Multicast Control Protocol

Like the PDCP, the BMC protocol exists only in the U-plane. This protocol provides a broadcast/multi-cast transmission service on the radio interface for common user data in TM or UM. It utilises UM RLC using the CTCH LoCH mapped onto the Forward Access Channel (FACH). The CTCH has to be configured and the TrCH used by the network has to be indicated to all UEs via RRC system information broadcast on the BCH [8].

2.4.2.6   Radio Resource Control (RRC) Protocol

RRC signalling is used to control the mobility of the UE in Connected Mode; to broadcast the information related to the NAS and AS; and to establish, reconfigure and release RBs. The RRC protocol is further used for setting up and controlling UE measurement-reporting criteria and the downlink outer-loop power control. Paging, control of ciphering, initial cell selection and cell reselection are also part of RRC connection management procedures. RRC messages carry all parameters required to set up, modify and release L2 and L1 protocol entities [9].

After power on, UEs stay in Idle Mode until a request to establish an RRC connection is transmitted to the network. In Idle Mode the connection of the UE is closed on all layers of the AS. In Idle Mode the UE is identified by NAS identities such as International Mobile Subscriber Identity (IMSI), Temporary Mobile Subscriber Identity (TMSI) and Packet-TMSI. The RNC has no information about any individual UE, and it can only address, for example, all UEs in a cell or all UEs monitoring a paging occasion [9]. The transitions between Idle Mode and UTRA Connected Mode are shown in Figure 2.16.

The UTRA Connected Mode is entered when an RRC connection is established. The RRC connection is defined as a point-to-point bidirectional connection between RRC peer entities in the UE and in the UTRAN. A UE has either none or a single RRC connection. The RRC connection establishment procedure can only be initiated by the UE sending an RRC connection request message to the RAN. The event is triggered either by a paging request from the network or by a request from upper layers in the

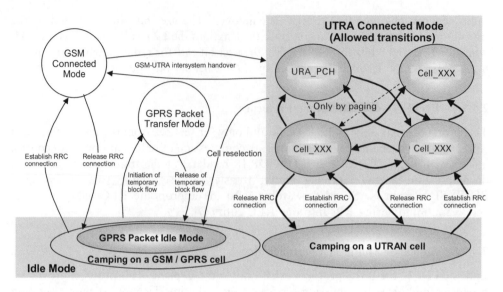

Figure 2.16 Radio resource control states and state transitions, including GSM Connected Mode for PSTN/ISDN domain services and GSM/GPRS Packet Modes for IP domain services.

UE. When the RRC connection is established, the UE is assigned a Radio Network Temporary Identity (RNTI) to be used as its own identity on CTCHs. When the network releases the RRC connection, the signalling link and all RBs between the UE and the UTRAN are released [9]. As depicted in Figure 2.16, the RRC states are as follows:

- *Cell_DCH*. In this state the Dedicated Physical Channel (DPCH), plus eventually the Physical Downlink Shared Channel (PDSCH), is allocated to the UE. It is entered from Idle Mode or by establishing a DTCH from the Cell_FACH state. In this state the terminal performs measurements according to the RRC MEASUREMENT CONTROL message. The transition from Cell_DCH to Cell_FACH can occur via explicit signalling – e.g., through expiration of an inactivity timer.
- *Cell_FACH*. In this state no DPCH is allocated to the UE; the Random Access transport Channel (RACH) and the FACH are used for transmitting signalling and a small amount of user data instead. The UE listens to the BCH system information and moves to the Cell_PCH substate via explicit signalling when the inactivity timer on the FACH expires.
- *Cell_PCH*. In this state the UE location is known by the SRNC on a cell level, but it can only be reached via a paging message. This state allows low battery consumption. The UE may use Discontinuous Reception (DRX), reads the BCH to acquire valid system information and moves to Cell_FACH if paged by the network or through any uplink access – e.g., initiated by the terminal for cell reselection (cell update procedure).
- *URA_PCH*. This state is similar to Cell_PCH, except that the UE executes the cell update procedure only if the UTRAN Registration Area (URA) is changed. One cell can belong to one or several URAs in order to avoid ping-pong effects. When the

number of cell updates exceeds a certain limit, the UE may be moved to the URA_PCH state via explicit signalling. The DCCH cannot be used in this state, and any activity can be initiated by the network via a paging request on PCCH or through uplink access by the terminal using RACH.

The understanding of RRC functions and signalling procedures is essential for radio network tuning and optimisation. Through RRC protocol analysis, it is possible to monitor the system information broadcast in the cell, paging messages, cell selection and reselection procedures, the establishment, maintenance and release of the RRC connection between the UE and UTRAN, the UE measurement reporting criteria and their control, and downlink open-loop and outer-loop power control.

2.4.3 Transport Channels

In UTRAN, data generated at higher layers is carried over the air interface using TrCHs mapped onto different physical channels. The physical layer has been designed to support variable bit rate transport channels, to offer bandwidth-on-demand services, and to be able to multiplex several services within the same RRC connection into one Coded Composite Transport Channel (CCTrCH). A CCTrCH is carried by one physical CCH and one or more physical data channels. There can be more than one downlink CCTrCH, but only one physical CCH is transmitted on a given connection [4].

In 3GPP all TrCHs are defined as unidirectional – i.e., uplink, downlink or relay link. Depending on services and state, the UE can have simultaneously one or several TrCHs in the downlink, and one or more TrCHs in the uplink.

As shown in Figure 2.17, for each TrCH, at any Transmission Time Interval (TTI) the physical layer receives from higher layers a TBS and the corresponding Transport Format Indicator (TFI). Then L1 combines the TFI information received from different TrCHs into one Transport Format Combination Indicator (TFCI). The TFCI is transmitted in the physical CCH to inform the receiver about what TrCHs

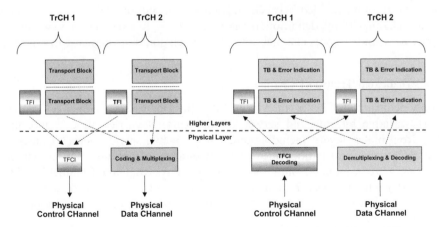

Figure 2.17 Interface between higher layers and the physical layer [19].

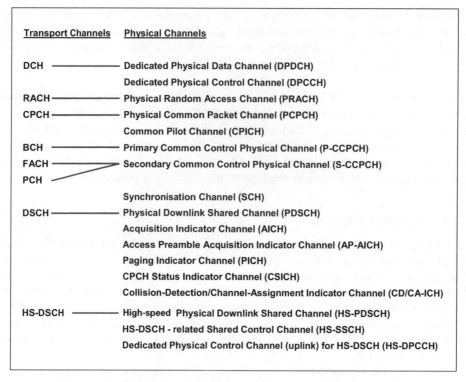

Figure 2.18 Mapping of transport channels onto physical channels.

are simultaneously active in the current radio frame. In the downlink, in the case of limited TFCSs the TFCI signalling may be omitted and Blind Transport Format Detection (BTFD) can be employed, where decoding of TrCHs can be done so as to verify which position of the output block is matched with the CRC results [4].

Two types of TrCHs exist: dedicated channels and common channels. A common channel is a resource divided between all users or a group of users in a cell, whereas a dedicated channel is by definition reserved for a single user. The connections and mapping between transport channels and physical channels are depicted in Figure 2.18.

2.4.3.1 Dedicated Transport Channels

The only dedicated TrCH specified in 3GPP is the Dedicated Channel (DCH), which supports variable bit rate and service multiplexing. It carries all user information coming from higher layers, including data for the actual service (speech frames, data, etc.) and control information (measurement control commands, UE measurement reports, etc.). It is mapped onto the Dedicated Physical Data Channel (DPDCH). The DPCH is characterised by closed-loop power control and fast data rate change on a frame-by-frame basis; it can be transmitted to part of the cell and supports soft/softer handover [4].

2.4.3.2   Common Transport Channels

The common TrCHs are a resource divided between all users or a group of users in a cell (an in-band identifier is needed). They do not support soft/softer handover, but some of them can have fast power control – for example, the Common Packet Channel (CPCH) and Downlink Shared Channel (DSCH). As depicted in Figures 2.15 and 2.18, the common TrCHs are as follows ([4], [2]):

- *Broadcast Channel (BCH).* This is used to transmit information (e.g., random access codes, cell access slots, cell-type transmit diversity methods, etc.) specific to the UTRA network or to a given cell; it is mapped onto the Primary Common Control Physical Channel (P-CCPCH), which is a downlink data channel only.
- *Forward Access Channel (FACH).* This carries downlink control information to terminals known to be located in the given cell. It is further used to transmit a small amount of downlink packet data. There can be more than one FACH in a cell, even multiplexed onto the same Secondary Common Control Physical Channel (S-CCPCH). The S-CCPCH may use different offsets between the control and data field at different symbol rates and may support slow power control.
- *Paging Channel (PCH).* This carries data relevant to the paging procedure. The paging message can be transmitted in a single cell or several cells, according to the system configuration. It is mapped onto the S-CCPCH.
- *Random Access Channel (RACH).* This carries uplink control information, such as a request to set up an RRC connection. It is further used to send small amounts of uplink packet data. It is mapped onto the Physical Random Access Channel (PRACH).
- *Uplink Common Packet Channel (CPCH).* This carries uplink packet-based user data. It supports uplink inner-loop power control, with the aid of a downlink Dedicated Physical Control Channel (DPCCH). Its transmission may span over several radio frames and it is mapped onto the Physical Common Packet Channel (PCPCH).
- *Downlink Shared Channel (DSCH).* This carries dedicated user data and/or control information and can be shared in time between several users. As a pure data channel, it is always associated with a downlink DCH. It supports the use of downlink inner-loop power control, based on the associated uplink DPCCH. It is mapped onto the Physical DL Shared Channel (PDSCH).
- *High-speed Downlink Shared Channel (HS-DSCH).* This downlink channel is shared between UEs by allocation of individual codes from a common pool of codes reserved for the HS-DSCH. The HS-DSCH is defined as an extension to DCH transmission. Physical channel signalling is used for indicating to a UE when it has been scheduled including the necessary signalling information for the UE to decode the High-speed Physical Downlink Shared Channel (HS-PDSCH) as well.

The common TrCHs needed for basic cell operation are RACH, FACH and PCH, while the DSCH, CPCH and HS-DSCH may or may not be used by the operator.

2.4.3.3 Formats and Configurations

In order to describe how the mapping of TrCHs is performed and controlled by L1, some generic definitions and terms valid for all types of TrCH are introduced in this section. Further information can be found in [4].

- *Transport Block (TB)* is the basic unit exchanged between L1 and MAC for L1 processing; a TB typically corresponds to an RLC PDU or corresponding unit. L1 adds a CRC to each TB.
- *Transport Block Set (TBS)* is defined as a set of TBs that are exchanged between L1 and MAC at the same time instant using the same TrCH.
- *Transport Block Size* is defined as the number of bits in a TB and is always fixed within a given TBS – i.e., all TBs within a TBS are equally sized.
- *Transport Block Set Size* is defined as the number of bits in a TBS.
- *Transmission Time Interval (TTI)* is defined as the inter-arrival time of TBSs, and is equal to the periodicity at which a TBS is transferred by the physical layer on the radio interface. It is always a multiple of the minimum interleaving period (i.e., 10 ms, the length of one radio frame, an exception is HS-DSCH with TTI = 2 ms as discussed in Section 2.4.5). MAC delivers one TBS to the physical layer every TTI.
- *Transport Format (TF)* is the format offered by L1 to MAC (and vice versa) for the delivery of a TBS during a TTI on a given TrCH. It consists of one *dynamic part* (TB Size, TBS Size) and one *semi-static part* (TTI, type of error protection– i.e., turbo code, convolutional code or no channel coding – coding rate, static Rate Matching parameter, size of CRC). An empty TF is defined as a TF that has a TBS size equal to zero.
- *Transport Format Set (TFS)* is a set of TFs associated with a TrCH. The semi-static parts of all TFs are the same within a TFS. TB size, TBS size and TTI define the TrCH bit rate before L1 processing. As an example, for a DCH, assuming a TB size of 336 bits (320 bits payload + 16 bits RLC header), a TBS size of 2 TBs per TTI, and a TTI of 10 ms, the DCH bit rate is given by $336 * 2/10 = 67.2$ kbps, whereas the DCH user bit rate, which is defined as the DCH bit rate reduced by the RLC headers, is given by $320 * 2/10 = 64$ kbps. Depending on the type of service carried by the TrCH, the variable bit rate may be achieved by changing between TTIs either the TBS size only, or both the TBS and TBS size.
- *Transport Format Combination (TFC)* is an authorised combination of the currently valid TFs that can be simultaneously submitted to L1 on a CCTrCH of a UE – i.e., containing one TF from each TrCH that is part of the combination. An empty TFC is defined as a TFC that is only made up of empty TFs.
- *Transport Format Combination Set (TFCS)* is defined as a set of TFCs on a CCTrCH and is produced by a proprietary algorithm in the RNC. The TFCS is what is given to MAC by L3 for control. When mapping data onto L1, MAC chooses between the different TFCs specified in the TFCS. MAC has only control over the dynamic part of the TFC, since the semi-static part corresponds to the service attributes (quality, transfer delay) set by the admission control in the RNC. The selection of TFCs can be seen as the fast part of the RRC dedicated to MAC, close to L1. Thereby the bit rate can be changed very quickly and with no need of L3 signalling. An example of

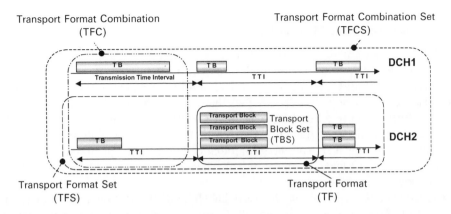

Figure 2.19 Example of data exchange between Medium Access Control and the physical layer when two Dedicated Channels are employed.

data exchange between MAC and the physical layer when two DCHs are multiplexed in the connection is illustrated in Figure 2.19.

- *Transport Format Indicator (TFI)* is a label for a specific TF within a TFS. It is used in the inter-layer communication between MAC and L1 each time a TBS is exchanged between the two layers on a TrCH.
- *Transport Format Combination Indicator (TFCI)* is used to inform the receiving side of the currently valid TFC, and hence how to decode, demultiplex and transfer the received data to MAC on the appropriate TrCHs. MAC indicates the TFI to L1 at each delivery of TBSs on each TrCH. L1 then builds the TFCI from the TFIs of all parallel TrCHs of the UE, processes the TBs appropriately and appends the TFCI to the physical control signalling (DPCCH). Through the detection of the TFCI the receiving side is able to identify the TFC.

The TFCS may be produced as shown in Figure 2.20 – i.e., as a Cartesian product between TFSs of the TrCHs that are multiplexed onto a CCTrCH, each considered as a vector. In theory every TrCH can have any TF in the TFC, but in practice only a limited number of possible combinations are selected.

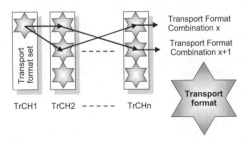

Figure 2.20 Relations of transport format, transport format set and transport format combination.

2.4.3.4 Functions of the Physical Layer

One UE can transmit only one CCTrCH at a time, but multiple CCTrCHs can be simultaneously received in the downlink direction. In the uplink one TFCI represents the current TFs of all DCHs of the CCTrCH. RACHs are always mapped one-to-one onto physical channels (PRACHs) – i.e., there is no physical layer multiplexing of RACHs. Further, only a single CPCH of a CPCH set is mapped onto a PCPCH, which employs a subset of the TFCs derived by the TFS of the CPCH set. A CPCH set is characterised by a set-specific scrambling code for access preamble and collision detection, and is assigned to the terminal when a service is configured for CPCH transmission [4].

In the downlink the mapping between DCHs and physical channel data streams works in the same way as in the uplink direction. The current configuration of the coding and multiplexing unit is either signalled (TFCI) to the UE, or optionally blindly (BTFD) detected. Each CCTrCH has only zero or one corresponding TFCI mapped (each 10 ms radio frame) on the same DPCCH used in the connection. A PCH and one or several FACHs can be encoded and multiplexed together forming a CCTrCH, one TFCI indicates the TFs used on each FACH and PCH carried by the same S-CCPCH. The PCH is always associated with the Paging Indicator Channel (PICH), which is used to trigger off the UE reception of S-CCPCH where the PCH is mapped. A FACH or a PCH can also be individually mapped onto a separate physical channel. The BCH is always mapped onto the P-CCPCH, with no multiplexing with other TrCHs [4].

The main functions of the physical layer are Forward Error Correction (FEC) encoding and decoding of TrCHs, measurements and indication to higher layers (e.g., BER, SIR, interference power, transmission power, etc.), macro-diversity distribution/combining and softer handover execution, error detection on TrCHs (CRC), multiplexing of transport channels and demultiplexing of CCTrCHs, rate matching, mapping of CCTrCHs onto physical channels, modulation/ demodulation and spreading/despreading of physical channels, frequency and time (chip, bit, slot, frame) synchronisation, closed-loop (inner-loop) power control, power weighting, combining of physical channels and RF processing.

The multiplexing and channel coding chain is depicted in Figures 2.21 and 2.22 for the uplink and downlink direction, respectively. As shown in these figures, data arrive at the coding/multiplexing unit in the form of TBSs once every TTI. The TTI is TrCH-specific from the set (10 ms, 20 ms, 40 ms, 80 ms) [12].

Error detection is provided on transport blocks through a CRC. The CRC length is determined by the admission control in the RNC and can be 24, 16, 12, 8 or 0 bits [12]. Regardless of the result of the CRC, all TBs are delivered to L2 along with the associated error indications. This estimation is then used as quality information for UL macro-diversity selection/combining in the RNC, and may also be used directly as an error indication to L2 for each erroneous TB in TM, UM and AM RLC, provided that RLC PDUs are mapped one-to-one onto TBs.

Depending on whether the TB fits in the available code block size (channel coding method), the TBs in a TTI are either concatenated or segmented to coding blocks of suitable size.

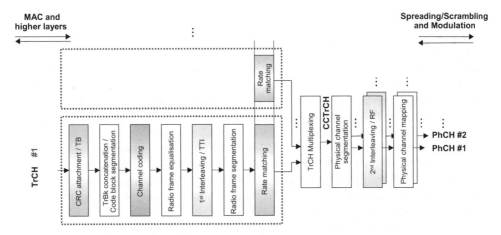

Figure 2.21   Uplink multiplexing and channel coding chain.

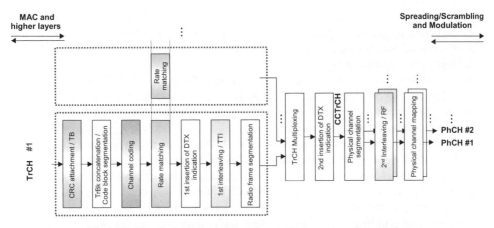

Figure 2.22   Downlink multiplexing and channel coding chain.

Channel coding and radio frame equalisation is performed on the coding blocks after the concatenation or segmentation operation. Only the channel-coding schemes reported in Table 2.2 can be applied to TrCHs – i.e., either convolutional coding, turbo coding or no coding (no limitation to coding block size).

Table 2.2   Transport Channel coding schemes.

Type of TrCH	Coding scheme	Coding rate
BCH, PCH, RACH	Convolutional coding	1/2
CPCH, DCH, DSCH, FACH	Convolutional coding	1/3, 1/2
	Turbo coding	1/3
	No coding	

Convolutional coding is typically used with relative low data rates – e.g., the BTFD using the Viterbi decoder is much faster than turbo coding – whereas turbo coding is applied for higher data rates and brings performance benefits when a large enough block size is achieved for a significant interleaving effect [19]. For example, the Adaptive Multi Rate (AMR) speech service (coordinated TrCHs, multiplexed in the FP) uses Unequal Error Protection (UEP): class A bits, strong protection (1/3 convolutional coding and 12 bits CRC); class B bits, less protected (1/3 convolutional coding); and class C bits, least protection (1/2 convolutional coding).

The function of radio frame equalisation (padding) is to ensure that data arriving after channel coding can be divided into blocks of equal length when transmitted over more than a single 10 ms radio frame. Such radio frame equalisation is only performed in the uplink, because in the downlink the rate matching output block length is already produced in blocks of equal size per frame.

The first interleaving (or the first radio frame interleaving) is used when the delay budget allows more than 10 ms of interleaving period. The first interleaving period is related to the TTI.

The rate matching procedure is used to match the number of bits to be transmitted to the number available on a single frame (DPCH), either by puncturing or by repetition. The amount of puncturing or repetition depends on the particular service combination and their QoS requirements.

Rate matching takes into account the number of bits of all TrCHs active in that frame. The admission control located in the RNC provides a semi-static parameter, the *rate matching attribute*, to control the relative rate matching between different TrCHs. The rate matching attribute is used to calculate the rate matching value when multiplexing several TrCHs for the same frame. With the aid of the rate matching attribute and TFCI, the receiver can back-calculate the rate matching parameters used and perform the inverse operation. By adjusting the rate matching attribute, admission control of the RNC fine-tunes the quality of different services in order to reach an equal or nearly equal symbol power-level requirement for all services.

Variable rate handling is performed after TrCH multiplexing for matching the total instantaneous rate of the multiplexed TrCHs to the channel bit rate of the DPDCH (when the TBSs do not contain the maximum number of DPDCH bits). The number of bits on a TrCH can vary between different TTIs.

In the downlink, transmission is interrupted if the number of bits is less than the maximum allowed by the DPDCH. As shown in Figure 2.23(a), a fixed position TrCH always uses the same symbols in the DPCH. If the transmission rate is below the maximum, Discontinuous Transmission (DTX) indication is then used for those symbols. The different TrCHs do not have a dynamic impact on the rate matching values applied for the other channel, and all TrCHs can use the maximum bit rate simultaneously (the space taken always depends on the maximum TF of the TFS). A fixed position TrCH allows easier blind detection. If TrCH positions were flexible when mapped onto the physical channel, as shown in Figure 2.23(b), the channel bits not being used by one service might be used by another. Blind detection is possible (for low data rates and for a few possibly higher data rates) but is not required by the specifications [19].

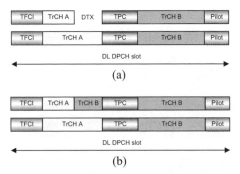

Figure 2.23   Example of (a) fixed position and (b) flexible position Transport Channels.

In the uplink, bits are repeated or punctured to ensure that the total bit rate after TrCH multiplexing is identical to the total channel bit rate of the allocated DPCHs. Rate matching is performed in a more dynamic way and may vary on a frame-by-frame basis.

Multicode transmission is employed when the total bit rate to be transmitted on a CCTrCH exceeds the maximum bit rate of the DPCH. Multicode transmission depends on the multi-code capabilities of the UE and Node B, and consists of several parallel DPDCHs transmitted for one CCTrCH using the same Spreading Factor (SF):

- In the downlink, if several CCTrCHs are employed for one UE, each CCTrCH can have a different spreading factor, but only one DPCCH is used for them in the connection.
- In the uplink, the UE can use only one CCTrCH simultaneously. Multicode operation is possible if the maximum allowed amount of puncturing has already been applied. For the different codes it is mandatory for the terminal to use $SF = 4$. Up to six parallel DPDCHs and only one DPCCH per connection can be transmitted.

The second interleaving is also called intra-frame interleaving (10 ms radio frame interleaving). It consists of block inter-column permutations, separately applied for each physical channel (if more than a single code channel is transmitted).

2.4.4   Physical Channels and Mapping of Transport Channels (FDD)

In this section the dedicated physical channel structure is described. Further explanation can be found in [11]. A physical channel is identified by a specific carrier frequency, scrambling code, channelisation code (optional), duration and, on the uplink, relative phase (0 or $\pi/2$). In UMTS the transmission of a physical channel in normal mode is continuous, but in compressed mode it is interrupted to allow the UE to monitor cells on other FDD frequencies and those from other radio access technologies, such as GSM.

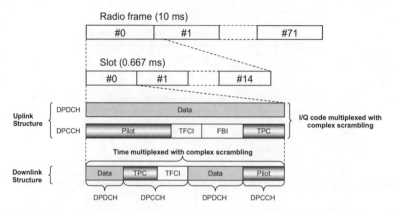

Figure 2.24 Structure of the dedicated physical channels, in the uplink and downlink directions.

2.4.4.1 Dedicated Physical Channel Structure

The dedicated physical channel structure is depicted in Figure 2.24. In this model each 2-bit pair represents an I/Q pair of Quaternary Phase Shift Keying (QPSK) modulation (symbol). As shown in the figure, the frame structure consists of a sequence of radio frames, one radio frame corresponding to 15 slots (10 ms or 38400 chips) and one slot corresponding to 2560 chips (0.667 ms), which equals one power control period.

2.4.4.2 Dedicated Uplink Physical Channel

The dedicated uplink physical channel structure for one power control period is shown in Figure 2.24. The dedicated higher layer information, including user data and signalling, is carried by the uplink DPDCH, and the control information generated at L1 is mapped onto the uplink DPCCH. The DPCCH comprises pre-defined Pilot symbols (used for channel estimation and coherent detection/averaging), power control commands, Feedback Information (FBI) for closed-loop mode transmit diversity and Site Selection Diversity Technique (SSDT), and optionally a TFCI. There can be zero, one or several uplink DPDCHs on each radio link, but only one uplink DPCCH is transmitted. DPDCH(s) and DPCCH are I/Q-code-multiplexed with complex scrambling. Further, as shown in Table 2.3, the uplink DPDCH can have a spreading factor from 256 (15 ksps) down to 4 (960 ksps), whereas the uplink DPCCH is always transmitted with a spreading factor of 256 (15 ksps). Table 2.3 also shows the uplink physical channel parameters for multiplexing of data, speech and Signalling Radio Bearer (SRB) [11].

Admission control in the RNC produces the TFCS and estimates the minimum allowed SF. As already pointed out, in the uplink for variable rate handling the DPDCH bit rate (spreading factor) may vary frame by frame. The parallel transmission of DPDCH and DPCCH, as depicted in Figure 2.25, allows continuous transmission regardless of the bit rate and data transmission (DTX). Audible interference to other equipment is then reduced without affecting spectral efficiency.

Table 2.3 Uplink Dedicated Physical Data Channel symbol rates and examples of services multiplexing.

SF	Channel symbol rate [ksps][a]	User bit rate [kbps]	Example of services multiplexing	Transport format (semi-static part)
256	15	3.4	Standalone mapping of DCCH 3.4 kbps	SRB (TTI 40 ms, CC coding rate 1/3)
128	30	—	—	—
64	60	12.2 + 3.4	AMR speech 12.2 kbps, DCCH 3.4 kbps	AMR (TTI 20 ms, CC 1/3 for TrCH #A and #B; CC 1/2 for TrCH #C) and SRB (as above)
32	120	28.8 + 3.4	Modem 28.8 kbps, DCCH 3.4 kbps	CS data (TTI 40 ms, turbo coding 1/3) and SRB (as above)
16	240	$(12.2)^{b}$ + 64 + 3.4	(AMR speech 12.2 kbps), packet data 64 kbps, DCCH 3.4 kbps	Packet data 64 kbps (TTI 20 ms, turbo coding 1/3), AMR and SRB (as above)
16	240	64 + 3.4	ISDN 64 kbps, DCCH 3.4 kbps	CS data (TTI 40 ms, turbo coding 1/3) and SRB (as above)
16	240	57.6 + 3.4	Fax 57.6 kbps, DCCH 3.4 kbps	CS data (TTI 40 ms, turbo coding 1/3) and SRB (as above)
8	480	(12.2) + 128 + 3.4	(AMR speech 12.2 kbps), packet data 128 kbps, DCCH 3.4 kbps	Packet data 128 kbps (TTI 20 ms, turbo coding 1/3), AMR and SRB (as above)
8	480	(12.2) + 144 + 3.4	(AMR speech 12.2 kbps), packet data 144 kbps, DCCH 3.4 kbps	Packet data 144 kbps (TTI 20 ms, turbo coding 1/3), AMR and SRB (as above)
4	960	(12.2) + 384 + 3.4	(AMR speech 12.2 kbps), packet data 384 kbps, DCCH 3.4 kbps	Packet data 384 kbps (TTI 20 ms, turbo coding 1/3), AMR and SRB (as above)

[a] In the uplink 1 symbol = 1 bit.
[b] AMR speech when shown in brackets does not affect the spreading factor.

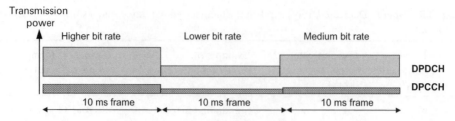

Figure 2.25   Parallel transmission of Dedicated Physical Data Channel and Dedicated Physical Control Channel.

2.4.4.3   Dedicated Downlink Physical Channel

In the downlink the downlink DPCH consists of a downlink DPDCH and a downlink DPCCH time-multiplexed with complex scrambling. Therefore the dedicated data generated at higher layers carried on DPDCH are time-multiplexed with pilot bits, TPC commands and TFCI bits (optional) generated by the physical layer. As pointed out in Section 2.4.3.4, the DPCH may or may not include the TFCI; if the TFCI bits are not transmitted, DTX is used in the corresponding field. The dedicated downlink physical channel structure for one power control period is shown in Figure 2.24. The I/Q branches have equal power and the SFs range from 512 (7.5 ksps) down to 4 (960 ksps) [11]. Examples of services multiplexing are shown in Table 2.4.

As introduced in Section 2.4.3.4, when the total bit rate to be transmitted on one downlink CCTrCH exceeds the maximum bit rate of the downlink physical channel, multi-code transmission is employed and several parallel code channels are transmitted for one CCTrCH using the same spreading factor. Different spreading factors can be used when several CCTrCHs are mapped onto different DPCHs transmitted to the same UE. As illustrated in Figure 2.26, the L1 control information is only transmitted on the first DPCH and the transmission is interrupted during the corresponding time period of the additional DPCHs [11].

2.4.4.4   Common Uplink Physical Channels

The common uplink physical channels are the PRACH and the PCPCH, which are used to carry RACH and CPCH, respectively. The RACH is transmitted using open-loop power control. The CPCH is transmitted using inner-loop power control and is always associated with a downlink DPCCH carrying power control commands [11].

Physical Random Access Channel (PRACH)
Random access transmission is based on a slotted ALOHA approach with fast acquisition indication. There are 15 access slots per two frames spaced 5120 chips apart, as shown in Figure 2.27. Information concerning which access slots are available in the cell for random access transmission is broadcast on the BCH [11].

Random access transmission consists of one or several preambles and a message part. The structure of the RACH transmission is illustrated in Figure 2.28. The preamble

Table 2.4 Downlink Dedicated Physical Data Channel symbol rates and examples of services multiplexing.

SF	Channel symbol rate [ksps][a]	User bit rate [kbps]	Example of services multiplexing (RBs and SRB)	Transport format (semi-static part)
512	7.5	—	—	—
256	15	3.4	Standalone mapping of DCCH 3.4 kbps	SRB (TTI 40 ms, CC coding rate 1/3)
128	30	12.2 + 3.4	AMR speech 12.2 kbps, DCCH 3.4 kbps	AMR (TTI 20 ms, CC 1/3 for TrCH #A and #B; CC 1/2 for TrCH #C) and SRB (as above)
64	60	28.8 + 3.4	Modem 28.8 kbps, DCCH 3.4 kbps	CS data (TTI 40 ms, turbo coding 1/3) and SRB (as above)
32	120	57.6 + 3.4	Fax 57.6 kbps, DCCH 3.4 kbps	CS data (TTI 40 ms, turbo coding 1/3) and SRB (as above)
32	120	$(12.2)^b$ + 64 + 3.4	(AMR speech 12.2 kbps), packet data 64 kbps, DCCH 3.4 kbps	Packet data 64 kbps (TTI 20 ms, turbo coding 1/3), AMR and SRB (as above)
32	120	64 + 3.4	ISDN 64 kbps, DCCH 3.4 kbps	CS data (TTI 40 ms, turbo coding 1/3), SRB (as above)
16	240	(12.2) + 128 + 3.4	(AMR speech 12.2 kbps), packet data 128 kbps, DCCH 3.4 kbps	Packet data 128 kbps (TTI 20 ms, turbo coding 1/3), AMR and SRB (as above)
16	240	(12.2) + 144 + 3.4	(AMR speech 12.2 kbps), packet data 144 kbps, DCCH 3.4 kbps	Packet data 144 kbps (TTI 20 ms, turbo coding 1/3), AMR and SRB (as above)
8	480[c]	(12.2) + 384 + 3.4	(AMR speech 12.2 kbps), packet data 384 kbps, DCCH 3.4 kbps	Packet data 384 kbps (TTI 20 ms, turbo coding 1/3), AMR and SRB (as above)
4	960	—	—	—

[a] In the downlink 1 symbol = 2 bits.
[b] AMR speech when shown in brackets does not affect the spreading factor.
[c] Or multicode 3 ∗ 240 ksps.

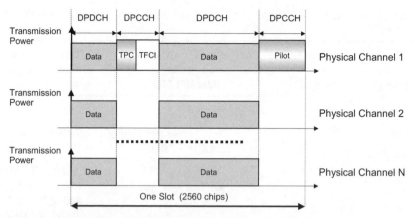

Figure 2.26 Downlink slot format in case of multicode transmission, showing N parallel physical channels.

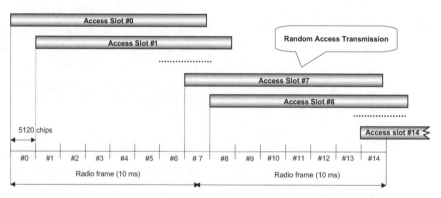

Figure 2.27 Random Access Channel access slot numbers and spacing between consecutive access slots.

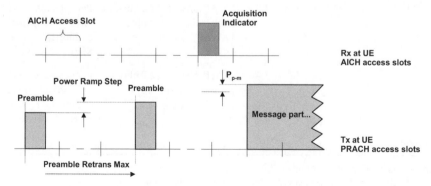

Figure 2.28 Physical Random Access Channel ramping and message transmission.

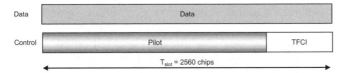

Figure 2.29   Structure of the random access message part radio frame.

comprises 4096 chips, being made up of 256 repetitions of a signature of length 16 chips ($256 * 16 = 4096$) [14].

The slot structure of the PRACH message is illustrated in Figure 2.29. It consists of two parts, a data part where the RACH transport channel is mapped and a control part where the L1 control information is carried. The data and control parts are transmitted in parallel. The SFs of the data part are 256, 128, 64 and 32. The control part consists of Pilot and TFCI bits and has a spreading factor of 256. The TFCI field indicates the TF of the RACH mapped onto the data part of the radio frame and is repeated in the second radio frame if the message part lasts for 20 ms [11].

A RACH sub-channel is defined as a subset of the total set of the uplink access slots. The 12 RACH sub-channels available for each cell can be found in [14].

Each cell is configured during radio network planning setting the preamble scrambling code, the message length in time (either 10 or 20 ms), the Acquisition Indicator Channel (AICH) Transmission Timing parameter (0 or 1, for setting the preamble-to-AI distance), the set of available signatures and the set of available RACH sub-channels for each Access Service Class (ASC).[2] As depicted in Figure 2.28, other essential parameters that need to be set during radio network planning are the power ramping factor ('Power Ramp Step'), the maximum number of preamble retransmissions ('Preamble Retrans Max'), and the power offset between the power of the last transmitted preamble and the control part of the PRACH message (Power offset $P_{p\text{-}m} = P_{message\text{-}control} - P_{preamble}$). The UE receives these data from the system information broadcast on the BCH, which may be updated by the RNC before any physical random access procedure is initiated. The physical random access procedure is illustrated in Figure 2.28 and may be summarised as follows (more information can be found in [14]):

- The UE derives the available uplink access slots (in the next full access slot set) from the set of available RACH sub-channels within the given ASC.
- The UE randomly selects one access slot from among those previously determined

---

[2] In order to provide different priorities of RACH usage when the RRC connection is set up, PRACH resources (access slots and preamble signatures) can be divided between eight different ASCs numbered from 0 (highest priority, used in case of emergency call or for reasons with equivalent priority) to 7 (lowest priority). The PRACH partitioning and the one-to-one correspondence (mapping) between the terminal Access Class (AC) and ASC are specified in [9]. If the UE is a member of several ACs, then it selects the ASC for the highest AC number. An ASC defines a certain partition of the PRACH resources and is always associated with a persistence value computed by the terminal as a function of a *dynamic persistence level* (1–8) and a *persistence-scaling factor* (seven values, from 0 to 1 for ASC 2-7) set during radio network planning.

and randomly selects a signature from the set of available signatures within the given ASC.

- The UE transmits the first preamble using the selected uplink access slot, signature and preamble transmission power, calculated as explained in Section 4.2.1.1.
- If no positive or negative Acquisition Indicator (AI $\neq$ +1 nor −1) corresponding to the selected signature is detected in the downlink access slot corresponding to the selected uplink access slot, then the terminal selects the next available access slot in the set of available RACH sub-channels within the given ASC, randomly selects a new signature from the set of available signatures within the given ASC and increases the preamble power by $\Delta P_0 =$ Power Ramp Step [dB].
- If the number of retransmissions exceeds the 'Preamble Retrans Max' value or if a negative AI corresponding to the selected signature is detected, then the UE exits the physical random access procedure. Otherwise, the UE transmits the random access message three or four uplink access slots after the uplink access slot of the last transmitted preamble, depending on the AICH transmission-timing parameter. The transmission power of the control part of the random access message is P_{p-m} [dB] higher than the power of the last transmitted preamble. The transmission power of the data part of the random access message is set according to the corresponding gain factor. The meaning of the gain factors is further explained in Section 2.4.7.

Physical Common Packet Channel (PCPCH)

The PCPCH is used to carry the CPCH TrCH. Briefly, CPCH is like RACH with fast power control and longer allocation time, and with the possibility of using higher bit rates to transfer larger amounts of data with a more controlled access method.

CPCH is intended to carry packet switched user data in the uplink direction. One of its main advantages is a short access delay with a high bit rate, which makes it especially suitable for bursty data. Compared with DCH, CPCH is a good alternative, because it can be better multiplexed in the time domain and it can also better adapt to data rate changes. On the other hand, CPCH may also degrade capacity, owing to its lack of soft handover. For longer uplink packet data transmission, it is better to use DCH. The lack of soft handover makes CPCH coverage inferior when compared with DCH. Since CPCH uses fast power control, it gives a better spectrum efficiency and thus a better capacity than RACH, which is not power-controlled. The effect of this advantage on overall network capacity depends on the extent to which these channels are used for data transmission.

If CPCH is used, it should be possible to use high bit rates. This means that CPCH can contribute to uplink noise rise. In that case, CPCH load should be taken into account in radio network planning.

CPCH transmission is based on the Collision Detection–Digital Sense Multiple Access (CD-DSMA) approach with fast AI. The UE can start transmission at the beginning of a number of well-defined time intervals. Access slot timing and structure are identical to those of RACH.

The structure of CPCH access transmission is shown in Figure 2.30. It consists of one or several Access Preambles (APs), one Collision Detection (CD) preamble, a PCPCH power control preamble and a message of variable length.

The structure of the PCPCH data part is shown in Figure 2.31.

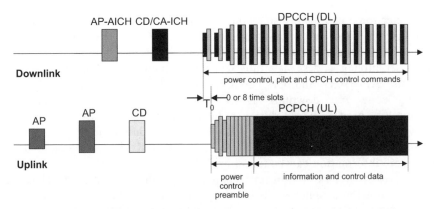

Figure 2.30 Structure of the Common Packet Channel access transmission.

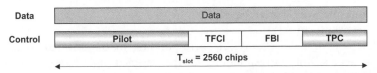

Figure 2.31 Structure of the Physical Common Packet Channel message part radio frame.

For the data part of the PCPCH message part, the permitted spreading factors may vary from 4 to 256, whereas the control part of the PCPCH message has a fixed spreading factor of 256. The spreading factor of the downlink DPCCH is fixed at 512. The maximum length of the message part – i.e., the maximum CPCH allocation time – can vary between 20 and 640 ms. It is a higher layer parameter and can be set by radio network planning as well as channel configurations including allowed spreading factors and bit rates.

The PCPCH AP part, the PCPCH collision detection/channel assignment preamble part and the PCPCH power control preamble part are UL physical signals associated with the PCPCH, which also carries CPCH transport channel data. A set of downlink physical channels are needed for the CPCH access procedure:

- CPCH Status Indicator Channel (CSICH);
- Access Preamble Acquisition Indicator Channel (AP-AICH);
- Collision Detection/Channel Assignment Indicator Channel (CD/CA-ICH).

Based on the availability information of each PCPCH that the CSICH indicates, the UE initiates the CPCH access procedure on an unused channel. A CSICH is always associated with an AP-AICH and uses the same channelisation code. The AP-AICH is used to carry access preamble acquisition indicators of the CPCH to the UE. The AP-AICH and the AICH are identical and may use the same channelisation code. The CD/CA-ICH is used to carry collision detection and channel assignment indicators to the UE.

The CPCH access procedure is fairly similar to the RACH access procedure. The main difference is the additional collision detection procedure. The extra step includes collision detection preamble transmission on PCPCH in the uplink, and transmission of collision detection and channel assignment on the CD/CA-ICH in the downlink.

Each cell is configured during radio network planning setting the AP and CD preamble scrambling codes, signature sets and sub-channels defining the available access slots, AP-AICH and CD/CA-ICH preamble channelisation codes, CPCH scrambling code and downlink DPCCH channelisation code. Other essential parameters that need to be set during radio network planning are the power ramp-up, access and timing parameters. The UE receives these data from the system information broadcast on the BCH. The CPCH access procedure may be summarised as follows; more information can be found in [14]:

- The UE selects a CPCH transport channel from the available CPCH set in the CSICH channel and builds a TB for the next TTI. The TB is sent to the physical layer, and the initial power value is set. The AP retransmission counter is set to its maximum value.
- The UE randomly selects a CPCH AP signature from the signature set of the CPCH channel and one available access slot.
- The UE transmits an AP.
- If the UE does not detect any AI corresponding to the selected signature in the downlink access slot corresponding to the selected uplink access slot, the UE selects the next available access slot and retransmits the AP.
- If the UE detects a negative acquisition indication in the AP-AICH in the corresponding slot with the selected signature, it aborts access.
- When the UE detects a positive acquisition indication in the AP-AICH, the contention segment starts. The UE randomly selects a CD signature and a CD access slot sub-channel, then transmits the CD preamble.
- If the UE does not receive the CD-AICH in the designated slot with the corresponding signature, it aborts access.
- If the UE receives the CD-AICH in the correct timeslot with the matching signature, it transmits the PC preamble; immediately thereafter data transmission starts.

The collision in the CPCH means that two UEs have selected the same access channel and preamble at the same time. After that it is unlikely, but not impossible, that they select again the same CD preamble. The Node B responds to only one CD preamble – i.e., the strongest. Although the channels are defined since Release '99 in 3GPP, another method High-speed Uplink Packet Access (HSUPA) is coming with Release 6 in 3GPP [37] as a more efficient and easy way to implement high bit rate packet data traffic access in the uplink.

2.4.4.5 Common Downlink Physical Channels

Most of the common downlink physical channels are used for transmitting signalling messages generated by the entity above the physical layer. The other common physical channels required for system operation are the physical layer control channels and the PDSCH, which is used for transmitting high peak rate data with a low activity cycle in the downlink like the HS-PDSCH.

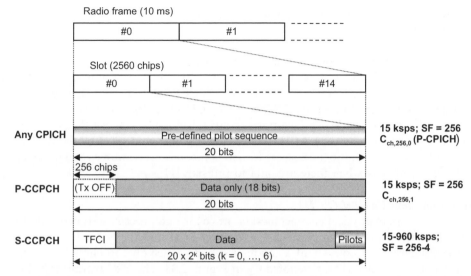

Figure 2.32 Slot structure of the Common Pilot Channel, Primary Common Control Physical Channel and Secondary Common Control Physical Channel.

Common Pilot Channel (CPICH)

There are two types of common pilot channels, the Primary and the Secondary CPICH. They are transmitted at a fixed rate (15 kbps, SF = 256) and carry only a pre-defined symbol sequence. The slot structure for the common pilot channels is illustrated in Figure 2.32.

The Primary Common Pilot Channel (P-CPICH) is characterised by a fixed channelisation code ($C_{ch,256,0}$) and is always scrambled using a primary scrambling code; see Section 2.4.7 for further explanation. There is one P-CPICH per cell and it is broadcast over the entire cell. The P-CPICH is the phase reference for the Synchronisation Channel (SCH), Primary Common Control Physical Channel (P-CCPCH), AICH, PICH, DL DPCCH for CPCH, S-CCPCH and by default for the DL DPCH [11].

The Secondary Common Pilot Channel (S-CPICH) is characterised by an arbitrary channelisation code with a spreading factor of 256, scrambled by either a primary or a secondary scrambling code. In a cell there may be no, one or several S-CPICHs. Each S-CPICH may be transmitted over the entire cell or over only a part of the cell [11].

If the P-CPICH is not used as a phase reference for the downlink DPCH, the UE is informed about it by the network. In that case for channel estimation it may use the S-CPICH or the pilot bits on the DL DPCCH [9].

Primary Common Control Physical Channel (P-CCPCH)

The P-CCPCH is a fixed rate (15 ksps, SF = 256) DL physical channel used to carry the BCH. It is a pure data channel characterised by a fixed channelisation code ($C_{ch,256,1}$). The P-CCPCH is broadcast over the entire cell and is not transmitted during the first

256 chips of each slot, where the Primary SCH and the Secondary SCH are transmitted instead (see Figure 2.32) [11].

Secondary Common Control Physical Channel (S-CCPCH)

The S-CCPCH is used to carry the FACH and PCH, which can be mapped onto the same S-CCPCH (same frame) or onto separate S-CCPCHs. The slot structure for the S-CCPCH is depicted in Figure 2.32. The S-CCPCH spreading factor ranges from 256 (15 ksps) down to 4 (960 ksps). Fast power control is not allowed, but the power of the S-CCPCH carrying only the FACH may be slowly power-controlled by the RNC. The S-CCPCH supports multiple transport format combinations (variable rate) using TFCI and it is on air only when there are data to transmit [11].

Synchronisation Channel (SCH)

The SCH is a pure physical channel used in the cell search procedure. It consists of two sub-channels transmitted in parallel, the Primary SCH and the Secondary SCH [11].

The Primary SCH consists of a modulated code of length 256 chips, the Primary Synchronisation Code (PSC), denoted C_p in Figure 2.33. The PSC is transmitted once every slot; it allows downlink slot synchronisation in the cell and is identical in every cell of the system.

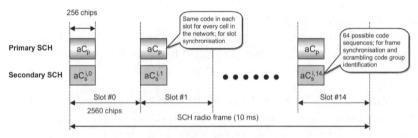

Figure 2.33 Structure of Synchronisation Channel (SCH); the symbol a indicates the presence or absence of Space Time Transmit Diversity (STTD) on the P-CCPCH; C_p and $C_s^{i,k}$ are the Primary and Secondary Synchronisation Codes (PSC and SSC), respectively.

The Secondary SCH consists of a sequence of repeatedly transmitted modulated codes of length 256 chips, the Secondary Synchronisation Codes (SSCs), denoted $C_s^{i,k}$ in Figure 2.33, where $i = 0, 1, \ldots, 63$ is the number of the scrambling code group, and $k = 0, 1, \ldots, 14$ is the slot number. This sequence permits downlink frame synchronisation and indicates from which of the code groups the cell got assigned its downlink primary scrambling code. This narrows down the search for the primary scrambling code to eight codes.

Physical Downlink Shared Channel (PDSCH)

The PDSCH is used to carry the DSCH TrCH. The DSCH offers fast power control and effective scheduling possibilities, but no soft handover.

The DSCH is targeted to transfer bursty non-real time packet switched data. The basic idea of the DSCH is to share a single downlink physical channel – i.e., orthogonal downlink channelisation code – between several users. DSCH scheduling can be considered as multiplexing of several DTCH logical channels of the same or different UEs to the DSCH transport channel in time division.

Faster allocation of the PDSCH will use potential capacity better than slower allocation of the DCH. As a result, QoS differentiation and prioritisation can be utilised effectively. From coverage point of view, the DSCH is not advantageous due to its lack of soft handover. The DSCH can be planned to be used over the whole cell, when hard handover is acceptable, or it can be planned not to cover the whole cell, in which case channel-type switching from the DSCH to the DCH is required when the DSCH coverage ends.

When data are transmitted with low activity on the DCH and inactive periods occur, a dedicated downlink channelisation code is still reserved, which may cause codes to run out. Since one code is shared between several users in the case of the DSCH, other users can take advantage of a user's inactive periods. Thus, downlink channelisation code usage is more efficient with the DSCH than with the DCH. Code blocking is less likely when the DSCH is used, and the capacity can be higher.

A PDSCH, which is used to carry the DSCH, corresponds to a channelisation code below or at a PDSCH root channelisation code. Figure 2.34 shows the PDSCH code resource allocation from the OVSF code tree.

A PDSCH is allocated on a radio frame basis to a single UE. Within one radio frame, the RAN may allocate different PDSCHs under the same PDSCH root channelisation code to different UEs based on code multiplexing. Within the same radio frame, multiple parallel PDSCHs with the same spreading factor may be allocated to a single UE. For the PDSCH the allowed permitted spreading factor may vary from 4 to 256.

For each radio frame, each PDSCH is associated with one downlink DPCH in order to support fast power control and to inform the UE of the arrival of data on the DSCH. The PDSCH and associated DPCH do not necessarily have the same spreading factor and are not necessarily frame-aligned. All relevant physical layer control is transmitted on the DPCCH part of the associated DPCH. The PDSCH itself does not carry any

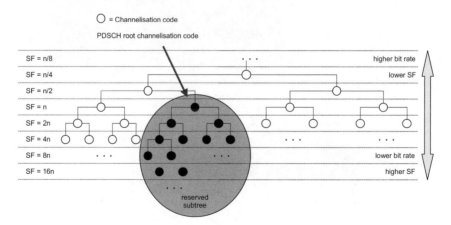

Figure 2.34 Physical Downlink Shared Channel code resource allocation from the orthogonal variable spreading factor code tree.

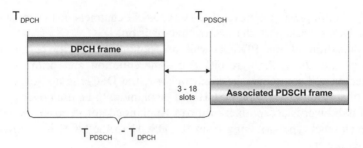

Figure 2.35 Dedicated Physical Channel and associated Physical Downlink Shared Channel timing relation.

physical layer control information but only channel-coded DSCH data. To indicate to the UE that there are data to decode on the DSCH the TFCI field of the associated DPCH is used. The TFCI informs the UE of the instantaneous bit rate as well as the channelisation code of the PDSCH.

Due to UE needing time for processing, there is a timegap between the DPCH and associated PDSCH frames, as illustrated in Figure 2.35. The associated PDSCH frame may start from 3 to 18 slots after the end of the DPCH frame.

PDSCH transmission power is controlled using the power offset between the PDSCH and the downlink DPCH. Figure 2.36 shows the power offset setting on the downlink DPCH and associated PDSCH. The power offsets between the DPCCH and DPDCH fields are denoted PO1, PO2 and PO3, referring to the TFCI, TPC and Pilot fields of the DPCCH, respectively. The power offset between the PDSCH and the downlink DPCH is defined as the offset relative to the power of the TFCI bits of the downlink DPCCH

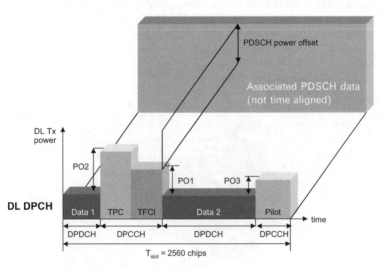

Figure 2.36 Downlink Dedicated Physical Channel, associated Physical Downlink Shared Channel structure and power setting.

directed to the same UE as the PDSCH. The RNC calculates the power offset and informs the Node B, which adjusts the PDSCH transmission power accordingly.

Although the PDSCH has existed in 3GPP since Release '99, the solution has not been implemented so far by any vendor. In the meantime, 3GPP Release 5 has introduced the HSDPA solution which possesses peak cell data throughput well in excess of the capabilities of the Release '99 solutions including PDSCH construction. HSDPA is discussed in more detail in Section 2.4.5.

Acquisition Indicator Channel (AICH)

The AICH is a fixed rate physical channel (SF = 256) used to indicate in a cell the reception by the Node B of PRACH preambles (signatures). Once the Node B has received a preamble, the same signature that has been detected on the PRACH preamble is then sent back to the UE using this channel. Higher layers are not involved in this procedure: a response from the RNC would be too slow to acknowledge a PRACH preamble. The AICH consists of a repeated sequence of 15 consecutive Access Slots (ASs) of length 5120 chips. Each AS includes an AI part of 32 real-valued symbols, as illustrated in Figure 2.37.

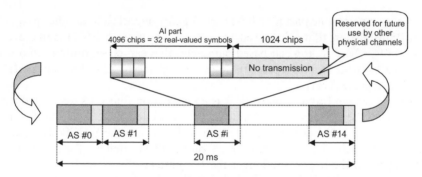

Figure 2.37   Acquisition Indicator Channel Access Slot structure.

As a function of the signature(s) detected on the PRACH preamble, the Node B derives the symbols of the AI part. The computation may result in a positive acknowledge, a negative acknowledge or no acknowledge at all if the detected signature is not a member of the set of available signatures for all the ASCs for the corresponding PRACH. Up to 16 signatures can be acknowledged on the AICH at the same time [11].

The UE receives the AICH information (channelisation code, STTD indicator and AICH transmission timing) from the system information broadcast on the BCH and accordingly starts receiving the AICH when the allocated PRACH is used. If AICH or PICH information is not present, the terminal considers the cell barred and proceeds to cell reselection, as specified in [9].

Paging Indicator Channel (PICH)

The PICH is a physical channel used to carry Paging Indicators (PIs). This channel is transmitted at a fixed rate (SF = 256) and is always associated with an S-CCPCH, where the PCH is mapped. As illustrated in Figure 2.38, a PICH radio frame

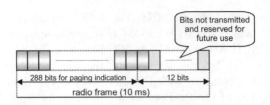

Figure 2.38 Paging Indicator Channel structure.

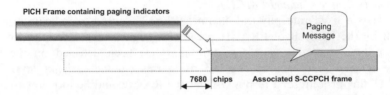

Figure 2.39 Relation between Paging Indicator Channel and a Secondary Common Control Physical Channel carrying a PCH.

consists of two parts, one part (288 bits) used for carrying PIs and another part (12 bits) with no transmission that is reserved for future use. In each PICH frame are transmitted N_p PIs, where N_p is a cell-based parameter that can be set during radio network planning to 18 (with 16 bits repeated), 36 (8 bits repeated), 72 (4 bits repeated) or 144 (only 2 bits repeated) [11].

If a PI in a certain frame is set to '1' it is an indication that UEs associated with this PI should read the corresponding frame of the associated S-CCPCH. As illustrated in Figure 2.39, once a PI has been detected, the UE decodes the S-CCPCH frame to see whether or not there was a paging message on the PCH intended for it. The less often the PIs appear in the frame, the longer the UE battery life.

2.4.5 High-speed Downlink Packet Access (HSDPA)

The HSDPA concept has been included by the 3GPP in the specifications of Release 5 as an evolution step to improve the WCDMA performance for downlink packet traffic. The feature improves downlink throughput and shortens the Round Trip Time (RTT) to below 100 ms. The feature however brings certain architectural changes into the 3GPP Release '99 protocol stack. Logical extension of HSDPA functionality in the uplink direction is the High-speed Uplink Packet Access (HSUPA) appearing in Release 6 of 3GPP [37]. To support the standardisation effort, and evaluate the performance of HSUPA, there are ongoing studies [38]. Currently, the HSUPA concept has been finalised in 3GPP [39].

The HSDPA concept consists of a downlink time-shared channel that supports a 2 ms TTI, Adaptive Modulation and Coding (AMC), multi-code transmission, and fast physical layer Hybrid ARQ (H-ARQ). The link adaptation and packet scheduling functionalities are controlled directly from the Node B, which enables them to acquire knowledge of the instantaneous radio channel quality of each user. Two of the main features of the WCDMA technology – closed-loop power control and variable spreading factor – have not been applied. Link adaptation of the system is performed by changing the modulation (QPSK or 16QAM) and the coding rate according to the instantaneous channel quality. Channel quality variations across TTI are minimised due to the reduction of TTI duration from the minimum 10 ms in WCDMA down to 2 ms.

2.4.5.1 High-speed Downlink Packet Access Architecture

To obtain recent channel quality information that permits the link adaptation and the packet scheduling entities to track the user's instantaneous radio conditions, the MAC functionality in charge of the HS-DSCH channel has been moved from the RNC to the Node B. Up-to-date channel quality information allows the packet scheduler to serve the user only when channel conditions are favourable. Thus, the HS-DSCH is directly terminated at the Node B. The MAC layer controlling the resources (called MAC-hs) is directly located in the Node B (Figure 2.40). Channel quality reports are provided by L1 signalling which support effective packet scheduling located directly in the Node B. This allows getting recent channel quality reports which make it possible to track and use the instantaneous signal quality for low-speed UEs. The location of the MAC-hs in Node B also enables execution of the H-ARQ protocol from the physical layer, which permits faster retransmissions.

The MAC-hs layer [5] is in charge of handling the H-ARQ functionality of every HSDPA user, distributing the HS-DSCH resources between all the MAC-d flows according to their priority (i.e., packet scheduling), and selecting the appropriate transport format for every TTI (i.e., link adaptation). The radio interface layers above the MAC are not modified from the Release '99 architecture because HSDPA

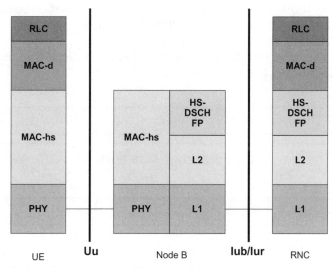

Figure 2.40 Radio interface protocol architecture of the High-speed Downlink Shared Channel Transport Channel (configuration without MAC-c/sh).

is intended for transport of LoCHs. Nonetheless, the RLC can only operate in either AM or UM, but not in TM due to ciphering [34]. For the RLC TM the ciphering is not done in RLC but in the MAC-d. However, neither MAC-c/sh nor MAC-hs support ciphering [5].

The MAC-hs also stores the user data to be transmitted across the air interface. This imposes some constraints on the minimum buffering capabilities of the Node B. Movement of the data queues to the Node B creates the need for a flow control mechanism (HS-DSCH FP) that aims at keeping the buffers full. The HS-DSCH FP handles the data transport from the SRNC to the Controlling RNC (CRNC) (if the Iur interface is involved) and between the CRNC and Node B.

For various practical reasons – like the complexity to synchronise the transmissions of various cells – the HS-DSCH does not support soft handover. Depending on coverage and Node B downlink power availability, the HS-DSCH may provide full or partial coverage in the cell.

2.4.5.2 High-speed Downlink Packet Access Channel Structure

The HSDPA concept relies on a new transport channel, the HS-DSCH, which can be seen as an evolution of the DSCH channel. The HS-DSCH is mapped onto a pool of physical channels (i.e., channelisation codes) denominated HS-PDSCHs to be shared among all the HSDPA users in a time-multiplexed manner. The spreading factor of the HS-PDSCHs is fixed at 16, and the MAC-hs can use one or several codes, up to a maximum of 15. Moreover, the scheduler may apply code-multiplexing by transmitting separate HS-PDSCHs to different users in the same TTI. The sub-frame and slot structure of HS-PDSCH are shown in Figure 2.41. The HS-PDSCH may use QPSK or 16QAM (16 State Quadrature Amplitude Modulation) modulation symbols. Thus, M in Figure 2.41 is the number of bits per modulation symbols – i.e., $M = 2$ for QPSK and $M = 4$ for 16QAM. Due to these two different modulation schemes, the HS-PDSCH raw bit rate (all bits in the HS-PDSCH sub-frame over 2 ms) on L1 is either 320 kbps or 960 kbps, respectively, for one code.

The uplink and downlink channel structure of HSDPA along with time relations is described in Figure 2.42 [36].

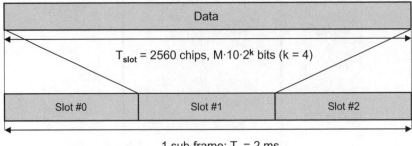

Figure 2.41 Sub-frame structure of the High Speed Physical Downlink Shared Channel ($M = 2$ for QPSK and $M = 4$ for 16QAM).

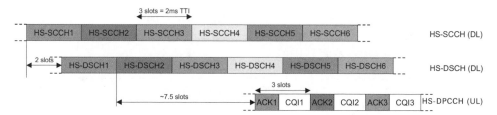

Figure 2.42 Uplink and downlink physical layer structure in High-speed Downlink Packet Access.

The HSDPA concept includes a High-speed Shared Control CHannel (HS-SCCH) in the downlink direction to signal the users when they are to be served and the necessary information for the decoding process. The HS-SCCH (60 kbps, SF = 128) carries the following information [33]:

- *UE Id Mask*: to identify the user to be served in the next TTI.
- *Transport-format-related information*: specifies the set of channelisation codes, modulation and modulation symbol constellation. The actual coding rate is derived from the TB size and other transport format parameters.
- *Hybrid ARQ-related information*: such as whether the next transmission is a new one or a retransmission and whether it should be combined, the associated ARQ process and information about the redundancy version.

The control information solely applies to the UE to be served in the next TTI, which permits this signalling channel to be a shared one. Figure 2.43 illustrates the sub-frame structure of the HS-SCCH.

The RNC can specify the power of the HS-SCCH (offset relative to the Pilot bits of the associated DPCH) [17]. The HS-SCCH transmit power may be constant or time-varying according to a certain power control strategy, though the 3GPP specifications do not set any closed-loop power control modes for the HS-SCCH. According to 3GPP [17] the RNC may set the maximum transmission power on all the codes of the HS-DSCH and HS-SCCH channels in the cell. Otherwise, the Node B may utilise all unused Node B transmission power for these two channels. The RNC determines the maximum number of channelisation codes to be used by the HS-DSCH channel.

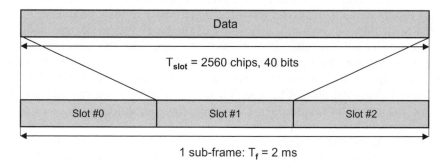

Figure 2.43 Sub-frame structure of the High-speed Shared Control Channel.

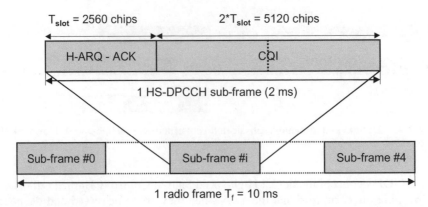

Figure 2.44 Frame structure of the uplink High-speed Dedicated Physical Control Channel.

A High-speed Dedicated Physical Control Channel (HS-DPCCH) carries the necessary control information in the uplink – namely, the ARQ acknowledgements, non-acknowledgements and the Channel Quality Indicator (CQI) reports. The HS-DPCCH can only exist together with an uplink DPCCH as a parallel code channel with a spreading factor of 256. The frame structure of the HS-DPCCH is shown in Figure 2.44.

2.4.5.3 Adaptive Modulation and Coding, Multicode Transmission and Link Adaptation

As mentioned, HSDPA utilises other link adaptation techniques to substitute power control and variable spreading factor. To cope with the dynamic range of the E_s/N_0 at the UE, HSDPA adapts the modulation, the coding rate and the number of channelisation codes to the instantaneous radio conditions. The combination of the first two mechanisms is denominated as AMC. Besides QPSK, HSDPA incorporates the 16QAM modulation to increase the peak data rates for users served under good radio conditions. The support for QPSK is mandatory for the UE. On the other hand, support of 16QAM is optional for the network and the UE [33]. The inclusion of this higher order modulation introduces some complexity challenges for the terminal receiver, which needs to estimate the relative amplitude of the received symbols, whereas it only requires the detection of the signal phase in the QPSK case.

Actually, the link adaptation functionality of the Node B is to change modulation, coding format and the number of multi-codes to cope with the instantaneous radio conditions represented by a certain E_s/N_0. The adaptation task is divided between the Node B having enough processing power for packet scheduling and the UE knowing the current radio conditions. The selection is thus based on mobile channel quality feedback and other information like UE capabilities, retransmissions, resource availabilities, status of buffers and priorities.

The CQI is the indicator reflecting the current downlink radio conditions. It is sent by the UE in the uplink HS-DPCCH every 2 ms if the HS-DPCCH is assigned. The report denominated CQI provides implicit information about the instantaneous signal quality received by the UE. The table including the set of reference CQI reports can be found in [14]. The CQI thus specifies the TB size, the number of codes and modulation from a set of reference ones that the UE is capable of supporting with a detection error no higher than 10% in the first transmission for a reference HS-PDSCH power. Node B is supposed to use the information for HS-PDSCH channel setting. The setting depends on the available power for HSDPA. In practice a certain combination of channel-coding robustness along with puncturing is set and this obviously impacts the throughput. In any case, a certain offset to decrease the reported values of CQI could be helpful for the case when the UE reports too optimistic CQIs. Too optimistic CQI estimation results in too many failed retransmissions and, thus, the system possesses a high Block Error Rate (BLER) and low throughput in the end.

The AMC together with multi-code transmission works as a tool for quite a wide link adaptation. If the user enjoys good channel conditions, the Node B can exploit the situation by transmitting multiple parallel codes, reaching significant higher peak throughputs. For example, with 16QAM modulation together with punctured convolutional coding and a set of 15 multi-codes, a maximum peak data rate of 10.8 Mbps can be obtained. With multi-code transmission, the overall dynamic range of the AMC can be increased by $10 \cdot \log_{10}(15) = 12$ dB as a difference between 1 code and 15 codes. The overall link adaptation dynamic range achieved with the combination of the AMC and multi-code transmission is thus around 30 dB [32].

2.4.5.4 Fast Hybrid ARQ (H-ARQ)

HSDPA incorporates a physical layer retransmission functionality that significantly improves robustness against link adaptation errors. Since H-ARQ functionality is located in the MAC-hs entity of the Node B, the transport block retransmission process is considerably faster than RLC-layer retransmissions because the RNC or the Iub are not involved. This benefit is directly reflected in a lower UTRAN transfer delay (both in terms of average and standard deviation). A low transfer delay has advantages for end-to-end level performance (e.g., for TCP, FTP).

The H-ARQ technique is further fundamentally different from WCDMA retransmissions because the UE decoder combines the soft information of multiple transmissions of a TB at the bit level. The retransmissions include additional redundant information that is incrementally transmitted if the decoding fails on the first attempt. This concept of Incremental Redundancy (IR) along with buffering capabilities at the UE causes the effective coding rate to increase with the number of retransmissions. Downsides of the technique are the memory and processing requirements for the UE, which must store the soft information of unsuccessfully decoded transmissions. The standard [35] classifies the UEs into different categories according to the soft memory and number of parallel spreading codes to support.

2.4.6 *Timing and Synchronisation in UTRAN (FDD)*

In UMTS, *network* synchronisation relates to the distribution of synchronisation references to the UTRAN nodes and the stability of the clocks in UTRAN, while *node* synchronisation concerns the estimation and compensation of timing differences between UTRAN nodes. For details of the two synchronisation mechanisms, see [18].

The *TrCH* synchronisation mechanism defines the synchronisation of the frame transport between the RNC and the BSs based on radio interface timing. *Radio interface* synchronisation relates to the timing of the radio frame transmission in the DL direction.

In radio network planning, understanding of timing and synchronisation in UTRAN is essential for synchronisation of DCH and CCH measurements – i.e., for the evaluation of the link-level performance of the UMTS network.

2.4.6.1 Timing Relationship between Physical Channels

The radio frame and access slot timing structure of the DL physical channels is illustrated in Figure 2.45.

As shown in the figure, the cell System Frame Number (SFN) is transmitted on the P-CCPCH which is used as the timing reference for all physical channels, since transmission timing in the uplink is derived from the timing of the downlink physical channels [11]. The SCH (primary and secondary), CPICH (primary and secondary), P-CCPCH and PDSCH have identical frame timing. The S-CCPCH timing may be different for different S-CCPCHs, but the offset from the P-CCPCH frame timing is a multiple of 256 chips. The PICH timing is 7680 chips prior to its corresponding S-CCPCH frame timing – i.e., the timing of the S-CCPCH carrying the PCH TrCH

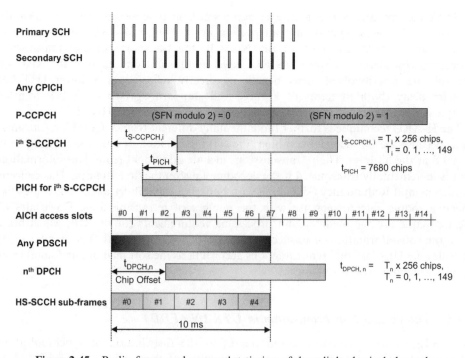

Figure 2.45   Radio frame and access slot timing of downlink physical channels.

with the corresponding paging information. AICH access slots #0 start at the same time as P-CCPCH frames with SFN mod 2 = 0. DPCH timing may be different for different DPCHs, but the offset from the P-CCPCH frame timing is always a multiple of 256 chips. The start of the DL HS-SCCH sub-frame #0 is aligned with the start of the P-CCPCH frames. HSDPA user data are sent via the associated HS-PDSCH channel with a sub-frame delayed 5120 chips after the start of the HS-SCCH sub-frame.

2.4.6.2 Transport Channel Synchronisation

TrCH synchronisation provides an L2 common frame numbering between UTRAN and UE.

The common frame reference at L2 is defined as the Connection Frame Number (CFN). The CFN is a unique number for each RRC connection, and is specified as the frame counter used for TrCH synchronisation between UE and UTRAN. A CFN value is associated with each TBS and is passed together with the TBS through the MAC L1 service access point. The duration of a CFN cycle (0–255 frames) is supposed to be longer than the maximum allowed transport delay between MAC and L1.[3] When used for PCH the range of the CFN is from 0 up to 4095 frames.

Other important (optionally frequency-locked) counters are the Node B Frame Number (BFN) and the RNC Frame Number (RFN). The BFN and RFN are, respectively, the Node B and RNC common frame number counters, which range from 0 up to 4095 frames.

The SFN counter ranges from 0 up to 4095 frames and is sent on the BCH. The SFN is used for scheduling the information transmitted in the cell. In FDD the SFN equals the BFN adjusted by the timing delay used for defining the start of the SCH, the CPICH and the downlink scrambling code(s) in the cell (T_cell in 3GPP). T_cell has been specified in 3GPP in order to avoid the overlapping of SCHs (collision of SCH bursts) in different cells belonging to the same Node B. In other words, the SFN in a cell is supposed to be delayed by T_cell chips with respect to the BFN. T_cell has a step size of 256 chips and ranges from 0 up to 9.

The CFN is not transmitted in the air interface, but is mapped by L1 to the SFN of the first radio frame used for the transmission of the TBS in question. As already mentioned in this section, the SFN is broadcast at L1 in the BCH and the mapping between the CFN and the SFN is performed as a function of a radio-link-specific parameter, denoted Frame Offset in 3GPP. The Frame Offset is computed by the SRNC and provided to the BS when the radio link is set up, where the mapping between L2 and L1 is performed as follows: SFN mod 256 = (CFN + Frame Offset) mod 256 (from L2 to L1); and CFN = (SFN − Frame Offset) mod 256 (from L1 to L2).

The TrCH synchronisation mechanism is valid for all DL TrCHs. In case of soft handover – i.e., only for DCHs belonging to radio links of different radio link sets – the Frame Offsets of the different radio links are selected by the SRNC in order to have a timed transmission of the diversity branches on the air interface. During soft handover

---

[3] Valid for the UTRAN side (between SRNC and Node B) only, where the L1 functions that handle TrCH synchronisation are in Node B.

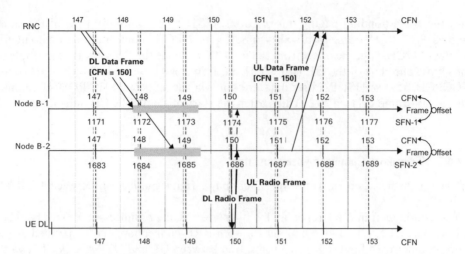

Figure 2.46   Transport Channel synchronisation during soft handover.

the CFN allows frame selection combining at L2 in the uplink and frame splitting in the downlink.

TrCH synchronisation during soft handover is illustrated in Figure 2.46, where the frame arrows represent the first chip or first bit in the frames. TTI and Chip Offset are 10 ms and 0 chips, respectively.

2.4.6.3   Radio Interface Synchronisation

Radio interface synchronisation ensures that the UE gets the correct frames while receiving from several cells. Figure 2.47 illustrates how offsets are signalled and used in different nodes when the (initial) radio link is set up and during a diversity handover [18].

When setting up the first radio link, the SRNC selects a default offset value for the dedicated physical channel, denoted by DOFF in Figure 2.47, which is then used to initialise the Frame Offset and Chip Offset in the Node B, and to inform the UE when the frames in the downlink are expected. In order to average out the Iub traffic and the Node B processing load, all services are scheduled by means of DOFF. In addition, DOFF is used to spread out the location of pilot symbols in the downlink in order to reduce the Node B peak power, since pilot symbols are always transmitted at the fixed location within a slot. Before any intra-frequency diversity handover the UE is supposed to measure the timing difference between the uplink DPCH and the target cell SFN and to report it to the SRNC. The SRNC breaks this time difference into two parameters (Frame Offset and Chip Offset) and forwards the computed values to the Node B. The Node B rounds the received Chip Offset to the closest 256 chip boundary value in order to maintain the downlink orthogonality in the cell (regardless of the spreading factor in use) and then uses it for the downlink DPCH transmission as an offset relative to the P-CCPCH timing, as illustrated in Figure 2.45.

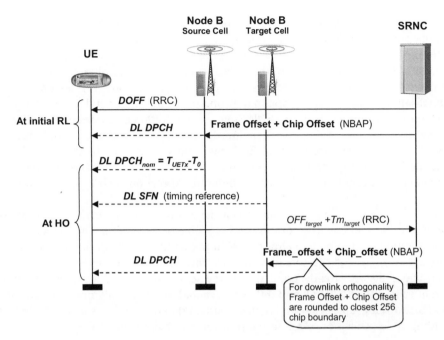

Figure 2.47   Usage of Offset values at initial radio link setup and at handover.

As shown in Figure 2.47, the handover reference is the time instant $T_{UETx} - T_0$, which is denoted *DL DPCH$_{nom}$* in the figure, where T_{UETx} represents the time when the UE transmits the uplink DPCH, and T_0 is the nominal difference between the first received DPCH finger (*DL DPCH$_{nom}$*) and T_{UETx} at UE (constant of 1024 chips). *OFF* and T_m are estimated by the UE according to the following equation: $OFF + T_m = (SFN_{target} - DL\ DPCH_{nom})\ \mathrm{mod}\ 256$ frames [chips], where *OFF* and T_m are expressed in frames and chips, respectively.

2.4.7   Spreading, Scrambling and Channelisation Concepts

A description of the WCDMA air interface principle and the concept of spreading the information in the DS-SS system have been presented in Section 2.1.

2.4.7.1   Scrambling and Channelisation Codes

The spreading concept is applied to physical channels and consists of two operations. The first is the channelisation operation, which transforms each data symbol into a number of chips, thus increasing the bandwidth of the signal. The number of chips per data symbol is called the spreading factor. The second operation is the scrambling operation, where a scrambling code is applied on top of the spread signal [13]. The spreading factor is variable with the exception of the High-speed Dedicated Physical Channel (HS-DPCH) where SF = 16 is the only option defined by the standard.

In the channelisation process, data symbols on I- and Q-branches are independently multiplied by an OVSF code, which is aligned in time with the symbol boundary. In 3GPP the OVSF codes used for different symbol rates are uniquely described as $C_{ch,SF,k}$, where SF is the spreading factor of the code and k is the code number ($0 \leq k \leq SF - 1$). Each level of the code tree defines the channelisation codes of length SF. The channelisation codes have orthogonal properties and are used for separating the information transmitted from a single source – i.e., different connections within one cell in the downlink, where own-cell interference is also reduced – and dedicated physical data channels from one UE in the uplink direction. In the downlink the OVSF codes in a cell are limited resources and need to be managed by the RNC, as explained in Section 4.5, whereas in the uplink direction such a problem does not exist.

The OVSF codes are effective only when the channels are perfectly synchronised at symbol level. The loss in crosscorrelation – e.g., due to multi-path – is compensated for by the additional scrambling operation. With the scrambling operation the real (I) and imaginary (Q) parts of the spread signal are further multiplied by a complex-valued scrambling code. As already pointed out, the scrambling codes are used to separate different cells in the downlink and different terminals in the uplink direction. They have good correlation properties (interference averaging) and are always used on top of the spreading codes, thus not affecting the transmission bandwidth [13].

2.4.7.2 Spreading on Uplink Dedicated Channels

The uplink spreading and modulation principle for the DPCCH and DPDCHs is illustrated in Figure 2.48.

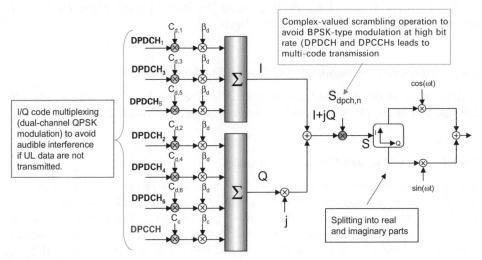

Figure 2.48 Spreading and scrambling for uplink Dedicated Physical Control Channels and Dedicated Physical Data Channels.

One DPCCH and up to six real-valued parallel DPDCHs can be spread and transmitted simultaneously. The DPCCH is always spread using the code $C_c = C_{ch,256,0}$, where $k = 0$. If only one DPDCH is transmitted in the uplink, the DPDCH$_1$ is spread using the code $C_{d,1} = C_{ch,SF,k}$, where SF is the spreading factor of DPDCH$_1$ and $k = SF/4$ is the OVSF code number. When more than one DPDCH needs to be transmitted, all DPDCHs have spreading factors equal to 4 – i.e., DPDCH$_n$ is spread by the code $C_{d,n} = C_{ch,4,k}$, where $k = 1$ if $n \in \{1, 2\}$, $k = 3$ if $n \in \{3, 4\}$ and $k = 2$ if $n \in \{5, 6\}$. In order to compensate for the difference between the spreading factors of the data and signalling parts, the spread signals are then weighted by the gain factors, denoted β_c for DPCCH and β_d for all DPDCHs in Figure 2.48. The gain factors are computed by the SRNC and forwarded to the terminal – e.g., when a radio link is set up or reconfigured [9]. The gain factors range from 0 up to 1, and at least one of the values β_c and β_d always has amplitude 1. The stream of chips on the I- and Q-branches are then summed and scrambled using a complex-valued scrambling code, denoted by $S_{dpch,n}$ in Figure 2.48. The scrambling code is applied aligned with the radio frames – i.e., the first scrambling chip corresponds to the beginning of a radio frame [13].

As explained in [19], any discontinuity in the uplink transmission may cause audible interference to audio equipment close to the terminal. A typical example is the interference of the frame frequency ($217\,\text{Hz} = 1/4.615\,\text{ms}$) caused by GSM terminals. To avoid this effect, the DPCCH and DPDCHs are not time-multiplexed, but I/Q code-multiplexed (dual-channel QPSK modulation) with complex scrambling. As seen in Figure 2.49, this modulation scheme allows continuous transmission even during the silent period where only L1 signalling information for link maintenance purposes (DPCCH) is transmitted.

As illustrated in Figure 2.50, the complex scrambling codes are formed in such a way that the rotation between consecutive chips within one symbol period is limited to $\pm 90°$. With this operation the efficiency of the power amplifier in the terminal remains constant irrespective of the power ratio β between the DPDCH and the DPCCH, thus enabling a backoff in the output power of the linear amplifier [19].

The DPCCH and DPDCHs may be scrambled by either long or short scrambling codes. The complex-valued short and long scrambling sequences are defined in [13]. There are 2^{24} long and 2^{24} short UL scrambling codes. Since millions of codes are available, no uplink code planning is needed. The uplink DPCH scrambling code type (short or long) and number $(0, \ldots, 16777215)$, together with the minimum allowed SF (4, 8, 16, 32, 64, 128 or 256) of the channelisation code for the data part, are assigned by higher layers when, for example, the RRC connection is set up or a handover to UTRAN is commanded [9].

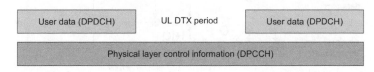

Figure 2.49 Uplink Dedicated Physical Control Channel and Dedicated Physical Data Channel transmission when data are present/absent (DTX).

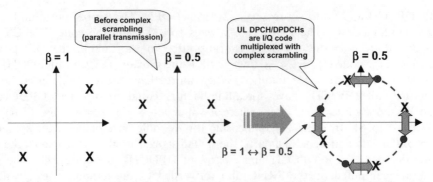

Figure 2.50 Signal constellation for I/Q codes multiplexed with complex scrambling; β denotes the power ratio between the Dedicated Physical Data Channel and Dedicated Physical Control Channel.

2.4.7.3 Spreading on Uplink Common Channels

This section describes the code allocation for the PRACH preamble and message parts. In addition, the channelisation code for the PCPCH power control preamble, the code allocation for the CPCH message part, preamble and power control preamble are pointed out.

Physical Random Access Channel Message Part Scrambling and Preamble Codes
The spreading and scrambling of the PRACH message part is illustrated in Figure 2.51. The control part of the PRACH message is spread with the channelisation code $C_c = C_{ch,256,m}$, where $m = 16 \cdot s + 15$ and s $(0 \leq s \leq 15)$ is the preamble signature; and the data part is spread using the channelisation code $C_d = C_{ch,SF,m}$, where SF is the spreading factor used for the data part and $m = SF \cdot s/16$ [13].

The PRACH message part is always scrambled with a long scrambling code. The length of the scrambling code used for the PRACH message part is 10 ms. There are 8192 scrambling codes given by [13].

Physical Common Packet Channel Message Part Scrambling and Preamble Codes
The spreading and scrambling of the PCPCH message part is similar to the one illustrated in the previous section for the PRACH. A comprehensive and detailed description can be found in [13].

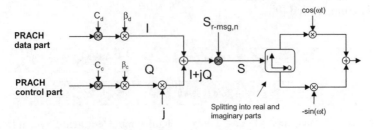

Figure 2.51 Spreading and scrambling of the Physical Random Access Channel message part.

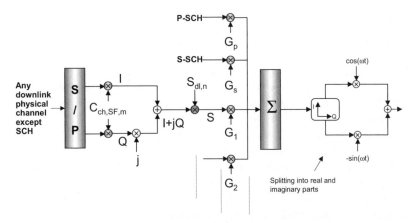

Figure 2.52 Spreading and scrambling scheme for all downlink physical channels.

2.4.7.4 Downlink Spreading and Modulation

The spreading and scrambling concept for all downlink physical channels is illustrated in Figure 2.52. Apart from SCHs, each pair of two consecutive symbols is first serial-to-parallel converted and mapped onto I- and Q-branches. The I- and Q-branches are then spread to the chip rate by the same channelisation code $C_{ch,SF,m}$. The sequences of real-valued chips on the I- and Q-branch are then scrambled using a complex-valued scrambling code, denoted $S_{dl,n}$ in Figure 2.52. The scrambling code is applied aligned with the scrambling code applied to the P-CCPCH, where the first complex chip of the spread P-CCPCH frame is multiplied by chip number 0 of the scrambling code [13].

After spreading, each physical downlink channel (except SCHs) is separately weighted by a weight factor, denoted G_i in Figure 2.52. The complex-valued P-SCH and S-SCH are separately weighted by weight factors G_p and G_s. All downlink physical channels are combined using complex addition, and the resulting sequence generated by the spreading and scrambling processes is then QPSK-modulated [13].

Downlink Spreading Codes
In the downlink the same channelisation codes as in the uplink (OVSF codes) are used. Typically only one code tree per cell is used and the code tree under a single scrambling code is then shared between several users. By definition, the channelisation codes used for P-CPICH and P-CCPCH are $C_{ch,256,0}$ and $C_{ch,256,1}$, respectively. The resource manager in the RNC assigns the channelisation codes for all the other channels, with some restrictions on the usage of SF = 512 in the case of diversity handover [13].

In compressed mode[4] there are three methods for generating gaps: rate matching, reduction of the spreading factor by a factor of 2 and higher layer scheduling. When the

---

[4] State where at least one transmission gap pattern sequence (Layer-1 parameterisation unit, which contains one or two transmission gaps within a set of radio frames) is active. The aim of the downlink and uplink compressed mode is to allow the UE to monitor cells on other FDD frequencies and on other modes and radio access technologies that are supported by the UE – i.e., TDD and GSM.

mechanism for opening the gap is to reduce the spreading factor by a factor of 2, the OVSF code used for compressed frames is $C_{ch,SF/2,\lfloor n/2 \rfloor}$ if an ordinary scrambling code is used, and $C_{ch,SF/2,n \bmod SF/2}$ if an alternative scrambling code is used (see next section), where $C_{ch,SF,n}$ is the channelisation code used for non-compressed frames.

In the downlink the spreading factor of the dedicated physical channel does not vary on a frame-by-frame basis. As illustrated in Figure 2.23, the data rate variation on the DPCH is managed either by a rate matching operation or by L1 DTX, where the transmission is interrupted during a part of the DPDCH slot. In case of multi-code transmission, the parallel code channels have different channelisation codes but the same spreading factor under the same scrambling code. Different spreading factors may be employed in case several CCTrCHs are received by the same UE.

The OVSF code may vary from frame to frame on the PDSCH. The rule is that the OVSF code(s) below the smallest spreading factor is from the branch of the code tree pointed at by the smallest spreading factor used for that connection. If the DSCH is mapped onto multiple parallel PDSCHs, the same rule applies, but all branches identified by the multiple codes, corresponding to the smallest spreading factor, may be used for higher spreading factor allocation.

Downlink Scrambling Codes

In the downlink only long scrambling codes are used. There are $2^{18} - 1 = 262143$ scrambling codes, numbered from 0 up to 262142. The scrambling code sequences, denoted by $S_{dl,n}$ in Figure 2.53, are constructed as segments of the Gold sequence, as specified in [13].

In order to speed up the cell search procedure only 8192 codes of those 262143 are used in practice, and the phase pattern from 0 up to 38399 is forcibly repeated, thus resulting in a periodical scrambling code of period 10 ms, which facilitates the UE in

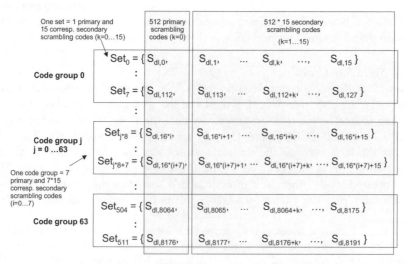

code number $n = 16*[(8*j+i]+k; j = 0...63; i = 0...7$ for each j and $k = 0...15$ for each j, i

Figure 2.53 Primary and secondary scrambling codes.

finding the correct code phase. As illustrated in Figure 2.53, only the scrambling codes with $n = 0, 1, \ldots, 8191$ can be employed. Those codes are divided into 512 sets. Each set consists of a primary scrambling code and 15 secondary scrambling codes, unambiguously associated in one-to-one correspondence – i.e., the mth primary scrambling code corresponds to the mth set of 15 secondary scrambling codes ($m = 0, \ldots, 511$). One set of scrambling codes is further divided into 64 groups of each 8 primary codes and $8 * 15$ corresponding secondary scrambling codes. Further, each scrambling code n is unambiguously associated with a left (denoted $n + 8192$) and a right alternative scrambling code (denoted $n + 16384$), which can be used for scrambling compressed frames during the downlink compressed mode [13].

Each cell is allocated one and only one *primary* scrambling code. The primary CCPCH, CPICH, PICH, AICH and S-CCPCH carrying the PCH are always transmitted using the primary scrambling code. The allocation of primary scrambling codes is a radio network planning task and can be done with the aid of any radio network planning or optimisation tool. Scrambling code planning and optimisation strategies are discussed in Section 4.5.2.4. The other downlink physical channels can be transmitted either with the same primary scrambling code or using a secondary scrambling code taken from the set of codes associated with it. For one CCTrCH it is allowed to use a mixture of primary and secondary scrambling codes. However, if the CCTrCH is of type DSCH, then all PDSCH channelisation codes received by a single UE must be under a single scrambling code [13].

2.5 WCDMA Radio Link Performance Indicators

This section deals with the WCDMA link performance indicators used in radio network dimensioning and planning. Typically, the link performance indicators are produced by a link-level simulator or in laboratory measurements with a real BTS and UE and a channel simulator. Optimally, link and network performance are measured in a live network.

Link performance indicators are used extensively in radio network dimensioning and planning, therefore the most realistic and consistent set of figures must be used covering all the relevant cases. The receiver and transmitter algorithms of the simulation models must be as realistic as possible. One way to achieve this is to require at least compliance with the link performance requirements given by 3GPP standards.

The link performance requirements are different for different services, for example, because of different channel-coding schemes and interleaving depths. Assumptions regarding the radio propagation channel must be carefully chosen, as the propagation channel has a significant effect on the link performance indicators. Also the speed of the UE must be taken into account.

In reality, channel conditions vary from cell to cell and even within cells. Thus, choosing a specific multi-path channel model, as is usually done in simulations, is not ideal. However, it is the only way to ensure consistency when comparing issues in the development phase, such as the performance of different receiver algorithms or of different network-level radio resource management algorithms, or even different approaches in cell deployment. Link performance figures can be classified to, for example:

- direction of transmission (uplink, downlink);
- UE speed;
- service, bit rate (speech 8 kbps, 12.2 kbps, packet data, circuit-switched data);
- multi-path channel (ITU models, 3GPP models);
- environment (dense urban, urban, suburban, rural);
- cell deployment (macro, micro, pico);
- diversity solution (1Rx, 2Rx, 4Rx, no Tx diversity, with Tx diversity).

A selected set of link performance indicators is defined in Section 2.5.1. Section 2.5.2 shows standardised multi-path channel models for which the link performance figures are determined by simulations. In addition, a classification according to services is discussed briefly. The basic principles of simulation are given in Section 2.5.3. In Section 2.5.4 there is a list of *physical-layer measurements* from the 3GPP standard, which support logging the link performance from a live network. The WCDMA link performance figures and their use in radio network dimensioning is described in Section 3.1 and [19].

2.5.1 Definitions

The most important link performance indicators or related quantities are introduced in this section. The selection is not the only one possible, but gives a sufficient set of variables for the modelling of link performance in radio network planning and dimensioning.

2.5.1.1 Block Error Rate (BLER)

BLER is the long-term average block error rate calculated for TBs. The TB is considered erroneous if it has at least one bit error. The system knows the correctness of the blocks with very high reliability through the CRC.

2.5.1.2 Bit Error Rate (BER)

BER refers here to the information bit error rate – i.e., for user bits after decoding. The BER of channel-coded bits is always higher. Note that [15] specifies *TrCH BER* and *Physical Channel BER* as BS measurements.

2.5.1.3 Bit Rate, R

The bit rate, R, used in link-level simulations refers to user information bits. This means that the overhead from L1, such as CRC bits, coding and DPCCH control bits, is added in the simulations, but this only increases the energy required to transport the information bits over the air with the required quality (BER, BLER) for the information bits. Retransmissions are typically not modelled in link-level simulations, unless some specific ARQ schemes are under testing.

2.5.1.4 E_b/N_0 and Orthogonality, α

Originally, E_b/N_0 meant simply bit energy divided by noise spectral density. However, over time the expression 'E_b/N_0' has acquired an additional meaning. One reason is the fact that in CDMA the interference spectral density is added to the noise spectral density, since the interference is noiselike, due to the spreading operation. Thus, N_0 can usually be replaced by I_0, interference plus noise spectral density. The performance indicator E_b/N_0 is always related to some quality (BLER) target.

Equation (2.7) is a basic example of how to calculate E_b/N_0 in the uplink. Suppose that the signal is received at constant power p_{rx} and the received interference power is I. Assume further that the user bit rate is R and the bandwidth is W. Now the interference is uniformly distributed over the frequency band of width W [Hz] (equal to the chip rate) and the bit energy is $p_{rx} \cdot (1/R)$ [Ws], thus:

$$\frac{E_b}{N_0} = \frac{p_{rx}/R}{I/W} = \frac{W}{R}\frac{p_{rx}}{I} \tag{2.7}$$

The target of the fast power control in WCDMA is to keep the received E_b/N_0 constant. Due to the fast feedback loop (1.5 kHz) this is fairly successful. It means that for a chosen service, chosen channel conditions and chosen required BLER, the received power on the traffic channel divided by the interfering power is approximately constant.

In the downlink the E_b/N_0 is defined in another way, because the synchronised orthogonal codes reduce the interference from the serving cell (or cells, in soft handover). In the downlink the E_b/N_0 is calculated by the model:

$$\frac{E_b}{N_0} = \frac{W}{R} \cdot \frac{p_{rx}}{I_{own} \cdot (1 - \alpha) + I_{oth} + P_N} \tag{2.8}$$

where I_{own} is the total power received from the serving cell; I_{oth} is the total power received from the surrounding cells; and P_N is the noise power (thermal and equipment). The factor α is the so-called orthogonality factor, which depends on the instantaneous multi-path conditions. The codes are fully orthogonal, thus in the case of no multi-path the interference from the serving cell is cancelled and $\alpha = 1$. If, for example, there are instantaneously two equally strong propagation paths, then only half of the interference is cancelled from the receiver point of view and $\alpha = 0.5$ at this instant. Notice that E_b/N_0 and α should always be kept together, because in calculating the downlink capacity or spectral efficiency these together are the inputs that model the link-level performance. Using several scrambling codes in one cell naturally destroys the orthogonality.

The applicability of the model of Equation (2.8) is explored further in Section 2.5.3.2.

2.5.1.5 E_c/I_0

E_c/I_0 is the received chip energy relative to the total power spectral density. In the uplink this is the same as E_b/N_0 divided by the processing gain – i.e., by W/R. In the downlink I_0 is the total received power spectral density, thus the orthogonality effects are not taken into account. Notice that the notation of CPICH E_c/N_0 in [15] is actually the E_c/I_0 of the CPICH. E_c/I_0 is typically used as a link performance indicator for the signals having no information bits in the usual meaning. Such signals are for example CPICH, AICH and PICH.

2.5.1.6 E_c/I_{or}

E_c/I_{or} is the transmitted energy per chip on a chosen channel relative to the total transmitted power spectral density at the BS. Notice that it is also the fraction of the power allocated to the channel from the total BS transmitted power used. It is often utilised in downlink performance requirements: see [20]. It is either an input parameter, in the context 'for a certain E_c/I_{or} of a channel the UE is able to decode the channel at the cell edge', or an output, as in 'the E_c/I_{or} was, on average, X decibels for the BLER of Y%'. E_c/I_{or} is always used together with the *geometry factor* – see Equation (2.10).

2.5.1.7 Average Power Rise

In the case of low UE speed, the fast power control is able to compensate fast fading fairly well within the power control range, so the average E_b/N_0 needed for the required BLER is low. However, due to the deep fades, especially in low-multi-path diversity channels, the transmitted power has a peaky behaviour modelled by average power rise: see Figure 2.54.

It is measured from the link-level simulations as the difference between the average transmitted power and the average received power, with the condition that the average channel gain is 1. In the uplink, the power rise effectively only increases the interference received from surrounding cells, and in dimensioning it is added during the interference

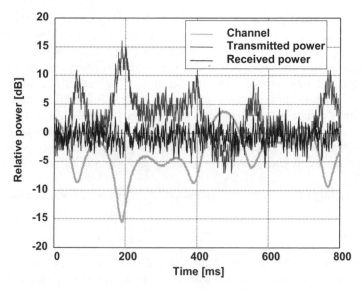

Figure 2.54 A snapshot of power control efficiency in a simulation with a low User Equipment speed.

calculation. In the downlink the average power rise is included in the basic E_b/N_0 figures. The *average power rise* is described in further detail in [22].

2.5.1.8   Power Control Headroom

Uplink coverage calculation starts from the required E_b/N_0. As explained in the previous section, the E_b/N_0 can be low for low UE speeds, but the cost is in the transmitted side due to the compensation of fading. When a low-speed UE is approaching the cell edge, the UE transmission power occasionally reaches its maximum value due to deep fades. Thus, a margin is required in the link budget for those low-speed UEs. This is called the *power control headroom* and is explained in further detail in [22].

2.5.1.9   Macro Diversity Combining Gain

The Macro Diversity Combining (MDC) gain is the reduction of the required E_b/N_0 per link in soft or softer handover when compared with the situation with one radio link only. Due to the power control, the gain is small when measured as the average required E_b/N_0. However, when measured as the required peak power – i.e., at cell edge – there is a substantial gain. In dimensioning, the reduction of the required peak power can be modelled by a reduced power control headroom.

Notice that typically the MDC gain is higher in the downlink than in the uplink. This does not mean that soft handover has bigger net gain in the downlink than in the uplink. It is merely a matter of definition –   namely, that each soft handover connection in the downlink adds the contribution of an additional transmitted channel to the total downlink transmitted power. As an example, if the MDC gain was 3 dB for a two-way soft handover in the downlink, then the net gain in total transmitted power would be 0 dB.

The MDC gain should not be confused with *multi-cell soft handover gain*, which results from the ability to keep the best cell always in the active set. A similar gain could also be achieved by a very fast hard handover. Uplink MDC gain is explained in further detail in [23].

2.5.1.10   Little i

The 'little i', i, is not actually a link-level performance indicator but is described here because it is one of the main issues affecting the spectral efficiency of a CDMA radio network. It stands for other-to-own-cell received power ratio and is calculated as the sum of received powers from the connections in surrounding cells divided by the sum of those from the connections in the serving cell, according to Equation (2.9):

$$i = \frac{I_{oth}}{I_{own}} \tag{2.9}$$

In the uplink, i affects all connections of one cell similarly, since it is calculated for the BS receiver. In the downlink it is calculated for each UE and thus is highly dependent on the UE location. However, it can be shown, assuming homogenous traffic and similar BSs in all cells, that the average i over all MSs in the network is the same as the average i over all BSs.

According to simulations, typical values for average i range from 0.15 (very well-isolated micro-cells) to 1.2 (poor radio network planning). Some simulations on i can be found in [24] and [25].

2.5.1.11   Geometry Factor, G

The geometry factor, G, used mostly in the downlink, is almost the same as the inverse of little i. G is defined as the ratio of the received power from the serving cell divided by the received power from surrounding cells plus thermal noise, i.e.:

$$G = \frac{I_{own}}{I_{oth} + P_N} \tag{2.10}$$

One can see that if the network is interference limited in the downlink – i.e., $P_N \ll I_{oth}$ – then approximately $G = 1/i$. The geometry factor reflects the distance of the UE from the BS antenna. A typical range is from $-3\,dB$ to $20\,dB$, where $-3\,dB$ is for the cell edge. G is usually an input parameter in the DL link-level simulations.

2.5.2   Classification according to Multi-path Channel Conditions and Services

There are two important sets of multi-path channel models in use. The first set is specified by 3GPP and used in UE and BTS performance requirements. These are collected here in Section 2.5.2.1. The second set, presented here in Section 2.5.2.2, has been recommended by ITU for the evaluation of the radio performance of different 3G system proposals [26].

2.5.2.1   3GPP Multi-path Channel Models

In 3GPP (see [20] and [21]) the UE and BS performance tests are defined in a 'static' propagation environment and in certain multi-path propagation conditions. The static environment means that there is one propagation channel – i.e., no multi-path, and no fading. AWGN is added to the signal before reaching the receiver. Table 2.5 shows the multi-path propagation conditions according to 3GPP. The fading models in all taps have a classical Doppler spectrum.

Case 1 in Table 2.5 is essentially a heavily fading one-tap channel, mostly similar to the ITU Pedestrian A model with a UE speed of 3 km/h. Cases 2 and 4 are also for low-speed UE but with higher multi-path diversity and thus less fading. Case 3 has four significantly strong channel taps, but the UE speed is 120 km/h. The channel is very like the Vehicular A channel at the same UE speed. Case 5 is the same as Case 1 but with a different UE speed, 50 km/h.

2.5.2.2   ITU Multi-path Models

Tables 2.6–2.8 show the multi-path propagation model recommended by International Telecommunication Union (ITU) in [26]. The 'indoor office' channel models have been

Table 2.5 Propagation conditions for multi-path fading environments [20].

Channel taps	Case 1, speed 3 km/h		Case 2, speed 3 km/h		Case 3, speed 120 km/h		Case 4, speed 3 km/h		Case 5, speed 50 km/h	
	Rel. delay [ns]	Av. power [dB]	Rel. delay [ns]	Av. power [dB]	Rel. delay [ns]	Av. power [dB]	Rel. delay [ns]	Av. power [dB]	Rel. delay [ns]	Av. power [dB]
1	0	0	0	0	0	0	0	0	0	0
2	976	−10	976	0	260	−3	976	0	976	−10
3			20 000	0	521	−6				
4					781	−9				

Table 2.6 Indoor office test environment tapped-delay-line parameters [26].

Tap	Channel A		Channel B		Doppler spectrum
	Relative delay [ns]	Average power [dB]	Relative delay [ns]	Average power [dB]	
1	0	0	0	0	Flat
2	50	−3.0	100	−3.6	Flat
3	110	−10.0	200	−7.2	Flat
4	170	−18.0	300	−10.8	Flat
5	290	−26.0	500	−18.0	Flat
6	310	−32.0	700	−25.2	Flat

Table 2.7 Outdoor-to-indoor and pedestrian test environment tapped-delay-line parameters [26].

Tap	Channel A		Channel B		Doppler spectrum
	Relative delay [ns]	Average power [dB]	Relative delay [ns]	Average power [dB]	
1	0	0	0	0	Classic
2	110	−9.7	200	−0.9	Classic
3	190	−19.2	800	−4.9	Classic
4	410	−22.8	1200	−8.0	Classic
5	—	—	2300	−7.8	Classic
6	—	—	3700	−23.9	Classic

Table 2.8 Vehicular test environment, high antenna, tapped-delay-line parameters [26].

	Channel A		Channel B		Doppler spectrum
Tap	Relative delay [ns]	Average power [dB]	Relative delay [ns]	Average power [dB]	
1	0	0.0	0	−2.5	Classic
2	310	−1.0	300	0	Classic
3	710	−9.0	8900	−12.8	Classic
4	1090	−10.0	12900	−10.0	Classic
5	1730	−15.0	17100	−25.2	Classic
6	2510	−20.0	20000	−16.0	Classic

recommended for use in modelling indoor systems. The 'outdoor-to-indoor and pedestrian' models are good for micro-cells, according to the ITU, as well as for indoor UEs served by an outdoor micro-cell. The 'vehicular' models can be used in modelling multi-path environments in macro-cells. Note that despite the name 'vehicular' these models have nothing directly to do with vehicles – i.e., although the UE may be in a car this does not mean that the multi-path channel is on average the 'Vehicular A' model. A similar possibility for misunderstanding exists with the name 'outdoor-to-indoor and pedestrian' test environment.

2.5.2.3 Reference Measurement Channels and the Link Performance Requirements in the Standard

As seen in Section 2.4 the standard offers huge flexibility for the system to select how to map the user information bits onto the air interface.

The easiest way to avoid endless simulation and measurement campaigns is to stick to a set of most typical cases and to use interpolation in planning and dimensioning other cases. An example of such a subset has been defined by 3GPP in [20] and [21] as *reference measurement channels*. The UE and Node B performance requirements have been defined for these channels. The reference measurement channels define the transport and physical channel parameters for speech (12.2 kbps codec rate) and for data channels with 64 kbps, 144 kbps, 384 kbps and 2048 kbps. Both uplink and downlink parameters were defined. As well as U-plane information bits, a *layer 3* signalling channel has been added.

Although the reference measurement channels may not be the most relevant in all cases, they are important, since the link performance requirements for the UE and BS have been given for these channels in the standard. Most of the standard requirements are, however, given in a form that is not directly suitable for use in radio network dimensioning and planning. This is because the power control is switched off in most test cases.

In the 3GPP standard (Release 5) the uplink performance requirements are given for dedicated channels as the average required E_b/N_0 values for fixed BLER points, without power control. In the downlink the link performance requirements are defined as the average required values of E_c/I_{or} for fixed BLER points at selected geometries (i.e., at assumed distances from the BS: see Section 2.5.1). In the downlink there is also one case with power control 'ON'. In this case it is required that the E_c/I_{or} stays below given thresholds for 90% of the time, assuming that the BS adjusts its transmit power according to the power control commands sent by the UE.

UE receiver performance requirements for the common channels are not defined with the exception of HSDPA by the standard Release 5.

2.5.3 Link-level Simulation Principles

2.5.3.1 Uplink Simulations

Uplink link-level simulations can be made according to the simulation chain depicted schematically in Figure 2.55.

The 'Data' block generates pseudo-random information bits that are input to the transmitter. The 'Transmitter' generates the physical channel according to the transport and physical channel parameters, performs the spreading and adjusts the transmitted power according to the power control commands received from the power control feedback loop. The transmitted signal is input to the 'Channel', which is the propagation channel, typically one of the channel models described in Section 2.5.2. The gains of the multi-path channel taps are scaled so that the average channel gain is 1 – i.e., large-scale propagation loss is excluded. After the channel, the pseudo-randomly generated 'AWGN' is added to the signal; this emulates thermal noise and interference. The composed signal is input to the 'Receiver' which despreads the signal and decodes the information bits according to the transport and physical channel parameters. The receiver also performs the SIR estimation which is input to the 'Power control' block for the inner-loop power control. The correctness of the bits and blocks is evaluated in the 'Evaluation' block. Information on the correctness of a block is further input to the (outer-loop) power control. The inner-loop power control compares the measured SIR with the SIR_{target} and sends the power control bits back to the 'Transmitter'. Random errors are generated to the power control bits usually with the probability of 4%. The transmitter changes the transmit power, usually by $\pm 1\,dB$, where the sign depends on

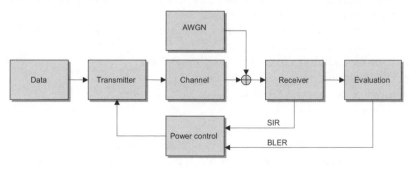

Figure 2.55 Uplink simulation chain.

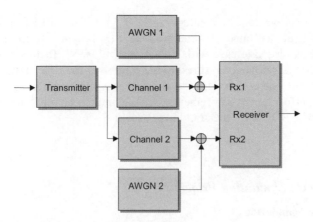

Figure 2.56 Receive diversity simulation chain.

the power control bits. The outer-loop power control controls the SIR$_{target}$ according to the BLER target and to the information received from the 'Evaluation' block.

When simulating receive diversity the 'Channel' and the 'AWGN' blocks are duplicated (see Figure 2.56). Different seeds of the random generators of the 'Channel' and 'AWGN' blocks are used, which emulates uncorrelated antenna diversity, but also a chosen correlation between the antennas can be arranged.

The most important outputs collected from the transmitter and receiver blocks are:
- transmitted power in each slot;
- received power in each slot;
- bit error statistics;
- BLER statistics.

From the transmitted and received power traces one can easily calculate the required link performance indicators, such as, for example, E_b/N_0 according to Equation (2.7), where I is the AWGN power, which is known. Notice that the inter-path interference from the own signal is not included in the N_0 in this calculation but is seen in the required E_b, since the delayed multi-path components interfere with the signal. This requires at least chip-level resolution in the simulation.

The following features of the simulator are crucial for reliability of the results:
- high time resolution (at least chip resolution);
- realistic channel estimation which uses as input only the received signal, not the known channel coefficients;
- power control, including closed-loop and outer-loop;
- realistic SIR estimation;
- delay in the power control feedback loop;
- power control bit errors.

Figure 2.57 shows an example of uplink simulation results. The information BER and BLER are plotted as a function of the received E_b/N_0 at the BS antenna. This simulation was performed with power control 'ON'. Simulations without power control

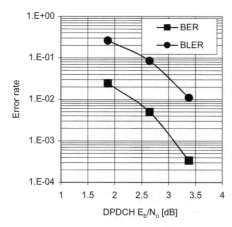

Figure 2.57 Example of uplink simulation output: information bit error rate and block error rate as a function of E_b/N_0 at the BS receiver: speech 12.2 kbps, Vehicular A multi-path channel, UE speed 3 km/h, power control 'ON'.

are also needed to estimate the power control headroom used in cell range calculation; for further detail see [22]. Table 2.9 shows some of the E_b/N_0 results in tabular form.

2.5.3.2 Downlink Simulations

Generally speaking, uplink simulation principles and the block diagram of Figure 2.55 apply to the downlink as well. There are, however, some major modifications needed in the transmitter model. In the downlink the interference originating from the same cell with the analysed signal must be included. This is because the own-cell interference propagates through the same propagation channel with the signal and thus the distribution of the total interference greatly varies as a function of distance from the BS.

Another difference from the uplink is that the UE uses the CPICH for channel estimation and this needs to be modelled. In simulations approximately 10% of the total Node B transmit power is allocated to the CPICH.

Figure 2.58 shows schematically how the channel of interest, CPICH, and a chosen number N (typically $N = 10, \ldots, 20$) of interfering channels are multiplexed (by summing the chips) to the Node B transmit signal.

In downlink simulations an additional input parameter is needed. This is the ratio of the power of the Node B own-signal to that of the AWGN source, which is actually the geometry factor explained in Section 2.5.1. In practice, the power of the interfering channels and the CPICH is kept constant while the AWGN power is varied to achieve different geometry factors, or vice versa. The transmit power of the analysed channel p_{tx}, the total Node B transmit power, I_{own}, and BER/BLER statistics are recorded from the simulations. Power control is 'ON' just as in uplink simulations.

When simulating downlink transmit diversity, the analysed signal and interference divides into two transmit branches in the 'Transmitter' block where the required antenna-specific coding and the required antenna-specific gains in the case of the feedback modes are applied on each branch. Similarly to the uplink, the transmitted

Table 2.9 Example of E_b/N_0 tables.

UL Rx E/N_0 [dB]	12.2 kbps voice, 20 ms interleaving			CS data 3 km/h, 40 ms interleaving		
	3 km/h	20 km/h	120 km/h	64 kbps	128 kbps	384 kbps
	4	4.5	5	2	1.5	1
	PS data 3 km/h, 10 ms interleaving			PS data 120 km/h, 10 ms interleaving		
	64 kbps	128 kbps	384 kbps	64 kbps	128 kbps	384 kbps
	2	1.5	1	3.3	3	2
DL Tx E_b/N_0 [dB]	12.2 kbps voice, 20 ms interleaving			CS data 3 km/h, 40 ms interleaving, BLER 1%		
	3 km/h	20 km/h	120 km/h	64 kbps	128 kbps	384 kbps
	6.5	6	6.5	5	5	5
	PS data 3 km/h, 10 ms interleaving, BLER 10%			PS data 120 km/h, 10 ms interleaving, BLER 10%		
	64 kbps	128 kbps	384 kbps	64 kbps	128 kbps	384 kbps
	5.5	5	4.5	5	4.5	4

signals from the two antennas are driven through independent fading channels, but now they are added together before the AWGN source. The power control feedback loop carrying the FBI bits should be implemented with random errors generated on the FBI bits. Also, realistic antenna verification performed by the UE should be used.

In Figure 2.59 there is an example of the output of a downlink simulation case using transmit diversity feedback mode 1. The example is used here also to show how the E_b/N_0 can be calculated from the simulated data. In the figure, the y-axis is the fraction of the traffic channel power from the total Node B power which is required for the average 1% BLER for this particular service and multi-path channel. The x-axis is the

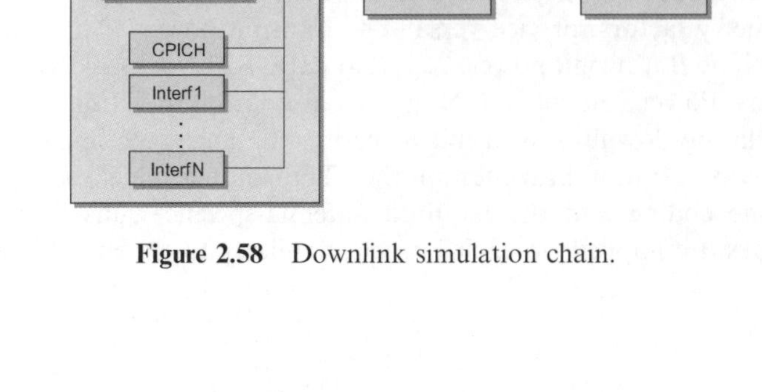

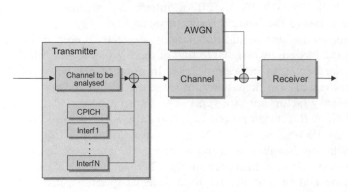

Figure 2.58 Downlink simulation chain.

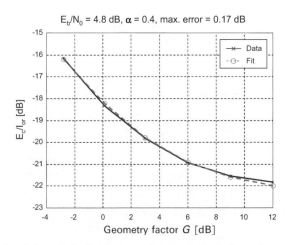

Figure 2.59 Example of downlink simulation output: required channel power relative to total BS power as a function of the geometry factor G: speech 12.2 kbps, BLER 1%, Vehicular A multipath channel, UE speed 3 km/h, transmit diversity feedback mode 1 used, power control 'ON'.

geometry factor. Now, we denote by ρ the required E_b/N_0 that is to be calculated. Using the definition of the geometry factor from Equation (2.10), the model (2.8) can be rewritten as:

$$\frac{W}{R} \cdot \frac{p_{tx}}{I_{own}} \cdot \frac{1}{(1-\alpha) + \dfrac{1}{G}} = \rho \tag{2.11}$$

Notice here that the received power from the own-cell in Equation (2.8) can be replaced by transmitted powers since the average channel gain is 1.

For each simulation point $(G = G_k)$ there is an output from the simulation – i.e., some $(p_{tx}/I_{own})_k$. Thus Equation (2.11) becomes a system of equations with two unknowns, α and ρ, which can be found by using, for example, least squares fitting. Another possibility is to calculate a fixed α based on the multi-path channel profile and to estimate only ρ from the simulations. The accuracy of the least squares fit can be seen as the difference between the 'Fit' and 'Data' curves of Figure 2.59. Thus the link-level performance in the downlink can be accurately modelled with two parameters, E_b/N_0 and α, at least in this case.

Also, with a known multi-path profile, using a fixed orthogonality provides accurate enough results according to experience from simulations.

Some downlink common channels, such as CPICH, SCH and P-CCPCH, typically do not use power control but use constant power instead. Thus the simulations for these channels are simpler. It is enough to simulate the geometry points near the cell edge $(G = -6, \ldots, -3 \text{ dB})$ and estimate the required E_c/I_0 for the chosen quality. The results can then be used directly in dimensioning and planning, as coverage thresholds and furthermore to estimate the fraction of the total Node B transmit power required for these channels.

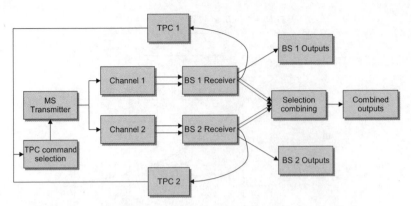

Figure 2.60 Simulation chain for uplink soft handover simulations.
Reproduced by permission of IEEE.

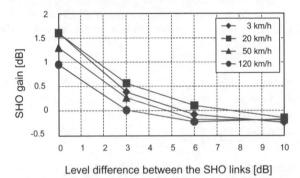

Figure 2.61 Example of uplink soft handover simulation results (3GPP channel model 4, antenna diversity).

2.5.3.3 Soft Handover Simulations[5]

In uplink soft handover simulations, two propagation channels, two BS receivers and two independent power control feedback loops need to be modelled and implemented, see Figure 2.60 (see [23]). Selection combining is performed at the combining block (RNC), based on CRC evaluation or, for equal CRC results, on quality estimates (SIR) from each branch.

In the simulation the relative channel gains between the two soft handover branches are varied. Figure 2.61 shows an example of soft handover simulation results. In the *x*-axis there is the relative gain difference between the soft handover branches and in the *y*-axis there is the gain in the average UE transmit power. In softer handover with MRC of the signals from different sectors, the gain is slightly bigger than in soft handover with selection combining.

---

[5] In this section soft handover gain refers to the 'Macro Diversity Combing (MDC) gain'. It is not the multi-cell gain, which is also often called 'soft handover gain'.

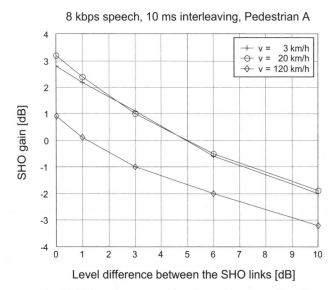

Figure 2.62 Example of downlink soft handover simulation results: ITU Pedestrian A channel model, no transmit diversity, soft handover gain in *total* transmit power.

In the downlink the UE typically combines all the signals in soft or softer handover in baseband, which is why a different model is used in the downlink. Actually, only one Node B transmitter block is needed, but the propagation channel must be duplicated. Similarly to the uplink, the relative channel gain difference between the soft handover branches is varied. Figure 2.62 shows an example from downlink soft handover simulations. On the *y*-axis is the soft handover gain in *total* transmit power (added from the two Node Bs), while the *x*-axis shows the average level difference between the soft handover links at the UE antenna. Note that the gain per radio link is 3 dB higher than in Figure 2.62, but because two transmissions are needed the net gain is clearly negative when the level difference between the soft handover branches increases.

2.5.4 Physical-layer Measurements Supporting the Measurement of Link-level Performance in a Live Network

The 3GPP standard specifies various *physical-layer measurements* that can be used as link-level performance indicators, either directly or indirectly. The most important physical-layer measurements from [15], which are related to link-level performance, are listed below. In most cases the analysis requires that a special *field measurement tool* is used to log the data, or that appropriate *performance indicators* are recorded from the network end. The standard measurement accuracy requirements are given in [16].

For the UE the measurements related to link-level performance that are supported by the standard (FDD) are as follows:

- CPICH RSCP – Received Signal Code Power. This is the received power of the CPICH as measured by the UE. It can be used to estimate the path loss, as the

transmit power of the CPICH is either known or can be read from the system information.
- UTRA carrier RSSI – Received Signal Strength Indicator, the wideband received power on the channel bandwidth in the downlink.
- CPICH E_c/N_0 – received energy per chip divided by the power density in the band. This is the most important UE measurement in WCDMA for network planning purposes, not only because it is typically the basic coverage indicator, but also because the accuracy is good, since the measurement can be done in baseband. Theoretically it is identical to CPICH RSCP/RSSI.
- TrCH BLER – the estimate for the BLER. It is based on the CRC evaluation of each TB after radio link combination.
- UE transmit power – this is very important as it can be used to verify power control performance, together with the SIR and BLER estimates at the Node B.

For the UTRAN the measurements related to link-level performance that are supported by the standard are as follows:
- Received total wideband power – this is simply the total received power (including noise generated in the receiver) within the bandwidth defined by the pulse-shaping filter. It is most important for estimating loading or some unexpected interference in the network, but not so important concerning link-level performance.
- SIR – defined as (RSCP/ISCP) ∗ SF, where RSCP is the Received Signal Code Power on the DPCCH, ISCP is the Interference Signal Code Power on the DPCCH and SF is the spreading factor on the DPCCH – i.e., 256. The reference point for the SIR measurement is the antenna connector. In compressed mode the SIR shall not be measured in the transmission gap. For one bearer service the E_b/N_0 of the TrCH can be estimated from SIR according to the equation, which includes both multi-code transmission and the overhead of the DPCCH in the energy per bit computation:

$$E_b/N_{0\,DCH} = SIR_{actual}^{UL,DPCCH} - 10 \cdot \log \left(\frac{R_{DCH}^{user}}{\left(N + \left(\frac{\beta_c}{\beta_d} \right)^2 \right) R_{DPDCH}} \right)$$

$$- 20 \cdot \log \left(\frac{\beta_c}{\beta_d} \right) - 10 \cdot \log \left(\frac{SF_{DPCCH}}{SF_{DPDCH}} \right) \tag{2.12}$$

where β_c (DPCCH gain factor), β_d (DPDCH gain factor), R_{DPDCH} (DPDCH bit rate) and SF_{DPDCH} are produced by admission control in the RNC considering the maximum bit rate of the transport channel; N is the number of DPDCHs used in multi-code transmission.
- SIR$_{error}$ – the difference between the SIR and the SIR$_{target}$ in the closed-loop power control algorithm. This can be used in analysing power control performance in the uplink.
- Transmitted carrier power – this is the ratio between the total and the maximum transmit power of a UTRAN access point (Node B). It can be used in analysing the downlink performance of DCHs, as it indicates the total BS power used.

- Transmitted code power – the transmitted power on one channelisation code on a given scrambling code on a given carrier. This can be used in analysing the downlink performance of DCHs.
- TrCH BER – an estimation of the average BER of the DPDCH.
- Physical channel BER – an estimation of the average BER on the DPCCH. This is often called 'raw BER' or 'uncoded BER'.

Note that BLER is not a mandatory measurement for the UTRAN.

References

[1] 3GPP, Technical Specification 23.101, General UMTS Architecture, v5.0.1.
[2] 3GPP, Technical Specification 25.301, Radio Interface Protocol Architecture, v5.3.0.
[3] 3GPP, Technical Specification 25.401, UTRAN Overall Description, v5.9.0.
[4] 3GPP, Technical Specification 25.302, Services Provided by the Physical Layer, v5.7.0.
[5] 3GPP, Technical Specification 25.321, MAC Protocol Specification, v5.10.0.
[6] 3GPP, Technical Specification 25.322, RLC Protocol Specification, v5.9.0.
[7] 3GPP, Technical Specification 25.323, PDCP Protocol Specification, v5.2.0.
[8] 3GPP, Technical Specification 25.324, BMC Protocol Specification, v5.5.0.
[9] 3GPP, Technical Specification 25.331, RRC Protocol Specification, v5.11.0.
[10] 3GPP, Technical Specification 23.107, QoS Concept and Architecture, v5.13.0.
[11] 3GPP, Technical Specification 25.211, Physical Channels and Mapping of Transport Channels onto Physical Channels (FDD), v5.6.0.
[12] 3GPP, Technical Specification 25.212, Multiplexing and Channel Coding (FDD), v5.9.0.
[13] 3GPP, Technical Specification 25.213, Spreading and Modulation (FDD), v5.5.0.
[14] 3GPP, Technical Specification 25.214, Physical Layer Procedures (FDD), v5.10.0.
[15] 3GPP, Technical Specification 25.215, Physical Layer Measurements (FDD), v5.5.0.
[16] 3GPP, Technical Specification 25.133, Requirements for Support of Radio Resource Management (FDD), v5.13.0.
[17] 3GPP, Technical Specification 25.433, UTRAN Iub Interface NBAP Signalling, v5.11.0.
[18] 3GPP, Technical Specification 25.402, Synchronisation in UTRAN, v5.3.0.
[19] Holma, H. and Toskala, A. (eds), *WCDMA for UMTS* (3rd Edition), John Wiley & Sons, Chichester, UK, 2004.
[20] 3GPP, Technical Specification 25.101, UE Radio Transmission and Reception (FDD), v5.13.0.
[21] 3GPP, Technical Specification 25.104, UTRA (BS) FDD Radio Transmission and Reception, v5.9.0.
[22] Sipilä, K., Laiho-Steffens, J., Jäsberg, M. and Wacker, A., Modelling the impact of the fast power control on the WCDMA uplink. *Proc. VTC 1999 Conf., Houston, Texas, May 1999*, pp. 1266–1270.
[23] Sipilä, K., Jäsberg, M., Laiho-Steffens, J. and Wacker, A., Soft handover gains in a fast power controlled WCDMA uplink. *Proc. VTC 1999 Conf., Houston, Texas, May 1999*, pp. 1594–1598.
[24] Wacker, A., Laiho-Steffens, J., Sipilä, K and Heiska, K., The impact of the base station sectorisation on WCDMA radio network performance. *Proc. VTC 1999 Fall Conf., Amsterdam, Netherlands, September 1999*, pp. 2611–2615.

[25] Laiho-Steffens, J., Wacker, A. and Aikio, P., The impact of the radio network planning and site configuration on the WCDMA network capacity and quality of service. *Proc. VTC 2000 Spring Conf., Tokyo, May 2000*, pp. 1006–1010.

[26] Guidelines for Evaluation of Radio Transmission Technologies for IMT-2000, Recommendation ITU-R M. 1225, 1997.

[27] Proakis, J.G., *Digital Communications* (3rd edn), McGraw-Hill, 1995, 927 pp.

[28] Peterson, R.L., Ziemer, R.E. and Borth, D.E., *Introduction to Spread Spectrum Communications*, Prentice-Hall, 1995, 695 pp.

[29] Noneaker, D.L. and Pursley, M.B., RAKE reception for a CDMA mobile communication system with multi-path fading. In: Glisic, S. G. and Leppänen, P. A. (eds), *Code Division Multiple Access Communications*, Kluwer Academic, 1995, pp. 183–201.

[30] Lee, J.S. and Miller, L.E., *CDMA Systems Engineering Handbook*, Artech House, 1998.

[31] Simon, M.K., Omura, J.K., Scholtz, R.A. and Lewitt B.K., *Spread Spectrum in Communications* (Vols 1–3), Computer Science Press, 1985.

[32] Gutierrez, P.J.A., Packet scheduling and quality of service in HSDPA. PhD thesis, University of Aalborg, October 2003.

[33] 3GPP, Technical Specification 25.308, High-speed Downlink Packet Access (HSDPA); Overall description, v5.7.0.

[34] 3GPP, Technical Specification 25.855, High-speed Downlink Packet Access; Overall UTRAN Description, v5.0.0 (currently withdrawn from the specifications).

[35] 3GPP, Technical Specification 25.306, UE Radio Access Capabilities, v5.9.0.

[36] 3GPP, Technical Specification 25.858, High-speed Downlink Packet Access: Physical Layer Aspects, v5.0.0.

[37] 3GPP, Technical Report 25.896, Feasibility Study for Enhanced Uplink for UTRA FDD, v6.0.0.

[38] Rosa, C., Outes, J., Dimou, K., Sørensen, T., Wigard, J., Frederiksen, F. and Mogensen, P., Performance of fast Node B scheduling and L1 H-ARQ schemes in WCDMA uplink packet access. *Proc. VTC 2004 Spring Conf., Milan, May 2004*, pp. 2071–2075.

[39] 3GPP, Technical Report 25.808, FDD Enhanced Uplink; Physical Layer Aspects (Release 6), v1.0.1.

3

WCDMA Radio Network Planning

Achim Wacker, Jaana Laiho, Tomáš Novosad, Terhi Rautiainen and Kimmo Terävä

In this chapter the pre-operational phase of the Wideband Code Division Multiple Access (WCDMA) Radio Network Planning (RNP) process, as depicted in Figure 3.1 in detail, is discussed. Like the planning process in second-generation (2G) systems, it can be divided into three phases. They are shown in Figure 3.1 and consist of initial planning (dimensioning), detailed planning and network operation and optimisation. Each phase requires additional support functions, such as propagation measurements, Key Performance Indicator (KPI) definitions, etc. In a cellular system where all the air interface connections operate on the same carrier, the number of simultaneous users directly influences the receivers' noise floors. Therefore, in the case of Universal Mobile Telecommunications System (UMTS) the planning phases cannot be separated into coverage and capacity planning. For post-2G systems data services start to play an important role. The variety of services requires the whole planning process to overcome a set of modifications. One of the modifications is related to the Quality of Service (QoS) requirements. So far it has been adequate to specify the speech coverage and blocking probability only, but it is increasingly necessary to consider the indoor and in-car coverage probabilities. In the case of UMTS the problem is slightly more multi-dimensional. For each service the QoS targets have to be set and naturally also met. In practice this means that the tightest requirement shall determine the site density. In addition to the coverage probability, the packet data QoS criteria are related to the acceptable delays and throughput. Estimation of the delays in the planning phase requires good knowledge of user behaviour and understanding of the functions of the packet scheduler.

There are also features common to 2G and third-generation (3G) coverage prediction. In both systems uplink (UL) as well as downlink (DL) have to be analysed. In current 2G systems the links tend to be in balance whereas in 3G systems one of the links can be loaded higher than the other, so that either link could be limiting the cell capacity or coverage. The propagation calculation is basically the same for all radio access technologies, with the exception that different

Radio Network Planning and Optimisation for UMTS Second Edition
Edited by J. Laiho, A. Wacker and T. Novosad © 2006 John Wiley & Sons, Ltd

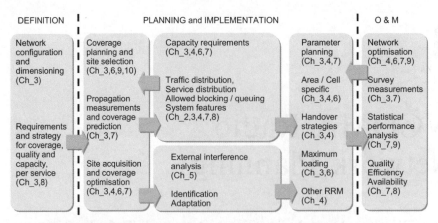

Figure 3.1 Radio network planning process for UMTS networks.

propagation models could be used. Another common feature is interference analysis. In the case of WCDMA this is needed for loading and sensitivity analysis; in the case of Time Division Multiple Access/Frequency Division Multiple Access (TDMA/FDMA) it is essential for frequency allocation. In order to fully utilise the capabilities of WCDMA, a thorough understanding of the WCDMA air interface is needed, from the physical layer to network modelling, planning and performance optimisation.

The purpose of the dimensioning phase is to estimate the approximate number of sites required, the Base Station (BS) configurations and the number of network elements, in order to forecast the projected costs and associated investments. Dimensioning is introduced in Section 3.1.

In Section 3.2 the detailed coverage and capacity planning with a static RNP tool is presented. The detailed planning takes into account the real site locations, the propagation conditions calculated on digital maps and real user distributions based on the operator's traffic forecasts. After the detailed planning has been performed, network coverage, capacity and other KPIs representing the network performance can be analysed.

In Section 3.3 the dimensioning and detailed planning are compared for a greenfield operator. The suitability of a static RNP tool for planning 3G systems has often been doubted and instead it has been proposed to use a fully dynamic simulator, which emulates moving mobiles and implements detailed Radio Resource Management (RRM) algorithms, such as power control and handovers. Owing to its complexity and need for computational power, however, it is not feasible for planning huge networks; rather it is a tool for benchmarking RRM algorithms. Nevertheless in Section 3.4 we demonstrate, for a small network, that the modelling in the static simulator presented is well in line with the outputs concerning the average network performance of the dynamic simulator. Sections 3.5 and 3.6 show the importance of interference control in a 3G network within one carrier as well as interference from the adjacent channel, emphasising that precautions should be taken to support this interference control as early as in the network planning phase. Bringing the chapter to a close in Section 3.7 we give an overview of how multi-layered network structures may be rolled out using one or several carrier frequencies.

3.1 Dimensioning

Initial planning (i.e., system dimensioning) provides the first and most rapid evaluation of the network element count as well as the associated capacity of those elements. This includes both the Radio Access Network (RAN) as well as the Core Network (CN). This section focuses upon the radio access part solely. The target of the initial planning phase is to estimate the required site density and site configurations for the area of interest. Initial planning activities include radio link budget and coverage analysis, capacity evaluation and lastly estimation of the amount of BS hardware and sites, Radio Network Controllers (RNCs), equipment at different interfaces and CN elements. The service distribution, traffic density, traffic growth estimates and QoS requirements are already essential elements in the initial planning phase. Quality is taken into account here in terms of blocking and coverage probability. Calculation of a radio link budget is carried out for each service and the tightest requirement determines the maximum allowed isotropic path loss.

3.1.1 WCDMA-specific Issues in Radio Link Budgets

In this section the WCDMA uplink and downlink radio link budgets are discussed. To estimate the maximum range of a cell, a radio link budget calculation is needed. In the radio link budget the antenna gains, cable losses, diversity gains, fading margins, etc. are taken into account. The output of the radio link budget calculation is the maximum allowed propagation path loss, which in return determines the cell range and thus the number of sites needed. There are a few WCDMA-specific items in the link budget, compared with TDMA-based radio access systems such as GSM. These include interference degradation margin, fast fading margin, transmit power increase and soft handover gain.

3.1.1.1 Uplink Radio Link Budget

The interference degradation margin is a function of cell loading. As more loading is allowed in the system, a larger interference margin is needed in the uplink and the coverage area shrinks. The uplink loading can be derived as follows; for simplicity the derivation is performed with service activity $\nu = 1$.

To find the required uplink transmitted and received signal power for Mobile Station (MS) k connected to a particular BS n, the basic CDMA E_b/N_0 Equation (3.1) is used. The usual, slightly idealistic, assumption in there is that I_{oth}, the power received from the MSs connected to other cells, is directly proportional (proportionality constant i) to I_{own}, the power received from the MSs connected to the same BS n as the desired MS. Assume that MS k uses bit rate R_k, its E_b/N_0 requirement is ρ_k and the WCDMA chip rate is W. Then the received power of the kth mobile, p_k, at the BS it is connected to, must be at least such that:

$$\frac{W}{R_k} \cdot \left(\frac{p_k}{I_{own} - p_k + I_{oth} + N} \right) = \frac{W}{R_k} \cdot \left(\frac{p_k}{I_{own} - p_k + i \cdot I_{own} + N} \right) \geq \rho_k, \quad k = 1, \ldots, K_n \quad (3.1)$$

where K_n is the number of MSs connected to BS n,

$$N = N_0 \cdot W = N_f \cdot \kappa \cdot T_0 \cdot W \tag{3.2}$$

is the noise power for an empty cell; N_f is the receiver's noise figure; κ is the Boltzmann constant $(1.381 \cdot 10^{-23}\,\text{Ws/K})$; and T_0 is the absolute temperature. For $T_0 = 293\,\text{K}$ $(20°\text{C})$ and $N_f = 1$, this results in $N_0 = -174.0\,\text{dBm/Hz}$ and $N = -108.1\,\text{dBm}$.

The inequalities in Equations (3.1) are slightly optimistic because it is assumed that there is no interference from the own signal, which is not exactly true in real multi-path propagation conditions. Equation (3.1) is, however, still chosen to avoid taking multi-path interference into account twice – i.e., the E_b/N_0 requirements determined from link-level simulations are presented so that N_0 means only noise and multi-path interference is visible in a higher E_b/N_0 requirement for a given Bit Error Rate (BER) performance. Solving the inequalities as equalities means solving for the minimum required received power (sensitivity), p_k:

$$p_k \cdot \left(1 + \frac{\rho_k \cdot R_k}{W}\right) = \frac{\rho_k \cdot R_k}{W} \cdot (1+i) \cdot I_{own} + \frac{\rho_k \cdot R_k}{W} \cdot N \quad \Rightarrow$$

$$p_k = \frac{1}{1 + \dfrac{W}{\rho_k \cdot R_k}} \cdot (1+i) \cdot I_{own} + \frac{1}{1 + \dfrac{W}{\rho_k \cdot R_k}} \cdot N, \quad k = 1, \dots, K \tag{3.3}$$

If the equations in (3.3) are summed over the MSs connected to BS n, then:

$$\sum_{k=1}^{K_n} p_k = \left[\sum_{k=1}^{K_n} \frac{1}{1 + \dfrac{W}{\rho_k \cdot R_k}} \cdot (1+i)\right] \cdot \sum_{k=1}^{K_n} p_k + \left[\sum_{k=1}^{K_n} \frac{1}{1 + \dfrac{W}{\rho_k \cdot R_k}}\right] \cdot N \quad \Rightarrow$$

$$\sum_{k=1}^{K_n} p_k \cdot (1+i) = \frac{N \cdot \left[\displaystyle\sum_{k=1}^{K_n} \frac{1}{1 + \dfrac{W}{\rho_k \cdot R_k}} \cdot (1+i)\right]}{1 - \left[\displaystyle\sum_{k=1}^{K_n} \frac{1}{1 + \dfrac{W}{\rho_k \cdot R_k}} \cdot (1+i)\right]} \tag{3.4}$$

since

$$I_{own} = \sum_{k=1}^{K_n} p_k \tag{3.5}$$

If we define uplink loading as:

$$\eta_{UL} = \sum_{k=1}^{K_n} \frac{1}{1 + \dfrac{W}{\rho_k \cdot R_k}} \cdot (1+i) \tag{3.6}$$

we can modify this to include the effect of sectorisation (sectorisation gain, ζ, number of sectors, N_S) and service activity, ν. Sectorisation gain values are listed in

Table 3.24:

$$\eta_{UL} = \sum_{k=1}^{K_n} \frac{1}{1 + \dfrac{W}{\rho_k \cdot R_k}} \cdot \nu_k \cdot \left(1 + i \cdot \frac{N_S}{\zeta}\right) \qquad (3.7)$$

In [1] uplink loading is estimated using Equation (3.8):

$$\eta_{UL} = \frac{1}{W} \cdot \sum_{j=1}^{m} R_j \cdot \nu_j \cdot \rho_j \cdot (1 + i) \qquad (3.8)$$

where m is the number of services used and each single user is counted as a separate service. The differences between Equations (3.7) and (3.8) are due to the fact that Equation (3.8) does not include sectorisation gain and that in the derivation starting from Equation (3.1) the denominator is $I_{own} - p_k + i \cdot I_{own} + N$ rather than $I_{own} + i \cdot I_{own} + N$, which is the case only when $p_k \ll I_{own}$.

3.1.1.2 Downlink Radio Link Budget

Downlink dimensioning follows the same logic as for the uplink. For a selected cell range the total BS transmit power needs to be estimated. In this estimation the soft handover connections must be included. If the power is exceeded, either the cell range should be limited, or the number of users in a cell should be reduced. For the downlink the loading (η_{DL}) is estimated based on:

$$\eta_{DL} = \sum_{i=1}^{I} \left[\frac{\rho_i \cdot R_i \cdot \nu_i}{W} \cdot \left((1 - \alpha_i) + \sum_{n=1,n \neq m}^{N} \frac{Lp_{mi}}{Lp_{ni}} \right) \right] \qquad (3.9)$$

where Lp_{mi} is the link loss from the serving BS m to MS i; Lp_{ni} is the link loss from another BS n to MS i; ρ_i is the *transmit* E_b/N_0 requirement for MS i, including the soft handover combining gain and the average power rise caused by fast power control; N is the number of BSs; I is the number of connections in a sector; and α_i is the orthogonality factor, which lies in the range from 0 up to 1 depending on multi-path conditions ($\alpha = 1$: fully orthogonal). The term:

$$i_{DL} = \sum_{n=1,n \neq m}^{N} \frac{Lp_{mi}}{Lp_{ni}} \qquad (3.10)$$

defines the other-to-own-cell-interference ratio in the downlink. A direct output of the downlink radio link budget is the single-link power required by a user at the cell edge. Total BS transmit power estimation must take into account multiple communication links with average ($\overline{Lp_{mi}}$) distance from the serving BS.

Furthermore, the multi-cell environment with orthogonalities α_i should be included in the modelling. For more on downlink loading and transmit power estimations, see [2].

In the radio link budget calculation in the uplink, the limiting factor is the MS transmit power; in the downlink the limit is the total BS transmit power. When balancing the uplink and downlink service areas both links must be considered.

The interference degradation margin, L, to be taken into account in radio link budgets due to a certain loading η (in either uplink or downlink) is:

$$L = 10 \cdot \log_{10}(1 - \eta) \tag{3.11}$$

The fast fading margin or power control headroom is another WCDMA-specific item in the radio link budget. Some margin is needed in the MS transmission power for maintaining adequate closed-loop fast power control in unfavourable propagation conditions such as near the cell edge. This applies especially to pedestrian users, where the E_b/N_0 to be maintained is more sensitive to the closed-loop power control. Power control headroom has been studied more in [3] and [4]; a summary can be found in Section 4.7. Another impact of fast power control is the greater average transmit power needed. In a slowly moving MS, the power control is able to follow the fading channel and the average transmitted power increases. In the own cell this is needed to provide adequate quality for the connection and causes no harm, since increased transmit power is compensated for by the fading channel. For neighbouring cells, however, this means additional interference because fast fading in the channels is uncorrelated. The transmit power increase (*TxPowerInc*) is used to reduce the reuse efficiency according to Equation (3.12). In Equation (3.6) i should be replaced with the term *TxPowerInc·i* in case MS transmit power increase is significant:

$$F_r = \frac{1}{1 + TxPowerInc \cdot i} \tag{3.12}$$

Soft handover gain has been discussed in [5]. Handovers – soft or hard – provide gain against shadow fading by reducing the required fading margin. Because slow fading is partly uncorrelated between cells and by making handovers, the mobile can select a better communication link. Furthermore, soft handover (macro-diversity) gives an additional gain against fast fading by reducing the required E_b/N_0 relative to a single radio link. The amount of gain depends on the mobile speed, the diversity combining algorithm used in the receiver and the channel delay profile. Soft handover gain is further discussed in Sections 3.1.3, 2.5.3.3 and 4.7.1.2.

3.1.2 Receiver Sensitivity Estimation

In the link budget the BS receiver noise level over one WCDMA carrier is calculated. The required Signal-to-Noise Ratio (SNR) at the receiver contains the processing gain and the loss due to the loading. The loading used is the total loading due to different services on the carrier in question. The required signal power, S, depends on the SNR requirement, receiver noise figure and bandwidth:

$$S = SNR \cdot N_0 \cdot W \tag{3.13}$$

where

$$SNR = \rho \cdot \frac{R}{W \cdot (1 - \eta)} \tag{3.14}$$

$N_0 \cdot W$ is the background noise introduced in Equation (3.2); R is the bit rate of the service used; ρ is its E_b/N_0 requirement; W is the WCDMA chip rate; and η is the loading of the cell. In some cases the basic noise/interference level is further corrected by applying a term that accounts for man-made noise.

3.1.3 Shadowing Margin and Soft Handover Gain Estimation

The next step is to estimate the maximum cell range and cell coverage area in different environments/regions. In the radio link budget the maximum allowed isotropic path loss is calculated and from that value a slow fading margin, related to the coverage probability, has to be subtracted. When evaluating the coverage probability, the propagation model exponent and the standard deviation for log-normal fading must be set. If the indoor case is considered, typical values for the indoor loss are from 15 to 20 dB and the standard deviation for log-normal fading margin calculation ranges from 10 to 12 dB. Outdoors, typical standard deviation values range from 6 to 8 dB and typical propagation constants from 2.5 to 4. Traditionally the area coverage probability used in the radio link budget is for the single-cell case [6]. The required probability is 90–95% and typically this leads to a 7–8 dB fading margin, depending on the propagation constant and standard deviation of the log-normal fading. Equation (3.15) estimates the area coverage probability for the single-cell case:

$$F_u = \frac{1}{2} \cdot \left\{ 1 - erf(a) + \exp\left(\frac{1 - 2 \cdot a \cdot b}{b^2}\right) \cdot \left[1 - erf\left(\frac{1 - a \cdot b}{b}\right) \right] \right\} \qquad (3.15)$$

where

$$a = \frac{x_0 - P_r}{\sigma \cdot \sqrt{2}}$$

and

$$b = \frac{10 \cdot n \cdot \log_{10} e}{\sigma \cdot \sqrt{2}}$$

where P_r is the received level at the cell edge; n is the propagation constant; x_0 is the average signal strength threshold; σ is the standard deviation of the field strength; and *erf* is the error function.

In real WCDMA cellular networks the coverage areas of cells overlap and the MS is able to connect to more than just one serving cell. If more than one cell can be detected, the location probability increases and is higher than that determined for a single isolated cell. Analysis performed in [7] indicates that if the area location probability is reduced from 96% to 90% the number of BSs is reduced by 38%. This number indicates that the concept of multi-server location probability should be carefully considered. In reality the signals from two BSs are not completely uncorrelated, and thus the soft handover gain is slightly less than estimated in [7]. In [5] the theory of the multi-server case with correlated signals is introduced:

$$P_{out} = \frac{1}{\sqrt{2\pi}} \int\limits_{-\infty}^{\infty} e^{-\frac{x^2}{2}} \cdot \left[Q\left(\frac{\gamma_{SHO} - a \cdot \sigma \cdot x}{b \cdot \sigma}\right) \right]^2 dx \qquad (3.16)$$

where P_{out} is the outage at the cell edge; γ_{SHO} is the fading margin with soft handover; σ is the standard deviation of the field strength and for 50% correlation of the log-normal fading between the mobiles and the two BSs $a = b = 1/\sqrt{2}$. With the theory presented, for example, in [6], this probability at the cell edge can be converted to the area probability. In the WCDMA link budget, soft handover gain is needed. The gain consists of two parts: combining gain against fast fading and gain against slow fading. The latter one dominates and is specified as:

$$G = \gamma_{single} - \gamma_{SHO} \qquad (3.17)$$

If we assume a 95% area probability, a path loss exponent of $n = 3.5$ and a standard deviation of the slow fading of 7 dB, the gain will be 7.3 dB − 4 dB = 3.3 dB. If the standard deviation is larger and the probability requirement higher then the gain will be more. Table 3.1 lists an example of a radio link budget for both uplink and downlink.

3.1.4   Cell Range and Cell Coverage Area Estimation

Once the maximum allowed propagation loss in a cell is known, it is easy to apply any propagation model for cell range estimation. The propagation model should be chosen so that it optimally describes the propagation conditions in the area. The restrictions on the model are related to the distance from the BS, the BS effective antenna height, the MS antenna height and the carrier frequency. One typical representative for the macro-cellular environment is the Okumura–Hata model (see Section 3.2.2.1), for which Equation (3.18) gives an example for an urban macro-cell with BS antenna height of 25 m, MS antenna height of 1.5 m and carrier frequency of 1950 MHz [8]:

$$Lp = 138.5 + 35.7 \cdot \log_{10}(r) \qquad (3.18)$$

After choosing the cell range the coverage area can be calculated. The coverage area for one cell in hexagonal configuration can be estimated with:

$$S = K \cdot r^2 \qquad (3.19)$$

where S is the coverage area; r is the maximum cell range; and K is a constant. Up to six sectors are reasonable for WCDMA, but with six sectors estimation of the cell coverage area becomes problematic, since a six-sectored site does not necessarily resemble a hexagon. A proposal for cell area calculation at this stage is that the equation for the 'omni' case is also used in the case of six sectors and the larger area is due to a higher antenna gain. The more sectors that are used, the more careful soft handover overhead has to be analysed to provide an accurate estimate. In Table 3.2 some of the K values are listed.

3.1.5   Capacity and Coverage Analysis in the Initial Planning Phase

Once the site coverage area is known the site configurations in terms of channel elements, sectors and carriers and the site density (cell range) have to be selected so that the traffic density supported by that configuration can fulfil the traffic requirements. An example of a dimensioning case can be seen in Section 3.3. The WCDMA radio link budget is slightly more complex than the TDMA one. The cell range depends on the number of simultaneous users – in terms of interference margin: see Equation (3.8). Thus the coverage and capacity are connected. From the beginning of network evolution the operator should have knowledge and vision of subscriber distribution and growth, since they have a direct impact on coverage. Finding the correct configuration for the network so that the traffic requirements are met and the

Table 3.1 Example of a WCDMA radio link budget.

	Uplink		Downlink	
Transmitter power	125.00	a	1372.97	mW
	20.97	$b = 10 \cdot \log_{10}(a)$	31.38	dBm
Transmitter antenna gain	0.00	c	18.00	dBi
Cable/body loss	2.00	d	2.00	dB
Transmitter EIRP (including losses)	18.97	$e = b + c - d$	47.38	dBm
Thermal noise density	−174.00	f	−174.00	dBm/Hz
Receiver noise figure	5.00	g	8.00	dB
Receiver noise density	−169.00	$h = f + g$	−166.00	dBm/Hz
Receiver noise power	−103.13	$i = 10 \cdot \log_{10}(W) + h$	−100.13	dBm
Interference margin	-3.01	j	−10.09	dB
Required E_c/I_0	−17.12	$k = 10 \cdot \log_{10}[E_b/N_0/(W/R)] - j$	−7.71	dB
Required signal power S	−120.26	$l = i + k$	−107.85	dBm
Receiver antenna gain	18.00	m	0.00	dBi
Cable/body loss	2.00	n	2.00	dB
Coverage probability outdoor (requirement)	95.00		95.00	%
Coverage probability indoor (requirement)	0.00		0.00	%
Outdoor location probability (calculated)	85.62		85.62	%
Indoor location probability (calculated)	32.33		32.33	%
Limiting environment	Outdoor		Outdoor	
Slow fading constant outdoor	7.00		7.00	dB
Slow fading constant indoor	12.00		12.00	dB
Propagation model exponent	3.50		3.50	
Slow fading margin	−7.27	o	−7.27	dB
Handover gain (including any macro-diversity combining gain at the cell edge	0.00	p	2.00	dB
Slow fading margin + Handover gain	−7.27	$q = o + p$	−5.27	dB
Indoor loss	0.00	r	0.00	dB
Power control headroom (fast fading margin)	0.00	s	0.00	dB
Allowed propagation loss	147.96	$t = e - l + m - n + q + r - s$	147.96	dB

Reproduced by permission of Group des Ecoles des Télécommunications.

Table 3.2 K values for the site area calculation.

Site configuration:	Omni	Two-sectored	Three-sectored	Six-sectored
Value of K:	2.6	1.3	1.95	2.6

Reproduced by permission of Groupe des Ecoles des Télécommunications.

network cost is minimised is not a trivial task. The number of carriers, number of sectors, loading, number of users and the cell range all affect the result.

3.1.6 Dimensioning of WCDMA Networks with HSDPA

In this section we describe the influence of the inclusion of High-speed Downlink Packet Access (HSDPA) transmission on the radio link budgets in both the uplink and downlink direction. The properties for HSDPA and the associated physical channels (HS-PDSCH, HS-SCCH in the downlink and the HS-DPCCH as a return channel in the uplink) have been described in Section 2.4.5. HSDPA dimensioning in this chapter assumes that dimensioning for Dedicated Channels (DCHs) ('Release '99 traffic') has already been done. The impact of the HSDPA can then be seen in following:

- In the uplink link budget an additional power margin is needed to be taken into account due to the introduction of the uplink High-speed Dedicated Physical Control Channel (HS-DPCCH: Section 2.4.5.2) transmitting ACK/NACK information and the Channel Quality Indicator (CQI).
- In the downlink direction the maximum power reserved for HSDPA transmission is constant, but it consists of two components that are time-variable. These two components are the powers of the High-speed Physical Downlink Shared Channel (HS-PDSCH) and the High-speed Shared Control Channel (HS-SCCH).
- In the downlink there is no soft handover, but the uplink return channel may or may not be in soft handover. In case soft handover is used, imperfect power control needs another margin in the link budget.

The main inputs for the dimensioning are the following:
- DCH traffic for the traditional link budgets;
- the desired HSDPA throughput in the downlink, either as average number for the cell or as average user throughput at the worst spot in the cell area (typically at the cell edge).

All three entities – i.e., cell range, coverage and throughput for HSDPA air interface – are then estimated. They are coupled together even more than for Release '99 data transmission on DCH. The behaviour can be understood as a consequence of there being more variables involved in HSDPA data transfer. On top of the usual WCDMA issues, in the HS-PDSCH there is the adaptive modulation switch between Quaternary Phase Shift Keying (QPSK) and 16 State Quadrature Amplitude Modulation (16QAM) working together with the Automatic Repeat reQuest (ARQ) scheme, 'fat pipe' scheduling, constellation and coding arrangement, which could change every Transmission Time Interval (TTI) – i.e., 2 ms. These features maximise air interface throughput and suppose there are no hardware-processing bottlenecks, the air interface is interference limited and the coverage for a certain capacity could be studied by connecting link-level simulations of the HSDPA 3GPP air interface with a power budget.

3.1.6.1 HSDPA Effects in Uplink Radio Link Budget

Although HSDPA is a downlink feature, there are additional effects on the uplink. The uplink HS-DPCCH, which provides the network with feedback from the MS (CQI and ACK/NACK) needs to be taken into account. The additional interference is not included in the original target E_b/N_0 values and a certain portion of the MS transmission power must be reserved for the additional traffic. This can be accounted for by including certain additional margins in the uplink link budget. As a result, the final uplink coverage is a bit worse compared with the Release '99 DCH. For more on the power offsets in the HS-DPCCH see Section 4.6.1. The additional margin depends on these power settings and on the bit rate of the uplink-associated DCH. Based on the default setting of the ratio of DPCCH over Dedicated Physical Data Channel Received Signal Code Powers (DPDCH RSCPs) ([9], table A.1) it may vary between 0.4 and 1.3 dB (see Table 3.3).

Table 3.3 Additional margin in uplink radio link budget due to uplink-associated DCH, CQI and ACK/NACK.

Uplink DCH bit rate	Margin
64 kbps	1.3 dB
128 kbps	0.6 dB
384 kbps	0.4 dB

Another additional margin that could be taken into account follows from the fact that the power control for HS-DPCCH is suboptimal for those HSDPA users applying soft handover on the HS-DPCCH [10]. To overcome this suboptimality a recommendation is to use the maximum possible HS-DPCCH power offset of 6 dB and an ACK/NACK repetition factor of 2. For this case, some applicable margin values are collected in Table 3.4.

Table 3.4 Additional margin in uplink radio link budget due to imperfect power control in soft handover.

UL DCH bit rate	Margin
64 kbps	2.70 dB
128/384 kbps	1.45 dB

However, considering the high data rate asymmetry for HSDPA, the main coverage limitation of the network will be on the downlink.

3.1.6.2 HSDPA Effects in Downlink Radio Link Budget

The main impact of the introduction of HSDPA will be visible in the downlink direction. The additional power needed for HSDPA transmission needs to be estimated and checked, whether this is compatible with DCH dimensioning. However, due to the physical properties of the HS-PDSCH as described above, the air interface cannot be fully described by E_b/N_0 and the BLER; therefore, we introduce another quantity instead into the link budget, which is the average HSDPA Signal-to-Interference-and-Noise Ratio (SINR). Additionally, one needs to keep in mind that there is no soft handover for the HS-PDSCH and therefore the appropriate gain in the radio link budget has to be removed.

Let's assume HSDPA transmission will use a certain portion of the cell power denoted by P_{HSDPA} that depends on the resource (power) management strategy used in the network. Typically, this part of the power is the remaining BS output power after deduction of both Release '99 traffic power and Common Control Channel (CCCH) power.

The power used for HSDPA will then impact the SINR as follows:

$$SINR = 16 \cdot \frac{P_{HSDPA} - P_{HS\text{-}SCCH}}{P_{tot} \cdot \left(1 - \alpha + \dfrac{1}{G}\right)} \tag{3.20}$$

where $P_{HS\text{-}SCCH}$ is the power of the HS-SCCH channel; α and G are the orthogonality and the Geometry factor explained in Section 2.5.1.11; P_{tot} is the total transmit power in the downlink including the HSDPA portion as multi-path propagation influences in the same way all downlink channels; and '16' (12 dB) is the fixed spreading factor for HSDPA as defined by Third Generation Partnership Project (3GPP) [11] and can be used directly in the radio link budget as the service processing gain for HSDPA users.

Next the relationship between achievable average throughput and the SINR present in the receiver environment needs to be established. Extended link-level simulations according to 3GPP specifications ([11] and [12]) have produced mapping tables between the two quantities. For five parallel codes and by simple second-order curve fitting the following approximate relationship can be derived:

$$Thr[Mbps] = 0.0039 * SINR^2 + 0.0476 * SINR + 0.1421, \quad -5\,\text{dB} \le SINR \le 20\,\text{dB} \tag{3.21}$$

where Thr is the average cell throughput in Mbps; and $SINR$ is the average SINR in dB in the cell. Equation (3.21) represents either the throughput of one user having a certain SINR or the combined cell throughput of several users having the same average SINR value together. More details can be found in [13] and [14].

The following process can now be identified for HSDPA downlink dimensioning. First the HSDPA throughput requirements need to be set by the operator and Equation (3.21) provides the needed SINR. With the additional inputs of the orthogonality and the G-factor at the cell edge (both could be results of simulations within the environment of the network or simple operator inputs), Equation (3.20) gives the power needed for HSDPA transmission (P_{HSDPA} and $P_{HS\text{-}SCCH}$). The power resulting from this calculation must be within the limits of the whole downlink loading. If violated, then additional sites or carriers need to be introduced to distribute the extensive load further. Finally, when the power used for HSDPA is known, one can estimate the cell capacity along with the downlink HSDPA coverage based on the power budget. HSDPA coverage (maximum path loss) is done in a similar way to the DCH case. HSDPA-

Table 3.5 Downlink High-speed Downlink Packet Access radio link budget example for 5 W of HSDPA power.

Service type: HSDPA

BS			
HSDPA power ($P_{HSDPA} + P_{HS\text{-}SCCH}$)	5.0	W	a
	37.0	dBm	$b = 10 * \log_{10}(a) + 30$
Receiver antenna gain	18.0	dBi	c
Cable/body loss	4.0	dB	$d = b + c - d$
Transmitter EIRP	51.0	dBm	e
MS			
Thermal noise	-108.0	dBm	f
Receiver noise figure	8.0	dB	g
Receiver noise power	-100.0	dBm	$h = f + g$
Downlink load	70.0	%	i
Interference margin	5.2	dB	$j = -10 * \log_{10}(1 - i/100)$
Interference plus noise	-94.8	dBm	$k = h + j$
Required SINR	5.3	dB	l
HSDPA processing gain	12.0	dB	$m = 10 * \log_{10}(16)$
Receiver antenna gain	0.0	dBi	n
Body/cable loss	0.0	dB	o
Receiver sensitivity	-101.5	dB	$p = k + l - m - n + o$
Power control headroom (fast fading margin)	0.0	dB	q
Soft handover gain	0.0	dB	r
Allowed propagation loss	152.5	dB	$s = e - p - q + r$

specific values are applied to the power budget. An example for such a power budget for HSDPA transmission is depicted in Table 3.5.

The allowed propagation loss is finally compared with the one from DCH dimensioning and, if compatible, HSDPA dimensioning can be accepted. Otherwise, it must be considered to add more sites or, if there is spectrum available, another carrier for HSDPA.

3.1.7 RNC Dimensioning

Mobile radio networks are too large for one RNC alone to handle all the traffic, so the whole network area is divided into areas each handled by a single RNC. In the rough dimensioning as described in this section it is normally assumed that sites are distributed uniformly across the RNC area and carry roughly the same amount of traffic. The purpose of RNC dimensioning is to provide the number of RNCs needed to support the estimated traffic. Several limitations on RNC capacity exist and at least the following must be taken into account, out of which the most demanding one has to be selected:

• maximum number of cells (a cell is identified by a frequency and a scrambling code);
• maximum number of BSs under one RNC;

- maximum Iub throughput;
- amount and type of interfaces (e.g., STm-1, E1).

Table 3.6 presents an example for the capacity of one RNC in different configurations. The number of RNCs needed to connect a certain number of cells can be simply calculated according to Equation (3.22):

$$numRNCs = \frac{numCells}{cellsRNC \cdot fillrate1} \tag{3.22}$$

where *numCells* is the number of cells in the area to be dimensioned; *cellsRNC* is the maximum number of cells that can be connected to one RNC; and *fillrate1* is a margin used as a backoff from the maximum capacity.

Next the number of RNCs needed according to the number of BSs to be connected must be checked with Equation (3.23):

$$numRNCs = \frac{numBSs}{bsRNC \cdot fillrate2} \tag{3.23}$$

where *numBSs* is the number of BSs in the area to be dimensioned; *bsRNC* is the maximum number of BSs that can be connected to one RNC; and *fillrate2* is a margin used as a backoff from the maximum capacity.

Finally, the number of RNCs to support Iub throughput has to be calculated with Equation (3.24):

$$numRNCs = \frac{voiceTP + CSdataTP + PSdataTP}{tpRNC \cdot fillrate3} \cdot numSubs \tag{3.24}$$

where *tpRNC* is the maximum Iub capacity; *fillrate3* is a margin used as a backoff from it; *numSubs* is the expected number of simultaneously active subscribers; and

$$\left. \begin{array}{l} voiceTP = voiceErl \cdot bitrate_{voice} \cdot (1 + SHO_{voice}) \\[4pt] CSdataTP = CSdataErl \cdot bitrate_{CSdata} \cdot (1 + SHO_{CSdata}) \\[4pt] PSdataTP = avePSdata/PSoverhead \cdot (1 + SHO_{PSdata}) \end{array} \right\} \tag{3.25}$$

are the throughputs for voice, Circuit Switched (CS) and Packet Switched (PS) data, respectively. *voiceErl* is the traffic of a single voice user; *CSdataErl* is the traffic from a

Table 3.6 Radio Network Controller capacity example.

Configuration	Iub throughput	Iub traffic capacity		Other interfaces	
		BSs	Cells	STm-1	E1
1	48 Mbps	128	384	4*4	6*16
2	85 Mbps	192	576	4*4	8*16
3	122 Mbps	256	768	4*4	10*16
4	159 Mbps	320	960	4*4	12*16
5	196 Mbps	384	1152	4*4	14*16

Table 3.7 Explanation of the parameters used in Equation (3.25).

voiceErl, CSdataErl	Expected amount of Erlangs per subscriber during busy hour in the RNC area.
avePSdata/PSoverhead (also called *FP_datarate* or *L2 data rate*)	This is the L2 data rate + overhead introduced by the Frame Protocol, including retransmission overhead (10%) and L2 + FP overhead (5%) – i.e., *L2 data rate = endUserDatarate*·1.1 · 1.05 (used only for PS data; for CS data there is no extra overhead).
SHOvoice, SHO$_{CSdata}$, SHO$_{PSdata}$	Overhead due to soft handover, typically 20–30% (i.e., 20–30% of MSs are connected to two or more BSs at the same time and this extra 20–30% of traffic is terminated in the RNC; therefore, transmission capacity is needed up to the RNC.

CS data user; and *avePSdata* is the average amount of PS data per user. *PSoverhead* takes into account 10% of retransmission as well as 5% of overhead from the Frame Protocol (FP) and L2 (RLC and MAC) overhead. The different SHOs are the overhead per service produced by soft handover. Note that in the case of asymmetric uplink and downlink the maximum number of both has to be taken and if there are several different services of one type (voice, CS or PS) summation has to be taken over all these services. The Erlang and kbps are measured as 'per area' values and are input data from the operator's traffic prediction, see Table 3.7.

Example of Radio Network Controller Dimensioning
In a certain area there are 800 BSs. Each BS has three sectors with two frequency carriers used per sector. If we assume a maximum capacity of *cellsRNC* = 1152 cells per RNC and a *fillrate1* of 90%, the number of RNCs needed is given by Equation (3.22):

$$\frac{800 \cdot 3 \cdot 2}{1152 \cdot 0.9} = 4.6 \ RNCs \tag{3.26}$$

If we assume that one RNC can support *bsRNC* = 384 BSs and take also 90% for *fillrate2*, Equation (3.23) leads to the following result for the number of RNCs needed:

$$\frac{800}{384 \cdot 0.9} = 2.3 \ RNCs \tag{3.27}$$

Finally, if we consider the following traffic profile:

- Voice service: *voiceErl* = 25 mErl/subs, *bitrate$_{voice}$* = 16 kbps,
- CS data service1:   *CSdataErl* = 10 mErl/subs, *bitrate$_{CSdata}$* = 32 kbps,
- CS data service2:   *CSdataErl* = 5 mErl/subs, *bitrate$_{CSdata}$* = 64 kbps,
- PS data services: *avePSdata* = 0.2 kbps/subs, *PSoverhead* = 15%,

with a soft handover factor for all services of 30%, a total of 350 000 subscribers, a maximum Iub capacity of *tpRNC* = 196 Mbps and a *fillrate3* of 90%,

Equations (3.24) and (3.25) yield:

$$\frac{(0.025 \cdot 16\,\text{kbps} + 0.010 \cdot 32\,\text{kbps} + 0.005 \cdot 64\,\text{kbps} + 0.2\,\text{kbps}/0.87) \cdot 1.3 \cdot 350000}{196\,\text{Mbps} \cdot 0.9}$$

$$= 3.3\ RNCs \quad (3.28)$$

Note that for the voice service above, the RNC input and output rates are assumed to be effectively 11.7 kbps (for EFR 12.2 kbps and 50% DTX), but 16 kbps is used for a voice channel in calculating the number of RNCs needed based upon the RNC processing limitation. For an Asynchronous Transfer Mode (ATM) switch-based RNC with no transcoding function, 11.7 kbps should be used. The reason for using 16 kbps is the estimate that a lower bit rate channel requires as much processing capacity (U- and C-plane) within an RNC as a 16 kbps channel.

We now take the maximum of the three results above, from Equations (3.26)–(3.28), for the number of RNCs needed, which in this example is 4.6 RNCs. In practice this would mean four RNCs with maximum capacity and one RNC with a smaller configuration.

It should be noted that using a typical three-sectored BS layout either the number of cells or the throughput is the limiting factor. In contrast, at the beginning of a typical network rollout, throughput is not a limiting factor. One RNC typically can support several hundred BSs. However, in a practical network, the number of BSs is expected to be significantly less (e.g., $32, \ldots, 64$), owing to the high capacity of each BS.

Based on the supported traffic or the actual expected traffic, there are the following different methods of RNC dimensioning (note that in any method, soft handover and air interface protocol overhead must be included):

- *Supported traffic (upper limit of RNC processing)*   This represents the planned equipment (and radio) capacity of the network. It is the upper limit of what RNC processing needs to support. Normally, the capacity is planned so that it is just slightly above the required traffic. However, in the case of data services, if the operator required a 384 kbps service, every cell would need to be planned for 384 kbps throughput. This usually gives too much data capacity, if averaged across the network. An RNC that is dimensioned based on supported traffic is able to offer 384 kbps throughput in every cell of the network at the same time.

- *Required traffic (lower limit of RNC processing)*   Based on the operator's prediction, this represents the actual traffic needs to be carried during the busy hour of the network and is an average value across the network. An RNC that is dimensioned based on required traffic can fulfil the mean traffic demand as predicted by the operator, but gives no room for dynamic variations in the data traffic (with the exception of buffering and increasing service delay). Therefore, it should be treated as the lower limit of the processing requirement. Note that:
  ○ RNC processing needs to include the overhead of soft handover;
  ○ voice traffic can be simply converted to kbps (1 voice channel = 16 kbps), for the purpose of calculating Iu interface loading.

- *RNC transmission interface to Iub*   If an RNC is dimensioned to support N sites, the total capacity for the Iub transmission interface must be greater than N times the transmission capacity per site, regardless of the actual load at the Iub interface.

- *RNC blocking principle* Normally, an RNC is dimensioned according to the assumed blocking at each BS (by Iub admission control or air interface admission control). Owing to allowed blocking at the BS, a certain proportion of subscriber peak traffic is never seen by the RNC. Consequently, we can convert the Erlangs per BS into physical channels per BS and use the result to calculate the number of RNCs needed. Similarly for NRT traffic, we can divide the average offered traffic by (*1-backoff_from_max_data_throughput*). In this way the RNC does not introduce any additional blocking to the offered traffic.
- An RNC can also be dimensioned directly according to the actual subscriber traffic in the area, and, for example, it can allow a similar amount of blocking as specified for the Iu interface. In this case, owing to the large amount of Erlangs per RNC area, the Erlang value can be used directly for calculating the number of RNCs needed.

3.2 Detailed Planning

In this section detailed planning with the help of a static radio network simulator is presented. Further information, together with a Matlab® implementation of such an example static simulator, can be acquired from the weblink at *www.wiley.com/go/laiho* and in [16]. This simulator was used in most of the studies presented in this book. It needs as inputs a digital map, the network layout and the traffic distribution in the form of a discrete user map. In a static simulator each of the users can have a different speed even though no actual mobility is modelled. How the MS speed is taken into account is described in Section 3.2.3. This speed and the service used (bit rate and activity factor, which can both be different for the uplink and downlink) together define the individual E_b/N_0 requirements, margins and gains imported from link-level simulations. Other static simulators are described, for example, in [17] and [18].

The simulator itself consists of basically three parts – initialisation, combined uplink and downlink analysis, and the post-processing phase (see Figure 3.2).

Following initialisation, both the uplink and downlink for all Mobile Stations (MSs) are analysed repeatedly in the main part of the tool. In the final step, after the iterations have fulfilled certain convergence criteria, the results of the uplink and downlink analyses are post-processed for various graphical and numerical outputs. On top of these results, for selected areas (which also can consist of the whole network), area coverage analyses for uplink and downlink DCHs, as well as for common channels (CPICH, BCCH, FACH and PCH on the P-CCPCH and/or S-CCPCH), can be performed.

In case a second carrier is present in the network area, used either by the same or by a different operator, Adjacent Channel Interference (ACI) can be taken into account. Only if the second carrier is assigned to the same operator can load be shared between the carriers by performing Inter-frequency Handover (IF-HO) according to different strategies.

This section is organised as follows. Section 3.2.1 lists general requirements for a planning tool. In Sections 3.2.2–3.2.5 the detailed processes and calculations in the three different phases of the analysis are presented. Section 3.2.2 describes the initialisation phase; Section 3.2.3 deals with the detailed iterations in the uplink and downlink;

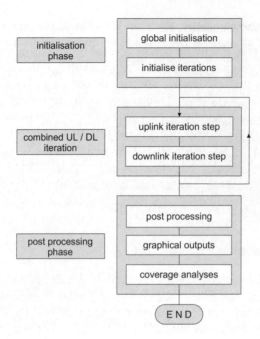

Figure 3.2   Static simulator overview.

and Section 3.2.4 shows how ACI can be modelled. Finally, Section 3.2.5 is concerned
with the post-processing phase.

3.2.1   General Requirements for a Radio Network Planning Tool

Planning tools (RNP tools) have always played a significant role in the daily work of
network operators. When business requirements for service demands are specified based
on business plans, the task of network planners is to fulfil the given criteria with
minimal capital investment. Typically, the input parameters include requirements
related to quality, capacity and coverage for each service. Most 2G networks have
only offered voice services. In 3G networks, there are various service types (voice
and data) and a multitude of different services, which may all have different requirements. Thus 3G planning tools play an even bigger role in the detailed network
planning phase than in the case of 2G networks. It is necessary to find an optimum
tradeoff between quality, capacity and coverage criteria for all the services in an
operator's service portfolio.

One or more tools should assist the network planner in the whole planning process,
covering dimensioning, detailed planning and, finally, pre-launch network optimisation.

Typically, a single tool alone cannot support all the phases of the planning process.
Instead, one tool is dedicated to dimensioning, another to network planning, a third to
optimisation. In modern applications, all the tools required are typically integrated
seamlessly into one package, which consists of a suite of tools. If this integration is

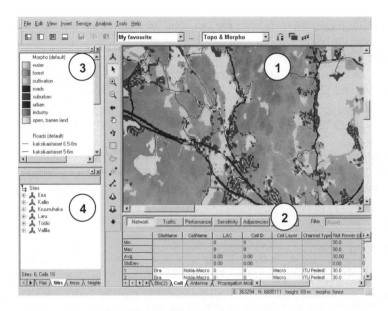

Figure 3.3   Example of the main user interface of an RNP tool.

performed properly, the end-user, here the network planner, is unaware of actually using several tools when performing the planning and optimisation activities.

This section gives the requirements for an RNP tool that will support the depicted phases of the planning process. The tool described is static, meaning that the simulator models one snapshot of time instead of dynamically modelling the active calls, for example.

Figure 3.3 shows an example of the main user interface of an RNP tool. It consists of:

1. map;
2. browser (table view);
3. legend dialog;
4. network element tree view.

The workflow supported by a typical RNP tool is presented in Figure 3.4. The given process is naturally part of the whole network planning process as set out in Figure 3.1. This section covers the workflow presented in Figure 3.4.

3.2.1.1   Preparations for Necessary Input Data

Digital Map

The most important basic preparatory requirement for an RNP tool is a geographical map of the planning area. The map is needed in coverage (link loss) predictions and subsequently the link loss data are utilised in the detailed calculation phase and for analysis purposes. For network planning purposes, a digital map should include at least

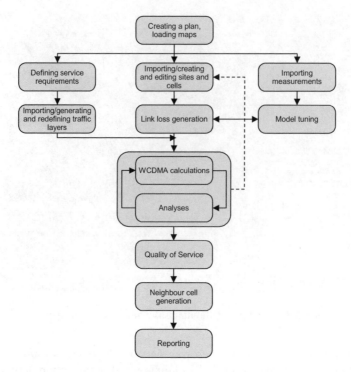

Figure 3.4 Example workflow supported by RNP tools.

topographic data (terrain height), morphographic data (terrain type, clutter type) and building location and height data, in the form of raster maps.

In addition, it is important to include vectorised data for building locations in digital maps. If available, road information (raster or vector) can also be used in certain operations, such as traffic modelling and coverage predictions.

A raster unit (map *resolution*) is usually in the range of 1 up to 200 m. Typically, in urban areas the minimum acceptable resolution is 12.5 m, whereas in rural areas up to 50–100 m resolution is common. However, as a rule of thumb, the more accurate map (finer resolution) that is available, the more precise calculation results can be achieved. Also, when considering 3G networks, a resolution as low as 5 m may be needed for dense urban areas, since geographical cell sizes will be small.

Other general requirements for RNP tool digital maps are the ability to support various projections, ellipsoids and coordinate systems – e.g., the Universal Transverse Mercator projection and the World Geodetic System 84 (WGS-84) ellipsoid.

Plan
A plan is a logical concept for combining various items of data into one 'package' that is understandable to the network planner. It is typically defined by the following items:
- digital map;
- map properties such as projection and ellipsoid;
- target planning area;

- selected radio access technologies;
- input parameters for calculations;
- antenna models.

A plan is always created and defined before the actual network planning activities are started. It will always contain all the configuration settings and parameter values for the planned network elements. In practice, the plan contains all the BS and cell data to be deployed finally in the real network. In modern tools, several radio access technologies are supported in one plan, thus providing a means of planning networks for both 2G and 3G systems simultaneously. An RNP tool should be able to create, define, save and retrieve several plans, so that different versions of the same target area can be compared in terms of which plan version best fulfils the given quality, capacity and coverage criteria. Naturally, an RNP tool should also provide means of assessing the differences between multiple plans: for example, by providing 'delta' reports of selected characteristics, such as coverage or planned network elements.

Antenna Editor

In RNP tools, 'antenna' is a logical concept that includes the antenna radiation pattern and parameters such as antenna gain and frequency band. Once 'antenna' is defined, it can then be assigned and used for selected cells and coverage predictions.

Typically, 'antenna' definition starts by importing radiation patterns into the RNP tool. Antenna vendors provide operators with data sheets that include the necessary radiation pattern information (direction and gain). Vendor-specific antenna data are converted and imported into the RNP tool and then logical antennas can be defined and antenna models stored in the RNP tool's database.

Modern RNP tools provide support for visualising antenna radiation patterns and also for editing patterns manually. Typically, two types of antenna models are supported: global and plan-specific. Global antenna models are available for all plans. If such models are modified, they are available to all new plans created subsequently. Plan-specific antenna models belong to individual plans and changes in them do not affect the global models.

Propagation Model Editor

Operators usually have separate regional and centralised planning organisations. One task of the central organisation is to provide templates and defaults for regional organisations. Having a 'default' coverage prediction model is one concrete example. Typically, a few propagation models are prepared for each area type for the regional organisations. The default model can then be tailored at the regional level according to local conditions.

An RNP tool should be able to support this facility and modern tools usually include so-called propagation model tuning or editing tools. The tuning itself is based on field measurements that provide basic signal strength data together with coordinates. Model tuning is described later in this section.

As with antenna models, two types of propagation models are available in modern RNP tools: global and plan-specific. Similar rules apply: if a global propagation model is modified, the changes are available to all new subsequently created plans.

RNP tools should also support different planning area characteristics and propagation environments. Therefore, various propagation models must be supported: Okumura–Hata, Walfisch–Ikegami and ray-tracing models are typically provided by RNP tools. The Okumura–Hata model is best suited for macro-cells and for small cells in which the antenna is located above the surrounding rooftop level. The Walfisch–Ikegami model is intended for small-cell planning where the maximum cell radius is 3–5 km.

Ray-tracing techniques are applied only in micro-cell environments in dense urban areas, since the necessary accurate map data are normally available only for such urban areas and calculation times are usually too long for planning a whole network. More about propagation models can be found in Section 3.2.2.1.

BS Types and Site/Cell Templates

Another example of the templates and defaults that should be provided by an operator's central planning organisation are network element parameter defaults and typical site configurations. An RNP tool should provide the functionality for defining and handling general hardware configurations and default configuration and parameter settings for network elements such as sites and cells. A typical example of a default hardware configuration is the BS hardware definition. In both 2G and 3G systems, network hardware vendors update their hardware regularly, usually adding more functionality and capacity in later hardware generations. In practice this means that more physical hardware can be installed in later hardware generations. Naturally, this is closely related to the actual number of needed BSs and sites in the planned network and this should be taken into account in the RNP tool when performing calculations and analyses. For WCDMA, the BS hardware template may include:

- maximum number of wideband signal processors;
- maximum number of channel units;
- noise figure;
- available transmit/receive diversity types.

Site templates may include default values for cell configuration, antenna directions, BS hardware capacity and propagation models used for cells, for example. Site templates are also defined by the central planning organisation.

When site deployment is being planned, the default values for almost all site and cell parameters come automatically from the site defaults. This can significantly reduce the time needed for manually entering these parameters, though in some cases manual editing of these parameters will still be required since the defaults cannot be used in all cases.

A site template may include general site information, BS information and cell template information for the site. A WCDMA cell template may include cell-layer type, channel model, transmit/receive diversity options, power settings, maximum acceptable load, propagation model used, antenna information and cable losses.

3.2.1.2 Planning

Importing Site Information

When planning 3G networks, a typical scenario is that an operator may wish to reuse existing 2G network sites as much as possible. Therefore, it is important for an RNP tool to provide support for importing 2G site locations and basic antenna data into a new plan, especially when making a combined network plan for both 2G and 3G networks.

Site import functionality automatically brings site and antenna information into an RNP tool plan. Naturally, such automatic importing of data saves network planner time. The imported information may include the site location, site ground height, number of cells and antenna directions.

Editing Sites and Cells

After existing site data are imported, it may still be necessary to add either sites or cells manually. Also manual modification of parameters and antenna information is typically needed during 'traditional' network planning operations.

RNP tools should provide various means to add and edit network elements manually, the most important being the manual addition of single elements and adding elements from templates.

When network elements are placed into planned geographical locations, their parameters should be checked before starting time-consuming calculations. Parameters are controlled by invoking individual network elements' dialogs or from specific browsers that usually list all the network elements from the current plan (or from the planning area). From these browsers, it is easy to see at a glance the data covering the whole network and any variations in parameter settings.

Defining Service Requirements and Traffic Modelling

Traffic modelling and service requirements form a basis for advanced RNP and for evaluating the interaction of coverage and capacity. Bearer service and traffic-modelling features should also enable flexible traffic forecast definitions. The more accurate the traffic estimate, the more realistic the results achieved.

In the service definition phase, the bit rate and bearer service type are assigned for each bearer service. For non-real time traffic it should also be possible to define the average packet call size and retransmission rate – i.e., to model packet data services in order to make it possible to calculate average throughputs for both uplink/downlink and delays.

In the traffic-modelling phase, it should be possible to create traffic forecasts in different ways. Busy-hour traffic can be given as input figures, or measured traffic data from measurement tools can be exploited. For example, knowledge of hotspot locations in the current network and traffic measurements from these locations are useful. Therefore, an RNP tool should be able to import traffic information from 2G network measurements, since traffic hotspots are often located in the same area independent of the radio access technology or method.

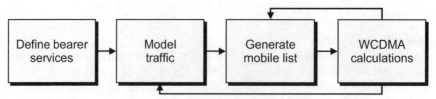

Figure 3.5   Iterative traffic-planning process for WCDMA networks.

Different weighting methods can be applied when assigning traffic amounts to areas. For example, uniform distribution or weighting based on clutter or road types can be used.

Traffic densities differ between services and therefore must be modelled separately for each service. Furthermore, traffic densities of different services can be combined and integrated concurrently. In an advanced 2G/3G RNP tool it must be possible to model a mixed bearer service situation, where there is both real time and non-real time traffic. Traffic forecasts can be utilised to realise a 'snapshot' of simultaneously active mobiles in the network. In the same context, a speed based on service and clutter information can be assigned to each MS. MS parameters – e.g., minimum and maximum transmission powers and speed – must also be modelled and specified.

MS lists including location, used bearer service and other MS parameters are used in WCDMA calculations, especially in assigning transmission powers. If an RNP tool is able to create several mobile lists, it is also possible to analyse the effect of varying mobile lists on network performance under unchanging traffic conditions – i.e., to analyse several snapshots and combine the results statistically. This method is one form of the so-called Monte Carlo analysis.

An RNP tool should be able to visualise traffic data at least in 2D and preferably also in 3D map view and to save different traffic scenarios and retrieve them for later usage.

The basic traffic-planning procedure is shown in Figure 3.5. The first task is to define bearer services and the second is to model traffic. Next, mobile lists are generated and, finally, WCDMA calculations are made. To perform WCDMA analyses with different traffic loads, several mobile lists with varying amounts of mobiles are needed. WCDMA analyses and iterations are carried out for each mobile list. Often, one representative mobile list is enough and WCDMA calculations need to be done only once. When changes are made in a network, for example, a site is relocated or its cell configuration is changed, then it is reasonable to make a WCDMA analysis only once with a representative mobile list. This is how 'what-if' trials can be evaluated rapidly.

Propagation Model Tuning
In the model-tuning phase, propagation models are tuned to match the propagation environment at hand as closely as possible. Therefore, several site locations must be selected for the measurements. Selected site locations should represent the whole planning area and the different propagation conditions inside this area. In other words, sites must be selected from all the different area types, including rural, suburban, urban and most of all dense urban areas. If necessary, for each area type a separate tuning process should be performed in order to get good accuracy. All selected sites must be visited and exact locations and hardware data must be

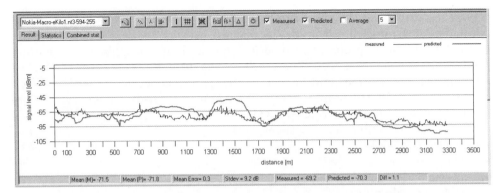

Figure 3.6   Example of propagation model-tuning application.

collected, if not known already. Site locations and sector bearings must be drawn on the map, or printed out from the RNP tool.

Measurement routes are planned so that the majority are inside the areas covered by antenna main lobes. Naturally, the routes are drawn on the map, so that driving (or walking) personnel can do the measurements as planned. Measurement equipment needs to be tested and calibrated before use. While making the measurements, log information is kept so that known anomalies and problems can be analysed after the measurements are done. Having made all the necessary measurements, the actual model tuning with the RNP tool can be started. Default propagation models are tuned to match actual signal strength values from the route. The RNP tool must provide support for comparing predicted and measured values and show the differences in graphical displays. Based on differences between the values at specific points on the measurement routes, the network planner can specify appropriate correction factors for different clutter types, for example. Naturally, the RNP tool must be able to check antenna and transmit parameters, such as tilt and EIRP.

After suitable propagation models are found and copied into relevant cells, link loss calculations can be started for the planning area at hand.

The RNP tool should provide support for tuning of different propagation models, such as Okumura–Hata and Walfisch–Ikegami. All tuning functionality must be available on a per-cell basis – i.e., it must be possible to tune one or more selected cells from the planning area. Naturally, the tool should be capable of tuning a model by several measurement routes even for the same physical cell.

Figure 3.6 shows an example screenshot of a model-tuning dialog from the measured route. This type of display can clearly indicate the problematic parts of the measured routes and the network planner is then able to modify clutter-type weightings, for example.

Perform Link Loss Calculations

When propagation models are tuned, the initial coverage plan is calculated – i.e., link losses from the BS towards the mobiles. Link loss calculations are used to obtain the signal level in each pixel in the given area.

Prior to starting link loss calculations, the RNP tool should automatically define a calculation area for each cell inside the planned network (in case it is not defined manually). The tuned propagation model(s) should always be utilised as a starting point. Furthermore, if needed, some cell-specific parameters can be adjusted – e.g., antenna tilt, transmit power and the propagation model that is used by a cell can be redefined, or the propagation model parameters can be fine-tuned. Factors affecting link loss calculation results include:

- Network configuration (sites, cells, antennas).
- Propagation model.
- Calculation area.
- Link loss parameters:
  - cable and indoor loss;
  - Line-of-Sight (LOS) settings;
  - clutter-type corrections;
  - topography corrections;
  - diffraction.
- Slow fading settings:
  - standard deviation;
  - weight factor for shadowing effect.

An RNP tool should be able to automatically provide combined coverage predictions for all the antennas belonging to the same cell.

After calculating link loss and investigating dominance areas from the map, either the predicted coverage is accepted or some RNP means should be performed. An RNP tool should provide easy coverage visualisation on a digital map, in either 2D or 3D displays. Visualisation must be possible for both single and multiple selected cells. When showing predictions for several cells, the results must be combined so that the highest signal strength is shown when there are several serving cells in the same location. An RNP tool should support different colour schemes for display purposes: for example, by using different colours for different signal thresholds, or by showing coverage areas simply by Serving RNC (SRNC) or cell colour.

Modern RNP tools provide means for distributing time-consuming link loss calculations among several workstations within the operator's Local Area Network (LAN).

Optimising Dominance
In addition to coverage area calculations and display functionality, an RNP tool should provide support for optimising cell dominance areas (best servers). 3G planning is more focused on interference and capacity analysis than on coverage area estimation alone, as was the case with 2G. During network planning, BS configurations need to be optimised: antenna selection and directions as well as the site locations need to be tuned as accurately as possible in order to meet the QoS and the capacity and service requirements at minimum cost.

Quite simple network planning solutions, such as antenna tilting, changing antenna bearing and correct antenna selection for each scenario, may already be sufficient to control interference and improve network capacity. In the initial planning phase (before WCDMA iterations) a good indicator of the interference situation is the dominance.

Each cell should have clear, not scattered, dominance areas. Naturally, since traffic is not distributed uniformly and propagation conditions vary, the cell dominance areas can never be exactly predicted and may also vary in size.

RNP tools should provide support for analysing cell dominance areas, and usually when performing the analyses it may be necessary to change some configuration settings. Facilities for rapid 'what-if' analysis when changing antenna direction, for example, offer network planners considerable time savings. An example of automated plan synthesis for interference limitation is presented in Section 7.3.

3.2.1.3   Simulating Link Performance

Link performance analysis forms the heart of the RNP tool. The 'calculation engine' must provide support for both 2G and 3G. In 2G it is enough merely to predict coverage, estimate the mutual interference between cells and perform frequency allocation. In WCDMA the analysis is more demanding. As described in Section 3.2.3 extensive uplink/downlink iterations must be conducted in order to find transmission powers for the MSs and BSs, respectively. After the RNP tool has calculated transmission powers, the number of served mobiles is also known and all the available information can then be used in further processing the data so that Key Performance Indicator (KPI) values can be generated, for example.

In estimating interference for WCDMA networks, modern RNP tools should also take adjacent channel interference into account. This is a basic requirement when more than one WCDMA carrier is used – e.g., for micro-cells. In traditional RNP tools for 2G it is also possible to estimate adjacent channel interference.

Figure 3.7 presents an example of an analysis hierarchy diagram for a modern RNP tool. Here only WCDMA-specific analysis examples are shown. It is also worth noting that Figure 3.7 shows the analysis for only one snapshot. Modern RNP tools can also

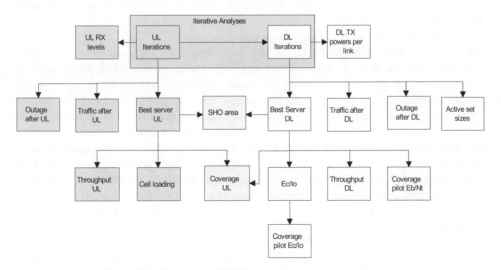

Figure 3.7   Example WCDMA analysis hierarchy diagram.

provide analysis results for several snapshots, therefore giving greater statistical reliability. This is depicted in the following sections.

Analysis of One Snapshot

Analysing only one snapshot is enough when a network planner wants to find out quickly whether current network deployment is feasible at all – for example, from the interference point of view.

With advanced RNP tools, the network planner should be able to perform single-snapshot analyses in at least two ways. In the first method, only a couple of iterations are performed for both the uplink and downlink, in order to find quickly those areas that are poorly covered and those that most likely experience heavy interference. The planner can then make the necessary RNP changes immediately, before starting more detailed calculations that require considerably more time and computing power.

The second method for analysing one snapshot takes much more information into account during the iterations, which naturally leads to longer calculation times than in the first method. For example, when performing full analysis for single-snapshot link loss calculations, a mobile distribution list and a traffic map are needed. During the iterative simulations, mobile users are put into outage until a steady state is reached. This means that the internal variables do not change more than by a predetermined small value. As a result, the indicators mentioned are calculated and ready for post-analysis treatment. However, it should be noted that a set of results is valid only for a given set of calculation parameters and input data, such as the mobile distribution at hand.

Advanced Analysis

The basic idea in advanced analysis is to automatically generate a multitude of snapshots, which are iterated accordingly, in order to generate a reliable set of WCDMA analyses from the current network deployment. A Monte Carlo simulation technique is used to verify changes in the network for varying mobile lists used under the same traffic conditions.

The implementation of advanced analysis in modern RNP tools is based on automatic generation of multiple mobile lists. The network planner can naturally also define the number of mobile lists required, in case more control of the analyses is desired. Each mobile list represents a snapshot of the traffic situation in the network – i.e., the locations of the mobile users at a given time. The WCDMA analysis results of each snapshot are combined to provide statistically relevant and reliable results. Because the same traffic conditions are used for a large number of generated mobile lists, the reliability of analysis results is improved due to the diminished randomness of the mobile locations. This is the more critical the fewer mobiles there are in the network: that is, for high bit rate services considerably more snapshots are needed to average out the dependence of the results on the mobile locations.

It is essential to verify that the planned coverage, capacity and QoS criteria can be met with the current network deployment and parameter settings. In order to make this crucial task easier, the RNP tool must provide support for performing a multitude of iterations automatically. If the calculated results show problem areas or cells in the planned scenario, it is extremely likely that the problems also occur in the real network.

Nowadays, in order to avoid too often modern RNP tools one can perform the above-mentioned analysis easily. The output results provided by the RNP tools usually consist of graphical plots based on all performed iterations and performance indicators that are relevant for the current analysis. All result values are provided with average, minimum, maximum and standard deviation figures with an overall summary, which enables quick and easy identification of possible problems and verification of overall network coverage, capacity and QoS. It must also be possible to show performance values like interference and throughputs for each cell.

An RNP tool should also provide support for analysing and studying information related to a particular iteration round and furthermore should provide the means to store this information for later use. This is necessary, since it might be the case that certain phenomena of a network's operating point can be revealed only from a specific iteration round – e.g., with certain locations of mobile users.

General requirements for advanced analysis are, for example, that users must be able to control the analyses. In RNP tools, the user can define a number of analysis-related settings, for example:

- number of iteration rounds;
- maximum calculation time;
- whether mobile lists are created automatically or existing lists are used;
- general calculation settings, such as pilot power allocation algorithm selection and checking of hardware capacity restrictions.

3.2.1.4 Analysing the Results

When calculations and simulations have been performed in the RNP tool, the next very important step is to verify and analyse whether the results are acceptable. RNP tools should provide support for post-processing, analysis and visualisations in different ways. All the phases mentioned are executed based on the results of the iterations saved previously. Naturally, if the coverage, quality and QoS targets are not met, normal network planning activities must be performed in order to change the network's operating point to the acceptable level. An RNP tool can show the necessary results and then it is the network planner's task to perform the actual optimisation. A modern RNP tool can show the results as *raster maps, numerical tables* or *histograms*.

Examples of the first format, raster maps, include the best server in the uplink and downlink, the uplink loading, pilot carrier-to-interference ratio, dominance and soft handover area plots on a digital map. Raster maps must be available for any calculated analysis result, but also for any KPI value that can then be shown for a cell dominance area with a specific threshold colour, for example. Advanced RNP tools can also show any kind of raster plots using 'transparent' colours so that the planning area can be seen together with the results. This makes pinpointing the real geographical areas from the map easier. An example of one type of raster plot is shown in Figure 3.8.

The second output format presents the results in the form of tables in which each row represents one cell (or any other network element) and each column represents a parameter value for this cell. The implementation in the RNP tool is done typically by a so-called browser, which is illustrated in Figure 3.9.

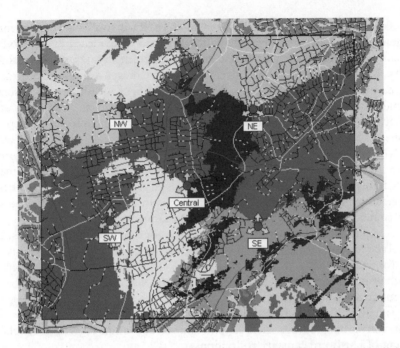

Figure 3.8 Cell loading (shading indicates the actual loading value in a certain threshold range).

	Site Name	Cell Name	Ach UL Throu	Ach DL Throu	SHO Through	# Best Server	# SHO Links
1	Central	North	848.0	848.0	384.0	42	14
2	Central	South-East	312.0	288.0	360.0	36	28
3	Central	South-West	272.0	272.0	264.0	17	32
4	NE	North	616.0	616.0	112.0	30	14
5	NE	South-East	768.0	392.0	424.0	32	6
6	NE	South-West	848.0	808.0	496.0	20	15
7	NW	North	520.0	520.0	200.0	31	7
8	NW	South-East	1056.0	1056.0	144.0	21	18
9	NW	South-West	1520.0	1520.0	240.0	41	12
10	SE	North	832.0	624.0	672.0	14	20

Figure 3.9 Example of a table view sheet.

The third output format presents the results as histograms or charts. Examples include active set size, soft handover probability for users, link transmit powers for each cell, etc.

3.2.1.5 Adjacent Cell Generation

An RNP tool must also provide the means for creating and managing adjacency relations between the cells. These so-called adjacent or neighbour cell lists contain definitions for neighbour cells for each cell in the RAN. Such information is necessary in order to ensure seamless mobility of the users in the network by performing cell changes and handovers between the cells successfully in a live network. Adjacency information is defined on a per-cell basis, but before performing adjacent cell list generation it is essential to have the right network element configuration and parameter settings. Therefore, adjacent cells are usually generated only after all other analyses have been successfully performed and the optimum configuration already achieved. With a contemporary RNP tool the inter-system (2G/3G) and 3G inter-frequency adjacencies can also be created. The possible relations between one cell and an adjacent cell are as follows:

- 2G–2G adjacency;
- 2G–3G adjacency;
- 3G–2G adjacency;
- 3G–3G inter-frequency adjacency (hard handover);
- 3G–3G intra-frequency adjacency (soft/softer handover).

After the adjacent cell lists have been created, it must be possible to view them and also to modify the adjacency parameters if necessary. The RNP tool must provide a means of visualising relationships between adjacent cells (incoming, outgoing) on a digital map. For large networks it is also very beneficial to have automated support for downlink scrambling code allocation for WCDMA cells after adjacent cell lists are generated or changed. In order to perform adjacency creation it must be possible to define at least the following items:

- radio access systems (2G/3G);
- target cells for adjacency creation (all cells, or only for cells without adjacencies);
- maximum number of neighbours per cell per adjacency type;
- field strength threshold.

In order to deploy the adjacencies and naturally all the other network element information as well, a functionality must be provided to transfer these data from the RNP tool to the network management system. This information download is described in Section 3.2.1.7.

3.2.1.6 Reporting

Reporting needs are various and, as a rule of thumb, it must be possible to print out or store for later use any output an RNP tool can provide. Therefore, RNP tools provide a rich set of reporting functionalities, usually including printouts of the following:

- raster plots from the selected area (and from the selected cells);
- network element configuration and parameter settings;
- various graphs and trends;
- customised operator-specific reports.

3.2.1.7 Inter-working with Other Tools

Every RNP tool must provide interfaces to several other tools. Operators typically have tools for managing business and customer information, dimensioning tools, transmission planning tools, measurement tools and network management systems in addition to an RNP tool. A very basic requirement is to provide data and information flow smoothly from every tool supporting the operator's whole working process.

As depicted in the planning process in Figure 3.1, the input data for an RNP tool come, for example, from network dimensioning. These data are derived from traffic and QoS requirements, which are input to a radio link budget. The output results from an RNP tool are needed in a network management system that provides plan provisioning and parameter control functionality. Therefore, an RNP tool should have a bidirectional interface for a network management system.

After the network plan is ready and its performance has been analysed by the RNP tool, the plan and configuration data from it can be exported into the operator's network management application. The exported data contain important network configuration and planned RRM parameters for network elements from a selected area or from the whole network.

Once the network is being operated and has been maintained for some period, there comes a need to replan network elements. For this purpose and naturally to save valuable time for the network planner, it should be possible to import valid network data and parameter values back to the RNP tool, so that planning and optimisation can continue there with the most up-to-date network configuration and real parameter values.

3.2.2 Initialisation: Defining the Radio Network Layout

In the global initialisation phase the network configuration is read in from parameter files for BSs, MSs and the network area. Some system parameters are set and propagation calculations are performed. In the following step, requirements coming from link-level performance are assigned to BSs and MSs. After some initialisation tasks for iterative analysis – setting default transmit powers and network performance – the actual simulation can start.

3.2.2.1 Path Loss Predictions and Propagation Models

In the network planning process, propagation models are used to predict the signal field strength of a given transmitter in the computation area. In macro-cells it is usually assumed that the transmitter is above the rooftops and the receiver is on ground level. The radio wave propagation from the transmitter to the receiver is typically impossible to compute analytically, because of different obstacles and complex scattering structures in the radio channel. However, by using a ray-optical way of thinking, we can assume that there are many different rays or wavepaths coming to the receiver. In micro-cells, the raypaths can be computed analytically because there are usually only a few strong ones.

In macro-cellular planning the propagation environment is much more complex because the distance from transmitter to receiver is larger and the propagation paths of the wave are therefore more difficult to determine. In such a situation, an empirical or semi-empirical model is more appropriate. Usually these models use free parameters and different correction factors that can be tuned by using measurements. The measurement data are obtained by receiving the signal from the BS at a number of receiver

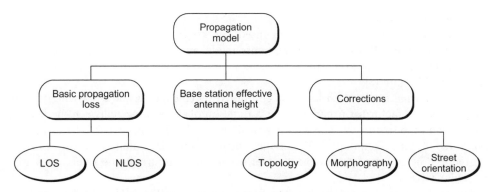

Figure 3.10 Propagation model components.

locations. These measurement samples are collected over different land use types, at different distances from the transmitter and at different topographical heights. The correction factors are tuned according to these measurements by comparing the modelled and measured signal strengths. If the location of the transmitter changes so that the statistical properties of the propagation environment change, the tuned parameters have to be changed with a new measurement set. The statistical propagation environment means that, in a certain area, the properties of the buildings, topography and vegetation are similar. The building height and its variations, as well as the distances between the buildings, are about the same. Additionally, radio wave propagation in macro-cells cannot be treated very reliably using ray theory, because the Fresnel zone in the case of large distances is too big. In such cases propagation of the waves is of a more statistical nature. These empirical models are usable in situations where the near environment of the transmitter has little effect on wave propagation.

The basic requirement for using any prediction model is to have a detailed digital map available in the simulator or in the RNP tool. Section 3.2.1 listed some general requirements for such a digital map.

In modern RNP tools the propagation models and path loss calculations consist of several components, as depicted in Figure 3.10: the basic path loss model, LOS checking, calculation of BS effective antenna height and corrections for topography, morphography and street orientation.

Most of these parts have a set of selectable correction functions with user-definable parameters. This, and the fact that each cell can have a unique model, enables the user to specify a suitable model for each propagation environment.

Correction factors are functions that are used to correct the basic propagation loss function in respect of certain site-specific features, such as large undulations in terrain. The user always defines the different correction factors in RNP tools.

3.2.2.2 Basic Propagation Loss

This section introduces the two most widely used propagation models – namely, the Okumura–Hata and Walfisch–Ikegami models. These models are the most typical means of calculating basic propagation loss.

Okumura–Hata Model

The Okumura–Hata model is widely used for coverage calculation in macro-cell network planning. Based on measurements made by Y. Okumura [19] in Tokyo at frequencies up to 1920 MHz, these measurements have been fitted to a mathematical model by M. Hata [8].

In the original model path loss was computed by calculating the empirical attenuation correction factor for urban areas as a function of the distance between the BS and the MS and the frequency. This factor was added to the free space loss. The result was corrected by the factors for BS antenna height and MS antenna height. Further correction factors were provided for street orientation, suburban and open areas, and irregular terrain.

Hata's formulas are valid when the frequency is 150–1000 MHz, the BS height is 30–200 m, the MS height is 1–10 m and the distance is 1–20 km. The BS antenna height must be above the rooftop level of the buildings adjacent to the BS. Thus, the model is proposed to be used in propagation studies of macro-cells. The original data on which the model was developed were averaged over a 20-m interval, being a kind of minimum spatial resolution of the model. Owing to frequency-band limitation, the original model was tailored by COST231 [20], resulting in a COST231–Hata model with a range of 1.5–2.0 GHz, which is also applicable to 3G radio networks. Of the available propagation models the Okumura–Hata model is most frequently referred to. It therefore became a reference with which other models are compared. Its range of usability with different land use and terrain types and for different network parameters has made the Okumura–Hata model very useful in many different propagation studies.

There are also several weaknesses in the empirical or semi-empirical models for propagation studies in micro-cellular environments. If the BS antenna height is below the rooftop level of the surrounding buildings, the nature of the propagation phenomena changes. This situation cannot be analysed with statistical methods because the individual buildings are too large compared with the cell size and the exact geometrical properties of the buildings can no longer be ignored as they can in macro-cellular models.

The Okumura–Hata equation ([8] and [19]) is in the form of propagation loss:

$$Lp = A + B \cdot \log_{10} f - 13.82 \cdot \log_{10} h_b - a(h_m) + (C - 6.55 \cdot \log_{10} h_b) \cdot \log_{10} d \quad (3.29)$$

where Lp is the path loss [dB]; f is the frequency [MHz]; h_b and h_m are the BS and MS antenna heights, respectively [m]; $a(h_m)$ is the mobile antenna gain function [dB]; and d is the distance [km].

The parameters A and B are set by the user according to Table 3.8. These values have been determined by fitting the model with measurements.

The parameter C gives the distance dependence of the model and is user-defined. C should be set using the appropriate measurement set, and it is possible to achieve a better fit in the model-tuning by changing this parameter. Its value is usually between 44 and 47 and the default value most often used, based on experience, is 44.9.

The constant term is specified in the slope part and the city type in the Okumura–Hata function. The city type specifies the function $a(h_m)$ for the mobile antenna gain for

Table 3.8   *A* and *B* constants for the Okumura–Hata model.

	150–1000 MHz	1500–2000 MHz
A	69.55	46.3
B	26.16	33.9

a medium or small city:

$$a(h_m) = (1.1 \cdot \log_{10} f - 0.7) \cdot h_m - (1.56 \cdot \log_{10} f - 0.8) \tag{3.30}$$

and for a large city:

$$a(h_m) = \begin{cases} 8.29 \cdot [\log_{10}(1.54 \cdot h_m)]^2 - 1.1 & f \le 200\,\text{MHz} \\ 3.2 \cdot [\log_{10}(11.75 \cdot h_m)]^2 - 4.97 & f \ge 400\,\text{MHz} \end{cases} \tag{3.31}$$

It should be noted that these functions do not usually have much meaning in practice because the MS antenna height used is almost always the same ($\approx$1.5 m). For this value, these functions are close to 0 and not very sensitive to small variations in MS antenna height.

Walfisch–Ikegami Model

The Walfisch–Ikegami model is based on the assumption that the transmitted wave propagates over the rooftops by a process of multiple diffraction. The buildings in the line between the transmitter and the receiver are characterised as diffracting half-screens with equal height and range separation [21] (Figure 3.11).

At the mobile terminal, the received field consists, e.g., of two rays as shown in Figure 3.11: (1) the direct diffracted ray and (2) the diffracted-and-single-reflected wave. The powers of these two components are combined together [22]. For the LOS situation, the original model was extended by the 'street canyon' model [23]. The resulting model is called the COST231–Walfisch–Ikegami model.

In RNP tools it is possible to define the LOS propagation as a two-slope function. This is based on the fact that, taking the Earth as flat, there are two main propagation paths from transmitter to receiver: a direct path and a ground-reflected path. When these two paths are combined coherently so that the phases of the waves are taken into account, it can be shown that there is a distance called the 'breakpoint' after which the slope is steeper than before. In RNP tools, this breakpoint effect is taken into account

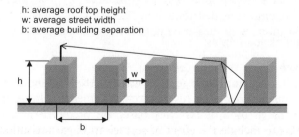

Figure 3.11   Definition of Walfisch–Ikegami model parameters.

by giving the user the possibility of changing the parameters of these two-slope functions. The distance of the breakpoint can be calculated from the following equation:

$$d_b = \frac{4 \cdot h_1 \cdot h_2}{\lambda}$$ (3.32)

where h_1 is the transmitter height; h_2 is the receiver height; and λ is the wavelength.

Although the Walfisch–Ikegami model is considered to be a micro-cell model, it should be used very carefully when the antenna of the transmitter is below the rooftops of the surrounding buildings. In such cases the transmitting wave is travelling through street canyons and not over the rooftops as is assumed in the model. For example, if the actual building size is large and the over-the-rooftop diffraction component is negligible, the Walfisch–Ikegami model overestimates path loss. The model implies, for obstructed paths in micro-cells, only a rough empirical function of BS antenna height. Thus, it must be applied very cautiously in this case and the result should be verified with measurements. The assumptions used in the Walfisch–Ikegami model restrict its usability in those cases where the dimensions of the buildings are identical and they are uniformly spaced. Also the terrain height must be constant across the cell calculation area.

The COST 231–Walfisch–Ikegami model is divided into two parts: (1) LOS and (2) Non-LOS (NLOS). Building height information is used to find out whether the path is in LOS or not. In this model path loss is calculated as follows:

$$Lp = \begin{cases} 42.6 + 26 \cdot \log_{10} d + 20 \cdot \log_{10} f & \text{when receiver is in LOS} \\ 32.4 + 20 \cdot \log_{10} d + 20 \cdot \log_{10} f + L_{rts} + L_{msd} & \text{when receiver is in NLOS} \end{cases}$$ (3.33)

where Lp is the total path loss [dB]; L_{rts} is the rooftop-to-street diffraction and scatter loss [dB]; and L_{msd} is the multi-screen diffraction loss [dB]. Note that this definition for LOS has no breakpoint, so it is valid only for relatively short distances. The breakpoint distance depends on the antenna height and distances and can be calculated with Equation (3.32).

When normal morphographic data are used the parameters h, w and b can be defined in the RNP tool for each cell independently of any digital map information.

Optionally, ray-specific parameters for street widths, building separations and building heights are calculated using the vector building map layer. This improves the accuracy of the model so that it can be also used when the building heights and the distances of the buildings are not uniform across the calculation area. However, the model is based on the assumption that the distances between the buildings and the building heights are uniformly distributed. So, care should be taken when this extension is used and verifications with measured data are recommended.

Ray-tracing Models

As frequencies got more and more scarce owing to the ever-tighter site distances, the importance of frequency planning rose as well. Frequency allocation, independent of the actual allocation method, is typically based on predicted propagation data, and therefore a need for more and more accurate propagation modelling has arisen. Examples of more accurate models are the ones based on ray-tracing. Some ray-tracing models can be found, for example, in [24]–[28]. With ray-tracing 2D and 3D modelling can be applied. Since a rather accurate environment description is required, ray-tracing is especially applicable for indoor propagation modelling, where the exact building structures are typically known, but ray-tracing has also found its application in outdoors and in outdoor-to-indoor propagation calculations.

3.2.3   Detailed Uplink and Downlink Iterations

In this section we introduce on a more detailed level the methods and example algorithms needed in iterative analysis during the detailed planning phase of a 3G radio network. Most of these arise due to the features that are typical of a 3G network. They include multiple services and their QoS requirements, fast transmit power control in the uplink and downlink, soft and softer handover and combinations thereof, multi-path profile of the propagation channel and speed of the terminal. To model the link-level requirements of different services in different multi-path channel conditions, five types of link-level simulation results can be identified and brought to a planning tool (these concepts were introduced briefly in Section 2.5):

- average received E_b/N_0 requirement;
- average power rise;
- multi-path fading margin (power control headroom);
- diversity combining gain in soft handover;
- orthogonality.

One possible way to bring the information from link-level simulation results into a planning tool is via so-called link performance tables. In these tables the most important numbers are the E_b/N_0 requirements for the services used and for the chosen MS speeds, in both the uplink and downlink and the orthogonality factor in the downlink. The numbers in the tables depend on the channel profiles and different tables should be generated for different channel profiles. In the same file there are also the required multi-path fading margins (headroom) in the uplink above the received E_b/N_0 as well as the average transmit power rise, as a function of MS speed. These are measured in decibels above the average received E_b/N_0. Soft handover diversity combining gains have been tabulated in the uplink and downlink as a function of MS speed and the level difference between the two best links. In addition to these parameters, the effective channel activity used in interference calculations is set in link performance tables.

Notice that link performance tables are not fixed; new values should always be used when there is more information available from the requirements in the standard, from the link-level simulations and, finally, from measurements during network operation. Examples of link performance tables can be found in the implementation of the static simulator (see weblink at *www.wiley.com/go/laiho*).

3.2.3.1   Uplink Iteration Step

The target in uplink iteration is to allocate MS transmit powers so that the (interference + noise) levels and thus BS sensitivity values converge. The average transmit powers of the MSs to each BS are estimated so that they fulfil the E_b/N_0 requirements of the BSs. MS average transmit powers are based on the sensitivity level of the BS, the service (data rate) and speed of the MS and link losses to the BSs. They are corrected by taking into account the activity factor, the soft handover gains and average power rise due to fast transmit power control. The impact of uplink loading on BS sensitivity (noise rise) is taken into account by adjusting it with $(1 - \eta)$. The loading η can be defined by Equation (3.7).

After the average transmit powers of the mobiles have been estimated, they are compared with the maximum allowed value. Mobiles exceeding this limit try IF-HO (if allowed) or are put to outage. Now the interference analysis can be performed again and the new loading and BS sensitivities are calculated until their changes are smaller than specified thresholds. Also, if the uplink loading of a cell exceeds specified limits, MSs are moved to another carrier (if allowed) (IF-HO), otherwise they are put to outage. A flowchart for the uplink iteration step is depicted in Figure 3.12.

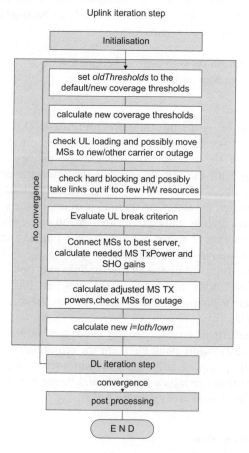

Figure 3.12   Flowchart for uplink iteration steps.

Reproduced by permission of Groupes des Ecoles des Télécommunications.

Selection of the Best Server in Uplink and Downlink

One way to determine WCDMA typical issues in the uplink iteration is to make them depend on how many and which BS(s) the MS in question is connected to. Therefore, it has to be decided how to determine the BSs that belong to the active set and which of them is the best server. In the example simulator, determination of the active set is based on the received signal strength of the P-CPICH. All BSs whose P-CPICHs are received within a certain window are included in the active set. In addition, a minimum required reception level could be considered. In the uplink then these BSs are ranked according to the power the MS needs to transmit so that it is received with the required signal quality. The calculation of the power needed is described below. The best server in the uplink then is simply selected as the BS requiring the minimum transmit power from the MS. In the downlink the BSs from the active set are simply ranked according to the level at which their P-CPICH is received by the MS.

A precondition for determining the active set is the allocation of the P-CPICH transmit powers of the individual cells. Several strategies can be applied here, for example:

- assume all cells use the same P-CPICH power;
- define the P-CPICH powers manually for each cell;
- use the maximum P-CPICH power for the lowest loaded cell (to make it more attractive) and scale other cells' P-CPICH powers by the loading relative to that cell.

Calculation of Transmit Powers Needed in Uplink

The transmit power [dBm] needed for MS n to transmit to BS k is determined from Equation (3.34):

$$neededMsTxPower(k, n) = bsSensitivity(k) + linklossUL(k, n) \qquad (3.34)$$

where $bsSensitivity(k)$ is the sensitivity of BS k [dBm]; and $linklossUL(k, n)$ is the total uplink link loss between MS n and BS k [dB]. The best server in the uplink for MS n is then determined as the BS that minimises Equation (3.34). Since only one sensitivity is calculated for each of the BSs, this is done for a reference service, which is defined by the data rate used and the speed of the terminal. For this reason, the transmit power needed for the MS is then corrected in the next step by the difference in sensitivity of the receiver for the different services, using Equation (3.35):

$$txPowerBase = minMsTxPower + deltaSensitivity \qquad (3.35)$$

where $minMsTxPower$ is the minimum power from Equation (3.34); and $deltaSensitivity$ is defined by Equation (3.36):

$deltaSensitivity(i)$ [dB]

$$= lin2log \left(\frac{\nu UL \cdot \left(1 + \dfrac{W}{\nu UL \cdot log2lin(refEbNo) \cdot refR}\right)}{\nu UL(i) \cdot \left(1 + \dfrac{W}{\nu UL(i) \cdot log2lin[msEbNoUL(i)] \cdot msRUL(i)}\right)} \right) \qquad (3.36)$$

where W is the chip rate; $\nu UL(i)$ is the activity factor in the uplink of MS i; νUL is the activity factor in the uplink of the reference service; $refEbNo$ is the E_b/N_0 of the reference service calculated from link-level performance tables using *refR* and *refSpeed* which are the reference data rate and the speed applied for calculating the sensitivity for the reference service; *msEbNoUL(i)* is the MS E_b/N_0 in the uplink; and *msRUL(i)* is the data rate of MS *i* in the uplink.

With the basic transmit power from Equation (3.35), all the MS transmit powers needed in the iterative calculation are calculated, as follows.

First, the gain coming from soft handover in the uplink is taken into account to estimate the power seen from the own cell:

$$msTxPower = txPowerBase - SHOgainRx \qquad (3.37)$$

where *SHOgainRx* is the gain due to soft handover taken from link-level simulations, depending on the level difference between the strongest and second strongest link in the active set.

Second, the power seen in the interference calculations of other cells is determined by:

$$msTxPowerRaised = txPowerBase + (msTxPowerRaise - SHOgainTx) \qquad (3.38)$$

where *msTxPowerRaise* is the average transmit power rise due to fast power control; and *SHOgainTx* is the reduction of this power rise due to soft handover gain. The first of these is calculated from link-level simulations using terminal speed and the second one using terminal speed and the level difference between the strongest and second strongest link in the active set.

Third, the transmit power including the fast fading margin is calculated according to Equation (3.39). This power is needed to check whether the MS has enough power or has to be put to outage:

$$msTxPowerPeak = txPowerBase + (msHeadRoom - SHOgainPeak) \qquad (3.39)$$

where *msHeadRoom* is the fast fading margin needed at the cell edge for the fast power control to follow the fading; and *SHOgainPeak* is the reduction of this margin due to soft handover gain. The first of these is calculated from link-level simulations using terminal speed and the second using terminal speed and the level difference between the strongest and second strongest link in the active set.

Estimation of BS Sensitivity, the Loading and Load Control in Uplink

The sensitivity of a BS – i.e., the required signal power at the receiver for the reference service – is calculated according to Equation (3.40):

$$bsSensitivity = BS\_noise\_power$$

$$+10 \cdot \log_{10} \left(\frac{1}{\nu UL \cdot \left(1 + \frac{W}{\nu UL \cdot log2lin(refEbNo)refR} \right) \cdot (1 - \eta)} \right) \qquad (3.40)$$

where *BS_noise_ power* is the receiver background noise [dBm] including the noise figure N_f; *vUL*, *refEbNo* and *refR* are the service activity, the E_b/N_0 requirement and the bit rate of the reference service, respectively; and η is the uplink loading calculated according to Equation (3.59).

The interference parts needed in the loading calculation are:

$$I_{own,m} = \sum_{k,bestServer(k)=m} \frac{\nu_k \cdot log2lin(msTxPower_k)}{Lp_{k,m}} \tag{3.41}$$

$$I_{oth,m} = \sum_{k=1,bestServer(k)\neq m} \frac{\nu_k \cdot log2lin(msTxPowerRaised_k)}{Lp_{k,m}} \tag{3.42}$$

$$I_{ACI,m} = \sum_{j} \frac{\nu_k \cdot max[log2lin(-acFilterUL) \cdot log2lin(msTxPowerRaised_k), log2lin(acMinPowUL)]}{Lp_{k,m}}$$

$$\tag{3.43}$$

When the uplink loading for each cell is calculated with the help of the above equations, the received loading can be checked against the maximum allowed loading in each cell. If it is exceeded, MSs can be put to outage directly, or after they have tried IF-HO if a second carrier is available. The strategy for selecting candidate mobiles can be chosen from among the following, for example:
- randomly from all mobiles in the network;
- randomly from all mobiles in the overloaded cell(s);
- according to their needed transmit powers;
- according to the service.

Finally, it should be checked whether the remaining mobiles can be served from a hardware point of view (hard blocking). Hardware in this connection means whether there are enough channel elements and codes available. Also, in this case mobiles are then taken to outage with or without trying IF-HO if a cell is running out of hardware resources. One way of implementing this is given in the documentation for the simulator downloadable from the weblink (see weblink at *www.wiley.com/go/laiho*).

3.2.3.2 Downlink Iteration Step

Similarly to the uplink, the goal of the downlink iteration is to assign the BS transmit powers for each link (including soft handover connections) in use by a MS, until all MSs receive their signal with the required Carrier-to-Interference ratio (C/I), defined by Equation (3.44):

$$targetCI = \frac{EbNo_{MS}}{W/R} \tag{3.44}$$

where $EbNo_{MS}$ is the received E_b/N_0 requirement of the MS depending on terminal speed and service. The actual received $(C/I)_m$ of MS m is calculated using Maximal Ratio Combining (MRC) according to Equation (3.45), summing the C/I values of all links k ($k = 1, \ldots, K$) in use by MS m:

$$\left(\frac{C}{I}\right)_m = \sum_{k=1}^{K} \frac{p_{km}/Lp_{km}}{(1-\alpha_k) \cdot P_k/Lp_{km} + I_{oth,k} + N_m} \tag{3.45}$$

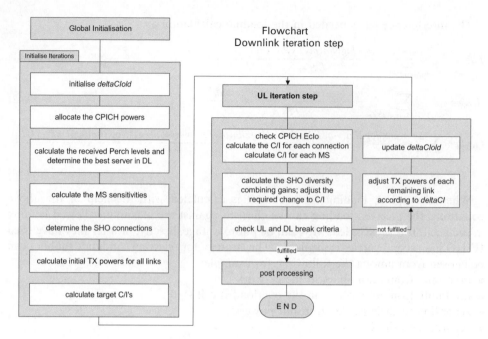

Figure 3.13   Flowchart for the downlink iteration steps.
Reproduced by permission of Groupes des Ecoles des Télécommunications.

where P_k is the total transmit power of the BS to which link k is established; Lp_{km} is the link loss from cell k to MS m; α_k is the cell-specific orthogonality factor; p_{km} is the power allocated to the link from BS k to MS m; $I_{oth,k}$ is the other-cell interference; and N_m is the background and receiver noise of MS m.

The initial transmit powers are adjusted iteratively according to the difference between the achieved and the targeted C/I value until convergence is achieved. The process requires iteration, since the C/I at each MS depends on all the powers allocated to the other MSs and it is not known *a priori* whether a link can be established or not. If either certain link power limits or the total transmit power of a BS is exceeded, MSs perform IF-HO (if allowed) or are taken out from the network randomly.

In a further step for each MS it is checked whether the received P-CPICH E_c/I_0 value is above a user-defined threshold so that the MS can reliably measure the BS and synchronise to it. Also in this case, if the threshold given is exceeded, the MS tries IF-HO or is put to outage. A flowchart for the detailed iteration steps is given in Figure 3.13.

Calculation of the Transmit Powers Needed in Downlink

The transmit powers needed for all soft handover links in the downlink are calculated by iteratively adjusting initially assigned link powers. The following steps must be done for initial transmit power allocation:

- based on the evaluation of a radio link budget for the downlink, MS sensitivity is calculated;
- initial transmit power for the best link in the active set is calculated;
- initial transmit power for other links in the active set is calculated.

MS sensitivity is simply calculated by adjusting the sensitivity of the BS, which is its best server, by the difference in the required E_b/N_0 values of the uplink and downlink according to Equations (3.46) and (3.47):

$$msSensitivity = bsSensitivity(bestServerDL) - EbNoCorrectionFactor \quad (3.46)$$

where $bsSensitivity(bestServerDL)$ is the sensitivity of the best serving BS in the downlink, and:

$$EbNoCorrectionFactor = -deltaSensitivity - (msEbNoDL - msEbNoUL) \quad (3.47)$$

Then the initial transmit power of each connection of MS m is assigned. First, the power needed for the connection at the best server is allocated. This transmit power is given by:

$$txPower_m \text{ [dBm]} = msSensitivity_m + linklossDL(bestServerDL, m) \quad (3.48)$$

where $msSensitivity_m$ is the sensitivity of MS m; and $linklossDL(bestServerDL, m)$ is the total link loss in the downlink between MS m and its best server.

After the transmit power of the dominant connection is assigned, the transmit powers for the other soft handover connections of each MS are allocated. Those transmit powers are adjusted by the difference between the P-CPICH power of the BS where the link is located and that of the best server, both in dBm, according to Equation (3.49):

$$txPower_{m,k} = txPower_m + CPICHPower(k) - CPICHPower(bestServerDL) \quad (3.49)$$

where $txPower_{m,k}$ is the transmit power of the link between cell k and MS m; $txPower_m$ is the transmit power of the best server to MS m – see Equation (3.48); $CPICHPower(k)$ is the P-CPICH power of BS k; and $CPICHPower(bestServerDL)$ is the P-CPICH power of the best server in the downlink (all values in dBm).

Estimation of C/I and Load Control in Downlink

When the transmit powers of all links in the downlink have been assigned, the C/I at each MS can be calculated. However, before this can be done it must be checked whether load conditions in the downlink have been fulfilled. Two criteria can be checked:

- The maximum power allowed for each link in the network. The exact definition of the maximum link power will be vendor-specific. A possible algorithm is presented in the documentation for the simulator (see weblink at *www.wiley.com/go/laiho*).
- The total transmit power (including the power for the common channels) in all cells.

If the first condition is violated, the link cannot be established and must be dropped. If the second condition is violated, enough links have to be dropped until the total transmit power is sufficiently low. Various strategies can be used; in the documentation for the accompanying simulator (see *www.wiley.com/go/laiho*) one possible implementation is given. After load control issues have been handled, the next major step is the estimation of interference and received levels for each connection. Interference from the own cell, *IownDL*, and from other cells, *IothDL*, is estimated separately. The latter includes both interference coming from other cells of the own operator and interference coming from cells of the other operator or carrier. Each soft handover connection is orthogonal only to the own cell. In the following, the calculations needed for one link – i.e., for MS k connected to BS m – are shown.

The received link power is given by Equation (3.50):

$$msRxPowerLin_{m,k} = \frac{P_{m,k}}{linklossDL_{m,k}} \tag{3.50}$$

and the own-cell interference by Equation (3.51):

$$IownDL_{m,k} = (1 - \alpha_k) \cdot \frac{P_{tot,m}}{linklossDL_{m,k}} \tag{3.51}$$

where α_k is the orthogonality factor of MS k; $P_{m,k}$ is the transmit power; $linklossDL_{m,k}$ is the link loss between BS m and MS k; and $P_{tot,m}$ is the total transmit power of BS m including P-CPICH and other common channels (all values are in linear scale). Note that Equation (3.51) assumes that downlink link-level simulations are made including own-cell as well as own-signal interference. If they have been done only against noise, the own-link signal has to be subtracted from $P_{tot,m}$.

When calculating the interference from other sources, $IothDL_{m,k}$, interference from other cells of the same frequency as well as from other frequencies (coming from either one's own network or a competitor's network), has to be taken into account. Equation (3.52) shows one possibility for a solution, where just one other frequency from a neighbouring frequency/operator is present. The other-cell interference for MS k at BS m of operator/carrier 1 thereby becomes:

$$IothDL_{m,k} = \sum_{i=indBStype1(1);i \neq m}^{indBStype1(numBStype1)} \frac{P_{tot,i}}{linklossDL_{i,k}}$$
$$+ \sum_{i=indBStype2(1)}^{indBStype2(numBStype2)} \frac{\max\{log2lin[-acFilterDL(channelOffset) \cdot P_{tot,i}, acMinPowDL]\}}{linklossDL_{i,k}} \tag{3.52}$$

where $P_{tot,i}$ is the total transmit power of one operator's BS_i, including P-CPICH and other common channels; $linklossDL_{i,k}$ is the link loss in the downlink between BS_i and MS_k; $acMinPowDL$ is the power level of interference which is minimal coming from other operators' BSs; $acFilterDL$ is specified by Equation (3.60) below; $numBStype1$ and $numBStype2$ are the number of BSs of operator/carrier 1 and 2, respectively; and $indBStype1$ and $indBStype2$ are the indices of BSs of operator/carrier 1 and 2, respectively.

Now the C/I matrix can be calculated, holding all C/I values for all links between MSs k and BSs m, by:

$$C\_over\_I\_all_{m,k} = \frac{msRxPowerLin_{m,k}}{IownDL_{m,k} + IothDL_{m,k} + MS\_noise\_power\_lin} \quad (3.53)$$

where

$$MS\_noise\_power\_lin = log2lin(Thermal\_noise\_density + MS\_noise\_figure) \cdot W \quad (3.54)$$

Finally, the C/I at MS k is the sum of the linear C/I values of all *connSHO* connections to that MS:

$$C\_over\_I_k = \sum_{i=1}^{connSHO} (C\_over\_I\_all_{i,k}) \quad (3.55)$$

where values are in linear scale.

Iterative Transmit Power Adjustments in Downlink

After the C/I has been calculated for each MS, the gain from soft handover diversity combining is calculated from the link-level performance table according to the service (terminal speed and data rate) the MS is using and the relative difference between the two strongest received P-CPICHs in the active set. The result is stored in vector *msSHOGainsDL* [dB].

$C\_over\_I_k$ is compared with the *targetCI$_k$*, from which the soft handover gain is subtracted:

$$deltaCI_k = (targetCI_k - msSHOGainsDL_k) - C\_over\_I_k \quad (3.56)$$

and the transmit power from BS m to MS k is corrected by the *deltaCI$_k$* (values are in logarithmic scale):

$$bsTxPower_{m,k} = bsTxPower_{m,k} + deltaCI_k \quad (3.57)$$

Calculations of *deltaCI* and power corrections are repeated until the maximum value of *deltaCI* is less than a specified threshold.

3.2.4 Adjacent Channel Interference Calculations

The influence of ACI, either from one's own network or from a competitor's network in the same area, is taken into account by filtering this interference with a filter that depends on channel separation. ACI is covered in more detail in Section 3.6. In both uplink and downlink, adjacent carrier filtering is implemented as a twofold process. In the uplink, one filter for the MSs has been implemented indicating its out-of-band radiation (*aciFilterUL*). This filter is used to indicate how much power the MS is leaking into the other carrier's receiving band (*Adjacent Channel Leakage power Ratio, ACLR*). For the BS in the uplink, another filter (*acpFilterUL*) has been implemented. This filter indicates the selectivity of the BS's receiver in a multi-carrier situation – i.e., how much of the adjacent channel power is received by the BS as ACI power (*Adjacent Channel Protection, ACP*). This filter setting also depends on carrier separation. The ACI situation in the uplink is shown in Figure 3.14.

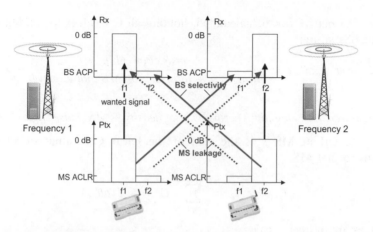

Figure 3.14   Uplink adjacent channel interference situation.
Reproduced by permission of IEEE and Groupe des Ecoles des Télécommunications.

In simulations these two filters are combined as a single filter by Equation (3.58):

$$acFilterUL = -10 \cdot \log_{10}\left(10^{-\frac{aciFilterUL}{10}} + 10^{-\frac{acpFilterUL}{10}}\right) \tag{3.58}$$

ACI (I_{ACI}) in the uplink is taken into account when calculating uplink load according to Equation (3.59):

$$\eta = \frac{I_{own} + I_{oth} + I_{ACI}}{I_{own} + I_{oth} + I_{ACI} + N} \tag{3.59}$$

where I_{own} is the interference from MSs of the own cell; I_{oth} is interference from MSs of other cells; I_{ACI} is interference from an adjacent carrier; and N is the receiver background noise. The calculation of the different quantities is introduced in Section 3.2.3.1.

Also in the downlink, a similar type of filtering is introduced as in the uplink. One filter for the BSs has been implemented indicating the out-of-band radiation of the BS (*aciFilterDL*). This filter is used to show how much power the BS is leaking into the other carrier's receiving band (ACLR). The filter setting depends on the separation between the carriers. For the MS another filter has been implemented (*acpFilterDL*). This filter indicates the selectivity of the MS's receiver in a multi-carrier situation – i.e., how much of the ACI power is received by the MS (ACP). Also this filter setting depends on carrier separation. The ACI situation in the downlink is depicted in Figure 3.15.

In simulations these two filters are combined as a single filter by Equation (3.60):

$$acFilterDL = -10 \cdot \log_{10}\left(10^{-\frac{aciFilterDL}{10}} + 10^{-\frac{acpFilterDL}{10}}\right) \tag{3.60}$$

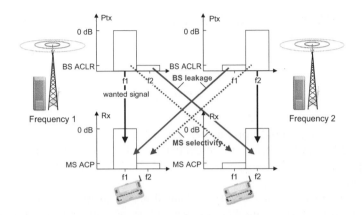

Figure 3.15 Downlink adjacent channel interference situation.

Reproduced by permission of IEEE and Groupe des Ecoles des Télécommunications.

ACI (I_{ACI}) in the downlink is taken into account when calculating the C/I for an MS according to Equation (3.61):

$$\left(\frac{C}{I}\right)_m = \sum_{k=1}^{K} \frac{p_{km}/Lp_{km}}{(1-\alpha_k) \cdot P_k/Lp_{km} + I_{oth,k} + I_{ACI} + N_m} \tag{3.61}$$

where variables are as defined in Equation (3.45); I_{oth} is the interference from other cells on the same carrier; and I_{ACI} is the ACI.

3.2.5 Post-Processing: Network Coverage Prediction and Common Channel Analysis

This section introduces the estimation of area coverage probabilities and explains the analyses for the following DCHs and CCHs in UMTS: uplink DCH, downlink DCH, P-CPICH, BCH, FACH and PCH.

In all analyses it is assumed that the interference situation is stable. This means that a certain traffic distribution has been assumed and the detailed uplink and downlink iterations have converged. A test mobile is then run through all pixels within the area of interest and all other MSs that could be served contribute to the interference. The test mobile has no influence on the interference situation; therefore, the other-to-own-cell-interference ratio will not change and also the total transmit powers of the serving BSs will be the same as they were after the iterations.

3.2.5.1 Uplink Dedicated Channel Coverage

In the uplink it can now be estimated whether or not this additional MS using a certain bit rate and having a certain E_b/N_0 requirement could get the service in the chosen geographical location – i.e., whether the maximum allowed transmit power of the test MS is enough to fulfil the E_b/N_0 requirement at the BS receiver. The needed transmit power for the MS, $P_{Tx,MS}$, is calculated using Equation (3.62) and compared with the maximum allowed:

$$P_{Tx,MS} = \frac{N_0 \cdot Lp}{\nu \cdot (1 - \eta) \cdot \left(1 + \dfrac{W}{R \cdot \rho \cdot \nu}\right)} \quad (3.62)$$

where N_0 is the background noise as defined in Equation (3.2); Lp is the propagation loss between MS and BS; R, ν, ρ are respectively the bit rate, service activity and uplink E_b/N_0 requirement of the chosen service; W is the WCDMA chip rate; and η is the uplink loading.

Finally, area coverage probability is defined as the proportion of the chosen area where the additional MS really gets the wanted service under that stable interference situation. The weakness in this approach is that in reality an additional MS would change the interference situation – e.g., some other mobiles could go to outage – but in the case of low bit rate services this effect can be neglected.

3.2.5.2 Downlink Dedicated Channel Coverage

In the downlink the coverage probability calculation is based on the transmit power limits per radio link. The main focus is on checking, pixel by pixel, whether or not there is enough transmit power per link from BSs in the active set, if there were an MS in the pixel, using a given service (bit rate) and having a given speed. Also, here it is assumed that the total transmit powers of BSs do not change from what was left after uplink/downlink iteration. In the iterations, however, there may or may not have been an MS in the pixel. The method described here for the calculation of the required downlink transmit power to each pixel is derived based on MRC. In the following, mathematical notation is used for convenience.

We assume that the transmit powers of the BSs in the active set of the studied MS pixel are equal, except that they will be scaled according to the possible differences in the limits for maximum powers per link at the different BSs in the active set. Then the condition of the needed transmit power per link, p_{tx} (maximum taken over the active set), for a sufficient connection is:

$$\frac{W}{R} \cdot p_{tx} \sum_{k \in AS} \frac{\beta_k}{Lp_k \cdot (I_{tot} - \alpha_k \cdot I_k + N_{ms})} \geq \rho \quad (3.63)$$

where ρ is the downlink E_b/N_0 requirement – i.e., the required energy per user information bit over the total in-band interference spectral density for the used service; R is the bit rate of the used service; W is the chip rate; N_{ms} is the background noise level at the MS; I_{tot} is the total wideband interference power received at the MS; I_k is the total wideband power received at the MS from BS k; Lp_k is the link loss from BS k to the MS; α_k is the orthogonality factor of cell k; and β_k are the scaling factors (relative maximum link powers) for different BSs in the active set. Thus, the equivalent requirement for p_{tx} is:

$$p_{tx} \geq \frac{\rho \cdot R/W}{\sum_{k \in AS} \dfrac{\beta_k}{Lp_k \cdot (I_{tot} - \alpha_k \cdot I_k + N_{ms})}} \tag{3.64}$$

For the whole selected area the inequality in Equation (3.64) is solved for each pixel for the studied service and MS speed. The result is compared with the maximum power allowed in each pixel; if it exceeds the limit in a pixel, the pixel is considered as outage. Finally, the coverage probability is calculated as the number of pixels *not* in outage in the wanted area divided by the total number of pixels in the wanted area. If the cumulative distribution of the required transmit power per pixel is drawn, then the maximum transmit power needed per link for a given coverage probability can be estimated.

3.2.5.3 Primary CPICH Coverage

During radio network planning the P-CPICH transmit power should be set as low as possible, while ensuring that the best cell and neighbour cells can be measured and synchronised to it and the P-CPICH can be sufficiently used as the phase reference for all other downlink physical channels. Typically this means that 5–10% of the total BS power is used for the P-CPICH. For each pixel in the chosen area the E_c/I_0 in that pixel is calculated using Equation (3.65):

$$CPICH_{ecio} = \frac{P_{CPICH}/Lp}{\sum_{i=1}^{numBSs} P_{tx,i}/Lp_i + I_{ACI} + N_0} \tag{3.65}$$

where P_{CPICH} is the P-CPICH power of the best server; Lp is the link loss to the best server; $P_{tx,i}$ is the total transmit power of BS i; Lp_i is the link loss to BS i; I_{ACI} is the adjacent channel interface; N_0 is the thermal noise of the default MS; and *numBSs* is the number of BSs in the network.

The E_c/I_0 achieved is then compared with a user-given threshold and the P-CPICH coverage is defined as the ratio of pixels where the threshold is exceeded compared with all pixels. The weakness of this modelling for the P-CPICH E_c/I_0 coverage is that it is done only for the best server. In an operating network, however, all neighbour cells must also be measured and, therefore, all neighbour cells' P-CPICH E_c/I_0 values should be analysed, too. This could be overcome by, for example, adding a threshold to the required P-CPICH E_c/I_0.

3.2.5.4 Primary and Secondary CCPCH Coverage

Of all common channels, the most important are the Broadcast Common Channel (BCCH) and the Paging Channel (PCH). In UMTS, the BCCH can be carried either by the BCH or (more seldom) by the Forward Access Channel (FACH). The BCH is then mapped onto the Primary Common Control Physical Channel (P-CCPCH), whereas the FACH and PCH are mapped onto either the same or different

Secondary Common Control Physical Channels (S-CCPCHs). To be able to decode the BCCH and PCH, certain requirements for the (narrowband) E_b/N_t must be fulfilled, but since the coding on P- and S-CCPCH are the same (apart from the fact that different spreading factors could be used), one generic analysis method for CCPCH can be used:

$$CCPCHebnt = \frac{P_{CCPCH}}{(1-\alpha) \cdot \dfrac{P_{tot,bs}}{Lp_{bs}} + \displaystyle\sum_{k,k\neq bs} \dfrac{P_{tot,k}}{Lp_k} + I_{ACI} + N} \cdot \frac{W}{R_{CCPCH}} \qquad (3.66)$$

where $CCPCHebnt$ is the narrowband E_b/N_t; P_{CCPCH} is the transmit power and R_{CCPCH} the bit rate of either primary or secondary CCPCH; $P_{tot,bs}$ is the total transmit power and Lp_{bs} the link loss of the best server; α is the orthogonality factor; I_{ACI} is the adjacent channel interface; N the background noise; and W is the chip rate. The summation in the denominator has to be taken over all BSs k in the own network, excluding the best server bs. After Equation (3.66) has been evaluated for all pixels in the test area, the coverage probability of the selected channel can be estimated by the ratio of pixels in which $CCPCHebnt$ exceeds the required level to all pixels in the test area. By evaluating the CDF one can identify the needed E_b/N_t requirements for a certain given coverage probability and thereby the transmit power required for the channel in question.

3.3   Verification of Dimensioning with Static Simulations

This section introduces in a very simple manner the initial steps of the radio network planning process in the case of a greenfield operator. The process starts with the traffic and QoS definition and continues with network dimensioning and radio network planning. Furthermore, the plan is analysed and the performance of the plan is compared with the set requirements.

Dimensioning provides a rough cell range estimation and radio network planning with the planning tool can start. The dimensioning and simulation methods used are as described in previous sections of this chapter. In this example case, only one network evolution phase is considered; in the real radio network planning case traffic growth should be more carefully considered. This study consists of two phases. In the first phase the operator's macro-cellular network is dimensioned and the cell range is estimated with the given input parameters. This study is based on the assumption that the traffic and QoS information is available from the operator. In the second phase the network is planned for the estimated site distance $(1.5 \cdot R)$ and the WCDMA analysis is performed for the radio network. From Figure 3.1 this planning case concentrates on the first half of the process: network dimensioning, network configuration definition and coverage/capacity planning.

In this study propagation for macro-cells has been calculated with the Okumura–Hata model, the average area-type correction in the simulations (excluding water areas) being 1.5 dB. Dimensioning was done with an area-type correction of 0 dB.

3.3.1 Macro-cellular Network Layout

In this study, an area of $9\,\text{km}^2$ of downtown Helsinki (Finland) is analysed. The result of the dimensioning proposes 13 sites (38 sectors) for the coverage and the required capacity. Radio network planning was done with 32 cells. The difference is due to the fact that in reality some 20% of the total area is water. In Figure 3.16 the network scenario is depicted. The traffic and quality requirements are collected in Table 3.9.

Table 3.10 gives an example of the uplink radio link budget used. From downlink dimensioning it follows that the 20 W BS, with 3.5 dB peak-to-average ratio, can serve the required number of users. The peak-to-average ratio is used to take account of the fact that not all MSs are located at the cell edge. Table 3.11 lists the cell ranges per service; as expected, the 384 kbps service limits the cell range to 610 m.

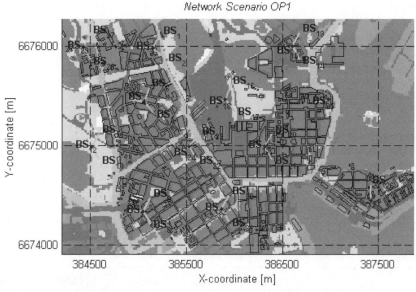

Figure 3.16 The selected network scenario, average site distance being roughly 910 m.

Table 3.9 Traffic requirements for the macro-cellular dimensioning and simulation case.

	Speech	64 kbps	144 kbps	384 kbps
Average traffic	45 mErl	6.5 mErl	2.2 mErl	2.2 mErl
		0.4 kbps/h	0.3 kbps/h	0.8 kbps/h
Subscribers	12091	955	636	318
Simultaneous users per cell	19	1.5	1	0.5
Blocking	2%	2%	2%	2%
Queuing time	0 s	5 s	5 s	5 s
Average call length	162 s	23.4 s	7.92 s	7.92 s
Coverage probability (outdoor)	98%	95%	90%	90%

Table 3.10 Example of a radio link budget.

Service	12.2 kbps	64 kbps	144 kbps	384 kbps	
MS transmit power	0.125	0.125	0.125	0.125	W
MS antenna gain incl. body loss	−3.00	−3.00	−3.00	−3.00	dBi
MS EIRP	17.97	17.97	17.97	17.97	dBm
Thermal noise density	−174.00	−174.00	−174.00	−174.00	dBm/Hz
BS receiver noise figure	5.00	5.00	5.00	5.00	dB
BS receiver noise density	−169.00	−169.00	−169.00	−169.00	dBm/Hz
BS noise power ($N_0 \cdot W$)	−103.16	−103.16	−103.16	−103.16	dBm
Interference margin	−6.02	−6.02	-6.02	−6.02	dB
Required BS E_c/I_0	−13.45	−7.75	-4.73	−0.97	dB
Required signal power	−116.61	−110.91	−107.89	−104.13	dBm
BS antenna gain	18.00	18.00	18.00	18.00	dBi
Cable losses	3.00	3.00	3.00	3.00	dB
Coverage probability outdoor	98	95	90	90	%
Outdoor point probability	93.82	81.42	70.58	70.58	%
Indoor point probability	62.07	57.08	54.30	54.30	%
Limiting environment	Outdoor	Outdoor	Outdoor	Outdoor	
Log-normal fading constant outdoor	5.00	5.00	5.00	5.00	dB
Log-normal fading constant indoor	12.00	12.00	12.00	12.00	dB
Propagation model exponent	3.50	3.50	3.50	3.50	
Log-normal fading margin	−6.88	−4.51	−2.33	−2.33	dB
Handover gain (incl. MDC gain at cell edge)	1.50	1.50	1.50	1.50	dB
Fading margin (incl. DHO gain)	−5.38	−3.01	−0.83	−0.83	dB
Indoor loss (−)	0.00	0.00	0.00	0.00	dB
Power control headroom	4.00	4.00	4.00	4.00	dB
Allowed propagation loss	−140.20	−136.86	−136.02	−132.26	dB

Table 3.11 Cell range per service according to dimensioning.

Service	12.2 kbps	64 kbps	144 kbps	384 kbps
Maximum cell range [km]	1.00	0.81	0.77	0.61
Selected cell range [km]	0.61	0.61	0.61	0.61
Site distance [km]	0.91	0.91	0.91	0.91

3.3.2 *Introduction to the Simulation Parameters*

The scope of this section is to simulate the network performance with a static simulator. The radio network plan is based on the dimensioning results of earlier sections. The cell range is based on the results of Table 3.11.

In these simulations, in the high-load case, users are randomly removed from the highly loaded cells, both in the uplink and in the downlink. During the simulations the total BS power and the uplink interference power relative to the noise floor were the indicators for the high load. During the simulations the MS transmit powers are

Table 3.12 Parameters used in the simulations.

Parameter	Value
Base station maximum transmit power	43 dBm
Mobile station maximum transmit power	21 dBm
Mobile station dynamic range	70 dB
Shadow fading correlation between BSs	50%
Indoor loss	12 dB
Standard deviation for the shadow fading	7 dB
Channel profile	ITU Vehicular A [29]
Mobile station speed	3 km/h for data, 50 km/h for speech
MS/BS noise figures	8 dB/5 dB
Soft handover window	−5 dB
P-CPICH power	30 dBm
Combined power for other common channels	30 dBm
Orthogonality	50%
BS antennas	65°/17.5 dBi
MS antennas	Omni/0 dBi
Cable losses	3 dB
Uplink loading limit	75%

corrected with voice activity factor, soft handover gain and average power rise due to fast power control for each MS. Table 3.12 presents the simulation parameters.

3.3.2.1 Macro-cellular Simulation Results

This section introduces the simulation results for the macro-cellular operator. As can be seen in Figure 3.17, users inside buildings will experience an additional indoor loss of 12 dB compared with the basic Okumura–Hata propagation. This additional loss was not introduced during dimensioning.

The dimensioning of this simulation case has succeeded rather well: only 2 users out of 704 are in outage. The average outage figure of 3 simulation runs is 4/704. So dimensioning and simulation show good agreement in terms of served users.

In the following tables the simulation results of the single operator are collected. The results in Table 3.13 indicate that the QoS criterion in terms of the uplink coverage probability is well met. The requirements were 98%, 95%, 90% and 90% (Table 3.9). The results after simulation were 99.2%, 96.3%, 93.6% and 89.1%, respectively (Table 3.13). Downlink coverage analysis is somewhat different from uplink. For the downlink the input is the actual coverage requirement and output is the link power limit for each service for which the coverage requirement is met. These results can be found in Tables 3.13 and 3.14.

In Tables 3.15 and 3.16 the MS and BS transmit power distributions are collected. In these simulations the window for soft handover was set to 5 dB (Table 3.12). In Figures 3.18 and 3.19 there are examples of soft handover probabilities and soft handover conditions.

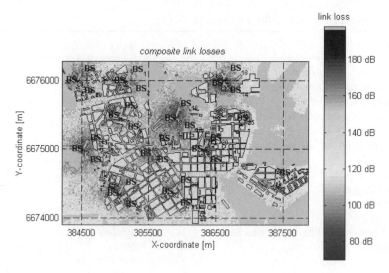

Figure 3.17 The link losses for the macro-cellular operator. An additional loss of 12 dB has been introduced for locations inside buildings.

Table 3.13 Uplink coverage probability results (area-based calculation in %).

	Speech	64 kbps	144 kbps	384 kbps
MS file 1	99.1	95.8	93.0	88.2
MS file 2	99.3	96.5	94.0	89.7
MS file 3	99.3	96.5	93.9	89.5
Average	99.23	96.27	93.63	89.13

Table 3.14 Downlink coverage analysis results (needed transmit power; area-based calculation). The coverage probability is shown in brackets.

	Speech	64 kbps	144 kbps	384 kbps
MS file 1	25.42 dBm	31.02 dBm (96.21%)	32.99 dBm (96.21%)	36.76 dBm (96.21%)
MS file 2	23.71 dBm	29.17 dBm (96.67%)	31.09 dBm (96.69%)	34.85 dBm (96.69%)
MS file 3	24.11 dBm	29.67 dBm (96.32%)	31.64 dBm (96.32%)	35.40 dBm (96.32%)
Average	24.42 dBm	29.95 dBm (96.40%)	31.91dBm (96.41%)	35.67 dBm (96.41%)

Table 3.15 Mobile station transmit power results [dBm].

	Max.	Q95	Q90	Q75	Q50	Min.
MS file 1	20.85	13.01	9.56	3.07	−5.70	−44.00
MS file 2	16.95	7.91	3.21	−3.75	−8.95	−44.00
MS file 3	15.65	7.25	4.75	−2.49	−8.93	−44.00
Standard deviation	2.71	3.15	3.31	3.63	1.87	0.00

Table 3.16 Base station transmit power results [dBm].

	Max.	Q95	Q90	Q75	Q50	Min.
Link power statistics						
MS file 1	38.67	31.10	28.52	24.10	23.04	17.50
MS file 2	34.41	28.34	26.07	21.54	20.25	15.92
MS file 3	35.40	29.20	26.49	22.17	20.99	15.65
Standard deviation	2.23	1.41	1.31	1.33	1.45	1.00
Total power statistics						
MS file 1	42.41	42.29	42.14	41.81	40.44	36.20
MS file 2	41.18	40.24	39.59	38.52	37.35	33.72
MS file 3	42.57	41.55	40.53	39.47	37.85	33.13
Standard deviation	0.76	1.04	1.29	1.69	1.66	1.63

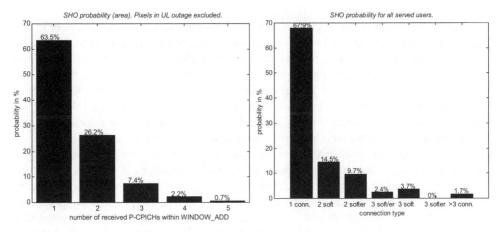

Figure 3.18 Soft handover situation in the network. In dimensioning 40% soft handover was assumed.

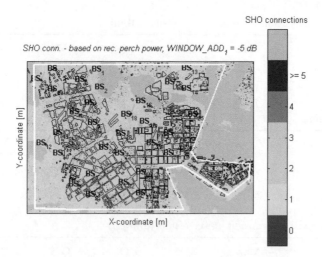

Figure 3.19   Number of connections in the network area. Main outage areas are inside buildings.

Table 3.17   Simulation results (average of three different MS location distributions).

	Total TCH power [W]	UL loading	UL users	DL users	SHO overhead	12.2 kbps links	64 kbps links	144 kbps links	384 kbps links
Average	7.94	0.54	21.90	21.27	0.47	26.83	2.16	1.38	0.71

Table 3.17 presents some tabular results. Of the served users, about 32% were in soft handover; from the whole network area roughly 36.5% were identified as in the soft handover area. Only few users/pixels had more than three connections, so little trouble should be expected from having too many received pilots.

In the exercise there were no limitations arising from site acquisition or transmission planning. Thus the positions of sites could be selected fairly optimally and the plan was able to support the required traffic. In a real planning process, such an idealised case would be too optimistic and planning compromises would be sought in order to optimise the radio network given the limitations on site selection. Furthermore, this plan was only done for the initial phase of network operation. If estimates of future traffic growth have been made, the evolution of the radio network plan should be carefully considered. Depending on traffic growth estimates, possible solutions are carrier additions, micro-cells and/or pico-cells, etc. In general it can be stated that dimensioning and static simulation results show good agreement. This has also been demonstrated in [30].

3.4 Verification of Static Simulator with Dynamic Simulations

As an alternative to the simulator described previously in this section one could think to use instead a fully dynamic simulator that implements all RRM functionality, such as power control, soft handover and packet scheduling, as well as genuinely moving users. However, the complexity of the necessary algorithms and the computational power requirements of such a tool make it rather inappropriate for planning larger networks, while it is an excellent tool for benchmarking RRM algorithms. A static simulator, however, has only a moderate processing power requirement, but its accuracy has been seriously questioned. The scope of this section is therefore to demonstrate that the accuracy of the modelling in the static simulator as described previously in this chapter is adequate for radio network planning purposes.

3.4.1 Introduction to the Dynamic Simulator

Typically, system-level simulators operate with a resolution determined by the feature that changes the interference situation most often. In WCDMA, fast closed-loop power control operating with 1.5 kHz frequency is the algorithm that has the highest frequency and therefore a frequency of 1.5 kHz is used in the system simulator for comparison purposes. Conventionally, the information obtained from the link-level tool is linked to the system simulation by using a so-called *average* value interface describing the BLER performance by average E_b/N_0 requirements. The average value interface is not accurate if there are rapid changes in interference due to, for example, high bit rate packet users. This kind of approach is well-suited to static snapshot simulations, but cannot be used when simulating systems with fast power control and high bit rate packet data. With the dynamic simulator presented, however, a so-called *Actual* Value Interface (AVI) is used that provides accurate modelling of fast power control and high bit rate packet data [31].

In the dynamic simulator, the users are making calls and transmitting data according to the traffic models. The call generation process for real time services, such as speech and video, follows a Poisson process ([32] and [33]). For speech, voice activity and discontinuous transmission (DTX) have to be considered. For CS data services, the traffic model is a constant bit rate model with 100% service activity.

The calculation of interference is an essential process of the system simulator. The better the interference modelling is, the more accurate are the results obtainable. The total interference power $I_{bs(k)}$ received by BS k is calculated as follows:

$$I_{bs(k)} = \sum_{n=1,n\neq m}^{N} \left[Lp_{n,k} \cdot \frac{\sum_{i=1}^{J} g_{i,n,k}}{\sum_{i=1}^{J} \hat{g}_{i,n,k}} \cdot p_{ms(n)} \right] \qquad (3.67)$$

where N is the total number of active MSs in the system; m is the index for the observed user; $Lp_{n,k}$ is the path loss (attenuation due to distance and slow fading) between BS k and MS n; $\sum g / \sum \hat{g}$ is the multi-path fading normalised to having a long-term average equal to 1; J is the number of multi-path components; and $p_{ms(n)}$ is the transmission power of mobile n. After the interference calculations, the uplink SNR, SNR_{UL}, can be calculated for user m connected to BS k as:

$$SNR_{UL(m,k)} = \sum_{i=1}^{J} \frac{G \cdot p_{ms(m)} \cdot a_i^2}{I_{bs(k)} + N} \tag{3.68}$$

where G is the processing gain; a_i is the amplitude attenuation of path i; and J is the number of allocated RAKE fingers. In Equation (3.68) it is assumed that the received signals are combined coherently with MRC. In the downlink the effect due to orthogonal codes has to be considered. Because of multi-path propagation, perfect orthogonality cannot be assumed. For optimal MRC, the downlink SNR, SNR_{DL}, for user m can be calculated as:

$$SNR_{DL(m)} = \sum_{k=1}^{M} \sum_{i=1}^{J_k} \frac{G \cdot p_{bs(m,k)} \cdot a_{k,i}^2}{I_{ms(m)} - P_{bs(k)} \cdot a_{k,i}^2} \tag{3.69}$$

where $I_{ms(m)}$ is the total interference power received by MS m; M is the number of BSs in the active set; $p_{bs(m,k)}$ is the transmit power to the observed user from BS k; $P_{bs(k)}$ is the total transmit power of BS k; $a_{k,i}$ is the amplitude attenuation of channel tap i; and J_k is the number of allocated RAKE fingers at BS k.

In the dynamic simulator, bad-quality calls, dropped calls and power outage were measured. *Bad-quality calls* are defined as calls having an average Frame Error Rate (FER) exceeding a threshold (usually 5% for speech). The minimum call duration is set to 7 s in order to increase the confidence of the averaging. The statistical data of these calls are recorded, such as coordinates, start and end time and call duration. *Dropped calls* are calls that have consecutive frame errors exceeding a threshold (usually 50 frame errors). They are usually considered as severely poor-quality calls. So bad-quality and dropped calls can be taken as one measure whose percentage is referred to the number of started calls after the warmup period. *Power outage*, for speech, is taken from active terminals including those that are in DTX. Therefore, it is slightly distorted due to the other half of the users that are in DTX. So the actual outage for terminals that are active is higher, roughly twice that of the output. There is no discrepancy for data. E_b/N_0 targets are taken from all active terminals, including those in soft handover. Finally, the soft handover probability histogram of the number of branches per user was collected.

From the static simulator for all uplink and downlink connections the histogram of the transmit powers and their cumulative distribution function are taken. Moreover, the pth percentiles Q_p for 0, 50, 75, 90, 95 and 100% are extracted and the number and type of soft handover links were gathered. For the whole simulated area the estimated active set size is collected, based on the received level of the P-CPICH. In each simulated case, the uplink loading level was also stored. In the final comparison the total traffic per cell, uplink power distribution [dBm], downlink total/link power distribution [dBm], soft handover statistics, soft handover areas, cell dominance areas and not served mobiles (static) versus dropped and bad-quality calls (dynamic) were of interest.

3.4.2   Comparison of the Results

In this section some of the comparison results are collected. The main conclusion is that the cell-level results (cell loading, for example) are in good agreement with both simulation methods. In Figure 3.20 the number of links is depicted cell by cell. It can be seen that the number of links per cell follows the same trend.

Figure 3.21 depicts the differences between the cell dominance areas as seen by the two simulators. It can be stated that the differences are minor. In 90–95% of all pixels, both simulators propose the same dominant cell.

Uplink power distribution statistics are collected for speech and data in Table 3.18. The maximum values do not differ significantly, but some of the percentile values are well apart. This could indicate different power distribution shapes for the two simulators. This cannot be avoided, due to the different nature of the simulators, since in the dynamic simulator DTX was modelled. Also the minimum allowed

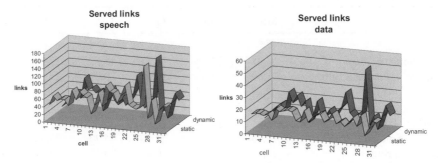

Figure 3.20   Number of links per cell for static simulator and dynamic simulator cell by cell.
Reproduced by permission of Groupe des Ecoles des Télécommunications.

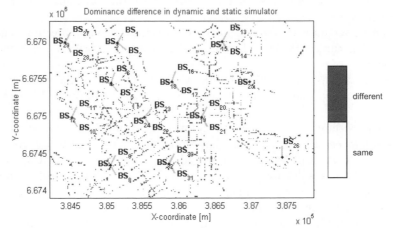

Figure 3.21   Dominance differences between the two simulators in the speech case.
Reproduced by permission of Groupe des Ecoles des Télécommunications.

Table 3.18 Uplink power distribution difference.

	Min.	Q50	Q75	Q90	Q95	Max.
Speech						
Static simulator [dBm]	−44.00	−10.38	−1.37	5.81	10.95	20.39
Dynamic simulator [dBm]	−49.00	−14.50	−7.00	0.00	4.50	20.00
Difference [dB]	−5.00	−4.12	−5.63	−5.81	−6.45	−0.39
Data						
Static simulator [dBm]	−41.79	−0.08	7.10	13.59	15.77	20.03
Dynamic simulator [dBm]	−44.00	−3.00	6.00	15.00	19.00	20.00
Difference [dB]	−2.21	−2.92	−1.10	1.41	3.23	−0.03

Reproduced by permission of IEEE.

Table 3.19 Downlink power distribution difference.

	Min.	Q50	Q75	Q90	Q95	Max.
Speech						
Static simulator [dBm]	8.25	13.58	16.12	18.37	18.98	24.14
Dynamic simulator [dBm]	−1.00	12.50	16.50	20.00	21.50	24.00
Difference [dB]	−9.25	−1.08	0.38	1.63	2.52	−0.14
Data						
Static simulator [dBm]	18.91	24.01	24.80	25.57	25.83	26.29
Dynamic simulator [dBm]	7.00	25.00	25.50	25.70	25.80	26.00
Difference [dBm]	−11.91	0.99	0.70	0.13	−0.03	−0.29

Reproduced by permission of IEEE.

transmit power of the MSs was different: −44 dBm for the static and −50 dBm for the dynamic simulator.

Transmit power statistics were also collected for the downlink direction. The results are presented in Table 3.19. The difference in the downlink shows a similar trend to that in the uplink.

Figure 3.22 depicts the downlink link power distributions for the two simulators in the speech case. The shapes of the distributions are similar. For data the variance is larger.

Another important result is the soft handover behaviour in the simulators. In Table 3.20 soft handover statistics from the data simulations have been collected.

In addition to the soft handover overhead, the differences in active set sizes were investigated. These results are shown in Figure 3.23. In radio network planning it is important to identify network outage areas. In this study, outage predictions were compared for the static and dynamic radio network simulator. The main conclusion is that problems tend to be distributed roughly in the same locations for either the static tool or the dynamic one. The numbers of problematic calls cannot be directly compared.

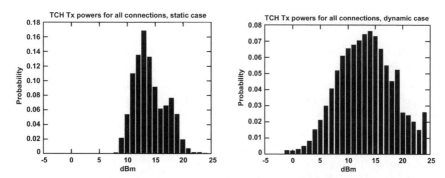

Figure 3.22   Link power distributions: left, static simulator; right, dynamic simulator.
Reproduced by permission of Groupe des Ecoles des Télécommunications.

Table 3.20   Handover comparison – 64 kbps circuit switched data case.

	No handover	Two-way handover	Three-way handover	Soft handover overhead
Static simulator [%]	83.7	14.9	1.4	17.7
Dynamic simulator [%]	72.0	23.0	5.0	33.0

Reproduced by permission of IEEE.

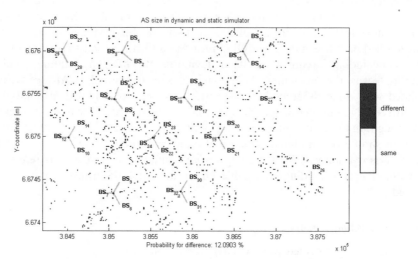

Figure 3.23   Active set size difference in the speech case.
Reproduced by permission of Groupe des Ecoles des Télécommunications.

3.5   Optimisation of the Radio Network Plan

In this section, two planning cases are presented with the aim of giving some guidance on starting optimisation as early as the planning phase. Since 3G systems and especially WCDMA systems are very sensitive to interference, it is of the utmost interest not to cause or receive too much of it. Means of controlling interference already in the radio network planning phase include, amongst others, optimising:

- site locations and configuration (sectorisation);
- height, main lobe direction, beamwidth and tilt of the antennas;
- antenna installations (cable losses);
- usage of Mast Head Amplifiers (MHAs).

The following two cases demonstrate the influence of the listed items on network capacity and coverage. The first case is an ideal case with a regular grid scenario. The second case makes use of digital topography and morphology information in propagation prediction. The studies presented in this section are carried out using the methods described in Section 3.2. A possible way of using these concepts in the initial automated optimisation with the help of a planning tool is presented in Section 7.3.1. An automated process and algorithm to optimise antenna tilting is presented in [34] together with some gain numbers.

3.5.1   Ideal Case

3.5.1.1   Simulation Scenario

The networks used in this ideal study [35] consisted of one central site surrounded by two tiers of sites on a regular hexagonal grid. Various different scenarios (number of sectors ∗ beamwidth) were studied. The site distance was 3 km in all cases. This distance was chosen so that the service probability exceeded 95% in all cases and the network was limited by capacity rather than coverage. The user profile used in the simulations was a homogeneous distribution of speech users (8 kbps) with a speed of 50 km/h.

The applied path loss model was Okumura–Hata for the urban environment overlaid with a slow fading having a standard deviation of 7 dB. The channel used in the simulations was the Vehicular A channel as specified in [29]. All scenarios were simulated with three different MS distributions to improve statistical reliability. For the relatively high number of speech users that can be served, three distributions proved to be enough. In the case of mobiles using higher bit rates, fewer users can be served and more runs should be simulated. The initial number of mobiles was set so high that the network was overloaded. Then an appropriate number of mobiles was removed until the required uplink loading was achieved or the total BS transmit power was not exceeded.

3.5.1.2   Simulation Results

The first analysis performed was the coverage probability for speech users. The results are collected in Table 3.21. It can be seen that in all cases the coverage probability was higher than the required 95%, increasing as the number of sectors increases.

Table 3.21 Service probabilities.

Number of sectors, beamwidth	Coverage probability [%]
1 sector, omni	96.6
3 sectors, 90°	97.6
3 sectors, 65°	98.6
6 sectors, 33°	99.6

Reproduced by permission of IEEE.

Table 3.22 Number of users per cell and per site.

	Users/cell		Users/site	
	Uplink	Downlink	Uplink	Downlink
1 sector, omni	59.2	43.0	59.2	43.0
3 sectors, 90°	50.7	35.5	152.2	106.4
3 sectors, 65°	56.7	42.6	170.1	127.9
3 sectors, 33°	55.7	40.3	167.0	120.8
4 sectors, 90°	46.0	30.7	183.9	122.7
4 sectors, 65°	53.2	39.9	212.8	159.7
4 sectors, 33°	50.9	35.7	203.5	142.8
6 sectors, 90°	39.4	24.5	236.7	147.1
6 sectors, 65°	46.4	32.4	278.6	194.2
6 sectors, 33°	49.6	36.3	297.4	218.0

Reproduced by permission of IEEE.

As a measure of capacity in all cases the number of users per cell and per site was of interest. The results collected in Table 3.22 demonstrate that the number of users per sector is decreasing when sectorisation is increasing.

The number of users per site, however, is increasing but not proportional to the number of sectors, because the overlap in the sectors is leaking interference from one sector to the other. Another result from Table 3.22 is that for each number of sectors an optimum beamwidth exists, optimum being when the number of users is at maximum. Another quantity of interest was the percentage of overhead due to soft handover connections. The soft handover overhead indicates the additional amount of hardware required and if the number of channel units is limited it increases the hard blocking probability.

As a last figure of merit, the other-to-own-cell-interference ratio, i, was collected. Both results, i and soft handover overhead, are shown in Table 3.23.

From Table 3.23 it can be seen that due to the increasing number of sectors the number of soft handover connections and therefore the overhead increases. Also the amount of interference leaking into neighbouring cells increases, but with proper choice of antenna beamwidth these effects can be controlled to an acceptable level. For the two most typical cases (three sectors, 65° antenna; six sectors, 33° antenna) it is important to

Table 3.23 Soft handover overhead and I_{oth}/I_{own}.

	Soft handover overhead [%]	$i = I_{oth}/I_{own}$ [%]
1 sector, omni	23	58
3 sectors, 90°	34	88
3 sectors, 65°	27	66
3 sectors, 33°	26	70
4 sectors, 90°	42	109
4 sectors, 65°	31	76
4 sectors, 33°	33	86
6 sectors, 90°	53	146
6 sectors, 65°	42	105
6 sectors, 33°	32	90

Reproduced by permission of IEEE.

note that there is only an increase of 5% in soft handover overhead when going to the higher sectorisation. If the beamwidth had been kept constant, this would have been 15% higher. This indicates that in the six-sector case the soft handover areas have to be planned more carefully than in the three-sector case. The smallest interference leakage occurred for the three-sector 65°-antenna case. With 66% it was in the order of the omni case, other scenarios have higher values of ~90%. This applies also to the six-sector 33° case, indicating that an even smaller beamwidth must be selected.

Finally the sectorisation gain, ξ, was estimated as the average number of simultaneous users relative to the average number of users of the omni-site configuration according to Equation (3.70):

$$\xi = \frac{Number\ of\ users\ of\ sectored\ site}{Number\ of\ users\ of\ omni\text{-}site} \tag{3.70}$$

The estimated numbers for ξ in the uplink and downlink are collected in Table 3.24. The simulation results were then compared with theoretically derived numbers from

Table 3.24 Simulated and theoretical sectorisation gains.

	UL gain	DL gain	Mean	Theoretical
1 sector, omni	1.00	1.00	1.00	1.00
3 sectors, 90°	2.57	2.56	2.56	2.51
3 sectors, 65°	2.87	3.00	2.94	2.90
3 sectors, 33°	2.82	2.85	2.84	2.90
4 sectors, 90°	3.11	2.97	3.04	2.86
4 sectors, 65°	3.59	3.71	3.65	3.59
4 sectors, 33°	3.44	3.41	3.42	3.79
6 sectors, 90°	4.00	3.64	3.82	3.22
6 sectors, 65°	4.70	4.62	4.66	4.40
6 sectors, 33°	5.02	5.17	5.09	5.60

Reproduced by permission of IEEE.

Equation (3.71), which represents the ratio of total received power to the power from within the sector:

$$\xi = N_S \cdot \int\limits_{-\pi}^{\pi} p(\vartheta) \cdot G(\vartheta)\, d\vartheta \Bigg/ \int\limits_{-\varphi/2}^{\varphi/2} p(\vartheta) \cdot G(\vartheta)\, d\vartheta \qquad (3.71)$$

where $G(\vartheta)$ is the antenna gain in direction ϑ; $p(\vartheta)$ is the received power in front of the antenna in direction ϑ; and $\varphi = 2\pi/N_S$ is the sector width in radians, N_S being the number of sectors. The numerical evaluation of Equation (3.71) with the antennas used in the simulations for a sectorisation with up to ten sectors can be seen in Figure 3.24.

The theoretical numerical values for the simulated cases are also listed in Table 3.24. Ideally, with sectorisation one would expect an increase of the capacity proportional to the number of sectors, but owing to overlap of the sectors the gain in capacity is smaller than this expected value. Also, the environment plays an important role. If there are obstacles in the vicinity of the antenna, the side lobe levels of the antennas are increased and the main beams are broadened. Both effects increase the amount of interference that is radiated to, or received from, other sectors, thus reducing the sectorisation gain.

It has been shown that higher sectorisation is giving higher capacity but the increase is not proportional to the number of sectors. The overlap in the antenna radiation patterns, as well as the influence of the environment on the shape of the patterns, makes it difficult to control interference leakage into neighbouring sectors and thus reduces the capacity of the network. By careful selection of the antenna beamwidth, however, the effect can be kept small and almost ideal sectorisation gains can be achieved. By the same careful selection, the increase of the soft handover areas can

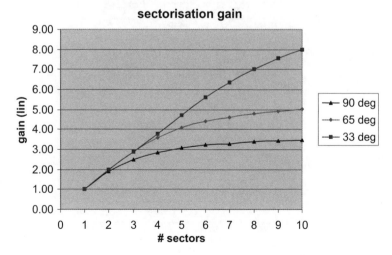

Figure 3.24   Theoretical sectorisation gains according to Equation (3.71).

Reproduced by permission of IEEE.

be kept at an acceptable level so that the additional overhead due to more soft handover connections, possibly resulting in hard blocking and also reducing system capacity, is small enough. However, control of the soft handover areas also requires careful selection of soft handover parameters. In addition, with the higher sectorisation, the orientation of the sectors – i.e., the bearing of the antennas – must be selected even more carefully, because the more sectors that are applied, the more of them will be pointing towards each other. All of these conclusions indicate that network performance can be significantly improved by higher sectorisation, but the more sectors that are applied the more care needs to be taken in network planning.

3.5.2 Shinjuku Case

3.5.2.1 Simulation Scenario

This study [36] was based in the Shinjuku area of Tokyo and assumed all users to be indoors. The system features used in the simulations were taken from [37]. For the multi-path channel profile the ITU Vehicular A channel from [29] was assumed. The 13.5 km^2 area was covered by ten sites. The selected antenna installation height was 50 m and the propagation loss was calculated with the Okumura–Hata model, with an average correction factor of −4.1 dB. The simulations used omni-, three-, four- and six-sector configurations and the site locations were kept fixed. The network scenario with six-sector implementation can be seen in Figure 3.25.

Five different antennas were used in the simulations with 3 dB beamwidths of 120°, 90°, 65° and 33° and additionally an omni-antenna. The gain of all antennas was set to 15 dBi and for the soft handover addition window a value of −4 dB was used – i.e., all

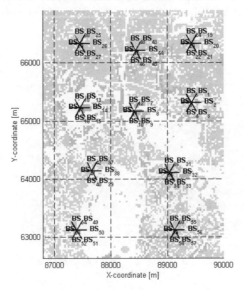

Figure 3.25 Example of the network scenario for the six-sectored BSs in the Shinjuku case.
Reproduced by permission of IEEE.

Table 3.25 Traffic information used in the studies.

Service	Service activity uplink [%]	Service activity downlink [%]	Users per service
8 kbps	50	50	720
64 kbps circuit switched data	100	100	240
144 kbps packet switched data	10	100	180

Reproduced by permission of IEEE.

sectors whose P-CPICH were received within -4 dB of the strongest P-CPICH were included in the active set. A service mix of voice users (8 kbps), circuit switched data users (64 kbps) and packet switched data users (144 kbps) was assumed. The exact traffic information used in this study is presented in Table 3.25. All simulations have been done with three different MS distributions and the results presented in Section 3.5.2.2 are averages over all distributions.

3.5.2.2 Simulation Results

This study was performed in three parts. In the first part, the impact of antenna tilting was of interest. Various antenna tilts were simulated to find the optimum. In the second part, the influence of MHA usage in the uplink was studied. For each sectorisation, simulations with and without MHA were compared. In the third part, capacity improvement as a function of sectorisation and antenna selection is illustrated. In Tables 3.26 through 3.28 the results in terms of the other-to-own-cell-interference ratio, served users, soft handover overhead and uplink coverage probability are collected.

In the antenna tilting study electrical tilting was applied and the results show that an optimum tilt angle can be found: both capacity and coverage probability have to be considered. The results of this study are collected in Table 3.26.

In these simulations the optimum tilting angle is from 7° to 10°. This is relatively high because of the large antenna installation height (50 m). The trend that can be seen from Table 3.26 is that by tilting the antennas the other-to-own-cell-interference ratio, i, decreases as the tilting is increased. This is because the antenna main beam is delivering less power towards the other BSs; therefore, most of the radiated power is going to the area that is intended to be served by this particular BS. At the same time the network could also serve more users than if the antennas were not tilted. There is always some optimum value for tilting, which depends on the environment, site and user locations, and the antenna radiation pattern. If the tilting angle is too big, the service area could decrease and the BS would be unable to serve such a large area. This is seen from the result for uplink coverage probability, which also has some optimum value. Owing to antenna radiation pattern side lobes and nulls, there could be some variations of i and coverage probability as a function of tilting angle.

In the second part of the study the usability of a low-noise MHA is demonstrated. The MHA is used in the uplink direction to compensate for cable losses, thus reducing the

Table 3.26 Examples of the impact of antenna tilt on network capacity. Mast head amplifier in use. Mobile station maximum transmit power 24 dBm. In the downlink, if base station maximum transmit power exceeded, connections were randomly put to outage.

Antenna tilt	Other-to-own-cell-interference ratio, i	Served users	Soft handover overhead	Uplink coverage probability (outdoor to indoor) for 8/64/144 kbps
Omni case				
0°	0.79	239	28%	70/32/40%
Three-sector case, 65° antenna				
0°	0.88	575	40%	86/59/62%
4°	0.75	624	39%	91/71/72%
7°	0.59	697	36%	92/76/76%
10°	0.37	856	30%	90/75/74%
14°	0.38	787	32%	81/62/61%
Four-sector case, 65° antenna				
0°	1.09	604	41%	92/70/71%
4°	0.94	707	30%	95/81/81%
7°	0.72	833	26%	96/84/83%
10°	0.47	959	21%	94/82/81%
14°	0.50	886	26%	86/69/68%
Six-sector case, 33° antenna				
0°	1.15	880	48%	93/76/76%
4°	1.03	946	49%	96/83/83%
7°	0.88	1037	45%	96/85/84%
10°	0.73	1054	41%	95/83/82%
14°	0.58	930	33%	86/70/69%

Reproduced by permission of IEEE.

required MS transmit powers. The three- and four-sectored scenarios have been simulated with the 65° antenna and the six-sectored case applied the 33° antenna. In all cases the antenna tilt used was 7° and the maximum MS power was 27 dBm. The MHA simulation results are collected in Table 3.27. The results indicate that by using an MHA the performance in the uplink can also be improved in WCDMA systems. In all cases the number of users that can be served in the uplink has been increased due to the increased sensitivity. Also the coverage probability is greater when deploying an MHA. In the six-sectored case, the influence of an MHA has also been greater when assuming bigger cable losses in the uplink (4 dB instead of 2 dB). Table 3.27 also shows, however, that the scenarios are downlink limited and having more MSs on the uplink actually decreases downlink performance. In all cases downlink capacity was smaller when using an MHA in the uplink. The reason could be that if more users can be served in the uplink, the transmit powers in the downlink are increased due to more soft handover links, thus reducing downlink capacity.

In the third case analysed, which illustrates the capacity improvement as a function of sectorisation, each BS was simulated as an omni-site and as a site with three, four or six

Table 3.27 Impact of mast head amplifier. MS maximum transmit power 27 dBm. Antenna tilt 7°. In the downlink, if BS maximum transmit power is exceeded, connections were randomly put to outage.

	Other-to-own-cell-interference ratio, i	Served users in uplink	Served users in downlink	Uplink coverage probability (outdoor to indoor) for 8/64/144 kbps
Three-sector case, 65° antenna				
No MHA	0.60	1038	807	93/78/78%
With MHA	0.61	1064	746	95/82/82%
Four-sector case, 65° antenna				
No MHA	0.73	1089	884	96/86/85%
With MHA	0.73	1107	846	98/89/89%
Six-sector case, 33° antenna				
No MHA	0.88	1124	1052	97/87/86%
With MHA	0.90	1132	1021	98/90/90%
No MHA, 4 dB cable losses	0.88	1109	1057	95/83/82%
With MHA, 4 dB cable losses	0.90	1132	1016	98/90/90%

Reproduced by permission of IEEE.

sectors. Furthermore, by simulating the scenarios with antennas having different beamwidths, the importance of correct antenna selection for a sectored configuration is emphasised with the help of some examples. For all scenarios the MHA was in use, the maximum MS transmit power was 24 dBm and antennas were not tilted. The results related to the sectorisation study are in Table 3.28. In the case of omni-sites, coverage was very poor and only 240 users could be served. Even in the uplink the network was already heavily overloaded. There were almost an equal number of MSs going to outage because of excessive load and because of MSs running out of power. In all the sectored cases the outage reason in the uplink was insufficient MS power, though the situation in the downlink was even more limiting, with more mobiles going to outage. Table 3.28 clearly indicates that, with higher sectorisation, more mobiles can be served. Another observation that can be made from the results is that for each sectorisation case the selection of antenna beamwidth is important. To achieve the highest number of served users it is crucial to control interference and soft handover overhead effectively. If the overlap of the sectors is too big, interference leaks from one sector to the other sector, directly reducing its capacity. Another effect of the antenna beam being too wide is the waste of hardware resources and increased downlink transmit powers due to the soft handover overhead being too high. In the simulations, the 65° antenna was optimal for the three-sectored case and the 33° antenna was best for the four- and six-sectored scenarios.

It can be stated that, with rather simple radio network planning techniques (antenna tilting and correct antenna selection for each scenario), interference can be controlled and the capacity of the network improved. In the antenna-tilting study, electrical tilting

Table 3.28 Impact of antenna selection in the sectorisation case. Mobile station maximum transmit power 24 dBm, mast head amplifier in use. No antenna tilt.

Antenna 3 dB beamwidth	Other-to-own-cell interference ratio, i	Served users	Soft handover overhead	Uplink coverage probability (outdoor to indoor) for 8/64/144 kbps
Omni case				
Omni	0.79	240	28%	70/32/40%
Three-sector case				
120°	1.33	441	39%	85/50/59%
90°	1.19	461	35%	87/55/62%
65°	0.88	575	34%	86/59/62%
Four-sector case				
120°	1.72	489	54%	90/62/68%
90°	1.49	510	51%	92/67/72%
65°	1.09	604	41%	92/70/71%
33°	0.92	691	40%	88/65/64%
Six-sector case				
120°	2.18	593	64%	95/75/79%
90°	1.97	627	59%	96/80/82%
65°	1.43	758	55%	96/80/81%
33°	1.15	880	48%	93/76/76%

Reproduced by permission of IEEE.

was applied and the results show that an optimum tilt angle can be found. In the simulations presented in the study each of the BSs was optimised in a similar manner. In reality BS antennas are not installed at equal heights and thus optimisation of the BSs should be performed cell by cell. In this study it has also been demonstrated that an MHA is also feasible in WCDMA networks, though the benefit is rather small when the system is strongly downlink limited and thus the uplink sensitivity improvement is less beneficial. The results in Table 3.27 indicate that the QoS can be improved in the uplink direction in lightly loaded networks with an MHA. In all of the simulated cases the coverage probability was increased when an MHA was in use. How much of the uplink capacity improvement can be utilised in the downlink direction when an MHA is in use depends naturally on the current downlink loading situation and the admission and load control strategies implemented in the network. The results of this study also clearly show that higher sectorisation offers more capacity to the network, but to achieve this the antenna selection is very crucial to control interference and soft handover overhead effectively. For each sectorisation case an optimum beamwidth exists.

3.6   Interference in WCDMA Multi-operator Environment

This section addresses the problems arising from Adjacent Cell Interference (ACI) when multiple WCDMA operators operate their networks in the same area on adjacent carriers. ACI from the own network has been shown to be not critical, since by making use of radio network planning techniques and by performing IF-HO the problems can be alleviated. In multi-operator environments ACI might cause problems, because large guard bands cannot be allowed owing to scarce frequency resources, which would mean a waste of frequency bands, especially recalling the large bandwidth of WCDMA systems. On the other hand, ACI problems could be avoided by putting tight requirements on transmit and receive filter masks of BSs and MSs. But high requirements on minimising leakage of interference power into neighbouring bands on the one side and high resistance towards interference from neighbouring bands on the other would make it almost impossible to build cheap, compact equipment for BSs as well as MSs. Considering only WCDMA internal interference, the following critical scenarios arise:

- MS from operator 1 is coming close to BS of operator 2 and is blocking this BS because it is transmitting with full power – e.g., when it is located at the cell edge of operator 1.
- BS from operator 2 transmits with high power and therefore blocks *all* MSs of operator 1 in a certain area around it, because of excessive interference power (dead zones) and/or the maximum allowed input power at the MS receiver being exceeded (blocking).

3.6.1 Sources of Adjacent Channel Interference

The following sources of ACI in WCDMA can be identified:

- *Out-of-band emissions*, which are unwanted emissions immediately outside the nominal channel resulting from the modulation process and nonlinearity in the transmitter but excluding spurious emissions. This out-of-band emission limit is specified in terms of a spectrum emission mask [38].
- *Spurious emissions*, which are emissions caused by unwanted transmitter effects such as harmonics emission, parasitic emission, intermodulation products and frequency conversion products, but exclude out-of-band emissions. The frequency boundary and the detailed transitions of the limits between the requirements for out-of-band emissions and spectrum emissions are based on [39].

The following important quantities characterising the influence of adjacent channels on each other can be identified:

- *Adjacent Channel Leakage power Ratio (ACLR)*, defined in [38] and [40] as the ratio of the transmitted power to the power measured in an adjacent channel. Both the transmitted power and the adjacent channel power are measured with a filter that has a root-raised-cosine filter response with a rolloff factor of $\alpha = 0.22$ and a bandwidth equal to the chip rate. If the adjacent channel power is greater than -50 dBm then the ACLR requirements are defined by [38] for MS and by [40] for BS, as shown in Table 3.29.
- *Adjacent Channel Selectivity (ACS)*, defined in [38] and [40] as a measure of a receiver's ability to receive a WCDMA signal at its assigned channel frequency in the presence of an adjacent channel signal at a given frequency offset from the centre frequency of the assigned channel. ACS is the ratio of the receive filter attenuation on

Table 3.29 Adjacent channel performance requirements for mobile station [38] and base station [40].

Adjacency	Channel separation	Max. allowed ACLR	
		MS[a]	BS
First adjacent carrier	5 MHz	33 dB	45 dB
Second adjacent carrier	10 MHz	43 dB	50 dB

[a] If adjacent channel power is bigger than −50 dBm.

Table 3.30 Adjacent channel selectivity requirements for BS [40].

Parameter	Value
Data rate	12.2 kbps
Wanted signal	−115 dBm
Interfering signal	−52 dBm
Fuw (modulated)	5 MHz

the assigned channel frequency to the receive filter attenuation on the adjacent channel(s). For the MS, ACS shall be better than 33 dB and for the BS, the requirements for speech channel and a BER of 10^{-3} are given by Table 3.30.

- *Min. UE output power during power ON state*; equal to or less than −50 dBm [38].
- *Max. UE output power during power OFF state*; equal to or less than −56 dBm [38].
- *Occupied bandwidth*, which is a measure of the bandwidth containing 99% of the total integrated power of the transmitted spectrum, centred on the assigned channel frequency. The occupied channel bandwidth shall be less than 5 MHz based on a chip rate of 3.84 Mcps [38] and [40].

3.6.2 Minimum Coupling Loss

The Minimum Coupling Loss (MCL) is defined as the smallest path loss that can occur between the transmitters and receivers of the BSs and MSs. It is encountered when the MS is coming as close as possible to the BS. If the MS is power-controlled by that BS it will consequently reduce its transmit power until it reaches the minimum possible value. If this stage is reached and the MS is still approaching the BS, it is transmitting with excessive power, increasing uplink interference beyond what is absolutely necessary. Here it is assumed that the MCL between a micro-BS and an MS is about 53 dB, the MCL between a macro-BS and an MS is about 70 dB and the minimum MS power is −50 dBm. Considering these values, the minimum received level at the (micro-) BS can be calculated as:

$$-50\,\text{dBm} - 53\,\text{dB} = -103\,\text{dBm} \tag{3.72}$$

which means that the MS generates only a very small noise rise compared with the noise floor of about -103.1 dBm (assuming a noise figure of 5 dB).

The MCL problem can naturally also be encountered when an MS of a second operator is coming too close to the first operator's BS. The difference, however, is that the MS is not power-controlled by the BS it is approaching. If the two operators have co-sited their BSs this is not critical, since then the second operator's BS will command the MS to lower its power. In an ideal case there would not be any problems, since the operators are using different frequency carriers and there would be no interference between them. In reality, however, there are only finite values for ACS and ACLR (see Section 3.2.4). Assuming values of 33 dB and 45 dB, respectively, the coupling, C, between the carriers becomes:

$$C = -10 \cdot \log_{10}(10^{-33/10} + 10^{-45/10}) \, \text{dB} = 32.7 \, \text{dB} \qquad (3.73)$$

This means that if the own MS and the other operator's MS are transmitting with the same power, the interference received from the latter is about 32.7 dB less than that generated by the MS of the own system. The worst case scenario in the MCL problem, however, happens when some MS of the second operator is transmitting with its maximum power at the MCL distance from the BS of the other operator. This happens, for example, when the sites are not co-located. In an extreme situation one site is at the border of a cell of the other operator's network. If then an MS is moving towards that border and in doing so it is approaching the first operator's BS, it is transmitting with full power in the near vicinity of the first operator's BS, as can be seen in Figure 3.26.

With a maximum MS power of 21 dBm, 53 dB for MCL to the micro-BS and coupling between the carriers of $C = 32.7$ dB, the received level at the micro-BS can be estimated as:

$$21 \, \text{dBm} - 53 \, \text{dB} - 32.7 \, \text{dB} = -64.7 \, \text{dBm} \qquad (3.74)$$

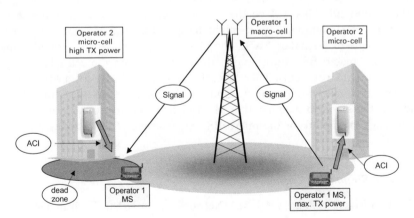

Figure 3.26 Worst case scenarios in intra-system ACI. Right part: uplink; left part: downlink with dead zone.

If the background noise level is -103.1 dBm, the micro-BS would suffer a 38.4 dB noise rise from one macro-user, which is located in the radio sense at the MCL distance from the micro-BS – i.e., such a macro-user would completely block the micro-BS.

Next we calculate the situation on the downlink: consider that the micro-BS is transmitting with even minimum power of 0.5 W (27 dBm); then the received interference at the MS in the adjacent channel is:

$$27\,\text{dBm} - 53\,\text{dB (MCL)} - 32.7\,\text{dB (ACS)} = -58.7\,\text{dBm} \qquad (3.75)$$

Assuming a speech service (processing gain of $G_p = 25$ dB) with an E_b/N_0 requirement at the MS of 5 dB and an allowed noise rise in the macro-cell of 6 dB, the maximum allowed propagation loss, Lp, to keep the uplink connection working is:

$$Lp = 21\,\text{dBm} - 5\,\text{dB} + 25\,\text{dB} - (-103\,\text{dBm} + 6\,\text{dB}) = 138\,\text{dB} \qquad (3.76)$$

Assuming a downlink transmit E_b/N_0 requirement of 8 dB, the transmit power, P_{tx}, would need to be:

$$P_{tx} = -58.7\,\text{dBm} + 8\,\text{dB} - 25\,\text{dB} + 138\,\text{dB} = 62.3\,\text{dBm} \qquad (3.77)$$

This simple example shows that clearly in these cases the downlink is the weaker link – i.e., before coming too close to a micro-BS, the connection of a macro-MS will be dropped due to insufficient downlink power and it cannot block the micro-BS.

3.6.3 Dead Zones

Dead zones are another problem that can occur due to MCL problems. A dead zone is an area in which either the BS in the downlink or the MS in the uplink does not have enough transmit power to maintain the QoS requirements of the other end. When entering such an area an existing connection is lost and it is not possible to establish a connection from that area. One possible scenario where a dead zone can arise is again in a multi-operator environment, if an MS from one operator is approaching at the cell edge a (micro-) BS from another operator that is transmitting with full power. Then the own BS does not have enough transmit power to overcome the interference generated from the second BS. This will be the case in a certain area around the second BS. Alternatively, or simultaneously, it might happen that the MS can no longer reach its own BS. Due to a smaller MCL, the problem is more severe around a micro-BS than around a macro-BS. Additionally, the link loss from the cell edge to the BS is bigger in macro-environments. Therefore, the most typical case for a dead zone will be for an MS of a macro-operator around the BS of a micro-operator. However, it depends on the scenario whether this MS will first lose its connection or whether it will first block the uplink of the micro-BS. An example of dead zones can be seen in Figure 3.27.

3.6.4 ACI Simulation Cases

3.6.4.1 Two Macro-cellular WCDMA Networks in an Urban Environment

In earlier work published in the field [41] and [42] the simulation scenario has been rather unrealistic. It is rather unlikely that in an (dense) urban area one operator would

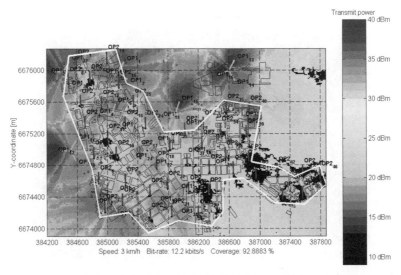

Figure 3.27 Example of downlink link power needed for a macro-operator's network. Also visible are some dead zones, where the maximum link power is not sufficient for good enough quality of service.

choose to employ a micro-cellular network modelled with a Manhattan grid, while another operator would see it feasible to provide services with a macro-cellular network.

This section describes the network simulation results of a study on the mutual influence of two macro-cellular WCDMA radio networks when operating in the same area. Both operators' networks were of macro-cellular type, located in an urban environment in the city centre of Helsinki (Finland). Both operators were assumed to have the same traffic and QoS requirements.

The first phase of the analysis considered the two operators' networks to be independent from each other – i.e., without experiencing the influence of external interference from the other operator's network. In the second phase, the influence of the interference leaking from one operator's network to the other's was taken into account by filtering the transmit powers from one operator to the other. In the whole study the two operators were considered to operate in immediately adjacent channels separated by 5 MHz. No other neighbouring channel interference was taken into account. The values of the minimum transmit power for the mobiles and the filter settings were chosen on a best guess basis, as their standardisation was not finished at the time of the study.

Urban Simulation Case
In the urban simulation case a 9 km^2 area in the city centre of Helsinki was analysed. The dimensioning proposed 13 sites (38 sectors) for the coverage and the required capacity. Because in reality some 20% of the total area is water, the actual network planning was done with 32 sectors, of which 31 used 65°/17.5 dBi sector antennas and one 11 dBi omni-antenna. The selected antenna installation height was from 16 m to

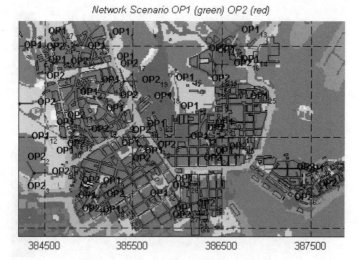

Figure 3.28   Used network scenarios in the urban case.

20 m and the propagation loss was calculated with the Okumura–Hata model, with an average area correction factor of −6.3 dB. For users inside the buildings an additional propagation loss of 12 dB was added. Two independent network layouts were created. The network scenarios can be seen in Figure 3.28.

The system features used in the simulations are from [37], except the chip rate which was modified to 3.84 Mcps. The multi-path channel profile was the ITU Vehicular A channel [29]. For the soft handover window a value of −5 dB was used – i.e., all sectors whose received P-CPICH are received within −5 dB of the strongest P-CPICH are in the active set. The maximum allowed uplink loading was set to 75%. Other relevant parameters applied in the simulations are listed in Table 3.31. The traffic requirements were as in Table 3.9.

Simulation Results

In this section results from the urban simulation case are collected. The numbers presented are averages over three different MS distributions following the traffic requirements of Table 3.9. Table 3.32 lists the uplink coverage probabilities. The requirements are well-met, except that the 384 kbps coverage is slightly too small. If a second operator is present, coverage does not drop significantly.

Table 3.33 gives an overview on the MS transmit powers in terms of maximum and minimum powers used, as well as the 50, 75 and 95 percentiles. In this case, too, no significant increase is noticed when introducing the influence of a second operator. Mobiles using their minimum allowed transmit powers indicate that there could be some problems in the network arising from excessive MCL, though no consequences, such as downlink dead zones, have been observed.

Table 3.34 shows the transmit powers in the downlink. Statistics from both the single-link powers and the total transmit powers are collected. If a second operator is

Table 3.31 Parameters used in the simulations.

Chip rate	3.84 Mcps
BS maximum transmit power	43 dBm
MS minimum/maximum transmit power	−44 dBm[a]/21 dBm
Shadow fading correlation between sites/sectors	50%/80%
Standard deviation for shadow fading	7 dB
Channel profile	ITU Vehicular A [29]
MS speed	3 km/h for data, 50 km/h for speech
MS/BS noise figures	8 dB/5 dB
P-CPICH power	30 dBm
Combined power for other common channels	30 dBm
Orthogonality	50%
MS antennas	Omni, 0 dBi
Cable losses	3 dB
Filter settings – Equations (3.58) and (3.60)	
aciFilterUL (BS selectivity, ACS)	45 dB
acpFilterUL (MS leakage, ACLR)	33 dB
aciFilterDL (MS selectivity, ACS)	33 dB
acpFilterDL (BS leakage, ACLR)	45 dB

[a] In this study, the minimum transmit power of the mobile station was −44 dBm. In 3GPP standards this value was adjusted later to −50 dBm.
Reproduced by permission of IEEE.

Table 3.32 Uplink coverage in urban case.

Uplink coverage	Speech	64 kbps	144 kbps	384 kbps
One operator	99.23%	96.27%	93.63%	89.13%
Two operators	99.19%	96.18%	93.52%	88.93%

Table 3.33 Mobile station transmit powers in the urban case.

MS transmit powers [dBm]	Max.	Q95	Q75	Q50	Min.
One operator	17.82	9.39	−1.06	−7.86	−44.0
Two operators	18.01	9.50	−0.90	−7.73	−44.0

Reproduced by permission of IEEE.

introduced, transmit powers increase slightly, though no dramatic effects could be noticed.

In Table 3.35 the average number of users per cell, the uplink load, the average number and type of links per cell and the soft handover overhead are given. Again, these results indicate that with the chosen filter values no significant influence from the neighbouring operator is experienced.

Table 3.34 Base station transmit powers in the urban case.

	Max.	Q95	Q75	Q50	Min.
Link power statistics [dBm]					
One operator	36.16	29.55	22.60	21.43	16.36
Two operators	35.85	29.76	22.90	21.63	16.52
Total power statistics [dBm]					
One operator	42.05	41.36	39.93	38.55	34.35
Two operators	42.30	41.75	40.04	38.74	34.67

Reproduced by permission of IEEE.

Table 3.35 Other results from the urban case.

	Users	Load	Links				Soft handover overhead
			12.2 kbps	64 kbps	144 kbps	384 kbps	
One operator	21.27	0.54	26.83	2.16	1.38	0.71	0.47
Two operators	21.44	0.55	27.18	2.18	1.43	0.66	0.47

Reproduced by permission of IEEE.

Conclusions

In this study the influence of two operators on each other in a macro-cellular environment was investigated for an urban area. Owing to the relatively tight filter settings describing the mutual influence, network performances did not suffer significant degradation. Almost the same performance with and without the second operator was achieved. The biggest degradation was observed for the outage probabilities, but the changes were not too dramatic as the outage was only slightly increased. In this urban study none of the so-called dead zones could be observed. One explanation for this could be that the link losses were calculated using an Okumura–Hata model without LOS check, so the minimum link losses were bigger than the minimum coupling loss required to avoid the problem. The result could, however, be different if an LOS check were used, especially in a scenario where there are BSs of two operators aligned along streets or even highways. The same reason lies behind the observation that there was no significant difference in performance whether cells of different operators were almost co-located or whether they were positioned at each other's cell edge. Another case in which networks are located in a suburban area can be found in [43]. Those results indicate the same behaviour in terms of ACI.

3.6.4.2 Macro- and Micro-cellular WCDMA Networks in an Urban Environment

In this ACI exercise the two networks comprised one macro- and one micro-cellular layout, operated on adjacent carriers servicing the same urban area (downtown

Table 3.36 Some general simulation parameters.

	Macro	Micro
Maximum BS power	43 dBm	36 dBm
Maximum downlink transmit power per link	40 dBm	33 dBm
P-CPICH power	30 dBm	23 dBm
Other common channel powers	30 dBm	23 dBm
Soft handover window	3 dB	3 dB
BS antenna height	25.0 m	10.0 m
MCL	70 dB	53 dB
BS selectivity/leakage	45 dB	45 dB
MS selectivity/leakage	33 dB	33 dB
Minimum MS transmit power	−44 dBm	−44 dBm
Shadowing standard deviation/correlation between BSs	7 dB/0.5	7 dB/0.5

Helsinki) as in the previous section with sufficient capacity and coverage. The dimensioning in this case suggested that the macro-operator has 32 cells and the micro-operator 46 cells in an area of about $4 \, km^2$. In the simulations the basic idea was that each operator optimises its network first so that the outage was below 2%, without considering the other operator. Therefore, the cell plans are totally independent. In the real case the parameters could be optimised in a more efficient way.

The propagation environments were calculated using a ray-tracing program for the micro-cell scenario and the Okumura–Hata model for the macro-cell scenario. In the study the micro-/macro-scenarios were first analysed independently. Then the scenarios were combined and the interaction of these two operators in the form of interference was deduced. Both network-based indicators and cell-based indicators were of interest.

The general simulation parameters are listed in Table 3.36. These serve as default values, if not stated otherwise, in the simulation cases.

3.6.4.3 Simulations in Helsinki with 32 Macro-cells and 46 Micro-cells

Figure 3.29 shows the cell plans used in the simulation together with the studied area. For each simulated case three snapshots with random positions of MSs were used. On average, 20, 25, 30 and 35 users per cell were input for the macro-operator and 55, 65, 75 and 85 users per cell on average for the micro-operator.

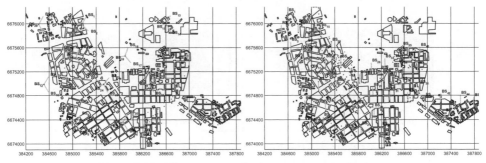

Figure 3.29 The macro- and micro-operators' cell plans.

Simulation Results

This section and the figures that follow give the main simulation results for macro- and micro-operators with and without the other operator present. Service probability (number of users served after iterations divided by initial number of users), uplink noise rise and BS total transmit power are shown. In addition, performance has been studied with two settings of the maximum traffic channel power for a single link in the downlink: 5.5 dB below CPICH (left diagrams) and 0 dB below CPICH (right diagrams). The latter corresponds to an aggressive parameter setting to avoid dead zones. All the curves show averages from all three snapshots and the powers averaged over the cells. The *x*-axis is always 'Number of users' or 'Number of served users': this means on average per cell, as the traffic was generated uniformly onto the area. For the macro-cells only 'inner cells' on the area were included in the cell-based analysis to avoid bias from border effects.

From the simulation results one can see that there is always a significant loss of downlink performance for the macro-operator. If the loading in the macro-operator's network is low, an aggressive parameterisation (allowing high transmit power for the traffic channels) may help slightly and make the micro-operator's life slightly more difficult, but for high loading it does not help. Also one can see that if the macro-operator uses aggressive parameterisation the micro-operator can suffer in the uplink because of a slightly bigger noise rise.

Simulation Results for the Macro-operator (Figures 3.30–3.32)

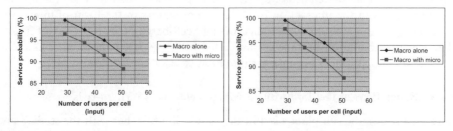

Maximum link power 5.5 dB below CPICH   Maximum link power equals CPICH

Figure 3.30 Service probability of the macro-operator when alone and with the micro-operator.

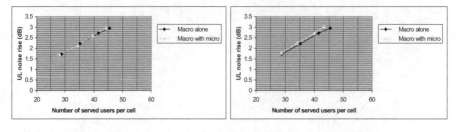

Maximum link power 5.5 dB below CPICH   Maximum link power equals CPICH

Figure 3.31 Uplink noise rise of the macro-operator when alone and with the micro-operator.

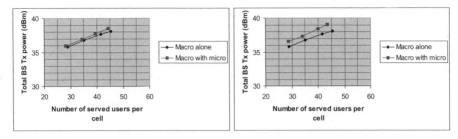

Maximum link power 5.5 dB below CPICH   Maximum link power equals CPICH

Figure 3.32   Total base station transmit power of the macro-operator when alone and with the micro-operator.

No pure capacity effects can be seen from these simulations – i.e., moving the pole capacity – but according to the results one could think of adding the effect of the adjacent carrier, if cell planning between the macro- and micro-layers is un-coordinated, as an offset to the noise level in dimensioning. In the optimisation process the other operator on the adjacent carrier should be taken into account to avoid local dead zones.

Simulation Results for the Micro-operator (Figures 3.33–3.35)

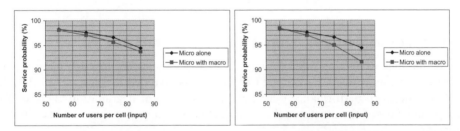

Maximum link power 5.5 dB below CPICH   Maximum link power equals CPICH

Figure 3.33   Service probability of the micro-operator when alone and with the macro-operator.

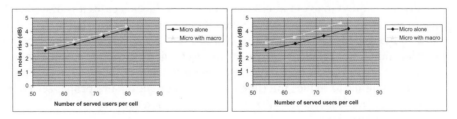

Maximum link power 5.5 dB below CPICH   Maximum link power equals CPICH

Figure 3.34   Uplink noise rise of the micro-operator when alone and with the macro-operator.

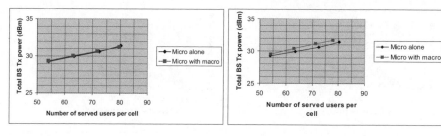

Maximum link power 5.5 dB below CPICH   Maximum link power equals CPICH

Figure 3.35 Total base station transmit power of the micro-operator when alone and with the macro-operator.

Conclusions

The macro-operator is more affected by the micro-operator than vice versa. The macro-operator can lose downlink coverage near the micro's BSs. The micro-operator's uplink noise rise can be slightly higher because of the macro's MSs if the macro-operator uses aggressive downlink power allocation (giving high power for a single MS). No clear capacity effects were found but only coverage effects. Downlink dead zones can occur in such places where the macro-cell boundary is close to the micro-operator's BS (the micro–micro case is probably easier, since in most cases the cell boundaries are inside buildings for both operators). The problem is made worse by a larger average path loss difference.

3.6.5   *Guidelines for Radio Network Planning to Avoid ACI*

The simulations in Section 3.6.4 prove that with proper radio network planning the severest problems with ACI within WCDMA can be avoided to such a level that the WCDMA network performance does not suffer significant degradation. This section gives a summary of the most popular radio network planning means to alleviate ACI problems.

- BS and antenna locations:
  - in macro-cellular-only environments, the natural distance between the MS and BS is normally large enough to provide sufficient decoupling. In mixed environments, however, when micro-cells and pico-cells are present, the minimum coupling loss is usually not enough to avoid interference problems. In such cases it is desirable that operators try to *co-locate BSs*, since then there is no possibility that an MS that is close to the cell edge of one operator comes close to the BS of the other operator;
  - if co-location is not achievable then one means to increase the MCL is to *deploy the antennas in a position as high as possible above the MS*;
  - other possibilities to reduce interference between operators are proper selection of the *antenna direction* and the *correct tilting of the antennas*.
- Base station configuration:
  - after selection of the correct sectorisation to meet the coverage and capacity requirements, for each configuration there exists an *optimum antenna beamwidth*.

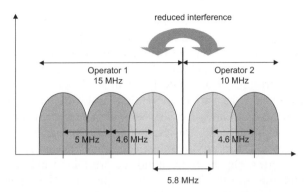

Figure 3.36 Reduction of ACI by creating a guard band with reduced carrier spacing.

Antennas that are too wide cause too much interference to adjacent sectors, naturally not only in the same frequency but also in adjacent ones;
- ○ in case other means are not possible or do not achieve the required coupling loss, it is still possible to reduce artificially the sensitivity of the BS receiver by increasing the noise figure. This technique, called *desensitisation*, reduces the effect of ACI but unfortunately also makes the receiver less sensitive to wanted signals, which results in reduction of coverage area and increased battery consumption in the MS. This approach, therefore, is normally applicable only in small micro- and pico-cells where coverage is not an issue.
- *Inter-frequency Handovers (IF-HOs)*:
  - ○ an operator can apply a second frequency in interference problematic areas and, for example, provide the possibility of *Inter-frequency Handover (IF-HO)* to the less interfered frequency, such as for services with especially high QoS requirements (service-based IF-HO).
- *Inter-system Handovers (IS-HOs)*:
  - ○ if there is a neighbouring system, such as a 2G GSM system, available, *Inter-system Handovers (IS-HOs)* can be performed in such areas where there are dead zones. Of course, this requires the affected mobiles to be multi-system-capable.
- Guard bands:
  - ○ the standards allow the centre frequencies of the different channels to be adjusted in a 200 kHz raster. If at least one operator has two or more frequencies available, he can decide to select a *different carrier spacing* than the nominal 5 MHz between at least the two frequencies closest to the other operator. By applying this method a guard band to the frequency band of the neighbouring operator's frequency band can be generated, which can help to alleviate ACI problems (see Figure 3.36).

3.7 CELL DEPLOYMENT STRATEGIES

As outlined in previous sections of this chapter, there are certain issues to be taken into account when deploying multiple frequencies and layers in a network. This section discusses tasks that need to be done for deploying operational 3G RANs, and the strategies for utilising frequencies if Hierarchical Cell Structures (HCSs) are used. Section 3.7.1 highlights the general process of rolling out a network and presents some differences between the strategies for an operator who is starting anew in an area (a 'greenfield' operator) and those for an operator already running a network from a previous generation – e.g., a GSM system.

In 3G systems, due to the variety of services and different capacities of different layers, an operator needs to have a clear vision about the deployment strategy of the different cell layers. Micro-cells, for example, may be necessary to accommodate hotspots with increased capacity requirements, but they may also be needed to support higher bit rates. On the other hand, before taking micro-cells into use, it is very likely a continuous macro-cell layer is already present. The simplest way to operate different cell layers is to have them on different frequency carriers, but this is not the only possible scenario that can be deployed. Section 3.7.2 discusses various issues of hierarchical cell structures and studies the influence of different scenarios on network performance when two frequencies are available with and without reusing them in different layers of a hierarchical WCDMA network.

3.7.1 Rollout

Rollout refers to a process that has to be completed in order to generate an operational network. 3G systems set high requirements for rollout, since effective and rapid rollout confers competitive advantage. The performance of UMTS must be at least as high as that provided by current systems. The services provided by UMTS must outperform the services provided currently. Therefore, effective means for integrating WCDMA networks are required. The prompt startup of network operation and aggressive introduction of new services could be the differentiating factor between two operators. Rollout and network development-related issues to be considered early in the business planning phase include:

- Services to be provided.
- Evolution of the services and the network (see Section 3.7.2):
  o usage of carriers;
  o usage of HCSs.
- Provisioning of indoor coverage and services.

Services to be provided will have a direct impact on site density. Furthermore, capacity limited networks should be planned with multiple carriers or with HCSs. The extension plan for the network must be considered so that new services will be introduced as seamlessly as possible, preferably without major changes in the network configuration.

For *3G greenfield operators*, rollout includes radio network planning, site acquisition, packet core network planning, construction work, commissioning and integration of the network elements. In the radio network planning phase, dimensioning and site acquisition information is combined with the traffic and service quality requirements, see Section 3.1. The site density and configuration for the network regions are determined, and the work schedule and instructions for civil engineering and equipment installation are generated for site deployment. Transmission requirements are estimated and transmission planning is performed. A part of the radio network and transmission planning is the preparation of parameter files and templates for the ATM layer and RNC. After installation, the sites can be commissioned with the parameter files and commissioning reports. The result of the installation and commissioning visit to each site is an operating network element with a connection to an Operation and Maintenance Centre (OMC), enabling effective networkwide mass operations for the radio network part. Now the radio parameters can be downloaded and the sites made operational. When the network plans are ready and the rollout project tasks are in place, effective tools are required to implement the plans quickly, cost-effectively and without manual errors. A successful rollout ends when the network is ready and operational, and the monitoring of the network performance can start. As the configuration of installed network elements is based on predicted network behaviour and default parameter settings, it is usually necessary not only to monitor and report on the actual performance but also to react fast with appropriate performance optimisation. Immediate feedback from network performance is also needed for providing information for network development tasks and plans. More about measurement-based configuration tuning in a Network Management System (NMS) can be found in Sections 7.3.3 and 9.3.

For *GSM operators* the radio network planning phase is slightly different. Information (location, height, possible antenna directions, etc.) on sites that will be reused is needed as input for 3G planning. Data from the existing GSM network can be effectively utilised. GSM traffic density information can be used to indicate traffic hotspots also in WCDMA. IS-HO (see Section 4.3.4) gives an opportunity to start WCDMA implementation selectively. GSM can be used to extend coverage, introducing WCDMA initially only in areas where service requirements so demand, such as city centres or high-density business areas. Furthermore, experience of the cell coverage areas and interference situation in GSM can be used in planning WCDMA. In the case of co-siting, GSM interference problems will indicate possible interference and thus also capacity problems in the WCDMA network. More about co-siting can be found in Chapter 5.

3.7.2 Hierarchical Cell Structures in WCDMA Networks

In most UMTS frequency allocations done until today, operators have been allocated two or more Frequency Division Duplex (FDD) carriers. Spectrum allocation affects the operators' WCDMA deployment scenarios, and the use of HCSs. In principle, an allocation of one pair of FDD carriers allows the operation of only a single network layer. Two paired carriers can cater for a two-layer structure, such as a macro-cell layer together with a micro-cell or pico-cell layer. A full hierarchical cell structure, with each layer operating on its own carrier, can be built with three carriers. With four or more carriers additional capacity and flexibility in network design is achieved. In hotspot areas highly loaded cells can be given extra capacity by adding another carrier to the cell, which would be more effective than increasing the BS transmission power (see Section 6.4). In order to support HCSs and handovers between carriers, IF-HOs are required. An example for a typical evolution path in a 3G network is presented in Figure 3.37.

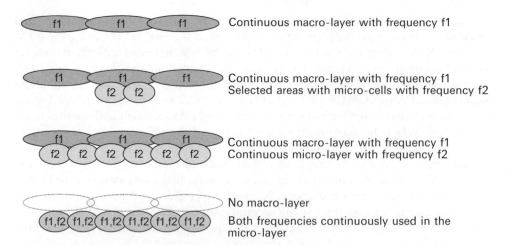

Figure 3.37   Example of WCDMA network evolution.

To start operating the network, one would begin typically with just one carrier in a macro-cellular layer to provide continuous coverage. This applies especially to a greenfield operator who cannot rely on an existing GSM network for coverage or cannot partner with an existing GSM operator. Later, a second carrier (and possibly more) is deployed to enhance capacity. This second carrier can then be added to the macro-cellular layer to create high-capacity sites or can be used to build a micro-layer. In its first phase, the micro-layer typically is deployed only in traffic hotspots or where high bit rates are needed. In a further stage of the network, then, both layers are giving continuous coverage in a specific area, and if further capacity is needed more carriers must be deployed. Again, the simplest way is to use a third frequency and assign it either to the macro- or the micro-layer. In cases, however, when an operator will be limited to two frequencies only, he will need to start to reuse a carrier that has already been used in another network layer.

The required capacity and coverage tradeoff needs to be carefully considered. Within an HCS in a WCDMA network, the micro-layer provides a very high capacity in a limited area, whereas the macro-layer can offer full coverage but with reduced throughput only. Typical air interface capacities are about 1 Mbps/carrier/cell for a three-sectored macro-BS and 1.5 Mbps/carrier/cell for a micro-BS.

Another important issue is whether to support mobiles moving at high speeds. If there is no such need, the easiest way to continue is to sacrifice the macro-layer and assign both frequencies to the micro-layer. This alternative might, however, result in increased investment, which has to be evaluated carefully. If high-mobility users have to be supported in a micro-cell layer there would be too many handovers between the cells, and it is therefore always beneficial to have an 'umbrella' macro-layer for those users. Then the strategy to further increase capacity is to reuse one frequency in the other layer. How the performance of a hierarchical WCDMA network is affected by reusing carrier frequencies in different layers is the subject of the study in Section 3.7.2.2.

3.7.2.1 Network Operation Aspects

There are certain aspects of WCDMA characteristics whose consideration is crucial from the point of view of frequency reuse. They are next recapped briefly.

Interference

It is impossible to consider any part of a WCDMA system in isolation. Changes to a part of the system may induce changes over a large area. For example, GSM systems are basically 'hard blocking' so that their ultimate capacity is limited by the number of channel elements, and blocking occurs when all frequencies and timeslots are fully occupied. WCDMA systems differ fundamentally from GSM in that the same spectrum is shared between all users.

In WCDMA, capacity limits can be reached before all channel elements in all cells are in use. The limit is reached when the QoS of the network degrades to a minimum acceptable level that depends on the interference levels in the system. In WCDMA, capacity and coverage can be limited by uplink and downlink interference. In the uplink the interference comes from other MSs, and in the downlink from adjacent BSs. Although the number of sources for downlink interference is low, the interfering power is relatively high. As the interference level experienced by a mobile depends on the path loss to all BSs, users suffer from different interference depending on their locations in the network [44]. Downlink interference levels are high even if cell load is low, because the BSs always have to transmit the downlink common channels.

In the downlink, the total transmitted power is shared between the users. In the uplink, there is a maximum interference level tolerable at the BS receiver. Each user contributes to the interference, and it is shared between the users in the cell. If the performance of some links can be improved, the power levels required in both the uplink and downlink and the interference generated are immediately reduced. With a common shared power resource, this results in reduced interference levels for all users, which can be further utilised as increased capacity and coverage, or improved link quality.

Soft Handover

In soft handover a mobile is located in an area where cell coverage of two (or more) sectors overlap, and the communication between the mobile and the BS occur via two (or more) air interface channels. Soft handover improves the performance of hard handover through the exploitation of macro-scopic diversity. In the downlink, signals received from different BS sectors are combined in the MS by MRC in RAKE processing. In the uplink, the signal from the MS is received at different sectors, which are combined in softer handover by using MRC and in soft handover by using selection combining.

Soft handover improves WCDMA system performance by minimising the received and transmitted powers when mobiles are close to cell boundaries. Typically, soft handover probability is targeted to keep below 20–30%, since excessive soft handover connections decrease downlink capacity. Each soft handover connection increases downlink interference to the network, and, if the increased interference exceeds the diversity gain, soft handover cannot provide any benefits for system performance [45].

Pilot Power Adjustment

P-CPICH power allocation is another important task in WCDMA network design. Optimum pilot powers ensure coverage with minimum interference to neighbouring cells. Excessive pilot powers will easily take too large a proportion of the total available BS transmission power so that not enough power is left for traffic channels. The cell can collect distant users whose mobile transmission power is not enough to connect to the BS, and which would more optimally be served by some other BS. On the other hand, pilot powers that are too low may not provide wide enough pilot coverage and result in very small dominance areas. Moreover, if link power limits are defined with respect to pilot power levels, low pilot powers also restrict link powers. Typically, approximately 5% of the total BS power is allocated to the pilot channel, and roughly the same amount to other common channels.

If the same carrier frequency is used at different network layers, a cell with higher pilot power easily blocks a nearby cell with lower pilot power. As the micro- and macro-BS total and pilot powers normally differ from each other by several decibels, micro- and macro-layer users on the same carrier may cause undesirable performance degradation – e.g., due to the near–far effect. This is shown in Figure 3.38.

The mobile is connected to the macro-cell BS with higher pilot power, although the path loss to the nearby micro-BS would be smaller. Higher transmission power is needed to compensate for the higher path loss, and the mobile introduces additional interference to the adjacent micro-cell (and the whole network). Therefore, it is not trivial to assign a carrier used in one network layer to another one.

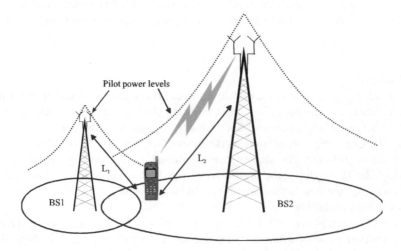

Figure 3.38   A mobile is connected to the macro-cell base station (BS2) with higher received pilot power, and increases uplink interference at the micro-cell base station (BS1).

If a mobile is in a location where numerous pilots are received with relatively equal signal strengths, it may happen that none of the pilot signals is dominant enough to enable the mobile to start a call. Pilot coverage from neighbouring BSs must overlap in cell border areas to accommodate handovers. However, each cell that has significant power in the soft handover area will increase I_0 and decrease E_c/I_0 (energy of the pilot signal divided by the total channel power). The total power in the channel includes the measured pilot signal, pilots from other BSs, traffic and other channels from BSs and thermal noise. Receiving too many pilot signals can degrade both capacity and quality, and can be prevented to a large extend by proper radio network planning. It is essential to create a network plan, where cells have clear dominance areas. Some pilot optimisation aspects are discussed in more detail in [44] and [46].

3.7.2.2   Case Study – Frequency Reuse in Micro- and Macro-cellular Networks

The basic issue in WCDMA network design is to determine the cell and carrier configurations at which the interference and QoS targets for given traffic are met. Since capacity and coverage in WCDMA networks are coupled with each other through interference, it is very difficult to consider any parts of a WCDMA network separately. Simple analytical studies, such as in [2] and [47], can be used to estimate asymptotic limits or study regular and simplified network scenarios, but have limited applicability in actual radio network planning. Such analyses often assume unrealistic assumptions or simplifications on traffic distributions, propagation models or cell patterns that do not reflect the complexity of real planning. In reality, uplink and downlink interference levels are affected by each mobile with different propagation conditions, service in use, E_b/N_0 requirements, soft handover situation, etc. Moreover, micro- and macro-cells and traffic distributions in urban areas do not readily form a regular pattern that could easily be handled by analytical means. Some factors, such as soft handover probabilities, are treated as input parameters for analytical approaches, although in reality one more often would expect them as outputs of the planning process, or factors to be optimised. Therefore, simulation methods often appear more appealing for network planning purposes. In the following section we have also adopted a simulation approach.

Network Configurations
In this study a static radio network simulator supporting IF-HOs between carriers was used to examine frequency reuse between micro- and macro-cellular layers in a WCDMA network. It is described in [16] and [48] and in more detail in its specifications at the weblink (*www.wiley.com/go/laiho*). The two-layered network this study is based on is shown in Figure 3.39. It consists of a micro-layer of 31 cells (sectors), and a macro-layer of 18 cells (six three-sectored sites). Micro- and macro-layers have been planned independently of each other without considering the other layer's site locations. The average micro- and macro-cell densities are ~ 8 and ~ 5 cells/km^2, respectively. Both network layers provide (nearly) continuous coverage, so that micro-cells are not used only as capacity fill-ins under the macro-cellular network, which could initially be planned to provide coverage (in GSM, for example). Hence the network can be considered to be in rather a mature deployment phase (see Figure

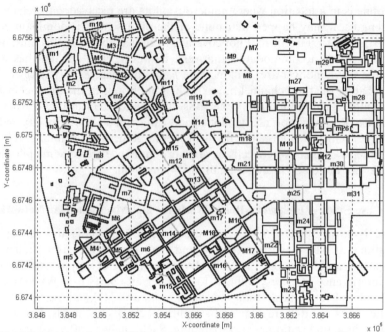

Figure 3.39   Micro (m) and macro (M) base station locations. Mobiles are uniformly distributed in the polygon in all cases.

3.37). In case of continuous micro-layer coverage, macro-cells serve more like umbrella cells, which are best suited for high-speed users to minimise the number of handovers. Alternatively they can fill micro-cell coverage holes or collect users who, for load reasons, for example, cannot be served by micro-cells. Propagation data for link loss tables for both micro- and macro-cells were calculated using a 3D ray-tracing model [49]. Initially all users were connected to (micro-) carrier 1. If not heard, mobiles were allowed to make an inter-frequency handover to carrier 2, if its pilot (P-CPICH) E_c/I_0 was sufficient. In this study no code limitation (hard blocking) was considered.

Initially, the micro- and macro-cellular networks were examined individually, and thereafter load balancing through inter-frequency handovers was allowed in a two-layer HCS. A key finding characterising the network operation in both cases was that micro-cells were first limited in the downlink by the total available BS power, whereas in macro-cells the uplink loading was the first factor restricting the performance. Figure 3.40 shows the reference scenario and the frequency reuse scenarios studied.

Performance of WCDMA networks where a macro-carrier is reused in micro-cells, and a micro-carrier is reused on macro-cells, are compared with that of a network with an HCS, where micro- and macro-layers operate on their own carriers. Tables 3.37 and 3.38 show parameters used for mobiles and BSs in basic micro- and macro-cell networks, respectively. When reusing a carrier on a different network layer, the pilot and total BS transmission powers were modified. Cases and modifications are listed separately in Table 3.39.

Reference scenario

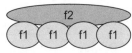

Continuous macro-layer with frequency f2

Continuous micro-layer with frequency f1

Reuse of micro-frequency in macro-layer

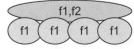

Continuous macro-layer with frequencies f1 and f2

Continuous micro-layer with frequency f1

Reuse of macro-frequency in micro-layer

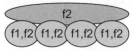

Continuous macro-layer with frequency f2

Continuous micro-layer with frequencies f1 and f2

Reuse of macro-frequency in selected micro-cells

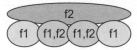

Continuous macro-layer with frequency f2

Continuous micro-layer with frequency f1
Selected micro-cells reusing macro-frequency f2

Figure 3.40   Hierarchical cell structures used in the study.

Table 3.37   Parameters for mobiles (common in all simulations).

Maximum transmission power	21 dBm
Minimum transmission power	−50 dBm
Service in use (uplink, downlink)	12.2 kbps
Mobile speed	3 km/h
Antenna	Omni, 1.5 dBi
Noise figure	8 dB
Adjacent channel leakage power ratio	33 dBc
Adjacent channel selectivity	33 dBc

Simulation Results

Figure 3.41 shows the service probabilities and Figure 3.42 the reasons for not serving mobiles. As such, the figures are not fully transparent regarding the feasibility of different carrier reuse cases.

User distributions among the carriers, other-to-own-cell-interference levels in the uplink, soft handover overheads, uplink loading and downlink transmission powers are shown in Figures 3.43–3.47 to give insight into the network operation in each case. They are presented as functions of users served per sector, 'sector' referring to both micro- and macro-sectors. To avoid confusion, the number of sectors remains unchanged throughout this study – i.e., 49 – although the number of cells changes when carriers are added to sectors. In Table 3.40 some cell-specific results are also given.

Table 3.38 Parameters used in the simulations for micro- and macro-cells.

	Micro-cell	Macro-cell
BS maximum transmission power	37 dBm	43 dBm
Pilot channel (CPICH) power	24 dBm	30 dBm
Power for other common channels	24 dBm	30 dBm
Cable losses	2 dB	2 dB
Multi-path channel profile	Two equal taps	Two equal taps
E_b/N_0 (uplink/downlink)	4/8.4 dB	4/8.4 dB
Soft handover addition window	3 dB	3 dB
Uplink loading limit	80 %	60 %
BS antennas	60°, 12 dBi	65°, 16 dBi
Average antenna height	10 m	32 m
Noise figure	5 dB	5 dB
Maximum single-link power below pilot power	5.5 dB	5.5 dB
Adjacent channel leakage power ratio	45 dBc	45 dBc
Adjacent channel selectivity	65 dBc	65 dBc

Table 3.39 Base station parameters in frequency reuse cases. Note that in reality the total base station power is pooled rather than split between the carriers.

	Micro-cell	Macro-cell
(a) Micro f1, macro f2 (reference case)		
Maximum transmission power	37 dBm	43 dBm
CPICH power	24 dBm	30 dBm
Power for other common channels	24 dBm	30 dBm
(b) Micro f1, macro f1 + f2		
Maximum transmission power	37 dBm	40 dBm (per carrier)
CPICH power	24 dBm	27 dBm (per carrier)
Power for other common channels	24 dBm	27 dBm (per carrier)
(c) Micro f1 + f2, macro f2		
Maximum transmission power	34 dBm (per carrier)	43 dBm
CPICH power	24 dBm (per carrier)	30 dBm
Power for other common channels	21 dBm (per carrier)	30 dBm
(d) Micro f1 + f2 on selected cells, macro f2		
Maximum transmission power	37 dBm (per carrier)	43 dBm
CPICH power	24 dBm (per carrier)	30 dBm
Power for other common channels	24 dBm (per carrier)	30 dBm

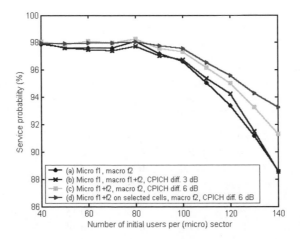

Figure 3.41 Service probabilities for the different base station and network configurations given in Table 3.39.

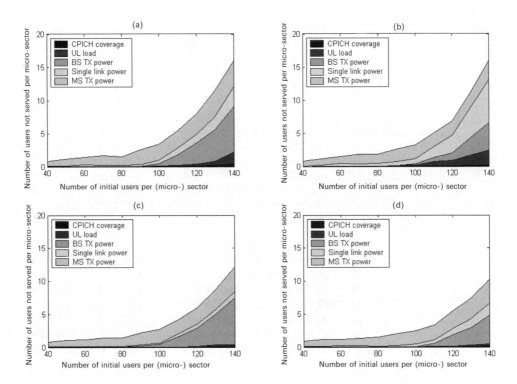

Figure 3.42 Reasons for not serving mobiles. (a), (b), (c) and (d) refer to the base station and network configurations given in Table 3.39.

Table 3.40 Served users, other-to-own-cell interferences, i, soft handover overheads, uplink loadings and base station transmission powers for the cases listed in Table 3.39. The initial network loading was 90 mobiles per micro-sector. In cells with two carriers f1 is on the left and f2 is on the right hand.

	BS ID	Downlink users	i	Soft handover overhead	Uplink loading	BS transmit power
(a) Micro f1, macro f2						
Micro	8	131.3	0.20	0.13	0.75	34.4
	9	100.0	0.17	0.06	0.60	32.2
	12	129.0	0.19	0.11	0.76	34.1
	13	57.3	0.54	0.20	0.47	31.7
	19	120.7	0.41	0.20	0.79	35.1
	20	65.7	0.20	0.15	0.47	32.6
	21	119.7	0.23	0.14	0.73	35.3
Mean (all cells)		*72.1*	*0.27*	*0.13*	*0.49*	*32.1*
Macro	12	23.3	0.43	0.42	0.19	35.3
	13	26.7	1.10	0.39	0.28	35.5
	14	44.7	0.83	0.28	0.41	36.9
	15	11.7	0.82	0.24	0.12	34.3
Mean (all cells)		*26.4*	*1.09*	*0.39*	*0.24*	*35.8*
(b) Micro f1, macro f1 + f2						
Micro	8	100.3	0.20	0.14	0.65	34.1
	9	83.0	0.19	0.07	0.53	32.2
	12	37.3	0.63	0.62	0.48	32.9
	13	23.3	0.97	0.26	0.29	30.1
	19	59.0	0.70	0.51	0.64	34.8
	20	34.7	0.43	0.23	0.32	30.6
	21	78.3	0.28	0.21	0.59	34.3
Mean (all cells)		*43.5*	*0.46*	*0.24*	*0.38*	*31.4*
Macro	12	73.0/25.0	1.05/0.41	0.57/0.31	0.59/0.19	38.0/33.2
	13	75.0/23.0	1.25/1.03	0.44/0.33	0.59/0.21	38.1/32.8
	14	46.7/10.7	1.21/2.22	0.78/1.30	0.51/0.24	37.6/32.5
	15	58.0/10.3	0.98/1.03	0.60/0.95	0.51/0.13	37.6/32.2
Mean (all cells)		*56.4/19.1*	*1.03/1.31*	*0.53/0.71*	*0.47/0.18*	*37.0/32.7*
(c) Micro f1 + f2, macro f2						
Micro	8	110.3/20.5	0.22/0.35	0.13/0.13	0.67/0.16	32.3/27.8
	9	98.2/0.7	0.17/2.88	0.06/0.33	0.58/0.03	31.6/26.1
	12	119.3/3.7	0.20/2.41	0.13/1.27	0.72/0.13	32.3/26.4
	13	61.3/0.2	0.41/1.00	0.19/0.0	0.46/0.02	30.8/25.9
	19	92.2/33.3	0.60/1.63	0.39/0.43	0.69/0.64	32.3/30.6
	20	65.8/1.0	0.15/6.75	0.10/0.50	0.43/0.08	31.0/26.2
	21	87.5/25.2	0.28/0.78	0.17/0.32	0.59/0.31	32.4/28.6
Mean (all cells)		*67.9/4.1*	*0.25/1.86*	*0.13/0.18*	*0.45/0.08*	*30.7/26.5*
Macro	12	32.8	0.51	0.25	0.23	36.0
	13	33.3	1.02	0.50	0.30	36.5
	14	33.8	1.54	0.78	0.39	37.8

	BS ID	Downlink users	*i*	Soft handover overhead	Uplink loading	BS transmit power
(c) Micro f1 + f2, macro f2 (cont.)						
Macro *(cont.)*	15	16.3	0.93	0.55	0.16	35.2
Mean (all cells)		*27.5*	*1.41*	*0.58*	*0.24*	*36.6*
(d) Micro f1 + f2 on selected cells, macro f2						
Micro	8	134.2/2.3	0.16/2.54	0.11/0.0	0.76/0.04	34.3/27.6
	9	96.8	0.19	0.06	0.59	32.2
	12	133.3	0.19	0.12	0.78	34.0
	13	61.0	0.44	0.20	0.49	31.6
	19	111.7/32.8	0.48/1.36	0.27/0.49	0.78/0.60	35.0/30.5
	20	64.0	0.20	0.12	0.45	32.3
	21	119.3/10.7	0.23/1.02	0.12/0.43	0.73/0.15	35.2/28.4
Mean (all cells)		*72.5/8.9*	*0.27/3.18*	*0.13/0.26*	*0.49/0.17*	*32.1/28.3*
Macro	12	26.5	0.33	0.23	0.19	35.8
	13	24.5	1.03	0.44	0.23	35.6
	14	31.0	1.28	0.55	0.33	36.4
	15	8,5	1.27	0.76	0.12	34.6
Mean (all cells)		*23.7*	*1.13*	*0.48*	*0.21*	*35.7*

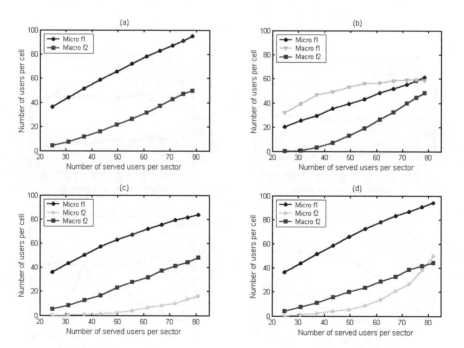

Figure 3.43   Served users on different cell layers. (a), (b), (c) and (d) refer to the base station configurations given in Table 3.39.

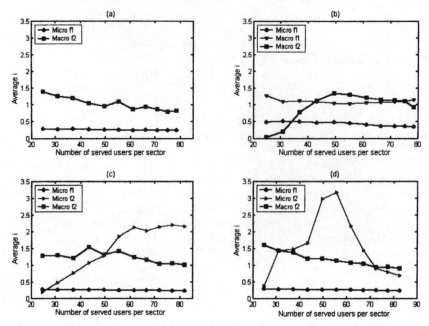

Figure 3.44 Other-to-own-cell-interference, *i*. (a), (b), (c) and (d) refer to the base station configurations given in Table 3.39.

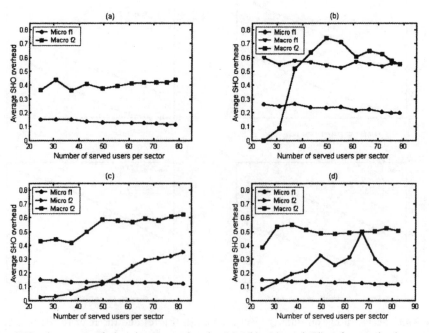

Figure 3.45 Average soft handover overheads. (a), (b), (c) and (d) refer to the base station configurations given in Table 3.39.

Micro f1, Macro f1 + f2

In the reference HCS case 95% service probability can be provided for up to 110 users per micro-sector. Reusing a micro-carrier on all macro-cells does not bring any improvements in network performance. Microcell users are mostly in the line of sight to the base station, and interference levels are lower than in macro-cells due to better physical cell isolation (Figure 3.44). Consequently, throughputs in micro-cells are greater than in macro-cells. When a micro-carrier is reused on macro-cells, the better capacity of micro-cells is sacrificed for a worse solution, since an additional carrier on macro-cells cannot compensate for the capacity reductions at micro-cells. Users can initially be connected also to macro-carrier 1 with higher pilot power, as depicted in Figure 3.38. Micro-carrier 1 now serves only ~50–65% (depending on the total network loading) of the users it serves in the reference HCS case (Figure 3.43). However, its uplink loading and downlink transmission power levels have not decreased in the same proportion as the number of users, as seen in Figures 3.46 and 3.47 and Table 3.40. Mobiles connected to macro-cells are required to transmit with higher power levels, as typically the minimum link losses to micro-cells are 53–55 dB, and to macro-cells over 70 dB. Higher transmission powers increase the uplink interference experienced at micro-cell BSs. In addition, micro-layer users are seen as additional uplink interference in macro-cells operating on carrier 1. As soon as carrier 1 macro-cells become fully loaded in the uplink (Figure 3.46), macro-cells operating on carrier 2 and micro-cells start to collect more users. Also, in the downlink the maximum transmission power is reached in many macro-cells (Figure 3.47), which pushes users to other carriers and layers. These can be seen in Figure 3.42 as major reasons for outages.

Another factor deteriorating network performance, if a micro-carrier is reused in macro-cells, is increased soft handover overhead (Figure 3.45). In this context soft handover overhead for a cell is defined as *Number_of_secondary_users/Number_of_primary_users*. Secondary users are those mobiles in soft handover to the sector, to which the sector is not the best server. Primary users refer to the users to whom the sector is the best server. Although in soft handover a mobile is using less transmit power and therefore introducing less uplink interference, the call is handled by two BSs. If used excessively, soft handovers decrease the overall capacity, as in the downlink the interfering power is increased. If a micro-carrier is reused in macro-cells, the soft handover overhead in macro-cells can be as high as 50–70%, and also micro-layer soft handover is increased to ~20–30%. The proportions have nearly doubled in comparison with the reference case. Also the single-link power in the downlink has become an important factor resulting in outages. The macro-cell pilot power is decreased by 3 dB when a micro-carrier is reused in macro-cells. As maximum link powers in our examples are defined with respect to the pilot power level, consequently the link powers are affected. In principle more power can be granted for a connection in the downlink than in the uplink, because BS transmission power is much higher than mobile output power. Therefore, services requiring high bit rates can be given better coverage in the downlink, if desired. By setting the link power limits properly, the uplink and downlink coverage areas can be balanced.

Micro f1 + f2, Macro f2

In our example micro-cells as such are inherently limited by the available downlink power earlier than by uplink loading. Also in the reference HCS case BS transmission

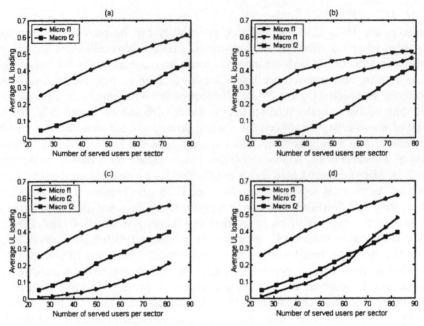

Figure 3.46 Average uplink loading. (a), (b), (c) and (d) refer to the base station configurations given in Table 3.39.

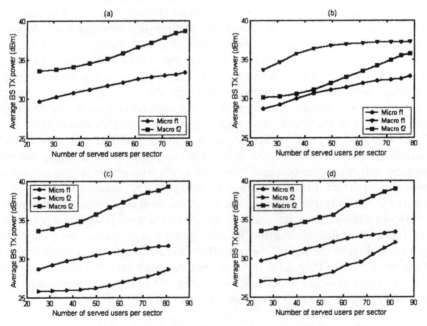

Figure 3.47 Average base station transmission powers. (a), (b), (c) and (d) refer to the base station configurations given in Table 3.39.

power is clearly the most limiting factor reducing the network performance, as seen in Figure 3.42. Therefore, sharing the total available BS transmission power between the micro-cell carriers (Table 3.39(c)) increased the number of outages due to downlink transmission power. On the other hand, the number of outages due to link power limitations and uplink loading was decreased. At a 95% service probability operating point, the network where a macro-carrier is reused on all micro-cells can serve ~10% more users than in the reference HCS case. The macro-layer is barely affected by the underlying micro-cells operating on the same carrier. Macro-cells serve more or less the same number of users, as in the reference HCS case (Figure 3.43), and the soft handover overhead and other-to-own-cell-interference levels have not drastically changed (see Figures 3.44 and 3.45).

Despite the users being handed over at a lower network loading to the other carrier, the reused macro-carrier on micro-cells is able to collect them, and the total number of users served on the micro-layer has slightly increased at a higher network loading compared with the reference HCS case, as can be seen in Figure 3.43 and Table 3.40.

The situation is somewhat different from the scenario where a micro-carrier is reused in macro-cells: now micro-cells operating on carrier 1, which is the network layer to which the users are initially connected, are not interfered by macro-cells. Therefore, their capacity can be fully exploited before inter-frequency handovers are made to carrier 2 micro- and macro-cells, which are more subject to increased interference levels due to carrier reuse. Micro-layer capacity could be best exploited if the total BS power is not shared between the carriers, but instead the power is doubled when another carrier is added. However, in terms of cost this is a much more expensive solution, as another power amplifier is required, which also affects the BS size and installation efforts. The problems of reusing a macro-carrier in a micro-cell are very like those involved in embedding micro-cells in hotspot areas under a macro-cell layer, as studied in [50] and [51].

Micro f1 + f2 on Selected Cells, Macro f2
Micro-cells do not benefit from the other carrier reused from macro-cells, if they still have unused capacity on their own carrier. In that case those cells with the reused carrier that do not collect traffic only generate additional downlink interference, since they still have to transmit the power for CPICH and other common channels. Figure 3.48 shows the users served, BS powers and uplink loadings at relatively high network loadings, when a macro-carrier is reused on all micro-cells. In this case the micro-cell power was doubled when the other carrier was added. We see that only about 20% of the micro-cells benefit from the other carrier. This is because inter-frequency handover is made to carrier 2 only when carrier 1 cannot serve users, and many cells shown in Figure 3.48 have unused capacity.

Micro-cells 8, 19, 21, 25, 28 and 31 (circulated in Figure 3.50) are selected to have a second carrier reused from the macro-layer. The BS configurations were given in Table 3.39. The selected cells can also collect traffic on carrier 2, and are not located in the immediate vicinity of a nearby macro-sector. Sufficient distance (attenuation) from a macro-sector is required to ensure that, if inter-frequency handover is made, the reused macro-carrier is likely to collect those mobiles. The selected cells can also establish reasonable dominance areas on carrier 2. Achieving clear cell dominance areas is

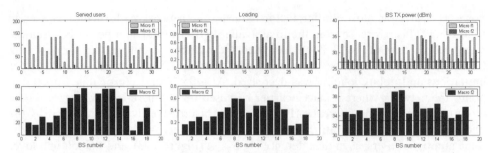

Figure 3.48 Served users, uplink loadings and downlink transmission powers for a network where a macro-carrier is reused in all micro-cells. The black line across the right-hand figure denotes the minimum power level required for pilot and other common channels. The network was initially loaded with 4340 users – i.e., 140 users per micro-sector. The service probability in this case was 93%.

essential for efficient WCDMA network operation. Dominance areas based on highest received pilot powers are shown in Figure 3.49 for micro-cells, and in Figure 3.50 for carrier 2 cells, when a macro-carrier is reused on micro-cells.

In the plain micro-cell network cells are easily distinguishable from each other, as the surrounding buildings provide good physical isolation. When a macro-carrier is reused on micro-cells, the cells are more fragmented in shape, have unequal sizes and may overlap each other. The effects of pilot power differences between micro- and macro-cells are distinguishable, for example, in micro-cells 7 and 12. They suffer from the nearby macro-sector, showing diminished dominance areas compared with the network

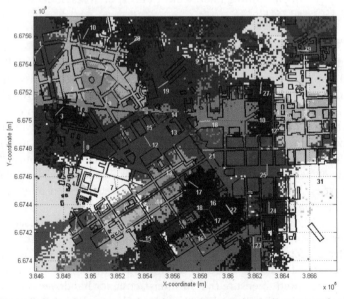

Figure 3.49 Downlink dominance areas based on highest received pilot power for f1 micro-cells in a hierarchical cell structure network.

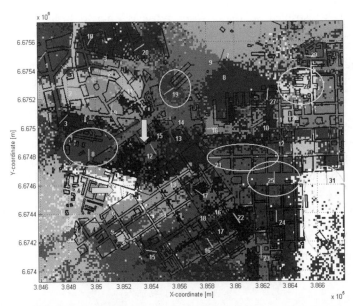

Figure 3.50   Downlink f2 dominance areas for f2 micro- and macro-cells, when macro-carrier f2 is reused on the micro-layer. Circulated cells are examples of micro-cells, which can establish clear dominance areas and collect traffic under the overlying macro-layer. The pilot power difference between micro- and macro-cells is 6 dB. Dominance areas of, for example, micro-cells 7 and 12 (shown with arrows) have shrunk considerably compared with Figure 3.49.

with HCS. It is noticeable that reusing a macro-carrier only on selected micro-cells hardly increases the soft handover and other-to-own-cell-interference levels in macro-cells compared with the HCS network, as seen in Figures 3.44 and 3.45. Micro-cells on carrier 2 are affected more than macro-cells, but they still have rather moderate soft handover overheads and other-to-own-cell-interference levels at higher network loading – i.e., over 50–60 served users per sector – when they start to collect traffic. If a micro-cell collects a lot of traffic, but is located very close to a macro-cell, macro-carrier reuse is not worthwhile. Micro-cell 12 is an example of such a situation. Although it collects plenty of users, the nearby macro-sectors 13 and 15 nearly fully disable carrier 2 on micro-cell 12, see Table 3.40. If BS transmit power limits micro-cell performance, decreasing the micro-cell pilot power could be a better solution instead. If the pilot power coverage is still adequate, the benefit is twofold. With smaller pilot power the micro-cell dominance area is reduced, and it attracts less users. Moreover, by decreasing pilot power, more power is left for traffic channels.

3.7.2.3   Concluding Remarks

This study indicates that the HCS of a WCDMA network can be divided in certain micro- and macro-cell scenarios, which could occur during WCDMA network deployment phases. The most important thing to avoid is excessive increase of interference levels in both the uplink and downlink, and it is essential to keep soft handover areas restricted so that carrier reuse is able to bring some performance improvements. Therefore, the selection of cells for carrier reuse may be limited in order to avoid increased interference levels in the network. In the reuse scenarios studied, the best capacity enhancements were achieved by reusing a macro-layer carrier in such highly loaded micro-cells, which were far enough away from macro-cells operating on the same carrier. By this, the interference levels could be kept reasonable, and the micro-cell load could be balanced between the two carriers. The results are based on snapshots of a static network simulator, and therefore do not take into account true mobility effects. In reality, problems may be encountered when, for example, a high-speed mobile connected to a macro-cell moves towards a micro-cell operating on the same carrier frequency. As the handover is a relatively slow process, the mobile may get deep into the micro-cell before the handover is completed. The fast and accurate operation of power control is crucial so that micro-cell users can quickly adjust their transmit powers to cope with the high-power interferer. In the radio network planning phase, operators can design different network layers to accommodate different types of traffic; handovers between cells can be directed, restricted or even completely prohibited.

References

[1] Sampath A. et al., Erlang capacity of a power controlled integrated voice and data CDMA system. *Proc. VTC 1997 Conf., May 1997, Phoenix, Arizona*, pp. 1557–1561.

[2] Sipilä, K., Honkasalo, Z., Laiho-Steffens, J. and Wacker, A., Estimation of capacity and required transmission power of WCDMA downlink based on a downlink pole equation. *Proc. VTC 2000 Spring Conf., Tokyo, Japan, May 2000*, pp. 1002–1005.

[3] Sipilä, K., Laiho-Steffens, J., Jäsberg, M. and Wacker A., Modelling the impact of the fast power control on the WCDMA uplink. *Proc. VTC 1999 Spring Conf., Houston, Texas, May 1999*, pp. 1266–1270.

[4] Sipilä, K., Jäsberg, M., Laiho-Steffens, J. and Wacker, A., Soft handover gains in fast power controlled WCDMA uplink. *Proc. VTC 1999 Spring Conf., Houston, Texas, May 1999*, pp. 1594–1598.

[5] Viterbi, A. J., *CDMA Principles of Spread Spectrum Communication*, Addison-Wesley, 1995, p. 198.

[6] Jakes, W.C., *Microwave Mobile Communications*, John Wiley & Sons, 1974, p. 126.

[7] Reunanen, J., Multiple server location probability in GSM/DCS1800 cellular systems, Master of Science in Engineering thesis, Helsinki University of Technology, 1997.

[8] Hata, M., Empirical formula for propagation loss in land mobile radio services. *IEEE Transactions on Vehicular Technology*, **VT-29**(3), August 1980, pp. 317–325.

[9] 3GPP, Technical Specification 25.141, Base Station (BS) conformance testing (FDD), v.5.9.0, September 2004.

[10] Pedersen, K.I., Toskala, A. and Mogensen, P.E., Mobility management and capacity analysis for high speed downlink packet access in WCDMA. *Proc. VTC 2004 Fall Conf., Los Angeles, California, September 2004*, pp. 3388–3392.

[11] 3GPP, Technical Specification 25.213, Spreading and Modulation (FDD), v.5.5.0, December 2003.

[12] 3GPP, TSG RAN Technical Report 25.858, High-speed Downlink Packet Access: Physical Layer Aspects 3G, v.5.0.0, March 2002.

[13] Holma, H. and Toskala A. (eds), *WCDMA for UMTS: Radio Access for Third Generation Mobile Communications* (3rd edn). John Wiley & Sons, 2004, chapter 11.

[14] Kolding, T.E., Pedersen, K.I., Wigard, J., Frederiksen, F. and Mogensen, P.E., High-speed downlink packet access: WCDMA evolution. *IEEE Vehicular Technology Society (VTS) News*, **50**(1), February 2003, pp. 4–10.

[15] *http://www.3gpp.org/*

[16] Wacker, A., Laiho-Steffens, J., Sipilä, K. and Jäsberg, M., Static simulator for studying WCDMA radio network planning issues. *Proc. VTC 1999 Spring Conf., Houston, Texas, May 1999*, pp. 2436–2440.

[17] Dehghan, S., Lister, D., Owen, R. and Jones, P., WCDMA capacity and planning issues. *IEE Electronics & Communication Engineering Journal*, June 2000, 101–118.

[18] Labedz, G. and Love, R., A new time-based outage criterion for the forward and reverse links of DS-CDMA cellular systems. *Proc. VTC 1998 Conf., Ottawa, Canada, May 1998*, pp. 2182–2186.

[19] Okumura, Y., Ohmori, E., Kawano, T. and Fukuda, K., Field strength and its variability in the VHF and UHF land mobile service. *Review Electronic Communication Laboratories*, **16**(9/10), 1968, pp. 825–873.

[20] Urban transmission loss models for mobile radio in the 900 and 1800 MHz bands. COST 231, TD(91)73, September 1991.

[21] Walfisch, J. and Bertoni, H.L., A theoretical model of UHF propagation in urban environment. *IEEE Transactions on Antennas and Propagation*, **AP-36**(12), December 1988, pp. 1788–1796.

[22] Ikegami, F., Yoshida, S., Takeuchi T. and Umehira, M., Propagation factors controlling mean field strength on urban streets. *IEEE Transactions on Antennas and Propagation*, **AP-32**(8), August 1984, pp. 822–829.

[23] Berg, J.E., Path loss and fading models for micro-cells at 900 MHz. COST 231, TD(92)95, Helsinki, September 1992.

[24] Wei, Q.X., Gong, K. and Gao, B.X., Ray-tracing models and techniques for coverage prediction in urban environments. *Proc. APMC 1999 Conf., Singapore, November/ December 1999*, pp. 614–617.

[25] Erricolo, D. and Uslenghi, L.E. Two-dimensional simulator for propagation in urban environments. *IEEE Transactions on Vehicular Technology*, **AP-50**(4), July 2001, pp. 1158–1168.

[26] Daniele, P., Frullone, M., Heiska, K., Riva, G. and Carciofi, C., Investigation of adaptive 3D micro-cellular prediction tools starting from real measurements. *Proc. ICUPC 1996 Conf., Cambridge, Massachusetts, September/October 1996*, pp. 468–472.

[27] Ji, Z., Li, B.H., Wang, H.X., Chen, H.Y. and Sarkar, T.K. Efficient ray-tracing methods for propagation prediction for indoor wireless communications. *IEEE Antennas and Propagation Magazine*, **AP-43**(2), April 2001, pp. 41–49.

[28] Rajala, J., Sipilä, K. and Heiska, K., Predicting in-building coverage for micro-cells and small macro-cells. *Proc. VTC 1999 Conf., Houston, Texas, May 1999*, pp. 180–184.

[29] Guidelines for Evaluation of Radio Transmission Technologies for IMT-2000, Recommendation ITU-R M. 1225, 1997.

[30] Laiho-Steffens, J., Sipilä, K. and Wacker, A., Verification of 3G radio network dimensioning rules with static network simulations. *Proc. VTC 2000 Spring Conf., Tokyo, Japan, May 2000*, pp. 478–482.

[31] Hämäläinen, S., Slanina, P., Hartman, M., Lappeteläinen, A., Holma, H. and Salonaho, O., A novel interface between link and system level simulations. *Proc. ACTS Summit 1997, Aalborg, Denmark, October 1997*, pp. 509–604.

[32] Hämäläinen, S., Holma, H. and Sipilä K., Advanced WCDMA radio network simulator. *Proc. PIMRC 1999, Aalborg, Denmark, October 1997*, pp. 509–604.

[33] Brady, P.T., A model for generating on–off speech patterns in two-way conversation. *Bell Systems Technical Journal*, **48**(9), September 1969, pp. 2445–2472.

[34] Wacker, A., Sipilä, K. and Kuurne, A., Automated and remote optimisation of antenna sub-system based on radio network performance. *Proc. WPMC 2002 Symposium, Waikiki, Hawaii, October 2002*, pp. 752–756.

[35] Wacker, A., Laiho-Steffens, J., Sipilä, K. and Heiska K., The impact of the base station sectorisation on WCDMA radio network performance. *Proc. VTC 1999 Fall Conf., Amsterdam, Netherlands, September 1999*, pp. 2611–2615.

[36] Laiho-Steffens, J., Wacker, A. and Aikio, P., The impact of the radio network planning and site configuration on the WCDMA network capacity and Quality of Service. *Proc. VTC 2000 Spring Conf., Tokyo, Japan, May 2000*, pp. 1006–1010.

[37] Toskala, A., Holma, H. and Muszynski, P., ETSI WCDMA for UMTS. *Proc. of ISSSTA 1998, South Africa, September 1998*, pp. 616–620.

[38] 3GPP, Technical Specification 25.101, UE Radio Transmission and Reception (FDD), v.5.13.0, December 2004.

[39] Spurious Emissions, Recommendation ITU-R SM.329-7.

[40] 3GPP, Technical Specification 25.104, UTRA (BS) FDD; Radio Transmission and Reception, v.5.9.0, September 2004.

[41] Hämäläinen, S., Lilja, H. and Hämäläinen, A., WCDMA adjacent channel interference requirements. *Proc. VTC 1999 Fall Conf., Amsterdam, Netherlands, September 1999*, pp. 2591–2595.

[42] Hämäläinen, S., Holma, H. and Toskala, A., Capacity evaluation of a cellular CDMA uplink with multiuser detection. *Proc. ISSSTA 1996, Mainz, Germany, September 1996*, pp. 339–343.

[43] Wacker, A. and Laiho, J., Mutual impact of two operators' WCDMA radio networks on coverage, capacity and QoS in a macro-cellular environment. *Proc. VTC 2001 Fall Conf., Atlantic City, New Jersey, October 2001*, pp. 2077–2081.

[44] Akl, R.G., Hedge, M.V., Naraghi-Pour, M. and Min P.S., Cell placement in a CDMA network. *IEEE Wireless Communications and Networking Conf., 1999*, Vol. 2, pp. 903–907.

[45] Gorricho, J.-L. and Paradells, J., Evaluation of the soft handover benefits on CDMA systems. *5th IEEE International Conf. on Universal Personal Communications, 1996*, Vol. 1, pp. 305–309.

[46] Love, R.T., Beshir, K.A., Schaeffer, D. and Nikides R.S., A pilot optimisation technique for CDMA cellular systems. *Proc. VTC 1999 Fall Conf., Amsterdam, Netherlands, September 1999*, pp. 2238–2242.

[47] De Hoz, A. and Cordier, C., WCDMA downlink performance analysis. *Proc. VTC 1999 Fall Conf., Amsterdam, Netherlands, September 1999*, pp. 968–972.

[48] Holma, H. and Toskala A. (eds), *WCDMA for UMTS: Radio Access for Third Generation Mobile Communications* (3rd edn). John Wiley & Sons, 2004, chapter 8.

[49] Wölfle, G., Gschwendtner, B.E. and Landstorfer, F.M., Intelligent ray tracing: A new approach for field strength prediction in micro-cells. *Proc. VTC 1997 Conf., Phoenix, Arizona, May 1997*, pp. 790–794.

[50] Wu, J.-S., Chung, J.-K. and Yang Y.-C., Performance study for a micro-cell hot spot embedded in CDMA macro-cell systems. *IEEE Transactions on Vehicular Technology*, **VT-48**(1), January 1999, pp. 47–59.

[51] Shapira, J., Micro-cell engineering in CDMA cellular networks. *IEEE Transactions on Vehicular Technology*, **VT-43**(4), November 1994, pp. 817–825.

4

Radio Resource Utilisation

Achim Wacker, Jaana Laiho, Tomáš Novosad, David Soldani, Chris Johnson, Tero Kola and Ted Buot

4.1 Introduction to Radio Resource Management

The expression Radio Resource Utilisation (RRU) covers all functionality for handling the air interface resources of a Radio Access Network (RAN). These functions together are responsible for supplying optimum coverage, offering the maximum planned capacity, guaranteeing the required Quality of Service (QoS) and ensuring efficient use of physical and transport resources. The Radio Resource Management (RRM) function consists of Power Control (PC), Handover Control (HC), congestion control – typically subdivided into Admission Control (AC), Load Control (LC) and Packet data Scheduling (PS) – and the Resource Manager (RM). PC is presented in Section 4.2. Since many users in a Wideband Code Division Multiple Access (WCDMA) network are operating on the same frequency, interference is a crucial issue and PC is responsible for adjusting the transmit powers in uplink and downlink to the minimum level required to ensure the demanded QoS. HC, presented in Section 4.3, takes care that a connected user is handed over from one cell to another as he is moving through the coverage area of a mobile network. AC and LC, together with PS, ensure that the network stays within the planned condition. AC, handled in Section 4.4.2, lets users set up or reconfigure Radio Access Bearers (RABs) only if these would not overload the system and if the necessary resources are available. LC takes care that a system temporarily going into overload is returned into a non-overloaded situation, and is the subject of Section 4.4.4. The main task of PS is to handle all NRT traffic – i.e., allocate optimum bit rates and schedule transmission of the packet data, keeping the required QoS in terms of throughput and delays. The functionality of PS is treated in Section 4.4.3. The RM is in charge of controlling the physical and logical radio resources under one Radio Network Controller (RNC). Its main tasks are to coordinate the usage of the available hardware resources and to manage the code tree. The RM is presented in Section 4.5. Most of the Release '99 RRM functionality is located in the RNC. Only part of the PC, LC and the RM can also be found in the Node B, and in the User Equipment (UE) only PC functionality is included. For Release 5 RRM involving High-speed Downlink Packet Access (HSDPA) transmission, big parts

Radio Network Planning and Optimisation for UMTS Second Edition
Edited by J. Laiho, A. Wacker and T. Novosad © 2006 John Wiley & Sons, Ltd

of the RRM (AC, LC and PS) have been located in Node B for speed and performance reasons. Enhancements of RRM due to HSDPA are introduced in Section 4.6.

4.2 Power Control

In mobile communication systems, such as 3G systems, which are based on the Code Division Multiple Access (CDMA) technique where all the users can share a common frequency, interference control is a crucial issue. This is especially important for the uplink direction, since one Mobile Station (MS or UE) located close to the base station and transmitting with excessive power can easily overshout others that are at the cell edge (the near–far effect) or even block the whole cell. In the downlink the system capacity is directly determined by the required code power for each connection. Therefore, it is essential to keep the transmission powers at a minimum level while ensuring adequate signal quality at the receiving end. In WCDMA a group of functions is introduced for this purpose. They are summarised as Power Control (PC) and consist of open-loop PC, inner-loop PC (also called fast closed-loop PC) and outer-loop PC in both the uplink and downlink and slow PC applied to some downlink common channels.

The following sections describe the radio network planning aspects of PC functionality and the parameters involved. The open-loop PC presented in Section 4.2.1 is responsible for setting the initial uplink and downlink transmission powers when a UE is accessing the network. The subject of Section 4.2.2 is the slow PC applied on some downlink common channels. The inner-loop PC adjusting the transmission powers dynamically on a 1.5 kHz basis and the differences between two algorithms specified in [1] are described in Section 4.2.3. The outer-loop PC estimates the received quality and adjusts the target Signal-to-Interference Ratio (SIR) for the fast closed-loop PC so that the required quality is provided. Outer-loop PC is the topic of Section 4.2.4. The PC procedure during Universal Mobile Telecommunications System (UMTS) Compressed Mode (CM) is presented in Section 4.2.5. A comprehensive and detailed analysis of the performance of the presented algorithms can be found in [2]. The PC for HSDPA is presented in Section 4.6.1.

4.2.1 Open-loop Power Control

Since the uplink and downlink frequencies of WCDMA are within the same frequency band, a significant correlation exists between the average path loss of the two links. This makes it possible for each UE, before accessing the network, and for each BS, when the radio link is set up, to estimate the initial transmit powers needed in the uplink and downlink based on the path loss calculations in the downlink direction. This function is denoted as open-loop PC.

4.2.1.1 Uplink Open-loop Power Control

The uplink open-loop PC function is located both in the UE and in the UTRAN and requires some control parameters being broadcast in the cell and the Received Signal

Code Power (RSCP) being measured by the UE on the active Primary Common Pilot Channel (P-CPICH). Based on the calculation of open-loop PC, the UE sets the initial powers for the first Physical Random Access Channel (PRACH) preamble and for the uplink Dedicated Physical Control Channel (DPCCH) before starting the inner-loop PC. During the random access procedure (Section 2.4.4), the UE sets the power of the first transmitted preamble as:

$$Preamble\_Initial\_power = CPICH\_Tx\_power - CPICH\_RSCP + UL\_interference$$

$$+ UL\_required\_CI \qquad (4.1)$$

where the transmit power of the P-CPICH, *CPICH_Tx_power*, and the required C/I (Carrier-to-Interference ratio) in the uplink, *UL_required_CI* (in 3GPP denoted as a constant value), are set during radio network planning. The *UL_interference* (in 3GPP called the 'receiver total wideband power'), measured at the BS, is broadcast on the Broadcast Channel (BCH). The same procedure is followed by the UE when setting the power level of the first Physical Common Packet Channel (PCPCH) access preamble.

When establishing the first DPCCH, the UE starts the uplink inner-loop PC at a power level according to Equation (4.2):

$$DPCCH\_Initial\_power = DPCCH\_Power\_offset - CPICH\_RSCP \qquad (4.2)$$

where the received signal code power of the P-CPICH, *CPICH_RSCP*, is measured by the UE; and the *DPCCH_Power_offset* is calculated by the AC in the RNC and provided to the UE – e.g., at RRC connection setup or during a radio bearer or physical channel reconfiguration – as:

$$DPCCH\_Power\_offset = CPICH\_Tx\_power + UL\_interference + SIR_{DPCCH}$$

$$- 10 \cdot \log_{10}(SF_{DPCCH}) \qquad (4.3)$$

where the SIR_{DPCCH} is the initial target SIR produced by the AC for that particular connection; and the SF_{DPCCH} is the spreading factor of the corresponding DPCCH.

4.2.1.2 Downlink Open-loop Power Control

In the downlink, the open-loop PC is used to set the initial power of the downlink channels based on the downlink measurement reports from the UE. This function is located in both UMTS Terrestrial Radio Access Network (UTRAN) and UE. A possible algorithm for calculating the initial power value of the DPDCH when the first radio link is set up is:

$$P_{Tx}^{Initial} = \frac{R \cdot (E_b/N_0)_{DL}}{W} \cdot \left(\frac{CPICH\_Tx\_power}{(E_c/N_0)_{CPICH}} - \alpha \cdot PtxTotal \right) \qquad (4.4)$$

where R is the user bit rate; $(E_b/N_0)_{DL}$ is the downlink planned E_b/N_0 value set during radio network planning for that particular bearer service; W is the chip rate; $(E_c/N_0)_{CPICH}$ is reported by the UE; α is the downlink orthogonality factor; and *PtxTotal* is the carrier power measured at the Node B and reported to the RNC. The algorithm for calculating the initial radio link power can be simplified when a diversity handover branch is set up or when a radio link is modified. For branch addition it is sufficient to scale the transmitted code power of the existing radio link(s) by the difference between the P-CPICH powers of the cell(s) with the existing link(s) and the P-CPICH power of the cell with the additional branch. For a modified radio bearer the scaling is done with the new user bit rate and new downlink E_b/N_0.

4.2.2 Power Control on Downlink Common Channels

From the downlink common channels, the only one that may be power-controlled is the S-CCPCH when it carries the Forward Access Channel (FACH). The transmit powers of the other downlink common channels are determined by the network. In general the ratio of the transmit powers between different downlink common channels is not specified in Third Generation Partnership Project (3GPP) and may even change dynamically. A feasible implementation solution is described in the following paragraphs and the typical values for the common channel power levels are reported in Table 4.1.

- The transmission powers of the P-CPICH, Primary and Secondary Synchronisation Channels (P-SCH and S-SCH) and Primary Common Control Physical Channel (P-CCPCH) are cell-specific configuration parameters, which are set during radio network planning to define the actual size of the cell. As a rule of thumb, for the P-CPICH a transmission power of about 5–10% of the total cell transmit power capability is allocated. Then the transmission powers of the other common channels mentioned are specified relative to this power as offsets in decibels.

- The transmission powers of the Acquisition Indicator Channel (AICH) and Paging Indicator Channel (PICH) are common TrCH configuration parameters set in the planning phase relative to the P-CPICH transmit power in order to have the same coverage over the cell. These parameters are sent to the Node B whenever the corresponding common TrCH is set up or reconfigured. The PICH transmission power depends on the number of Paging Indicators (PIs) per frame, Np, and is always provided to the Node B together with the power of the S-CCPCH carrying the Paging Channel (PCH). The more PIs per frame, the less often the bits are repeated per frame and the higher is the PICH power needed relative to the P-CPICH. Typical values for the offsets are $-10\,dB$ ($Np = 18$ or $Np = 36$), $-8\,dB$ ($Np = 72$) and $-5\,dB$ ($Np = 144$).

Table 4.1 Typical downlink common channel power levels.

Downlink common channel	Typical power level	Note
P-CPICH	30–33 dBm	5–10% of the maximum cell transmit power (20 W)
P-SCH and S-SCH	−3 dB	Relative to P-CPICH power
P-CCPCH	−5 dB	Relative to P-CPICH power
PICH	−8 dB	Relative to P-CPICH power and for $Np = 72$
AICH	−8 dB	Power of one Acquisition Indicator (AI) compared with the P-CPICH power
S-CCPCH	−5 dB	Relative to P-CPICH and for SF = 256 (15 ksps)

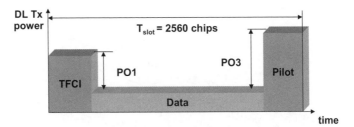

Figure 4.1 Downlink transmit power on secondary common control physical channel.
PO3 and PO1 denote the power offsets of Pilot and TFCI symbols, respectively.

- According to the 3GPP standard, when the S-CCPCH – i.e., FACH and PCH – is set up or reconfigured, Node B is given the S-CCPCH Power Offset information (PO1 for TFCI bits, PO3 for the Pilot bits: see Figure 4.1), FACH parameters, maximum FACH power, PCH parameters and PCH power. On the FACH slow PC can be applied based on the serving P-CPICH E_c/N_0 reported by the UE and other relevant control parameters with a proprietary algorithm in order to improve the cell downlink capacity. In this case, the indicated value is the negative offset relative to the maximum power configured for the S-CCPCH where the corresponding FACH is mapped.
- If we assume the same power for all TrCHs multiplexed onto the same physical channel, typical power values for the S-CCPCH relative to the P-CPICH are +1 dB for SF = 64 (60 ksps), −1 dB for SF = 128 (30 ksps) and −5 dB for SF = 256 (15 ksps). For the power offsets of the Pilot/TFCI symbols relative to the power of the S-CCPCH data field, typical values can be 2 dB for 15 ksps, 3 dB for 30 ksps and 4 dB for 60 ksps. During the communication the power offsets may vary according to the bit rate used.
- The Physical Downlink Shared Channel (PDSCH) supports inner-loop PC based on the Transmit Power Control (TPC) commands sent by the UE. PDSCH power can be further adjusted during the connection using the Iub interface user plane (U-plane) Control Channel Frame Protocol (CCH-FP) [3]. In this case a proprietary slow-PC algorithm is used to indicate the PDSCH transmission power level to Node B as an offset relative to the power of the TFCI bits of the downlink DPCCH transmitted to the same UE as the Downlink Shared Channel (DSCH).

4.2.3 Inner-loop Power Control

Inner-loop power control (fast closed-loop PC) relies on the feedback information at layer 1 from the opposite end of the radio link. This allows the UE/Node B to adjust its transmitted power based on the received SIR level at the Node B/UE for compensating the fading of the radio channel. The inner-loop PC function in UMTS is used for the Dedicated Channels (DCHs) in both the uplink and downlink directions and for the Common Packet Channel (CPCH) in the uplink. In WCDMA fast PC with a frequency of 1.5 kHz is supported. A graphical representation of the procedures further described in the following sections is shown in Figure 4.2.

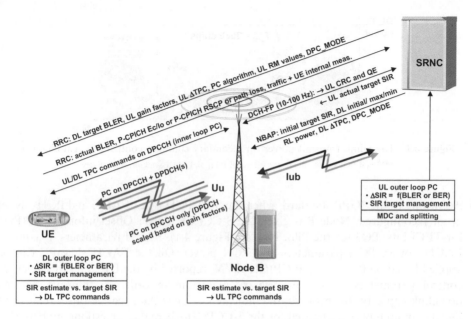

Figure 4.2   Uplink/downlink inner- and outer-loop power control.

4.2.3.1   Uplink Fast Closed-loop Power Control

The uplink inner-loop PC is used to set the power of the uplink DPCH and PCPCH as follows. The Node B receives the target SIR from the uplink outer-loop PC (see Section 4.2.4.1) located in the RNC and compares it with the estimated SIR on the pilot symbol of the uplink DPCCH once every slot. If the received SIR is greater than the target, the Node B transmits a TPC command 'down' to the UE on the downlink DPCCH. If the received SIR is below the target, a TPC command 'up' is sent.

In [4] it is shown that the optimum PC step size varies depending on the UE speed. For a given quality target, the best uplink PC step size is the one that results in the lowest target SIR. With an update rate of 1.5 kHz, a PC step size of 1 dB can effectively track a typical Rayleigh fading channel up to a Doppler frequency of about 55 Hz (30 km/h). At higher speeds, up to about 80 km/h, a PC step size of 2 dB gives better results. It is also shown in [5] and [4] that, for UE speeds greater than 80 km/h, the inner-loop PC can no longer follow the fades and just introduces noise into uplink transmission. This adverse effect on uplink performance could be reduced if a PC step size smaller than 1 dB was employed. Also, for UE speeds lower than about 3 km/h, where the fading rate of the channel is very small, a smaller PC step size is more beneficial.

In [1] two alternative algorithms, denoted as 'Algorithms 1 and 2', were specified for the UE for interpreting the TPC commands sent by the Node B. Algorithm 1 is used when the UE speed is sufficiently low to compensate for the fading of the channel. The PC step size is set during radio network planning and can have a value of either 1 or 2 dB. Algorithm 2 was designed for emulating the effect of using a PC step size smaller than 1 dB or to switch off the PC by sending an alternating series of PC commands.

Algorithm 2 can be used to compensate for the slow fading trend of the propagation channel rather than rapid fluctuations. It performs better than Algorithm 1 when the UE moves faster than 80 km/h or slower than 3 km/h. In Algorithm 2 the PC step size is fixed at 1 dB. The UE does not change its transmission power until it has received five consecutive TPC commands from slots aligned to the frame boundaries with no overlap between each other. At the end of the fifth slot, based on hard decisions, the UE adjusts its transmission power according to the following rules:

- If all five estimated TPC commands are 'down' the transmit power is reduced by 1 dB.
- If all five estimated TPC commands are 'up' the transmit power is increased by 1 dB.
- Otherwise the transmit power is not changed.

Before starting the uplink Dedicated Physical Data Channel (DPDCH) a UE may be instructed by the network to use an uplink DPCCH PC preamble while receiving the downlink DPCCH. The length of the uplink DPCCH PC preamble is a parameter that can be set during radio network planning and ranges from 0 up to 7 frames [6]. During the uplink DPCCH PC preamble the TPC commands sent by the Node B are always followed according to Algorithm 1, enabling the uplink transmit power to converge more rapidly and to stabilise faster before the beginning of normal PC.

In UMTS the diversity schemes are applied only to DCHs. After Layer 1 (L1) synchronisation is achieved, one or more cells participating in the diversity HO start uplink inner-loop PC. Each cell the UE is connected to measures the uplink SIR and compares the estimated value with the target SIR level. A TPC command is generated by each cell participating in the diversity HO and sent to the UE. The original idea was that the UE increases its transmit signal level only if all cells the UE is communicating with request an increase in the uplink power level. If only one cell requests the power to be decreased, the UE lowers its transmit signal level by the defined PC step size.

When the UE is in *softer* HO it is signalled by the serving cell that the TPC commands it receives are coming from the same radio link set and have to be combined into one TPC command according to Algorithm 1 or 2. If the UE is additionally in *soft* HO (SHO), this TPC command is then further combined by the UE with the TPC commands coming from other radio links of different radio link sets. The procedures for combining TPC commands from radio links of different radio link sets are defined in [1] and here illustrated in Figure 4.3.

If the TPC commands are from different cells and Algorithm 1 is used, the UE derives a combined TPC command based on soft decisions taken on each TPC command received from the different cells, and accordingly changes its transmit power by the pre-defined PC step size. If Algorithm 2 is employed, the UE makes a hard decision on the value of each TPC command from radio links of different radio link sets for five consecutive and aligned slots as described in the non-Diversity Handover (non-DHO) case. This results in as many hard decisions (temporary TPC commands) as there are cells participating in SHO. Finally, the UE derives a combined TPC command for the fifth slot according to the following rule:

- If any of the temporary TPC command is 'down', the transmit power is decreased by 1 dB.

- If all of the temporary TPC commands are 'up' ('1') or 'no change' ('0') and the mean of the estimated temporary TPC commands is bigger than 0.5, the transmit power is increased by 1 dB.
- Otherwise the transmit power is not changed.

During the communication, after applying DPCCH power adjustments and gain factors, the UE is not allowed to exceed the maximum transmit power value set during radio network planning. Furthermore, when transmitting on a DPCH the standard [7] requires that the UE must be able to reduce its transmit power at least to $-50\,$dBm. Assuming a maximum transmit power of the UE of 21 dBm (250 mW) this results in a dynamic range of about 70 dB.

4.2.3.2 Downlink Fast Closed-loop Power Control

The downlink inner-loop power control sets the power of the downlink DPCH. As illustrated in Figure 4.2, the UE receives from higher layers the Block Error Rate (BLER) target set by the RNC for downlink outer-loop PC together with other control parameters, and estimates the downlink SIR from the pilot symbols of the downlink DPCH. This SIR estimation is compared with a target SIR. If the estimate is greater than the target, the UE transmits the TPC command 'down' to the Node B, otherwise the UE transmits the TPC command 'up'. If DPC_MODE = 0 the UE sends a unique TPC command in each slot, otherwise the UE repeats the same TPC command over three slots. The TPC commands are sent on the uplink DPCCH and simultaneously control the power of a DPCCH and its corresponding DPDCHs in the downlink by the same amount. The relative power difference between the DPCCH

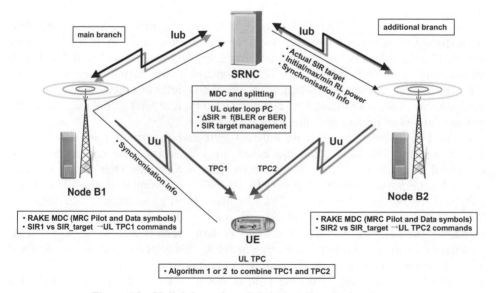

Figure 4.3   Uplink inner-loop PC during diversity handover.

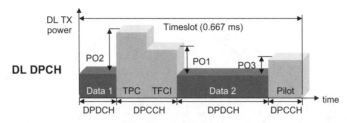

Figure 4.4 Power offsets for improving the downlink signalling quality.

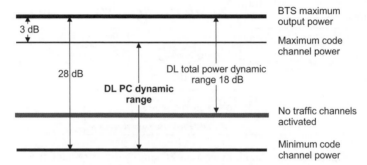

Figure 4.5 Downlink power control dynamic range.

and the TFCI, TPC and pilot fields of the downlink DPCCH are determined by PO1, PO2 and PO3, respectively, as illustrated in Figure 4.4. In SHO the UE transmit power is reduced if the PC signalling quality is improved by setting a higher power for the DPCCH than for the DPDCH in the downlink.

Depending on the DPC_MODE the Node B estimates the transmitted TPC command TPC_{est} over one or three slots to be '0' or '1', and updates the radio link power every slot or every third slot accordingly [1]. The downlink PC step size is a parameter in radio network planning that can take the values 0.5, 1, 1.5 or 2 dB. The only mandatory step size to be supported by the Node B is 1 dB; the other step sizes are optional [1]. However, if a UE is in SHO, all cells the UE is connected to must use the same PC step size to avoid power drifting (see Section 4.2.3.4). In case of congestion the Node B may be commanded by the RNC to ignore the TPC 'up' commands from the UE. The downlink PC dynamic range required by the standard is illustrated in Figure 4.5.

The inner-loop PC on downlink DPCHs during *softer* HO acts in the same way as in the single-link case. Only one DPCCH is transmitted in the uplink, and the signalling and data part of the received signals from different antennas are combined at symbol level in the Node B. On the downlink the Node B simultaneously controls the power of the radio link set and splits the data stream received from the DCH-FP to all cells participating in softer HO.

In *soft* HO the inner-loop PC on the downlink DPCHs has two major issues that differ from the single-link case: drifting in the cell powers, and reliable detection of uplink TPC commands. The inner-loop PC function during SHO is illustrated in Figure 4.6, while power drifting is described in more detail in Section 4.2.3.4.

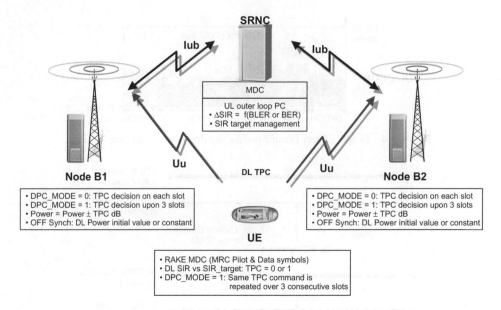

Figure 4.6   Downlink inner-loop PC during diversity handover.

4.2.3.3   Downlink Limited Power Increase

The downlink limited power increase algorithm defined in [1] operates in the Node B together with the downlink fast closed-loop PC. The algorithm takes into account only the downlink power increase due to TPC commands and is employed when the service being carried by the DPCH is a Non-Real Time (NRT) packet data transmission. Its purpose is to limit variations in the power levels that would degrade the quality of Real Time (RT) connections.

After estimating the kth TPC command TPC_{est} (see Section 4.2.3.2), the Node B adjusts the current downlink DPCH power according to Equation (4.5) as:

$$P(k) = P(k-1) + P_{TPC}(k) \tag{4.5}$$

where $P_{TPC}(k)$ [dB] is the kth power adjustment due to the downlink inner-loop PC. For the first $(DL\_Power\_Averaging\_Window\_Size - 1)$ power adjustments after the activation of this method, $P_{TPC}(k)$ is calculated with Equation (4.6):

$$P_{TPC}(k) = \begin{cases} +\Delta_{TPC} & \text{if } TPC_{est}(k) = 1 \\ -\Delta_{TPC} & \text{if } TPC_{est}(k) = 0 \end{cases} \tag{4.6}$$

After that, the following Equation (4.7) is used:

$$P_{TPC}(k) = \begin{cases} +\Delta_{TPC} & \text{if } TPC_{est}(k) = 1 \text{ and } \Delta_{sum} + \Delta_{TPC} < Power\_Rise\_Limit \\ 0 & \text{if } TPC_{est}(k) = 1 \text{ and } \Delta_{sum} + \Delta_{TPC} \geq Power\_Rise\_Limit \\ -\Delta_{TPC} & \text{otherwise} \end{cases} \tag{4.7}$$

where

$$\Delta_{sum}(k) = \sum_{i=k-DL\_Power\_Averaging\_Window\_Size+1}^{k-1} P_{TPC}(i) \tag{4.8}$$

Power_Rise_Limit and *DL_Power_Averaging_Window_Size* are parameters set during radio network planning. *Power_Rise_Limit* can range between 0 and 10 dB with a step size of 1 dB, the value of *DL_Power_Averaging_Window_Size* is allowed to be between 1 and 60 inner-loop PC adjustments [8]. Typical values are 3 dB and 30 inner-loop PC adjustments, respectively.

4.2.3.4 Downlink Power Drifting

When the UE is in SHO only a single TPC command is sent on the uplink to all cells participating in the SHO. Since it would introduce too much delay to combine all the received TPC commands in the RNC and send one combined command back, each cell detects the TPC command independently. Due to, for example, signalling errors in the air interface it is possible that each cell interprets this TPC command differently. As a consequence one cell lowers its transmission power to that UE while the other cell might increase it, and therefore the downlink powers drift apart.

Since power drifting degrades the performance in the downlink, methods to combat this effect are needed. The easiest possibility is just to limit the PC dynamic range, but this has the negative effect that also the gain from SHO is reduced. A more complex but more effective method is proposed in [8]. The transmission code power levels of the connections from the cells in SHO are forwarded to the RNC after they have been averaged. A typical time used for averaging is the *Measurement Reporting Period* set during radio network planning equal to, for example, 500 ms, corresponding to 750 TPC commands. From these measurements the RNC derives a reference power value, P_{ref}, which is sent to the cells. This is then used to periodically calculate a small power adjustment, P_{bal}, towards the reference value, which balances the link powers of the SHO connections and thereby reduces power drifting according to the following equation:

$$P(k) = P(k-1) + P_{TPC}(k) + P_{bal}(k) \tag{4.9}$$

4.2.4 Outer-loop Power Control

The aim of the outer-loop PC algorithm is to maintain the quality of the communication at the level defined by the quality requirements of the bearer service in question by producing an adequate target SIR for the inner-loop PC. This operation is done for each DCH belonging to the same Radio Resource Control (RRC) connection. As shown in [2], the SIR target needs to be adjusted when the UE speed or the multipath propagation environment change. The higher the variation in the received power, the higher the SIR target needs to be. If a fixed SIR target was selected the resulting quality of the communication would be too low or too high, causing an unnecessary power rise in most situations.

The outer-loop PC is shown in Figure 4.2 for a single link and in Figure 4.3 during SHO. In the latter, uplink quality is observed after macro-diversity selection combining in the RNC and the target SIR is provided to all cells participating in the SHO. The frequency of the outer-loop PC ranges typically from 10 to 100 Hz. As illustrated in Figure 4.3, during SHO the Iub and Iur DCH data streams coming from the different cells are combined in the Serving RNC (SRNC) to one data stream in the uplink. In the downlink the DCH data stream is split between the Node Bs. This combining and splitting in the RNC is performed by the Macro-Diversity Combiner (MDC). The MDC in the RNC is based on information received from the Node B in FP frames – namely, transport block-specific CRC results and possibly estimated quality information. Reliable SHO is based on the CFN information included in the Iub/Iur data streams. At the UE Maximal Ratio Combining (MRC) of the received signals is performed at symbol level (data and pilots). Only one DPCH is transmitted in the uplink direction. In the following sections a few aspects of the outer-loop PC applied to both the uplink and downlink directions are described.

4.2.4.1   Uplink Outer-loop Power Control

The uplink outer-loop PC operates within the SRNC and is responsible for setting a target SIR in the Node B for each individual uplink inner-loop PC. This target SIR is then updated on an individual basis for each UE according to the estimated uplink quality – e.g., BLER or BER – for that particular RRC connection. An algorithm, based on a Cyclic Redundancy Check (CRC) of the data stream as the quality measure, is presented in [2]. If the CRC is OK, the SIR target is lowered by a certain amount, otherwise it is increased. Typical values for the step size in the SIR adjustment are in the order from 0.1 to 1.0 dB.

A logical architecture of the uplink outer-loop PC function valid for the multi-bearer service case is illustrated in Figure 4.7.

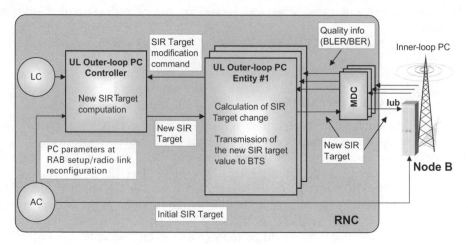

Figure 4.7   Uplink outer-loop power control algorithm – logical architecture.

There is one outer-loop power controller for each RRC connection and one uplink outer-loop PC entity for each DCH within the same RRC connection. The uplink outer-loop PC entities calculate the required change in the target SIR according to the corresponding uplink quality estimates. Under the same RRC connection, one of the uplink outer-loop PC entities – e.g., the signalling link (DCCH) – is selected to transmit the new common target SIR to the Node B. The new target SIR is computed by the uplink outer-loop power controller based on the changes in the target SIR received from the PC entities and other configuration parameters (e.g., initial/maximum/minimum target SIR) provided by the AC at RAB setup or radio link reconfiguration. The DCH-FP is used for the interactive communication between the RNC and the Node Bs.

Each uplink outer-loop PC entity receives uplink quality information from the MDC unit, where the incoming data from different SHO branches is combined (selection and combining procedure). Depending on the type of radio bearer, the PC entity receives either a BLER estimate, computed in the MDC according to the CRC bits of the selected frame, and/or a BER estimate, calculated in the Node B. If the CRC is not OK, the MDC may select the best of the BER estimates. At any Transmission Time Interval (TTI), one or several PC entities may contribute to the new SIR target computation when, for example, the difference between the BLER/BER estimate and the target BLER/BER multiplied by a step size is greater than 0.1 dB (accuracy of the Iub interface target SIR information element).

4.2.4.2 Downlink Outer-loop Power Control

The downlink outer-loop PC function is implemented in the UE. It adjusts the target SIR value for the downlink inner-loop PC using a proprietary algorithm that provides the same measured quality (BLER) as the quality target set by the RNC. If the CPCH is employed in the communication, the quality target signalled by the RNC is the downlink DPCCH BER, otherwise a BLER target value is provided to the UE. In addition, when the TrCH BLER is employed as the target in the communication, the downlink outer-loop PC in the UE ensures that the quality requirement is maintained for each TrCH to which a BLER target is assigned. On the other hand, if the BER of the downlink DPCCH is sent as a quality target, the control loop in the UE will keep the quality requirement for each CPCH to which a downlink DPCCH BER target is assigned.

The value of the downlink outer-loop PC quality target in the UE is controlled by the AC in the RNC, which determines the value of the downlink BLER target for each DCH mapped on the Coded Composite Transport Channel (CCTrCH). The downlink BLER target for each TrCH is then received by the UE on RRC messages as specified in [6].

4.2.5 Power Control during Compressed Mode

The UMTS PC function in downlink or/and uplink Compressed Mode (CM) provides the mechanisms to speed up the convergence of the SIR close to the target SIR after each transmission gap as quickly as possible.

The mandatory physical layer procedures for the transmit PC in CM can be found in [1], where the differences between the algorithms specified for the uplink and downlink directions are also pointed out.

Neither the uplink nor the downlink outer-loop PC functions are affected by CM implementation. However, the SIR target needs to be adjusted in the Node B and in the UE during the compressed frames compared with Normal Mode; that is:

$$SIR_{cm\_target} = SIR_{target} + \Delta SIR_{pilot} + \Delta SIR_{coding} \qquad (4.10)$$

where ΔSIR_{pilot} and ΔSIR_{coding} take respectively into account the reduction of pilot symbols in CM and the mechanism for generating the gaps. A comprehensive and detailed description of the SIR target computation during CM can be found in [1] and in [6] for the uplink and downlink, respectively.

4.2.6 Power Control with Transmit Power Control Command Errors

At higher speeds, the problem in the fast closed-loop PC is the high error rate in the TPC commands. This is especially true for bearers running at low SIRs. To maintain the same BLER at the receiver, the outer-loop PC will compensate the PC imperfection by maintaining a higher target E_b/N_0. Table 4.2 presents simulation results for target E_b/N_0 values as a function of the error probability in the TPC commands.

Table 4.2 E_b/N_0 targets as a function of error rate of transmit power control commands (UE speed 3 km/h).

Error rate [%]	Target E_b/N_0 uplink [dB]		Target E_b/N_0 downlink [dB]	
	Average	Standard deviation	Average	Standard deviation
5	4.05	0.27	6.37	0.48
10	4.08	0.3	6.44	0.5
20	4.33	0.49	6.74	0.47
30	5.36	0.64	7.51	0.57
40	9.5	1.51	10.4	1.38

4.2.7 Fast Power Control and User Equipment Speed

The SIR target for the closed-loop PC is dependent on the E_b/N_0 requirement of the requested service (target E_b/N_0) which in turn is dependent on the speed of the UE. Therefore, when setting the maximum allowed target E_b/N_0 one should also take into account UEs at higher speeds. There are two reasons why this is necessary. First, at lower speed, the fast PC is close to perfect and the target E_b/N_0 can be set to smaller values. At higher speeds fast PC becomes imperfect and at, for example, 120 km/h little correlation is left between the channel and TPC commands. To still achieve the required received BLERs a slightly higher average target E_b/N_0 is required. Second, there is also more variance in the target E_b/N_0 for higher UE speeds. As a result a higher maximum

Table 4.3 E_b/N_0 targets at various user equipment speeds (Target BLER = 1%, 64 kbps circuit switched data traffic).

UE speed [km/h]	Target E_b/N_0 uplink [dB]		Target E_b/N_0 downlink [dB]	
	Average	Standard deviation	Average	Standard deviation
3	3.75	1.00	5.72	0.78
20	3.75	1.10	5.09	0.93
50	3.7	1.46	5.87	1.47
120	4.1	1.86	5.99	1.90

target SIR setting in the outer-loop PC is necessary if the cell is to support UEs at higher speeds as well. Examples for target E_b/N_0 values as a function of the UE speed are shown in Table 4.3.

4.3 Handover Control

The HC of UTRAN supports different types of HOs and HO procedures. The following sections give an introduction to the most common types and procedures supported by the UTRAN. The HC can be divided into the following HO types:

- *Intra-system HO* occurring within a WCDMA system. It can be further subdivided into:
  - o *Intra-frequency HO* between cells belonging to the same WCDMA carrier;
  - o *Inter-frequency HO (IF-HO)* between cells operated on different WCDMA carriers.
- *Inter-system HO (IS-HO)* taking place between cells belonging to two different Radio Access Technologies (RATs) or different Radio Access Modes (RAMs). The most frequent case for the first type is expected between WCDMA and GSM/EDGE (Global System for Mobile communications/Enhanced Data rates for GSM Evolution) systems. However, IS-HO to other CDMA systems (e.g., cdmaOne) as well can be imagined. An example for inter-RAM HO is between UTRA FDD (Frequency Division Duplex) and UTRA TDD (Time Division Duplex) modes. Furthermore the following HO procedures can be identified:
- *Hard Handover (HHO)*, a category of HO procedures in which all the old radio links of a UE are released before the new radio links are established. For RT bearers it means a short disconnection of the bearer, for NRT bearers HHO is lossless.
- *Soft Handover (SHO)* and *softer HO* are categories of HO procedures in which the UE always keeps at least one radio link to the UTRAN. During *soft* HO the UE is *simultaneously* controlled by two or more cells belonging to different Node Bs of the same RNC (intra-RNC SHO) or different RNCs (inter-RNC SHO). In *softer* HO the UE is controlled by at least two cells under one Node B. SHO and softer HO are only possible within one carrier frequency.

Depending on the participation in SHO, the cells in a WCDMA system are divided into the following sets:

- *Active Set*, including all cells currently participating in a SHO connection of a UE.
- *Neighbour Set/Monitored Set*. Both terms are used synonymously. This set includes all cells being continuously monitored/measured by the UE and which are not currently included in its active set.
- *Detected Set*. This set includes the cells the UE has detected but are neither in the active set nor in the neighbour set.

4.3.1   Intra-system–Intra-frequency Soft Handover

SHO is a general feature in systems, such as WCDMA, in which neighbouring cells are operated on the same frequency. When in Connected Mode, the UE continuously measures serving and neighbouring cells (cells indicated by the RNC) on the current carrier frequency. The UE compares the measurement results with HO thresholds provided by the RNC, and sends a measurement report back to the RNC when the reporting criteria are fulfilled. SHO therefore is a Mobile Evaluated Handover (MEHO). The decision algorithm of SHO, however, is located in the RNC. Based on the measurement reports received from the UE (either periodic or triggered by certain events), the RNC orders the UE to add or remove cells from its active set (Active Set Update, ASU). UTRAN HC supports all types of SHO described, for both RT and NRT RABs.

In the WCDMA system the vast majority of HOs are intra-frequency HOs. Different types of intra-frequency HOs can take place simultaneously. For example, the RAN is able to perform SHOs (intra-RNC as well as inter-RNC) and softer HOs at the same time. The main objectives of soft/softer HO are the following:

- Optimum fast closed-loop PC, as the UE is always linked with the strongest cells.
- Seamless HO without any disconnection of the RAB.
- To enable a sufficient reception level for maintaining communications by combining the received signals (macro-diversity) at symbol level from multiple cells in cases when the UE moves to cell boundary areas, and cannot obtain a sufficient reception level from a single cell.
- Furthermore, the macro-diversity gain achieved by combining the received signal in the Node B (softer HO) or in the RNC (SHO) improves the uplink signal quality and thus decreases the required transmission power of the UE.

All RRM functions participate in the SHO process: HC processes the measurement reports from the UE and makes the final decision based on this information. HC also updates the reference transmit power used by the power-drifting-prevention algorithm during SHO (see Section 4.2.3.4). AC is needed first for the downlink admission decision for RT RABs and for queuing of an HO branch at a cell in case of congestion. When the new branch has been allocated, the AC is needed for downlink power allocation to the HO branch. If the HO branch addition request was rejected, AC may initiate either forced call release, forced IF-HO or forced IS-HO. The RM allocates the downlink spreading code to a new HO branch, and releases the spreading code when the HO branch is removed from the cell. The task of the RM is to activate the radio link for a new HO branch and to release the radio link when the HO branch is removed from the cell. LC updates the downlink load information of the cell when a new RT RAB (HO branch for SHO) is admitted or when an RT RAB (old HO branch) is removed from the cell. For packet switched data traffic the packet scheduler releases a downlink spreading code for the HO branch in the case of an NRT RAB and schedules the HO branch addition request for an NRT RAB with the highest priority.

4.3.2   Intra-system–Intra-frequency Hard Handover

Intra-frequency HHO is needed when cells participating in the HO are controlled by different RNCs in situations when the inter-RNC HO cannot be executed as an SHO or if SHO is not allowed. Intra-frequency HHO causes temporary disconnection of the RT RAB but is lossless for NRT bearers. Its decisions are made by the RNC based on the intra-frequency measurement results the UE is sending *periodically* after it has reported an intra-frequency triggering event and the active set could not be updated, and relevant control parameters. The reports are usually applied to the SHO procedure, so intra-frequency HHO is a MEHO. A simple algorithm for HHO could be based on the averaged P-CPICH E_c/I_0 values of the serving cell and the neighbouring cells, and an HO margin that is used as a threshold to prevent repetitive HHOs between cells. Before the intra-frequency HHO is possible in this case the measurement results of the neighbouring cell must satisfy Equation (4.11):

$$AveEcIoDownlink + EcIoMargin(n) < AveEcIoNcell(n) \qquad (4.11)$$

where *AveEcIoDownlink* is the averaged P-CPICH E_c/I_0 of the best serving cell; *AveEcIoNcell(n)* is the averaged P-CPICH E_c/I_0 of the neighbouring cell *n*; and *EcIoMargin(n)* is the margin by which the E_c/I_0 of the neighbouring cell *n* must exceed the E_c/I_0 of the best serving cell before HO is possible. By performing an HHO when SHO is not possible, excessive interference can be avoided. During the HHO procedure all links in the active set are replaced simultaneously by one new link.

4.3.3   Intra-system–Inter-frequency Handover

IF-HO is an HHO between different WCDMA carriers required to ensure an HO path from one cell to another cell in situations when different carriers have been allocated to the cells in question. Also, HHO here means that IF-HO causes temporary disconnection of the RT RAB and is lossless for NRT bearers. IF-HO also enables HOs between separate layers of a multi-layered cellular network – e.g., a network consisting of macro- and micro-cells where the cell layers are using different carriers. The RAN HC should support the following types of IF-HO:

- intra-BS HHO (to control the load between carriers);
- intra-RNC HHO;
- inter-RNC HHO.

IF-HO is a Network Evaluated Handover (NEHO) since its evaluation algorithm is located in the RNC. The RNC recognises the possibility of an IF-HO based on the configuration of the radio network (frequency/carrier allocation, neighbour cell definitions, cell layers etc.). When a UE is located where an IF-HO is possible and needed, the RNC commands the UE to start inter-frequency measurements and to report the results *periodically*. HO decisions are then made by the RNC based on those measurement results (inter- and intra-frequency) and relevant control parameters.

To be able to execute the inter-frequency measurements, the UE must be equipped with a second receiver tuned to the neighbouring frequency or it has to support so-called Compressed Mode (CM). More about CM can be found in Section 4.3.6.

4.3.4   Inter-system Handover

As the name indicates, IS-HO is an HO between UTRA FDD and a neighbouring system using a different RAT, or – within WCDMA – if the other system uses a different RAM – i.e., UTRA TDD. IS-HO is required, for example, to complement the coverage areas of WCDMA and a neighbouring system with each other when the coverage area of WCDMA is limited only to certain areas. When the coverage areas of WCDMA and the neighbouring system overlap each other, an IS-HO can be used to control the load between the systems as well. For example, speech connections can be handed over to a neighbouring second-generation (2G) system and data connections handled within the WCDMA system. IS-HO is an HHO – that is, it causes temporary disconnection of the RT RAB. When an RT RAB is handed over from one system to another, the Core Network (CN) is responsible for adapting the QoS parameters included in the RAB attributes according to the new system. For more about QoS see Chapter 8. Also IS-HO is a NEHO, since the evaluation is done in the RNC. The decision algorithm of the IS-HO is located in the RNC, but the UE must support IS-HO and its measurements completely before the feature can be used.

When the UE is located where an IS-HO is possible and needed, the RNC commands the UE to start inter-system measurements and to report the results *periodically*. The HO decisions made by the RNC are based on the results of these measurements (inter- and intra-system) and relevant control parameters.

The RNC recognises the possibility of IS-HO based on the configuration of the radio network (neighbour cell definitions and relevant control parameters). In case the second system is a GSM system, the decision algorithm of the IS-HO from GSM to WCDMA is located in the GSM Base Station Controller (BSC). From the viewpoint of the RNC an IS-HO from GSM to WCDMA does not differ from the inter-RNC HHO. Correspondingly, an IS-HO from WCDMA to GSM does not differ from the inter-BSC HO from the viewpoint of the GSM BSS.

As with inter-frequency measurements, the UE must be either equipped with a second receiver or support CM (see Section 4.3.6) to execute inter-system measurements.

4.3.5   Handover Measurement Reporting

HO measurement reporting can be divided into the following stages:
1. Neighbour cell definitions.
2. Measurement reporting criteria.
3. Reporting of measurement results.

4.3.5.1   Neighbour Cell Definitions

For each cell in the UTRAN an own set of neighbouring cells must be defined in the radio network configuration database, typically located in the RNC. Since a neighbouring cell may be located in the same network on the same frequency, on a different frequency or in any neighbouring Public Land Mobile Network (PLMN), the following neighbour lists need to be defined for each cell in case the corresponding HO needs to be supported [9]:

- *Intra-frequency neighbour cell list.* The UE must be able to monitor at least 32 cells on the same WCDMA carrier frequency as the serving cell.
- *Inter-frequency neighbour cell list.* The UE must be able to monitor at least 32 cells on a maximum of two WCDMA carrier frequencies in addition to the serving cell's frequency.
- *Inter-system neighbour cell lists.* For each neighbouring PLMN an own list is needed. In total a maximum of 32 inter-frequency neighbours must be supported by the UE.

The RAN broadcasts the initial neighbour cell list(s) of a cell in the system information messages on the BCCH (Broadcast CCH). In case a required ASU has been performed, a new neighbour list is combined in the RNC based on the neighbour lists of the cells in the new active set and then is sent to the UE on the DCCH.

To identify a WCDMA neighbour cell, this list includes the following information:

- UTRAN Cell Identifier:
  - Global RNC identifier (PLMN identifier MCC and MNC);
  - Cell Identifier (CI).
- Location Area Code (LAC).
- Routing Area Code (RAC).
- UTRA Absolute Radio Frequency Channel Number (UARFCN).
- Scrambling code of the P-CPICH.

For a GSM neighbouring cell, the following information is sent:

- Cell Global Identification, CGI = MCC + MNC + LAC + CI;
- BCCH frequency;
- Base Station Identity Code, BSIC = BCC + NCC.

Neighbour Cell Search on Current Carrier Frequency
In idle as well as in connected mode the UE continuously searches for new cells on the current carrier frequency. Details of the cell search procedure can be found in Section 4.5.2.1. If the UE detects a candidate cell that has not been defined as a neighbouring cell, it has to decode the cell's BCCH to identify the cell before it can report the measured E_c/I_0 of the detected neighbouring cell to the RNC. In this case the following Information Elements (IEs) are used to identify the undefined neighbouring cells: the downlink scrambling code, LAC and CI. When reporting the measurement result, the UE may or may not include this information in the measurement report.

4.3.5.2   Measurement Reporting Criteria

Depending on the HO type (MEHO or NEHO), different measurement reporting criteria can be used. The RNC may request the UE to execute and report the following different types of basic HO measurements:
- intra-frequency measurements (MEHO);
- inter-frequency measurements (NEHO);
- inter-system measurements (NEHO);
- UE internal measurements.

All HO measurement types are controlled independently of each other and are defined on a cell-by-cell basis, with the exception of UE internal measurements, which are partly controlled by parameters common to all cells under the same RNC. Two or more HO measurement types can be active simultaneously – e.g., intra- and inter-frequency measurements. Typically, in a RAN separate measurement parameter sets for RT and NRT bearers and for users applying HSDPA can be defined. Control of the HO measurements is explained in detail in the following sections in connection with the relevant HO types.

Intra-frequency Handover Measurements
The RAN broadcasts the measurement reporting criteria (measurement parameters) for intra-frequency measurements on the BCCH. When the criteria are fulfilled, the UE reports the results of its measurements to the RNC. The RNC in turn makes the HO decision. If the ASU could not be executed, the UE continues to measure the neighbouring cells but changes to *periodic* reporting of the results (more on this event-triggered periodic reporting can be found in Section 4.3.5.3). For this type of measurements the UE uses separate measurement reporting criteria transmitted to the UE.

Inter-frequency and Inter-system Handover Measurements
Inter-frequency and inter-system measurements are both made only when ordered by the RNC. They use separate measurement reporting criteria transmitted to the UE. When they are initiated, the UE *periodically* reports the results to the RNC. The measurements are controlled by two parameters: reporting duration and the reporting interval.

User Equipment Internal Measurements
UE internal measurement reporting criteria are controlled partly on a cell-by-cell basis and partly by parameters common to all cells in the whole RNC. The measurement information for UE internal measurements is not included in the system information on the BCCH but transmitted to the UE on a Dedicated CCH (DCCH). When the measurement-reporting criteria are fulfilled, the UE reports the results of its measurements to the RNC.

4.3.5.3 Reporting of Measurement Results

When the UE reports the measurement results from the intra- or inter-frequency measurements of the neighbouring cells back to the UTRAN, the following IEs are included to identify the neighbours:

- *P-CPICH information* (downlink scrambling code) identifies active and monitored cells when the UE reports intra-frequency or UE internal measurement results to the RNC.
- *P-CPICH information and UTRA RF (Radio Frequency) channel number* identifies neighbouring cells when the UE reports IF measurement results to the RNC.
- *BCCH frequency* identifies neighbouring GSM cells when the UE reports IS (GSM) measurement results to the RNC. The BSIC can be used additionally to verify identification if two or more neighbouring GSM cells have the same BCCH frequency. The RNC always applies the BSIC verification for the target cell before the execution of IS-HO so that the UE can synchronise with the GSM cell before HO execution. The UE reports the BSIC information only if it is requested by the RNC.

The UE generates at least the following event-triggered and periodic measurement reports:

- event-triggered intra-frequency measurement report;
- periodic intra-frequency measurement report;
- inter-frequency measurement report;
- inter-system measurement report;
- measurement reports on common channels;
- traffic volume measurement report;
- UE internal measurement report;
- quality deterioration report.

Reporting of Intra-frequency Measurements

Intra-frequency measurement reporting can be either event-triggered or periodic. During connected mode, the UE constantly monitors the P-CPICH E_c/I_0 of the cells defined by the intra-frequency neighbour cell list and evaluates the reporting criteria. If one of the reporting events is fulfilled, the UE sends an event-triggered measurement report. Before the P-CPICH E_c/I_0 of a cell is used by the HO algorithm in the UE, an arithmetic mean of a certain number of the latest measured values is taken. The number of the values taken into account is a UE performance specification parameter. The average is taken over the linear values of E_c/I_0, not the dB values.

For intra-frequency measurement criteria, the following reporting events are defined by [6]:

- **Event 1A**: *A P-CPICH enters the reporting range.* A report is triggered when Equation (4.12) is fulfilled:

$$10 \cdot \log_{10} M_{New} + CIO_{New} \geq W \cdot 10 \cdot \log_{10} \left(\sum_{i=1}^{N_A} M_i \right)$$

$$+ (1 - W) \cdot 10 \cdot \log_{10} M_{Best} - (R_{1a} - H_{1a}/2) \quad (4.12)$$

where M_{New} is the measurement result of the cell entering the reporting range; CIO_{New} is the cell-individual offset of the cell entering the reporting range; M_i is a measurement result of a cell in the active set not forbidden to affect the reporting range; N_A is the number of cells in the current active set not forbidden to affect the reporting range; M_{Best} is the measurement result of the strongest cell in the active set; W is a weighting parameter sent from the RNC to the UE; R_{1a} is the reporting range constant for event 1A sent from the RNC to the UE; and H_{1a} is the hysteresis parameter for event 1A. The hysteresis parameter together with the reporting range constant is usually called the *addition window*.

- **Event 1B**: *A P-CPICH leaves the reporting range*. A report is triggered when Equation (4.13) is fulfilled:

$$10 \cdot \log_{10} M_{Old} + CIO_{Old} \leq W \cdot 10 \cdot \log_{10} \left(\sum_{i=1}^{N_A} M_i \right)$$

$$+ (1 - W) \cdot 10 \cdot \log_{10} M_{Best} - (R_{1b} + H_{1b}/2) \quad (4.13)$$

where R_{1b} is the reporting range constant for event 1B sent from the RNC; M_{Old} is the measurement result of the cell leaving the reporting range; CIO_{Old} is the cell-individual offset of the cell leaving the reporting range; and H_{1b} is the hysteresis parameter for event 1B. The hysteresis parameter together with the reporting range constant is usually called the *drop window*.

- **Event 1C**: *A non-active P-CPICH becomes better than an active one*. A report is triggered when Equation (4.14) is fulfilled – i.e., when a P-CPICH that is not in the active set gets better than the worst P-CPICH from the active set when the active set is full. Used to replace the cell with the worst P-CPICH:

$$10 \cdot \log_{10} M_{New} + CIO_{New} \geq 10 \cdot \log_{10} M_{InAS} + CIO_{InAS} + H_{1c}/2 \quad (4.14)$$

where M_{InAS} is the measurement result of the cell in the active set with the lowest measurement result; CIO_{InAS} is the cell-individual offset for the cell in the active set that is becoming worse than the new cell; and H_{1c} is the hysteresis parameter for event 1C. The hysteresis parameter is usually called the *replacement window*.

- **Event 1D**: *Change of best cell*. A report is triggered when any P-CPICH in the reporting range becomes better than the current best plus an optional hysteresis value.

- **Event 1E**: *A P-CPICH becomes better than an absolute threshold*. A report is triggered when a new cell plus its cell-individual offset becomes better than an absolute threshold plus an optional hysteresis value.

- **Event 1F**: *A P-CPICH becomes worse than an absolute threshold*. A report is triggered when a new cell plus its cell-individual offset becomes worse than an absolute threshold minus an optional hysteresis value.

Figure 4.8 shows an example of the general WCDMA HO algorithm involving all reporting events 1A–1F. In the example, the reports are sent as soon as the event is triggered – i.e., the time-to-trigger method (see below) is not used. Also there are no hysteresis values involved and the weighting coefficients, W, are assumed to be zero.

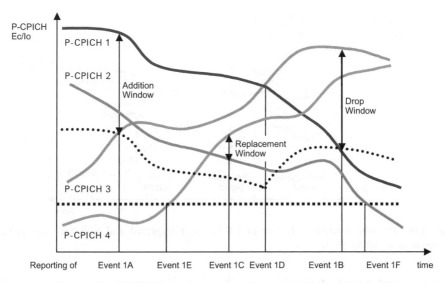

Figure 4.8 WCDMA handover algorithm reporting events 1A–1F.

Time-to-trigger Mechanism

The abundance of possible neighbouring cells together with the variety of triggering events could result in quite frequent reporting. To protect the network from an excessive signalling load, each of the reporting events can be connected with a timer. Only if the measurement criteria have been fulfilled during the whole period until the timer expires is the event reported to the network. Figure 4.9 shows an example of the time-to-trigger mechanism in case of event 1A. On the first two occasions when the event occurs, no report is triggered, since P-CPICH 3 does not stay within the reporting range for a long enough time. Only the third occurrence triggers the reporting of event 1A.

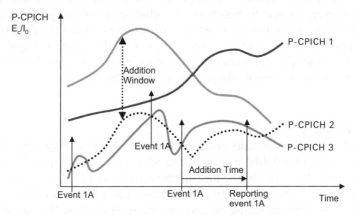

Figure 4.9 Example for time-to-trigger (Addition Time) in the case of event 1A.

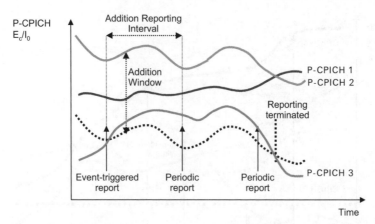

Figure 4.10   Periodic reporting after event 1A has been triggered and active set update failed (hysteresis value set to zero).

Event-triggered Periodic Reporting

Reporting one of the above-mentioned events typically results in an active set update. However, if the active set update cannot take place, owing to lack of capacity or hardware resources, for example, the UE changes to periodic reporting. In this case it sends a measurement report every reporting interval until the active set update has taken place, the measurement criteria are no longer fulfilled or the maximum number of measurement reports have been sent. Figure 4.10 shows an example of periodic reporting. After the first reporting of event 1A, the active set could not be updated, so the UE starts to report periodically. In this case periodic reporting is terminated when P-CPICH 3 is no longer within the reporting range.

Mechanism for Forbidding a Neighbouring Cell to Affect the Reporting Range

In case of events 1A and 1B when the weighting coefficient, W, is non-zero, all cells in the active set are used to evaluate whether or not the measurement criteria are fulfilled. In a RAN, however, it could be beneficial to exclude a specific neighbouring cell – i.e., its P-CPICH – from this active set weighting: for example, when the P-CPICH of that cell is very unstable within the reporting range. For this case, a neighbouring cell parameter can be specified for each cell, indicating whether or not this cell is allowed to affect the reporting range calculation when it is in the active set.

Cell-individual Offsets

To have an efficient means of reporting a monitored cell individually, a P-CPICH offset can be assigned to each neighbouring cell. The offset can be either positive or negative. The UE then adds this offset to the measurement quantity (E_c/I_0, *path loss* or *RSCP*) before it evaluates whether a reporting event has occurred. An example of the use of this mechanism for load balancing between neighbouring cells in the network together with an explanatory figure can be found in Section 9.3.6.3.

Reporting of Inter-frequency and Inter-system Measurements

Inter-frequency and inter-system measurement reports are always periodic. The events triggering them are not part of the standards. The RNC may initiate inter-frequency and/or inter-system measurements in various circumstances, for example:

- average downlink transmission power of a radio link as it approaches its maximum power level;
- uplink transmission power reaches a threshold or its maximum (events 6A/6D, see below);
- quality deterioration report from uplink outer-loop PC from the RNC;
- quality deterioration report from the UE;
- unsuccessful SHO (branch addition) procedure;
- unsuccessful RAB setup;
- UE located within cell where SHO capability is restricted;
- UE located within cell where admitted user bit rate is lower than requested;
- frequent SHOs (cell size and UE speed do not match);
- radio network recovery management initiates forced HO procedure;
- UE located within an area where cell structure is hierarchical (inter-frequency);
- UE located within an area where hierarchical network structure is composed of WCDMA and GSM systems (inter-system only);
- IMSI-based HO is needed;
- UE located within a cell with restricted intra-system HO capability (inter-system only).

UE Internal Measurements

UE internal measurements can be divided into two groups. The first group is used to indicate to the network the status of the UE transmit power. The reports may be used by the RNC to trigger off inter-frequency or inter-system measurements. The second group is the UE Rx–Tx (Receiver–Transmitter) time difference measurement. It is used to adjust the downlink DPCH air interface timing when the difference in time between the UE uplink DPCCH/DPDCH frame transmission and the first significant path of the downlink DPCH frame from a measured active set cell (UE Rx–Tx time difference) becomes too large.

The following events are specified in [6]:

- Event 6A: UE transmit power becomes larger than an absolute threshold.
- Event 6B: UE transmit power becomes less than an absolute threshold.
- Event 6C: UE transmit power reaches its minimum value.
- Event 6D: UE transmit power reaches its maximum value.
- Event 6E: UE RSSI (Received Signal Strength Indicator) reaches the UE receiver dynamic range.
- Event 6F: UE Rx–Tx time difference for a radio link included in the active set becomes larger than an absolute threshold.
- Event 6G: UE Rx–Tx time difference for a radio link included in the active set becomes less than an absolute threshold.

Node B Measurements

The Node B measurement report can be used to trigger off inter-frequency or inter-system (GSM) measurements, and to balance the PC (uplink and downlink) of the diversity branches during SHO. The Node B sends the measurement report to the RNC on a radio link by radio link basis at regular (e.g., 500–1000 ms) intervals. The measurement report from the Node B includes the following radio link measurement results:

- average downlink transmission power of the DPCH;
- average measured uplink SIR of the DPCH;
- uplink SIR target currently used on the DPCH.

4.3.5.4 Filtering of Measurement Results

In building the optimum active set, it is very important to have accurate enough P-CPICH E_c/I_0 measurements. Since the path loss is overlaid by fast fading, the SHO algorithm measures the P-CPICH E_c/I_0 by means of an SHO filter which is able to filter out the fast fading component of the channel depending on the UE speed. For low speeds in the range of 0–6 km/h, the fast fading component may not disappear, but for speeds in the range of 50 km/h, it can be filtered out completely. This also depends on the channel profile. Channels with one dominant path can produce large fast fading fluctuations that are clearly visible after filtering. To see the differences, Figure 4.11 compares a slow and a fast-moving UE.

In the high-speed UE, we can see that the fast fading rate around 60 km/h can be easily filtered out by filters of 50–200 ms in length. However, in the case of 6 km/h there exist deep fades with long fade duration that may be practically impossible to eliminate

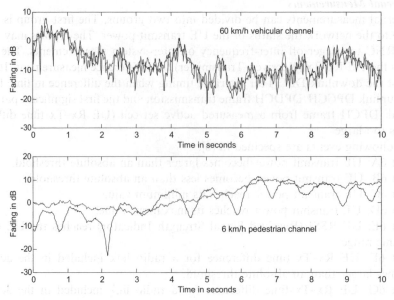

Figure 4.11 Fast fading filtering for different terminal speeds.

by just filtering. This will cause the SHO algorithm to result in measurement errors leading to unnecessary HOs, short active set update periods and increased signalling load. A longer filtering time – e.g., 1 s – could help in this situation, but the drawback in this case is the delay introduced to the HO. Delayed HOs, however, lead to degradation in network performance, since the UE might be connected for too long to the wrong cell. This is especially critical in situations where the best server is changing quickly – e.g., in micro-cell and with fast-moving UEs.

4.3.6 *Compressed Mode*

Intra-frequency neighbours can be measured simultaneously with normal transmission by the UE using a RAKE receiver. Inter-frequency and inter-system measurements, however, require the UE to measure on a different frequency. This can be done by incorporating multiple receivers in the UE. A second possibility that avoids receiver multiplicity is stopping the normal transmission and reception for a certain time period, enabling the UE to measure on the other frequency. To achieve this gap and not lose any information, the data sent have to be compressed in time – i.e., the transmission and reception enter *Compressed Mode* (CM), earlier also called *Slotted Mode*. The RNC determines which frames are compressed, and sends the information both to the Node B and to the UE. There are three methods to generate the gaps to use CM:

- reducing the data rate used in the upper layers (higher layer scheduling);
- reducing the symbol rate used in the physical layer (rate matching and/or puncturing).
- spreading factor splitting (halving the spreading factor doubles the available symbol rate).

The standard [10] allows CM to be applied in one direction only or in both directions simultaneously. Normally it is needed only in the downlink. However, to prevent the UE from interfering with itself by the effects described in Section 5.4, CM should also be used in the uplink and then must be synchronous with downlink CM. However in uplink CM, puncturing cannot be used. Different CM schemes have been specified in [10], Figure 4.12 illustrates the parameters in connection with CM. The most important parameters involved in CM are:

- Transmission Gap starting Slot Number (TGSN), the first slot in the transmission gap. All values from 0 to 14 are allowed.
- Transmission Gap Lengths (TGL1/TGL2), expressed in slots. Allowed lengths are 3, 4, 5, 7, 10 or 14 slots.
- Transmission Gap start Distance (TGD), indicating the number of slots between the starting slots of two consecutive transmission gaps within a transmission gap pattern. Allowed values are from 15 to 269.
- Transmission Gap Pattern Lengths (TGPL1/TGPL2), expressed in frames from 1 up to 144.
- Transmission Gap Pattern Repetition Count (TGPRC), the number of TGPs within one TGP sequence.
- Transmission Gap Connection Frame Number (TGCFN). This is the CFN of the first radio frame of the first pattern 1 within the transmission gap pattern sequence.

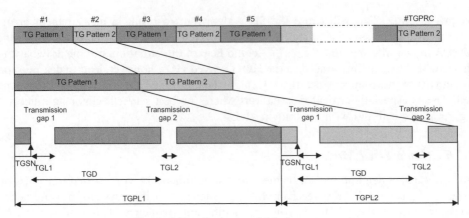

Figure 4.12   Compressed Mode pattern.

Naturally CM has a certain impact on RRM procedures. Especially crucial, for example, is the PC during CM. More on this topic is in Section 4.2.5. Also, network performance is affected by CM, see Section 4.7.1.4.

4.3.7   Inter-system Handover Procedure

IS-HO and inter-system cell reselection provide the ability to move connections belonging to dual-mode UE from one RAT to another. The two RATs are usually UMTS and GSM. IS-HO from the UMTS system to the GSM system is applicable when the UE is in the Cell_DCH RRC Connected Mode state. IS-HO from the GSM system to the UMTS system is applicable when the UE is in GSM Connected Mode. The division between IS-HO and inter-system cell reselection is presented in Table 4.4. This section focuses on IS-HO.

IS-HO from the UMTS system to the GSM system is applicable to both circuit switched services and packet switched services. IS-HO from the GSM system to the UMTS system is only applicable to circuit switched services. Packet switched services are moved from the General Packet Radio Service (GPRS) system to the UMTS system using cell reselection.

Table 4.4   Applicability of inter-system cell reselection and inter-system handover.

System	Cell reselection	Handover
UMTS	RRC Idle RRC Connected: Cell_FACH RRC Connected: Cell_PCH RRC Connected: URA_PCH	RRC Connected: Cell_DCH
GSM/GPRS	GSM Idle Mode GPRS Packet Idle Mode GPRS Packet Transfer Mode	GSM Connected Mode

The UMTS IS-HO procedure can be divided into five phases: triggering; GSM RSSI measurements; RNC evaluation of the GSM RSSI measurements; GSM BSIC verification; and the HO command or cell change order. The HO command is relevant to IS-HO for circuit switched services whereas the cell change order is relevant to IS-HO for packet switched services. The GSM IS-HO procedure can be divided into four phases: triggering; WCDMA measurements; BSC evaluation of the WCDMA measurements; and the HO command.

4.3.7.1 Triggering Mechanisms

There are numerous strategies possible for the triggering of IS-HO. A common strategy is to keep all dual-mode UEs connected to the UMTS system for as long as possible. In this case, IS-HO from the UMTS system is only triggered when the UE is approaching the edge of UMTS service coverage. It is important that the IS-HO procedure is triggered before the UE reaches the edge of UMTS coverage. If the procedure is triggered too late the UE will not have sufficient time to complete the HO and the connection will drop. This means that the duration of the IS-HO procedure must be accounted for when defining the thresholds used to trigger the procedure. If the IS-HO procedure is triggered too early then the procedure will be completed more frequently and the UE will start to complete unnecessary HOs. Once a circuit switched service connection has been moved onto the GSM system it is common to allow that connection to remain on the GSM system until the UE moves to GSM idle mode. The UE can then return to the UMTS system using inter-system cell reselection. Alternatively, IS-HO from the GSM system could be triggered as soon as the UE returns to an area of UMTS service coverage. In the case of packet switched services it is usual to move the connection back to the UMTS system as soon as possible. This is because the higher data rates provided by the UMTS system have a significant impact upon user experience. It is also possible to define IS-HO strategies based upon system load and service type. For example, IS-HO from the UMTS to the GSM could be triggered when the load of the WCDMA system exceeds a specific threshold, or IS-HO from the GSM system could be triggered whenever the UE initiates use of a specific service.

Coverage-based triggering mechanisms for the UMTS system include uplink transmit power, downlink transmit power, CPICH E_c/I_0, CPICH RSCP and uplink quality. In general, two or three of these triggering mechanisms are selected and subsequently optimised in the field. A common approach is to focus on the CPICH RSCP-triggering mechanism. This mechanism provides a means to trigger IS-HO based upon link loss – i.e., the fixed CPICH transmit power minus the measured CPICH RSCP indicates the link loss. Link budget calculations can be completed to determine an appropriate threshold for the maximum allowed link loss and thus the minimum allowed CPICH RSCP. The drawback of this approach is that it relies upon the accuracy of the link budget calculation. In practice, the link budget for one cell is likely to be different from the link budget for another cell. This makes it difficult to define a single CPICH RSCP threshold which is applicable to all cells. Nevertheless, as long as the link budget differences between cells are relatively small then the CPICH RSCP-triggering mechanism provides a practical approach to triggering IS-HO.

The uplink and downlink transmit-power-triggering mechanisms do not rely upon link budget assumptions. These mechanisms directly identify when the maximum uplink or downlink transmit powers are being approached. If service coverage is uplink limited then the uplink transmit-power-triggering mechanism would dominate. Likewise, if service coverage is downlink limited then the downlink transmit-power-triggering mechanism would dominate. The uplink transmit-power-triggering mechanism relies upon measurement reporting from the UE. This can be based upon the 3GPP measurement reporting events 6A and 6B. These events can be used to inform the RNC when the UE transmit power exceeds or drops below an absolute threshold. It is possible that not all UEs support these measurement reporting events. This would mean that the CPICH RSCP-triggering mechanism would be required as an alternative solution. A UE measures its transmit power on a per-slot basis – i.e., once every 0.67 ms. This represents a much faster rate than its CPICH measurements which are filtered on the basis of a 200 ms measurement interval. This means that UE transmit power measurements can be filtered at layer 3 using a much larger filter coefficient – e.g., a filter coefficient of 8 rather than 3. If the filter coefficient is not increased then the reporting events 6A and 6B may be triggered by very short-term increases or decreases in transmit power. The downlink transmit-power-triggering mechanism relies upon measurement reporting from Node B. Node B is able to report its radio link transmit powers to the RNC using NBAP: Radio Link Measurements Report message. It is common for Node B to periodically send this message to the RNC – e.g., every 500 ms. In general Node B filters the radio link transmit powers prior to signalling the result to the RNC. This means that the RNC may trigger the IS-HO procedure upon the basis of a single measurement report indicating that the downlink transmit power has exceeded its threshold. The thresholds associated with the downlink transmit-power-triggering mechanism should be defined on a per-service basis. This is because each service is likely to have a different maximum downlink transmit power assignment.

The CPICH E_c/I_0 IS-HO-triggering mechanism can be used in combination with other triggering mechanisms. Similar to the CPICH RSCP and uplink transmit-power-triggering mechanisms, this mechanism relies upon measurement reporting from the UE. This can be based upon the 3GPP measurement reporting events 1E and 1F. These events can be used to inform the RNC when the CPICH E_c/I_0 exceeds or drops below an absolute threshold. This could be an important triggering mechanism in areas of poor dominance although care should be taken to ensure that IS-HO is not triggered unnecessarily by short-term changes in radio conditions. The filter coefficient and time to trigger for events 1E and 1F can be used for this purpose.

The uplink quality IS-HO-triggering mechanism is less common. This mechanism can be justified when the uplink outer-loop PC has a limited dynamic range for SIR target changes. In this case there may be circumstances where the RNC has increased the uplink SIR target to its maximum allowed value but the uplink quality remains relatively poor. This means that the quality of service requirements for the connection cannot be achieved and it may be preferable to move the connection to the GSM system.

If there is a requirement to trigger IS-HO from the GSM system then it could be implemented using a combination of load and coverage-based triggering mechanisms.

If the GSM load exceeds a specific threshold then the BSC could start to evaluate IS-HO for dual-mode UE. The IS-HO load threshold should be coordinated with the intra-system HO load threshold. It may be desirable to balance the load within the GSM system prior to moving dual-mode UE onto the UMTS system. If there is a requirement to move dual-mode UE onto the UMTS system as soon as possible then the load threshold should be configured with a low value. The GSM IS-HO procedure requires the UE to measure the UMTS system. These measurements can be triggered by the UE using a coverage-based threshold. It is possible to trigger measurements when the GSM RSSI of the existing connection drops below a specific threshold. It is also possible to trigger measurements when the GSM RSSI exceeds a specific threshold. Alternatively the UE can be instructed to continuously measure or never measure the WCDMA system. If there is a requirement to move dual-mode UE onto the UMTS system as soon as possible then the UE should be instructed to continuously measure the WCDMA system.

4.3.7.2 Circuit Switched Services

It is usual to support IS-HO for the speech service but it is less common to support it for circuit switched data services. This is because the GSM system is not designed to support the circuit switched data services offered by UMTS – e.g., peer-to-peer video calls.

Once IS-HO has been triggered by the UMTS system the UE is instructed to complete a set of GSM RSSI measurements. These measurements are based upon an IS-HO list provided by the RNC using a dedicated measurement control message. The definition of inter-system neighbour lists forms part of the radio network planning process. If these neighbour lists are too short then missing neighbours may lead to failed IS-HOs. If the neighbour lists are too long then the UE measurement time increases and important neighbours may be removed from the list when the UE is in SHO. If the UE is in SHO when IS-HO is triggered then the neighbour lists belonging to each of the active set cells should be combined to form a composite neighbour list. 3GPP [6] limits the maximum number of inter-system neighbours to 32. This means that if there are more than 32 different neighbours belonging to active set cells then some of them must be removed. Each inter-system neighbour is specified in terms of its RF carrier and its BSIC. It is also possible to define neighbour-specific measurement offsets to make a neighbour appear either more or less attractive as an HO candidate.

Once the UE has received the inter-system neighbour list then it is able to start GSM RSSI measurements. These measurements are usually completed while in CM. The requirement for CM and the associated CM configurations are discussed in Section 4.3.7.4. GSM RSSI measurements are relatively straightforward for the UE because they do not require synchronisation with the GSM system. The UE simply has to tune to the appropriate RF carrier and then measure the power. It is usual for the RNC to configure periodic measurement reporting for the GSM RSSI measurements. This means that it is common to observe the UE sending a number of empty measurement reports to the RNC while it completes its first set of measurements. Once the UE has reported a set of measurements the RNC has the responsibility of evaluating whether or not an HO should be completed. The RNC should be configured with a minimum acceptable GSM RSSI threshold. This threshold is usually configured to be similar to the threshold used within the GSM system for allowing a UE to camp upon it. If none of the GSM RSSI measurements exceeds the threshold then IS-HO is not permitted. The RNC may base its decision upon an average of multiple UE measurements rather than a single UE measurement. This is possible if the UE is sending periodic GSM RSSI measurement reports. The benefit of using an average is that the HO decision is likely to be more reliable than when a single measurement is used – i.e., a single measurement provides a relatively short-term representation of the radio conditions. The drawback of using an average is that there is an increased delay because the RNC has to wait for multiple measurement reports. If the UE is using CM then this increased delay means spending more time in CM. If the delay becomes too great then the UE may move completely out of UMTS coverage before IS-HO has been completed. This would then result in a dropped connection.

Once the RNC has received one or more GSM RSSI measurements which achieve the minimum requirement the UE is instructed to complete BSIC verification. At this stage the RNC may reduce the size of the neighbour list to only a single inter-system neighbour or if the neighbour list includes multiple neighbours on the same RF carrier then it may reduce the neighbour list to only those on the relevant RF carrier. BSIC verification is more difficult than measuring the GSM RSSI. The UE must synchronise with the GSM system and manage to capture a BCCH slot which includes the BSIC. The BSIC is broadcast periodically within the SCH of the BCCH. There are 51 GSM frames within a GSM multiframe and the SCH is transmitted five times within each multiframe. Assuming that frame numbering starts at 0 then the frames occupied by the SCH are numbers 1, 11, 21, 31 and 41. The UE has no prior knowledge of the GSM system timing and must capture nine slots' worth of GSM data to be sure of capturing the BCCH.

If the UE manages to verify the BSIC then the result is provided to the RNC in a dedicated measurement report message. The RNC is then able to make the final IS-HO decision and subsequently make a corresponding CN request. The CN request is made using the RANAP: Relocation Required message. When the CN receives this message the MSC forwards a BSSMAP: Handover Request message to the relevant BSC. The BSC acknowledges the request and the RNC is sent a RANAP: Relocation Command message. Finally, the RNC sends an RRC: Handover from UTRAN Command message to the UE. The UE then moves onto the GSM system and sends an RR: Handover Complete message to the BSC.

If IS-HO is triggered by the GSM system then there is a requirement for the UE to start measuring the UMTS system. These measurements are usually made in terms of the CPICH E_c/I_0. In this case there is not a requirement for CM because the TDMA nature of GSM inherently provides transmission gaps which may be used for both intra-system and inter-system measurements. The UE is typically instructed the maximum number of UMTS neighbour measurements it is allowed to include within a single measurement report. Allowing more inter-system neighbour measurements means that there is less room for intra-system neighbour measurements. This could lead to a reduction in the performance of the GSM system. It is common to allow approximately one-third of the neighbour cell measurements to belong to the UMTS system. Once the UE has reported a set of measurements to the BSC then the BSC has the responsibility of evaluating whether or not an HO should be completed. The BSC should be configured with a minimum acceptable CPICH E_c/I_0 threshold. Configuring this threshold should account for the CPICH E_c/I_0 IS-HO-triggering threshold within the UMTS system. If the minimum acceptable CPICH E_c/I_0 threshold is defined to be less than the level which triggers IS-HO from the UMTS system then there are likely to be successive IS-HOs between the two systems. If none of the CPICH E_c/I_0 measurements exceed the threshold then IS-HO is not permitted. The BSC may base its decision upon an average of multiple UE measurements rather than a single UE measurement. Assuming that the UE reports a UMTS cell which exceeds the minimum quality requirement then the BSC is able to make the final IS-HO decision and subsequently make a corresponding CN request. The CN request is made using the BSSMAP: Handover Required message. When the MSC receives this message the CN forwards a RANAP: Relocation Request message to the relevant RNC. The RNC acknowledges the request and the BSC is sent a BSSMAP: Handover Command message. Finally, the BSC sends an RR: Handover Command message to the UE. The UE then moves onto the UMTS system and sends an RRC: Handover Complete message to the RNC.

4.3.7.3 Packet Switched Services

It is usual to support IS-HO from the UMTS to the GPRS for packet switched data services. Movement of the UE from the GPRS to the UMTS is completed as an inter-system cell reselection. Once IS-HO has been triggered by the UMTS system the UE is instructed to complete a set of GSM RSSI measurements. Similar to the circuit switched IS-HO procedure, these measurements are based upon an inter-system neighbour list provided by the RNC. The measurements are usually completed while in CM. The requirement for CM and the associated CM configurations for packet switched services are discussed in Section 4.3.7.4. The general GSM RSSI measurement procedure is the same as that which is used for circuit switched services. Evaluation of the measurement results against a minimum GSM RSSI threshold is also the same. If none of the GSM RSSI measurements exceeds the threshold then IS-HO is not permitted. Once the RNC has received one or more GSM RSSI measurements which achieve the minimum requirement the UE may be instructed to complete BSIC verification. In the case of IS-HO for packet switched services, BSIC verification is only required if there are multiple GSM neighbours using the RF carrier which has been reported to have the greatest RSSI. If BSIC verification is required then it is completed in a similar way to that for circuit switched services. If the UE manages to verify the BSIC then the result is provided to the RNC in a dedicated measurement report message. The RNC is then able to make the final IS-HO decision. In this case, the RNC does not have to complete any signalling towards the GPRS system prior to issuing the HO command in the form of an RRC: Cell Change Order from UTRAN message. Once the UE has received this message then it moves onto the GPRS system and requests a Routing Area (RA) update. The RA update triggers a transition of the data flow from the 3G SGSN (Serving GPRS Support Node) to the 2G SGSN and down to the BSS. The packet switched data service is then able to continue on the GPRS system.

4.3.7.4 Compressed Mode Configurations

Section 4.3.6 introduced CM and the most important parameters. It also introduces the three methods that may be used to generate CM transmission gaps – i.e., puncturing, spreading factor division by 2 (SF/2) and Higher Layer Scheduling (HLS).

Uplink CM methods are limited to SF/2 and HLS. SF/2 CM does not necessitate the use of a lower transport format combination. The physical layer is reconfigured to convey the same quantity of data within a shorter period of time. The slots transmitted with a halved spreading factor require approximately 3 dB greater transmit power. SF/2 CM is applicable to all services except those that use a spreading factor of 4. HLS necessitates the use of a lower transport format combination. The Medium Access Control (MAC) layer within the UE is responsible for selecting an appropriate lower transport format combination. The result of using a lower transport format combination is a reduction in service throughput. Release '99 3GPP specifications do not support HLS for TrCHs using fixed starting positions. HLS is not appropriate for RT data services because of their sensitivity to variations in throughput. HLS can be applied to the Adaptive Multi-Rate (AMR) speech service if a lower AMR bit rate can be used.

Downlink CM methods are puncturing, SF/2 and HLS. Release '99 3GPP specifications do not support puncturing for TrCHs using flexible starting positions. Puncturing removes L1 bits after channel coding. Typically, the maximum total puncturing applied to a Traffic Channel (TCH) is such that approximately 70% of the coded bits remain after puncturing. Puncturing for the AMR speech service may necessitate the use of a lower AMR bit rate. Similar to the uplink, downlink SF/2 CM requires the physical layer to be reconfigured to convey the same quantity of data within a shorter period of time. SF/2 in the downlink direction may require the use of an alternative scrambling code. There is a one-to-one relationship between the original channelisation code and the channelisation code with half the spreading factor. If the channelisation code with half the spreading factor is blocked by another connection then it must be applied with an alternative scrambling code. The use of an alternative scrambling code means that downlink orthogonality is lost for that radio link and downlink transmit powers for all radio links will increase. Similar to the uplink, downlink HLS necessitates the use of a lower transport format combination. The MAC layer within the RNC is responsible for selecting an appropriate lower transport format combination. Release '99 3GPP specifications do not support HLS for TrCHs using fixed TrCH starting positions. The AMR speech service uses fixed TrCH starting positions in the downlink if the UE is required to complete blind detection of the AMR bit rate.

One of the most important CM parameters is TGL. TGL defines the duration of the CM transmission gap – i.e., the quantity of time that the UE has available to move onto the GSM system, complete its measurements and then move back to the UMTS system. Selecting an appropriate TGL involves balancing the efficiency with which measurements can be made with the impact upon L1 and L2 performance. Large TGLs are more efficient in terms of the number of measurements that can be completed per unit time but have a greater impact upon L1 and L2. L1 performance is affected in terms of inner-loop PC. Larger TGLs result in less correlation between the propagation conditions before and after the transmission gap. This makes it more difficult for the

Table 4.5 The impact of transmission gap length on GSM received signal strength indicator measurements.

TGL [slots]	No. of GSM RSSI samples	No. of GSM RSSI samples per slot	Time to complete eight GSM RSSI measurements (three samples per measurement)
3	1	0.33	960 ms
4	2	0.50	480 ms
5	3	0.60	320 ms
7	6	0.86	160 ms
10	10	1.00	120 ms
14	15	1.07	80 ms

inner-loop PC to recover. In the case of CM by HLS, larger TGLs require the use of lower transport format combinations and result in lower L2 throughput. In the case of CM by SF/2, larger TGLs require the use of the double-frame approach meaning that two radio frames rather than a single radio frame have their spreading factor reduced.

Table 4.5 presents the relationship between TGL and the minimum requirement for the UE's ability to sample GSM RSSI. These figures have been extracted from [9]. The third column shows the efficiency with which measurements are made. Also included in the table is the equivalent time required to complete eight GSM RSSI measurements based upon three samples per measurement and a TGPL of four radio frames.

GSM RSSI measurements are made without acquiring GSM synchronisation and do not require the CM transmission gap to coincide with a particular section of the GSM radio frame. The measurement efficiency becomes relatively poor for TGLs of less than seven slots. A TGL of seven slots balances the efficiency but with an impact on the inner-loop PC.

In the case of BSIC verification, the frame structure and timing of the GSM system has a more significant impact on the required TGL. The GSM system is based on an eight-slot radio frame structure with a duration of 4.615 ms. The first slot of each frame is dedicated to the BCCH. The BSIC is broadcast periodically within the SCH of the BCCH. The UE has no knowledge of the timing of the GSM system and must capture 9 slots' worth of GSM data to be sure of capturing the BCCH. A CM TGL of 7 slots is equivalent to 4.667 ms and provides a high probability of capturing the BCCH. The fact that the BSIC is broadcast 5 times per 51 frames means that multiple transmission gaps are likely to be required. Table 4.6 presents the relationship between the TGL and the BSIC identification time that guarantees the UE at least two attempts at decoding the BSIC. These figures have been extracted from [9].

In practice BSIC identification times may be less than those presented in Table 4.6. It is possible that the UE manages to identify the BSIC within the first transmission gap. Longer TGLs and shorter TGPLs result in more rapid BSIC identification times.

The TGPL provides a tradeoff between the time spent in CM and the potential impact on L1 and L2 performance. Long TGPLs increase the time spent in CM. This means that CM must be triggered relatively early to prevent radio-link failure occurring prior to completing a successful IS-HO. Triggering CM relatively early means

Table 4.6 The impact of the transmission gap length on GSM BSIC verification.

TGL [slots]	Transmission gap pattern length	No. of transmission gap patterns	Equivalent time
7	3 frames	51	1.5 s
7	8 frames	65	5.2 s
10	12 frames	23	2.7 s
14	8 frames	22	1.8 s
14	24 frames	21	5.0 s

that it will also be triggered more frequently. TGPL should be defined such that CM can be triggered relatively late and less frequently. The benefit of using a long TGPL is that the inner-loop PC has more time to recover between transmission gaps. Throughput reductions caused by higher layer scheduling and L2 retransmissions will also be less frequent and thus will have lower average impact.

CM may be configured such that the UE has a fixed number of radio frames within which to complete its GSM RSSI measurements and a fixed number of radio frames to complete BSIC verification. The drawback of this approach is that the UE may complete its RSSI measurements very rapidly and subsequently have to wait until it can start BSIC verification. Alternatively the UE may not manage to complete its RSSI measurements within the fixed time and would then be forced to start BSIC verification without successful RSSI measurements. In this case, BSIC verification would have to be completed using the entire GSM neighbour list and the UE would have to report the GSM RSSI at the same time as reporting the BSIC. A different approach is to allow the UE to remain in CM for GSM RSSI measurements until instructed otherwise by the RNC. The RNC would be able to reconfigure the CM measurements for BSIC verification once the UE has provided sufficient RSSI measurements. In this case, BSIC verification could be completed using only the best GSM neighbour.

4.3.7.5 Common Issues

The definition of good inter-system neighbour cell lists is essential for reliable IS-HO performance. If neighbour lists are too short then missing neighbours may lead to failed IS-HOs. If the neighbour lists are too long then the UE measurement time increases and important neighbours may be removed from the list when the UE is in SHO. The initial definition of inter-system neighbour lists is part of the radio network planning process. The initial definition should be refined during pre-launch optimisation when, for example, RF scanner measurements or network performance statistics can be used to detect missing neighbours (see also Section 9.3.4.1).

If the RNC has reduced the GSM neighbour list to a single neighbour for BSIC verification then it is possible that the single neighbour is no longer available – i.e., the UE has moved out of its coverage area. This is more likely if the RNC has based its decision of which is the best GSM neighbour upon a single measurement report. Otherwise the UE may have difficulties synchronising and extracting the BSIC within the CM transmission gap. When GSM RSSI measurements or BSIC verification fail then the UE is unable to complete an IS-HO. It is then likely that the UE will trigger a further CM cycle and reattempt the HO procedure. Otherwise the UE may have moved back into good coverage or moved completely out of coverage and dropped the connection.

Once the HO command or cell change order has been issued by the RNC then the UE has a limited period of time to successfully connect to the GSM system. If connection is not achieved within this limited period of time then the UE returns to the UMTS system and issues a failure message. In the case of packet switched data services, GSM cell reselection after receiving the cell change order can slow down the IS-HO procedure. This may occur if the UE has moved onto a non-ideal GSM neighbour.

4.4 Congestion Control

In WCDMA it is of the utmost importance to keep the air interface load under pre-defined thresholds. The reasoning behind this is that excessive loading prevents the network from guaranteeing the needed requirements. The planned coverage area is not provided, capacity is lower than required and the QoS is degraded. Moreover, an excessive air interface load can drive the network into an unstable condition. Three different functions are used in this context, all summarised here under *congestion control*:

- *Admission Control* (AC), handling all new incoming traffic. It checks whether a new packet or circuit switched RAB can be admitted to the system and produces the parameters for the newly admitted RABs.
- *Load Control* (LC), managing the situation when system load has exceeded the threshold(s) and some countermeasures have to be taken to get the system back to a feasible load.
- *Packet Scheduling* (PS), which handles all the NRT traffic – i.e., packet data users. Basically, it decides when a packet transmission is initiated and the bit rate to be used.

4.4.1 Definition of Air Interface Load

Since WCDMA systems have the possibility of uplink and downlink being asymmetrically loaded, the tasks of congestion control have to be done separately for both links. Two different approaches can be used for measuring the load of the air interface. The first defines the load via the received and transmitted wideband power; the second is based on the sum of the bit rates allocated to all currently active bearers. The quantities have already been introduced in Chapter 3 and are thus only summarised here.

Wideband Power-based Uplink Loading

In this approach the Node B measures the total received power, *PrxTotal*, which can be split into three parts:

$$PrxTotal = Iown + Ioth + P_N \qquad (4.15)$$

where *Iown* is the received power from users in the own cell; *Ioth* comes from users in the surrounding cells; and P_N represents the total noise power, including background and receiver noise as well as interference coming from other sources (see Section 5.4). Two quantities representing the uplink loading can be derived from Equation (4.15). The first is called the uplink load factor, η_{UL}, and is defined as:

$$\eta_{UL} = \frac{Iown + Ioth}{PrxTotal} \tag{4.16}$$

The second quantity is called the uplink noise rise, *NR*, and can be derived as follows:

$$NR = \frac{PrxTotal}{P_N} = \frac{1}{1 - \eta_{UL}} \tag{4.17}$$

Throughput-based Uplink Loading
The definition of uplink loading follows the derivation in Section 3.1.1.1 and is based on the sum of the individual load factors of each user k:

$$\eta_{UL} = \sum_k \frac{1}{1 + \dfrac{W}{\rho_k \cdot R_k \cdot \nu_k}} \cdot (1 + i) \tag{4.18}$$

where W is the chip rate; and ρ_k, R_k and ν_k are the E_b/N_0 requirement, the bit rate and the service activity of user k, respectively.

Wideband Power-based Downlink Loading
One method of defining the air interface loading in the downlink direction is simply by dividing the total currently allocated transmit power at the Node B, *PtxTotal*, by the maximum transmit power capability of the cell, *PtxMax*:

$$\eta_{DL} = \frac{PtxTotal}{PtxMax} \tag{4.19}$$

Throughput-based Downlink Loading
The first way to define the downlink loading based on throughput is similar to that used in the wideband power-based approach: The loading is the sum of the bit rates of all currently active connections divided by the specified maximum throughput for the cell:

$$\eta_{DL} = \frac{\displaystyle\sum_{k=1}^{N} R_k}{Rmax} \tag{4.20}$$

where R_k is the bit rate of connection k; and N is the total number of connections. Note that in the summation the bit rates from the common channels also have to be included.

Alternatively, downlink loading can be defined as derived in Section 3.1.1.2 and simplifying Equation (3.9) by introducing an average orthogonality $\bar{\alpha}$ and an average downlink other-to-own-cell-interference ratio i_{DL}:

$$\eta_{DL} = [(1 - \bar{\alpha}) + i_{DL}] \cdot \sum_{k=1}^{N} \left(\frac{\rho_k \cdot R_k \cdot \nu_k}{W} \right) \tag{4.21}$$

where W is the chip rate; and ρ_k, R_k and ν_k are the E_b/N_0 requirement, the bit rate and the service activity of connection k, respectively.

4.4.2   Admission Control

This section describes the tasks performed in AC and the parameters involved. AC is the main location that has to decide whether a new RAB is admitted or a current RAB can be modified. Because of the different nature of the traffic, AC consists of basically two parts. For RT traffic (the delay-sensitive conversational and streaming classes) it must be decided whether a UE is allowed to enter the network. If the new radio bearer would cause excessive interference to the system, access is denied. For NRT traffic (less delay-sensitive interactive and background classes) the optimum scheduling of the packets (time and bit rate) must be determined after the RAB has been admitted. This is done in close cooperation with the packet scheduler (Section 4.4.3). The AC algorithm estimates the load increase that the establishment or modification of the bearer would cause in the RAN. Separate estimates are made for uplink and downlink. Only if both uplink and downlink admission criteria are fulfilled is the bearer setup or modification request accepted, the RAB established or modified, or the packets sent. Load change estimation is done not only in the access cell, but also in the adjacent cells to take the inter-cell interference effect into account, at least in the cells of the active set. The bearer is not admitted if the predicted load exceeds particular thresholds either in the uplink or downlink. In the decision procedure, AC will use thresholds produced during radio network planning and the uplink interference and downlink transmission power information received from the wideband channel. To be able to decide whether AC accepts the request, the current load situation of the surrounding cells in the network has to be known and the additional load due to the requested service has to be estimated. Therefore, AC functionality is located in the RNC where all this information is available.

4.4.2.1   Wideband Power-based Admission Control

The uplink admission decision is based on cell-specific load thresholds given during radio network planning. An RT bearer will be admitted if the non-controllable uplink load, $PrxNC$, fulfils Equation (4.22) and the total received wideband interference power, $PrxTotal$, fulfils Equation (4.23):

$$PrxNC + \Delta I \leq PrxTarget \tag{4.22}$$

$$PrxTotal \leq PrxTarget + PrxOffset \tag{4.23}$$

where $PrxTarget$ is a threshold and $PrxOffset$ is an offset thereof, defined during radio network planning. For NRT bearers only the latter condition is applied. The non-controllable received power, $PrxNC$, consists of the powers of RT users, other-cell users, and noise. ΔI is the increase of wideband interference power that the admission of the new bearer would cause. For its estimation in [2] two methods are proposed. The first is called the derivative method and defines the power increase as:

$$\Delta I \approx \frac{PrxTotal}{1 - \eta} \cdot \Delta L \qquad (4.24)$$

where η is calculated with Equation (4.16). The second approach is called the integration method. Here the power increase is estimated to be:

$$\Delta I \approx \frac{PrxTotal}{1 - \eta - \Delta L} \cdot \Delta L \qquad (4.25)$$

In both Equations (4.24) and (4.25) the fractional load ΔL of the new user can be calculated as derived in Section 3.1.1:

$$\Delta L = \frac{1}{1 + \dfrac{W}{\rho \cdot R \cdot \nu}} \qquad (4.26)$$

where W is the chip rate; ρ the required E_b/N_0; and ν the service activity of the new bearer.

For the downlink direction a similar admission algorithm as in the uplink is defined. An RT bearer will be admitted if the non-controllable downlink load, $PtxNC$, fulfils Equation (4.27) and the total transmitted wideband power, $PtxTotal$, fulfils Equation (4.28).

$$PtxNC + \Delta P \le PtxTarget \qquad (4.27)$$

$$PtxTotal \le PtxTarget + PtxOffset \qquad (4.28)$$

where $PtxTarget$ is a threshold; and $PtxOffset$ is an offset thereof defined during radio network planning. For NRT bearers only the latter condition is applied. The non-controllable transmitted power, $PtxNC$, consists of the powers of RT users, other-cell users and noise. ΔP can be based on the initial transmit power estimated by the open-loop PC as specified in Section 4.2.1.

4.4.2.2 Throughput-based Admission Control

The throughput-based AC is pretty simple by nature. The strategy is simply that a new bearer is admitted only if the total load after admittance stays below the thresholds defined during radio network planning. In the uplink this means that:

$$\eta_{oldUL} + \Delta L \le \eta_{thresholdUL} \qquad (4.29)$$

must be fulfilled, and in the downlink:

$$\eta_{oldDL} + \Delta L \le \eta_{thresholdDL} \qquad (4.30)$$

where η_{oldUL} and η_{oldDL} are the network load before the bearer request, estimated with Equations (4.20) and (4.21); and ΔL is the load increase calculated with Equation (4.26).

4.4.3   Packet Scheduling

4.4.3.1   Packet Data Characteristics

The RAN provides a capability to allocate RAB services for communication between the CN and the UE. RAB services realise the RAN part of end-to-end QoS. They have different characteristics according to the demands of different services and applications. In the UMTS QoS concept, RAB services are divided into four traffic classes, according to the delay sensitivity of the traffic. These traffic classes are:

- conversational class;
- streaming class;
- interactive class;
- background class.

Conversational class is meant for traffic that is very delay-sensitive, while background class is the most delay-insensitive traffic class. Conversational and streaming classes are intended to carry RT services between the UE and either a circuit or packet switched CN. Typical examples of packet switched RT services are Voice over IP (VoIP) and multimedia streaming of audio, video or data. Interactive and background classes are intended to carry NRT services between the UE and a packet switched CN. The characteristics of interactive and background class bearers are that they do not have transfer delay or guaranteed bit rates defined. Due to looser delay requirements, compared with conversational and streaming classes, both NRT classes provide better error rate by means of channel coding and retransmission. Retransmissions over the radio interface allow the use of a much higher BLER for NRT packet data on the radio link, while still fulfilling the residual BER target that is part of the QoS definition.

Typical characteristics of NRT packet data are the bursty nature of traffic. A packet service session contains one or several packet calls depending on the application. The packet service session can be considered as an NRT RAB duration and the packet call as an active period of packet data transmission. During a packet call several packets may be generated, meaning that the packet call constitutes a bursty sequence of packets. UMTS QoS classes and traffic modelling are described in more detail in Chapter 8.

PS can be considered as the scheduling of data of the NRT RABs – i.e., interactive and background class bearers over the radio interface in both the uplink and downlink. Conversational and streaming classes are delay-sensitive and require dedicated resources for the whole duration of the connection. Radio resource allocation for RT packet switched bearers is an AC function and thus not considered in this section.

4.4.3.2   WCDMA Packet Access

WCDMA packet access is controlled by the packet scheduler, which is part of the RRM functionality in the RNC. The functions of the packet scheduler are:

- to determine the available radio interface resources for NRT radio bearers;
- to share the available radio interface resources between the NRT radio bearers;
- to monitor the allocations for the NRT radio bearers;

- to initiate TrCH-type switching between common, shared and dedicated channels when necessary;
- to monitor the system loading;
- to perform LC actions for the NRT radio bearers when necessary.

As shown in Figure 4.13, AC and the packet scheduler both participate in the handling of NRT radio bearers.

AC takes care of admission and release of the RAB. Radio resources are not reserved for the whole duration of a connection but only when there is actual data to transmit. The packet scheduler allocates appropriate radio resources for the duration of a packet call – i.e., active data transmission. As shown in Figure 4.13, short inactive periods during a packet call may occur, due to bursty traffic.

PS is done on a cell basis. Since asymmetric traffic is supported and the load may vary a lot between the uplink and downlink, capacity is allocated separately for both directions. However, when a channel is allocated to one direction, a channel has to be allocated in the other direction as well, even if the capacity need was triggered only for one direction. The packet scheduler allocates a channel with a low data rate for the other direction, which carries higher layer (TCP) acknowledgements, data link layer (RLC) acknowledgements, data link layer control and PC information. This low bit rate channel is typically referred as the 'return channel'.

Packet scheduler functionality consists of UE- and cell-specific parts. The main functions of the UE-specific part are traffic volume measurement management for each UE TrCH, taking care of UE radio access capabilities and monitoring allocations for NRT radio bearers. SHO is also possible for the DCHs allocated to NRT radio bearers. During SHO, PS is done in every cell in the active set, and the UE-specific part of the PS function is the controlling entity between the cell-specific functions.

The cell's radio resources are shared between RT and NRT radio bearers. The proportions of RT and NRT traffic fluctuate rapidly. It is characteristic of RT traffic that the load caused by it cannot be controlled efficiently. The load caused by RT traffic, interference from other-cell users and noise together is called

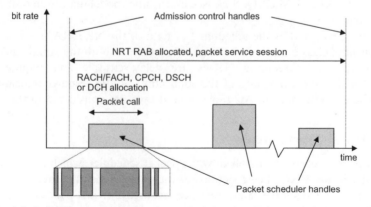

Figure 4.13 Admission control and packet scheduler handle non-real time radio bearers together.

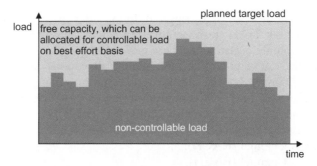

Figure 4.14   Capacity division between non-controllable and controllable traffic.

the non-controllable load. The available capacity that is not used for non-controllable load can be used for NRT radio bearers on a best effort basis, as shown in Figure 4.14. The load caused by best effort NRT traffic is called controllable load.

PS as well as RRM in general can be based on, for example, powers, throughputs and spectrum efficiency. Figure 4.15 shows the input measurements for a packet scheduler.

The Node B performs received uplink total wideband power (RSSI) and downlink transmitted carrier and radio link power measurements, and reports them to the RNC over the Iub interface using the NBAP signalling protocol. Throughput measurements can be performed in the RNC. If spectrum efficiency is taken into account, the P-CPICH E_c/I_0 measurement can be used to estimate transmission power. Traffic volume measurements can trigger radio resource allocation for NRT radio bearers. Traffic volume measurements are controlled by the RNC. The UE measures uplink TrCH traffic volumes and sends measurement reports to the RNC. Measurement

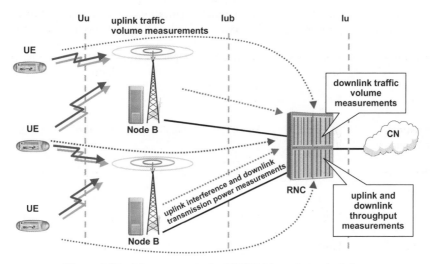

Figure 4.15   Measurements for WCDMA packet scheduler.

reporting can be periodical or event-triggered. In the latter case the measurement report is sent when the uplink TrCH traffic volume exceeds the threshold given by the RNC. Downlink traffic volume measurements are performed by the RNC.

According to the UE state and current channel allocations, system load, the radio performance of different TrCHs, the load of common channels and TrCH traffic volumes the packet scheduler selects an appropriate TrCH for the NRT radio bearer of the UE. The following TrCHs are applicable for packet data transfer:

- Dedicated transport Channel (DCH);
- Random Access Channel (RACH);
- Forward Access Channel (FACH);
- Common Packet Channel (CPCH);
- Downlink Shared Channel (DSCH).

Table 4.7 shows the key properties of these TrCHs. Applicable TrCH configurations for packet data in the uplink/downlink are DCH/DCH, RACH/FACH, CPCH/FACH, DCH/DSCH. A comparision of DSCH and HS-DSCH can be found in Table 4.8.

Table 4.7 Properties of WCDMA transport channels applicable for packet data transfer (HS-DSCH see Table 4.8).

TrCH	DCH	RACH	FACH	CPCH	DSCH
TrCH type	Dedicated	Common	Common	Common	Shared
Applicable UE state	Cell_DCH	Cell_FACH	Cell_FACH	Cell_DCH	Cell_DCH
Direction	Both	Uplink	Downlink	Uplink	Downlink
Code usage	According to maximum bit rate	Fixed code allocations in a cell	Fixed code allocations in a cell	Fixed code allocations in a cell	Codes shared between several users
Power control	Fast closed-loop	Open-loop	Open-loop	Fast closed-loop	Fast closed-loop
SHO support	Yes	No	No	No	No
Targetted data traffic volume	Medium or high	Small	Small	Small or medium	Medium or high
Suitability for bursty data	Poor	Good	Good	Good	Good
Setup time	High	Low	Low	Low	High
Relative radio performance	High	Low	Low	Medium	Medium

4.4.3.3 Packet Scheduling Methods

The principle of load distribution in a WCDMA cell, which RRM functionality controls, is that load targets for total load in a cell for the uplink and downlink are set during radio network planning so that those will be the optimal operating points of the system load. In wideband power-based RRM the uplink total RSSI and downlink transmitted carrier power are the quantities measured by the Node B that are planned to be below the target values. Instantaneously these targets can be exceeded due to changes of interference and propagation conditions. If the system load exceeds the load threshold in either the uplink or downlink that are set during radio network planning, an overload situation occurs and LC actions are applied to return the load to an acceptable level.

The flow chart in Figure 4.16 shows the basic functionality of the packet scheduler. In addition to load target and overload threshold, the maximum allowed load increase and decrease margins are important parameters, to avoid peaks in interference and to maintain system stability.

Usually NRT users use the resources left from RT users, since the scheduling of NRT radio bearers happens on a best effort basis. It is, however, possible to configure dedicated resources for the NRT radio bearers, by using separate load targets for RT and NRT users, which are considered in AC.

When the NRT radio bearer is set up, the applicable TrCH configurations are determined. The possibility of using CPCH and DSCH channels depends on the UE radio access capability definitions. The CPCH and DSCH are both optional, whereas RACH, FACH and DCH are mandatory and always supported.

When data arrive at the RLC buffer, the TrCH type to be used has to be decided. Uplink TrCH-type selection between RACH, CPCH and DCH is performed by the

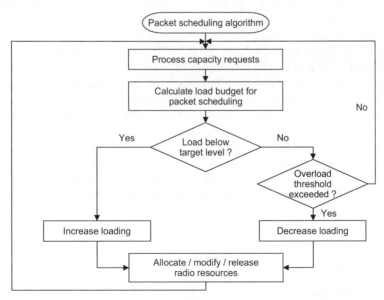

Figure 4.16   Flow chart of packet scheduling basic functionality.

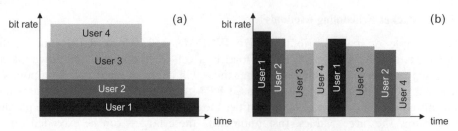

Figure 4.17   Basic packet scheduling approaches: (a) code division; (b) time division.

UE, based on the radio network planning parameters sent by the RNC. The parameters may include different thresholds for TrCH data volume that trigger the traffic volume measurement reporting or data transmission on RACH or CPCH. The RNC performs downlink TrCH-type selection between FACH, DSCH and DCH, which is also controlled by radio network planning parameters. The selection of the channel type used can be based on thresholds for TrCH traffic volume, system and common channel load, taking into account the performance over the radio interface.

The packet scheduler decides the bit rate and length of the allocation to be used. Several PS approaches can be utilised. Figure 4.17 illustrates the two basic approaches, which are:

- time division scheduling;
- code division scheduling.

In time division scheduling the available capacity is allocated to one or very few radio bearer(s) at a time. The allocated bit rate can be very high and the time needed to transfer the data in the buffer is short. The allocation time can be limited by setting the maximum allocation time, which prevents a high bit rate user from blocking others. Scheduling delay depends on load, so that the waiting time before a user can transmit data is longer when the number of users is higher. Time division scheduling is typically used for DSCH, where the scheduling of PDSCH can happen at a resolution of one 10 ms radio frame, but it can be also utilised for DCH scheduling.

In code division scheduling the available capacity is shared between a large number of radio bearers, allocating a low bit rate simultaneously for each user. Allocated bit rates depend on load, so that the bit rates are lower when the number of users is higher.

In practice, PS is a combination of these two approaches. When the packet scheduler decides the order of radio bearers to be allocated, different QoS differentiation methods can be utilised. The simplest is to use only arrival time as input (First In, First Out – FIFO) but also other factors – such as traffic classes, priorities of the bearers and spectrum efficiency – can be used. Since the spectrum is used more efficiently with higher bit rates, the bit rates allowed for PS can also be configured according to the network operator's preference.

4.4.4   Load Control

The main functionality of LC can be divided into two tasks. In normal circumstances LC takes care that the network is not overloaded and remains in a stable state. To achieve this, LC works closely together with AC and PS. This task is called 'preventive load control'. In very exceptional situations, however, the system can be driven into an overload situation. Then overload control is responsible for reducing the load relatively quickly and thereby bringing the network back into the desired operating area defined during radio network planning. LC functionality is distributed between Node B and RNC. The following list of actions can be performed to reduce the load:

- Fast LC actions located in Node B:
  ○ deny downlink or overwrite uplink TPC 'up' commands;
  ○ use a lower SIR target for the uplink inner-loop PC.
- LC actions located in the RNC:
  ○ interact with the packet scheduler and throttle back packet data traffic;
  ○ lower the bit rates of RT users – i.e., speech service or circuit switched data;
  ○ make use of WCDMA IF-HO or GSM IS-HO.
  ○ drop single calls in a controlled manner.

In wideband power-based LC, the measures to decide whether some LC action has to be taken are the total received interference power per cell, $PrxTotal$, in the uplink and the total transmission power per carrier, $PtxTotal$, in the downlink. It is a task during radio network planning to set the maximum allowed values for those quantities. For both links two thresholds can be defined:

- In the uplink:
  ○ $PrxTarget$, the optimal average of $PrxTotal$;
  ○ $PrxOffset$, the maximum margin by which $PrxTarget$ can be exceeded.
- In the downlink:
  ○ $PtxTarget$, the optimal average of $PtxTotal$;
  ○ $PtxOffset$, the maximum margin by which $PtxTarget$ can be exceeded.

If either of the first thresholds ($PrxTarget$ or $PtxTarget$) is exceeded, the cell enters the state where preventive LC actions are initiated. If either ($PrxTarget + PrxOffset$) or ($PtxTarget + PtxOffset$) is exceeded, the cell is moved to an overload state and overload control actions kick in. Figure 4.18 presents an overview of the inter-working actions of AC, PS and LC in the different load states defined by the above parameters.

The AC and PS functions together perform preventive LC actions, LC working as mediator between these two functions. LC updates the cell load status based on radio resource measurements and estimations provided by AC and PS. If the cell is in the normal load state, AC and PS can work normally. If the loads exceed the targets but are less than the specified overload thresholds, only preventive LC actions are performed. AC only admits new RT bearers if the RT load is below $PrxTarget$ or $PtxTarget$. The packet scheduler does not further increase the bit rate of the admitted NRT bearers. If the cell moves to an overload state, the packet scheduler starts to decrease the bit rates, for example, of randomly selected NRT bearers, taking into account the bearer classes and the priorities set by the operator within the same traffic class. However, the bit rate should not be reduced below the minimum allowed bit rate assigned during radio network planning to the selected bearer(s). Another possible way to reduce the load is to try to move NRT traffic from the DCH to FACH in case the FACH is not overloaded. In the most extreme case RT and NRT bearers might even be dropped.

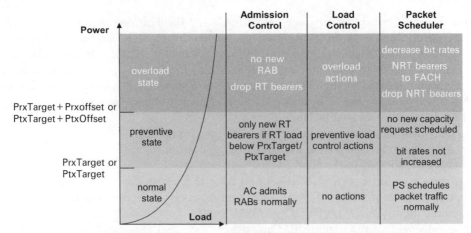

Figure 4.18 Example of inter-working actions of admission control, packet scheduler and load control to control system load if high-speed downlink packet access is not present.

4.5 Resource Management

The main function of the Resource Management (RM) is to allocate physical radio resources when requested by the RRC layer. To be able to do this the RM has to know all the necessary radio network configuration and state data, including the parameters affecting the allocation of logical radio resources.

The RM is located partly in the RNC and partly in Node B. It works in close cooperation with AC and PS: the actual input for resource allocation comes from AC/PS and the RM informs the packet scheduler about the resource situation.

The RM only sees the logical radio resources of a Node B and thus the actual allocation means that the RM reserves a certain proportion of the available physical radio resources according to the channel request from the RRC layer for each radio connection. In the channel allocation the RM attaches a certain spreading (or channelisation) code for each connection in the downlink direction. The length of the spreading code depends on the available codes at that moment and the requirement for a data rate in the channel request: the higher the rate the shorter the code. The RM has to be able to switch codes and code types for different reasons – e.g., SHO, defragmentation of the code tree, etc. The RM is also responsible for the allocation of scrambling codes for uplink connections. And obviously the RM has to be able to release the allocated resources as well.

4.5.1 The Tree of Orthogonal Channelisation Codes in Downlink

Orthogonal channelisation codes are used in WCDMA for channel separation within the same cell. If unshifted – i.e., channels are perfectly synchronised on a symbol level – the codes are perfectly pairwise orthogonal. Unfortunately, this assumption is not wholly justified due to multi-path propagation (delay spread). Consequently, there is mutual interference between different code channels on the receiving (UE) end.

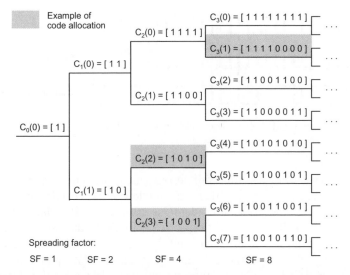

Figure 4.19 The tree of orthogonal short codes. High-speed downlink packet access-related issues with respect to the scrambling and spreading codes are introduced in Section 4.6.4.2.

The concept of parallel use of different codes is mainly used in the downlink. The uplink is connected with a single user, thus normally one code at a time is used.

The codes are just rows from a Hadamard matrix. They are based on Hadamard's work dating from the end of the 19th century. Orthogonality is preserved across different symbol rates (i.e., different spreading factors give different user data rates in parallel), but the selection of one short code will 'block' the sub-tree in both directions. This has an impact in the following ways:

- Codes must be allocated in the RNC.
- The code tree may become 'fragmented', so that code reshuffling is needed (arranged by the RNC).
- The allocation of codes is completely under the control of the RNC. A network planner or optimiser might have to interfere only in the case of constantly occurring problems – e.g., when a Node B is permanently running out of codes, which could happen with very high data rates typical of indoor applications – i.e., with low spreading factors. Nevertheless, in most cases AC or LC will take action first in the form of (soft) blocking.

An example of codes and code allocation policy can be seen in Figure 4.19. To maintain orthogonality a hierarchical selection of short codes from a code tree must be made.

4.5.2 Code Management

The WCDMA system divides spreading and scrambling (randomisation) into two steps. The user signal is first spread by the channelisation code and then scrambled by the scrambling code. This is similar to IS-95, but as 3G's WCDMA system is asynchronous,

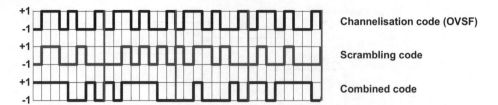

Figure 4.20 Spreading (SF = 8) and scrambling for all downlink physical channels except the synchronisation channel.

scrambling codes are not just time-shifted replicas of the same sequence, but the codes are really different from each other, having low cross-correlation properties. The scrambling code of the downlink identifies a whole cell, while in the uplink a scrambling code is call- or transaction-specific. In IS-95 the same (long) PN code is used in all cells as the scrambling code and they are separated with phases of the same code. This is possible since the BSs are synchronised. The planning of phaseshifts ensures that phaseshifts are longer than propagation delays, so that UEs do not hear any two cells having the same code phase. Such long code planning is definitely easier than frequency planning, but it is necessary and mistakes done could be a source of interference problems in some cases. The overall spreading and scrambling scenario is shown in Figure 4.20.

The basic assumption for good performance of a spread spectrum system with direct spreading such as WCDMA is for the UE to have a strong ability for fast synchronisation. There are two basic issues supporting each other:

- Implementation of the code acquisition strategy in the UE. The requirements are given by 3GPP [9]; the strategy and its implementation are specific to phone manufacturers.
- Scrambling code planning in the network. This task is carried out during radio network planning and described together with scrambling code optimisation in detail in Section 4.5.2.4.

4.5.2.1 Cell Search Procedure

The purpose of the cell search procedure is to find a suitable cell and to determine the downlink scrambling code and frame synchronisation of that cell. The cell search is typically carried out in the following three steps [1], also illustrated in Figure 4.21:

- **Step 1: Slot synchronisation.** During the first step of the cell search procedure the UE uses the SCH's primary synchronisation code to acquire slot synchronisation for a cell. This is typically done with a single matched filter (or any similar device) matched to the primary synchronisation code which is common to all cells. Slot timing of the cell can be obtained by detecting peaks in the matched filter output.
- **Step 2: Frame synchronisation and code group identification.** During the second step of the cell search procedure, the UE uses the SCH's secondary synchronisation code to find frame synchronisation and identify the code group of the cell found in the first step. This is done by correlating the received signal with all possible secondary

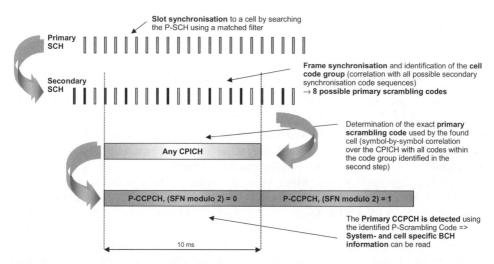

Figure 4.21 Example of the cell search procedure. If the user equipment has received information about which scrambling codes to search for, Steps 2 and 3 above can be simplified.

synchronisation code sequences and identifying the maximum correlation value. Since the cyclic shifts of the sequences are unique, the code group as well as frame synchronisation are determined.

- **Step 3: Scrambling code identification.** During the third and last step of the cell search procedure, the UE determines the exact primary scrambling code used by the cell found. The primary scrambling code is typically identified through symbol-by-symbol correlation over the P-CPICH with all codes within the code group identified in the second step. After the primary scrambling code has been identified, the P-CCPCH can be detected and the system- and cell-specific BCH information can be read.

4.5.2.2 Scrambling and Spreading Code Allocation for Uplink

In the uplink the spreading operation in WCDMA is done in two phases. The first is the channelisation operation, which transforms every data symbol into a number of chips. This increases the signal bandwidth. The number of chips per data symbol is called the Spreading Factor (SF). After this the scrambling operation is performed, meaning that a scrambling code is applied to the spread signal.

In channelisation the I- and Q-branches are independently multiplied by an orthogonal spreading code. The resulting signals are then scrambled by multiplying them by a complex-valued scrambling code.

Uplink channels are scrambled with a complex-valued scrambling code. There are 2^{24} long and 2^{24} short (length 256 chips) uplink scrambling codes. Either long or short scrambling codes can be used to scramble the DPCCH and DPDCH. In the uplink both the channelisation and the scrambling codes are allocated by the system and require little action during radio network planning. Uplink scrambling codes are call-specific and are allocated in connection establishment by the RNC. The uplink scrambling code space is divided between RNCs. Each RNC has its own planned range. The UE can use the same allocated code as long as it is connected to the 3G network.

4.5.2.3 Scrambling and Spreading Code Allocation for Downlink

In the downlink the symbols (non-spread physical channel) of the P-CCPCH, Secondary CCPCH (S-CCPCH), P-CPICH, PICH and DPCH are first converted and mapped onto I- and Q-branches. These branches are then spread by the same real-valued channelisation code. As a result the signal has its final chip rate. Then these chip sequences are scrambled by a complex-valued scrambling code. The channelisation codes in the downlink are the same as in the uplink. The channelisation codes for the P-CPICH and P-CCPCH are fixed; those for all other physical channels are assigned by the UTRAN. A total of $2^{18} - 1 = 262143$ long scrambling codes can be generated, but not all of them are used. The codes are divided into 512 sets each consisting of a primary scrambling code and 15 secondary scrambling codes. Furthermore, the set of primary scrambling codes is divided into 64 scrambling code groups, each consisting of 8 primary scrambling codes.

Each cell is allocated one and only one primary scrambling code. The P-CCPCH and P-CPICH are always transmitted using the primary scrambling code. The other downlink physical channels except the SCHs can be transmitted with either the primary or a secondary scrambling code from the set associated with the primary scrambling code of the cell. In case of parallel multi-code transmission, the mixture of primary scrambling code and secondary scrambling code for one CCTrCH is allowable. But, in the case of the CCTrCH of type DSCH then all the PDSCH channelisation codes that a single UE may receive have to be under a single scrambling code (either the primary or a secondary scrambling code). The same is applied for the case of CCTrCH of type HS-DSCH. Here all the HS-PDSCH channelisation codes and the HS-SCCH that a single UE may receive shall be under a single scrambling code.

The SCHs are under no scrambling code. They are formed by hierarchical Golay sequences to have optimal aperiodic autocorrelation properties to support fast slot boundary acquisition.

4.5.2.4 Downlink Scrambling Code Planning and Optimisation

The downlink *channelisation* codes are allocated by the UTRAN. Allocating the downlink *scrambling* codes and code groups to the cells is part of radio network planning.

As previously described, from 262143 possible long downlink scrambling codes a total of only 512 codes is used, subdivided into 64 groups each of 8 codes. All the cells a UE is able to measure in one location should have different scrambling codes. The simplest method is to use different scrambling code groups in neighbouring cells. This would ensure the previous requirement in most cases. The reuse could be 64, as there are 64 code groups. Another method that allocates as many codes as possible from the same code group to neighbouring cells could bring an advantage from the system point of view in the form of a less complex code search procedure for the UE.

In general, the speed of the code acquisition process depends on the match between scrambling code allocation in the network and the acquisition strategy applied in the mobile, which is manufacturer-specific. Nevertheless, any UE shall perform as required for any scrambling code allocation strategy. Both strategies are likely to have on average a similar performance. A discussion of both strategies can be found in [11] and [12]. A few planning rules that are recommended to keep in mind can be formulated as:

- A UE should never receive the same scrambling code from more than one cell. This can be achieved by explicitly specifying a minimum difference in received signal levels from the cells in question or – easier – by a minimum reuse distance.
- In no case can the same scrambling code be reused within one neighbour cell list.
- No repetition of one cell's scrambling code in any neighbour cell list of any neighbouring cells. Otherwise duplicated scrambling codes will arise when neighbour cell lists are combined during SHO.
- When inserting a new cell in the network plan, its scrambling code must be different in all neighbour cells and also in the neighbours' neighbours. Otherwise a neighbouring cell will have duplicate scrambling codes in its neighbour cell list.
- If network evolution must be considered in an early planning phase, a certain number of codes may be excluded from the initial planning and allocated during a second network rollout phase.

Scrambling code group planning for different RF carriers can be done independently. However, if the operator deploys Node Bs equipped with a second or more RF carriers, reusing the same scrambling code plan in all carriers is possible. This reduces the complexity of the network and eases the planning and optimisation work. A precondition for this strategy is obviously that all carriers also have the same neighbour cell definitions. *It should be noted that both neighbour cell definition and primary scrambling code planning are closely related and should always be done in conjunction.* The high number of codes enables code planning even manually; although this could be a very time-consuming task in large networks manual allocation is recommended only for small clusters.

Some special care needs to be taken in 3G networks in the area of international borders. Operators on both sides may use the same RF carrier and using then the same scrambling codes may result in problems. Limiting both sides to disjoint sets of scrambling codes in this case is the easiest way out. Regulatory organisations could be consulted in case the operators cannot achieve an agreement on the usage of scrambling codes. In Europe the ERC has issued a recommendation for operators following the above rule [13].

Code planning in WCDMA resembles frequency planning in the GSM. However, it can be seen that scrambling code planning in WCDMA is not such a key performance factor as is frequency planning in frequency division systems. In contrast to frequency planning, in scrambling code planning it is not crucial from the interference or synchronisation point of view which scrambling codes are allocated to neighbours as long as they are not the same.

4.6 RRU for High-speed Downlink Packet Access (HSDPA)

HSDPA is one of the major enhancements of the 3G cellular system introduced in Release 5 and is a high-speed version of the downlink shared channel known from earlier releases. The physical properties of HSDPA were introduced in Section 2.4.5. This section is devoted to the impacts of HSDPA on RRU procedures in the RAN. The main motivation was to account for the generally acknowledged asymmetry in uplink and downlink data transmission and its bursty nature. The main characteristics therefore are a short, fixed packet TTI, Adaptive Modulation and Coding (AMC) and a fast L1 retransmission (H-ARQ) based on feedback in the uplink direction (ACK/NACK and CQI). A short but comprehensive introduction to HSPDA can be found, for example, in [2] or [14]. The main differences to the DSCH introduced in Sections 2.4.3.2 and 4.4.3 are summarised in Table 4.8.

Table 4.8 Fundamental differences between Release '99 DSCH and Release 5 HS-DSCH.

Feature	DSCH	HS-DSCH
Variable spreading factor	Yes (4–256)	No (fixed at 16)
Fast power control	Yes	No
Adaptive coding and modulation	No	Yes
Fast L1 retransmission	No	Yes (H-ARQ)
Multicodes	Yes	Yes
Location of control	RNC	Node B

4.6.1 Power Control for High-speed Downlink Packet Access

In principle, for HSDPA, there is no 'classical' WCDMA PC at all. The radio resource allocation policy uses rather the maximum available HSDPA power for a certain short time for a certain connection and maximises the data throughput for that period. The available power for HSDPA is a radio network parameter and can be set per Node B.

The HSDPA channel is accompanied by relevant control channels, which may or may not be power-controlled. There are two HSDPA channels on the downlink direction: the High-speed Physical Downlink Shared Channel (HS-PDSCH) carrying the user data and the High-speed Shared Control Channel (HS-SCCH) carrying control information. The third HSDPA-specific channel is used in the uplink direction for feedback information from the UE: High-speed Dedicated Physical Control Channel (HS-DPCCH). The behaviour of the channels is defined by [1] as follows.

High-speed Shared Control Channel
The HS-SCCH PC is under the control of Node B. It may, for example, follow the PC commands sent by the UE to Node B or any other PC procedure applied by Node B and based on feedback information. Another possibility would be to simply apply an offset to the power of the downlink DCH. As can be concluded, the PC behaviour of the channel is thus vendor-specific.

High-speed Physical Downlink Shared Channel

The HS-PDSCH power setting is also under the control of Node B. When the HS-PDSCH is transmitted using 16 State Quadrature Amplitude Modulation (16QAM), the UE may assume that the power is kept constant during the corresponding HS-DSCH sub-frame. In case of multiple HS-PDSCH transmissions to one UE (multi-code transmission), all the HS-PDSCHs intended for that particular UE will be transmitted with equal power.

The sum of the powers used by all HS-PDSCHs and HS-SCCHs in a cell cannot exceed the maximum value of the *HS-PDSCH and HS-SCCH total power* signalled by higher layers [8]. Instead of using PC on the HS-PDSCH, the modulation and coding scheme is changed based on the channel conditions (Link Adaptation, LA). Dependent on the uplink feedback information and a proprietary algorithm, Node B selects the best suited modulation from the available Quaternary Phase Shift Keying (QPSK) and 16QAM and the best code rate, together denoted as Transport Format and Resource Combination (TFRC). The allowed combinations of TFRCs can be found in [15] and [16], a selection with the corresponding throughput is collected in Table 4.9.

Table 4.9 Example transport format and resource combinations and theoretically achievable throughput [2].

TFRC	Modulation	Code rate	Max. throughput [Mbps] (15 codes)
1	QPSK	1/4	1.8
2	QPSK	2/4	3.6
3	QPSK	3/4	5.3
4	16QAM	2/4	7.2
5	16QAM	3/4	10.7

Dedicated Physical Control Channel/High-speed Dedicated Physical Control Channel in Uplink Direction

For the uplink direction, a power difference between DPCCH/HS-DPCCH could be applied to adjust the high-speed feedback channel performance. This difference is independent from the inner-loop PC. When an HS-DPCCH is active, the relative power offset $\Delta_{HS-DPCCH}$ between the DPCCH and the HS-DPCCH for each HS-DPCCH slot is set. The offset could be different for HS-DPCCH slots carrying Hybrid Automatic Repeat reQuest (H-ARQ) Acknowledgement (ACK) $\Delta_{HS-DPCCH} = \Delta_{ACK}$ or Negative Acknowledgement (NACK) $\Delta_{HS-DPCCH} = \Delta_{NACK}$ and for HS-DPCCH slots carrying a Channel Quality Indicator (CQI) $\Delta_{HS-DPCCH} = \Delta_{CQI}$ (see Figure 4.22). The values for Δ_{ACK}, Δ_{NACK} and Δ_{CQI} are parameters set by higher layers, which can be quantised into nine steps $(0, \ldots, 8)$. Mapping onto amplitude ratios can be found in [17, table 1A]; for other details see also [1].

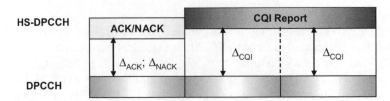

Figure 4.22 HS-DSCH–DPCCH power offsets.

4.6.2   Congestion Control for High-speed Downlink Packet Access

4.6.2.1   Admission Control for High-speed Downlink Packet Access

In case HSDPA transmission is supported in a Node B, then the AC has to be modified to take the power resources of the HSDPA channels into account. How much power will be allowed to be used is based on proprietary algorithms. One example would be that the RNC informs Node B in certain periods about the allowed power. Another could be that Node B is allowed to use any unused power for HSDPA. Whether or not to allow HSDPA transmission to be started, similar targets and thresholds as introduced in Section 4.4.2 could be used; one example can be seen in Figure 4.23.

The admission decision for the first HSDPA user could follow Equation (4.31):

$$PtxTotal \leq PtxTargetHSDPA \qquad (4.31)$$

where *PtxTotal* is the sum of the controllable and non-controllable instantaneous power measured by Node B.

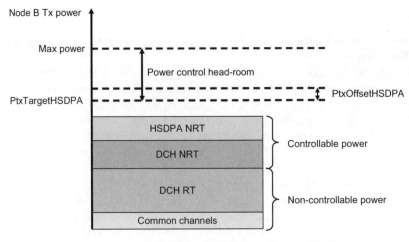

Figure 4.23 Downlink power budget for cells with HSDPA.

4.6.2.2 Load Control for High-speed Downlink Packet Access

Overload control actions are required for similar reasons to those discussed earlier in Section 4.4.4 and shall include the strategies introduced there. An additional requirement in the case of HSDPA would simply be that if, for example, Equation (4.32) is fulfilled, then HSDPA transmission is stopped and only resumed in case Equation (4.31) is again satisfied:

$$PtxNonHSDPA \geq PtxTargetHSDPA + PtxOffsetHSDPA \qquad (4.32)$$

where *PtxNonHSDPA* is the transmit power allocated to connections not applying HSDPA. Which type of NRT traffic, HSDPA or non-HSDPA, will first be restricted may be fixed in the implementation or left for the operator to choose according to their own strategy to prioritise either HSDPA or DCH NRT.

4.6.2.3 Packet Scheduling for High-speed Downlink Packet Access

The computational effort, the shortness of the allocation period and the fast H-ARQ transmission make it necessary for the packet scheduler for HSDPA to be located in Node B with its own MAC-hs. Another reason is the high number of AMCs, which should allow for rapid adjustments of the transmission formats to the current channel conditions. On top of this comes the fact that HSDPA uses the concept of a shared channel, so that in total this makes a very efficient means to serve high bit rates to individual users. The following inputs can be seen to have an impact on PS strategies:

- available system resources;
- data amount to be scheduled;
- instantaneous channel conditions of each user;
- QoS requirements (delay, throughput) of each user;
- capability classes supported by different UEs;
- SHO condition of the connections.

Various allocation strategies have been investigated (e.g., [18] and [19]) and since 3GPP does not require a certain one, the choice is on the vendors' or operators' sides. The main representatives are the Round Robin (Fair Resource) and the Proportional Fair algorithms. The Round Robin method shares the available resources (codes and powers) equally amongst all UEs – i.e., without exploiting any *a priori* knowledge of the channel conditions – while the better the channel conditions are for the UE, the higher the capacity allocated to the Proportional Fair algorithm. The first one guarantees a solid 'best effort' throughput on a low-complexity basis, the latter maximises cell throughput at the cost of much higher complexity.

4.6.3 *Handover Control and Mobility Management for High-speed Downlink Packet Access*

Compared with the DCHs in Release '99, the fundamental difference between the HC and mobility management involving cells where HSDPA is enabled comes from the issue that downlink channels involved in the HSDPA transmission (HS-PDSCH and

HS-SCCH) can neither be in soft nor in softer handover – i.e., they can only belong to one link in the active set of a UE. The cell to which this link belongs is called the 'serving HS-DSCH cell'. In case a certain UE also has DCHs allocated, those DCHs may or may not be in SHO. In order to make full mobility possible between cells supporting HSDPA or not, the following procedures have been specified in 3GPP [16] and they are explained in detail in [2, §11.7]:

- *High-speed Downlink Shared Channel to High-speed Downlink Shared Channel HO*, where an HSDPA connection is changed from one cell supporting HSDPA directly to another. This event is further refined so that it is possible:
  o without simultaneous update of the active set – i.e., for Release '99 DCHs; or
  o in combination with 'regular' HHO or SHO of existing DCHs.

Depending on whether or not the source and the target cells belong to the same Node B, the event is called *intra-* or *inter-Node B* HS-DSCH to HS-DSCH HO. In the latter case, the source and target cell may even belong to different RNCs. In any case, the procedure must be transparent to the UE – i.e., it must not be aware whether or not the source and target cell are within the same Node B.

- *High-speed Downlink Shared Channel to Dedicated Channel HO*, which is required in case the coverage of HSDPA ends and the target cell does not support HSDPA.
- *Dedicated Channel to High-speed Downlink Shared Channel HO*, in case the UE moves from a source cell not supporting HSDPA to a target cell that does.

The measurements and the reporting thereof to determine the active set of a UE were described in Section 4.3. In general, the RNC is in charge of determining which cells to include or exclude from the active set. Also in the event of HO in the HSDPA case, the UE is responsible for making the appropriate measurements. Also here the decision to which cell of the active set an HSDPA connection is established is in the responsibility of the SRNC, based on the measurement reports of the UE and some, in general, proprietary algorithm. It could be simply the best cell (based on P-CPICH E_c/I_0 or $RSCP$) within the current active set or from a subset of the cells of the candidate set (see Section 4.3) fulfilling a certain window criteria and supporting HSDPA. In case AC prohibits the selection, the next best cell can be chosen.

One possibility for initiating a serving HSDPA cell change or HO could be simply to exploit event 1D (change of best server, based in P-CPICH E_c/I_0 or $RSCP$, see Section 4.3.5.2), which can also be enhanced by the mechanisms described earlier (hysteresis, time-to-trigger mechanism, cell-individual offsets, etc.), but also decisions involving other reporting events as defined by 3GPP can be applied. Periodical reporting may be especially attractive in this case or in general any active set update can also trigger re-evaluation of the best candidate for the serving HS-DSCH cell.

In case the HSDPA coverage of a cell is smaller than the DCH coverage, another mechanism denoted as 'HS-DSCH–DCH fallback' is initiated. Reasons to trigger such a procedure may be, for example, event 1F (a P-CPICH becomes worse than an absolute threshold) or UE-related events 6A (UE transmit power becomes bigger than an absolute threshold) or 6D (UE transmit power reaches its maximum).

4.6.4 Resource Manager for High-speed Downlink Packet Access

This section introduces the additions to the RM due to HSDPA transmission. They can be seen mainly in managing the code tree – i.e., allocation of the channelisation codes (power allocation was handled in Section 4.6.1). In case of HSDPA the same principles for code allocation are applied as for the Release '99 channels introduced in Section 4.5 with the exceptions or restrictions described in the following sections.

4.6.4.1 Scrambling and Spreading Code Allocation in Uplink for High-speed Downlink Packet Access

For HSDPA-enabled cells in the uplink direction, the same *scrambling* code as for the other uplink Release '99 channels shall be applied.

The *spreading* code, C_{ch}, applied for the spreading of the HS-DPCCH, is dependent on the number of maximum available DPDCHs, N_{max}, in that cell. Three different fixed values are specified in [17] and collected in Table 4.10.

Table 4.10 Channelisation codes for high-speed dedicated physical control channel.

Number of maximum available DPDCHs, N_{max}	Channelisation code C_{ch}
1	$C_{ch,256,64}$
2, 4, 6	$C_{ch,256,1}$
3, 5	$C_{ch,256,32}$

4.6.4.2 Scrambling and Spreading Code Allocation in Downlink for High-speed Downlink Packet Access

Also in the downlink direction in HSDPA-enabled cells the same *scrambling* code as for the Release '99 channels shall be used for both HSDPA channels, HS-PDSCH and HS-SCCH.

For the spreading codes, the spreading factors are fixed. For HS-PDSCH, the spreading factor is always 16 and for the HS-SCCH, the spreading factor has a mandatory value of 128 [17]. The channelisation codeset information is reported via the HS-SCCH. Orthogonal Variable Spreading Factor (OVSF) codes must be allocated in such a way that they are positioned in sequence in the code tree. That is, for P multi-codes starting at offset O the following codes are allocated:

$$C_{ch,16,O} \cdots C_{ch,16,O+P-1}$$

The number of multi-codes and the corresponding offset for HS-PDSCHs is signalled in the HS-SCCHs. The controlling RNC is responsible for the allocation of the spreading codes.

4.7   Impact of Radio Resource Utilisation on Network Performance

4.7.1   Impact of Fast Power Control and Soft Handover on Network Performance

The results presented in this section are based on [20] and [21]. Simulations have been performed with parameters that are not fully compatible with the current 3GPP specifications, but the trends visible in the results do also apply to the current standard.

4.7.1.1   Impact of Fast Power Control

In WCDMA radio network dimensioning and planning the link level performance should be modelled as accurately as possible. Various services must be taken into consideration, with different bit rates, multiplexing and channel-coding schemes. In this section only one of the fundamental issues is discussed: how to model the effects of the fast PC in the uplink. Accurate PC is one of the basic requirements for high WCDMA system capacity. The transmit powers must be kept as low as possible in order to minimise interference, and just high enough to ensure the required QoS. Even though a relatively slow PC algorithm would be able to compensate for large-scale attenuation, distance attenuation and shadow fading, fast PC is needed for multi-path fading, in the case of slowly moving UEs. This is because for low-speed UEs interleaving does not provide enough diversity. In this section, first the statistics of transmit powers are analysed in the case of ideal PC. Ideal PC keeps the received SIR constant over time. The deviation of the statistics of a real PC from the ideal PC is shown with the help of uplink link level simulations. A method is proposed for taking the effects of fast PC into account in interference estimation. This is clarified by a numerical example. The effects of fast PC on WCDMA cell ranges are discussed. Link level simulation results with a very limited PC range are presented. Furthermore, a definition of the fast PC headroom to be used in cell range calculation is suggested.

The WCDMA reference system studied here is based on [22] in which a fast closed-loop PC is specified for both the uplink and downlink. The system operates around the 2 GHz frequency band with a 4.096 Mcps chip rate. 3GPP-compatible E_b/N_0 values, including PC errors, can be found in Table 4.2. The numbers presented in this section are only indicative and, therefore, should be seen as certain trends only, but should not be taken as absolute estimates of the performance.

Ideal Power Control
The instantaneous transmit power of the UE is denoted by p_t. In ideal PC p_t is set so that the received bit-energy to interference-spectral-density ratio (E_b/I_0) is constant just ensuring the desired quality. Here it is assumed that interference is close to Additive White Gaussian Noise (AWGN), which is a reasonable assumption in CDMA.

The ideal PC equation can be written as:

$$G \cdot \frac{p_t \cdot X}{I} = \rho \tag{4.33}$$

where I is the interference power at the Node B; G is the processing gain; ρ is the required bit-energy-to-noise-spectral-density ratio; and X is the instantaneous channel gain varying under multi-path conditions. It can be assumed that the expectation value of X is 1, $E(X) = 1$, since only fast PC effects are studied here. As the PC keeps the SIR constant, p_t can be solved from (4.33):

$$p_t = \frac{\rho \cdot I}{G} \cdot \frac{1}{X} \tag{4.34}$$

Thus the statistics of the transmit power are those of the inverse channel gain Y, $Y = 1/X$. In the following the expectation value of Y is calculated for some special cases. This is called here the average transmit power rise caused by fast PC. Assuming the signal is received by an ideal RAKE receiver using ideal maximal ratio combining of L multi-paths, X and its Probability Density Function (PDF) f_X can be written as (see, e.g., [23, p. 802]):

$$\left. \begin{array}{ccc} X = X_1 + \cdots + X_L, & E(X_k) = \bar{\gamma}_k, & k = 1, \ldots, L \\ f_X(x) = \sum_{k=1}^{L} \frac{\pi_k}{\bar{\gamma}_k} e^{-x/\bar{\gamma}_k}, & \pi_k = \prod_{j=1, j \neq k}^{L} \frac{\bar{\gamma}_k}{\bar{\gamma}_k - \bar{\gamma}_j} \end{array} \right\} \tag{4.35}$$

and for L equally strong Rayleigh-distributed paths, the average transmit power rise is:

$$E(Y) = \frac{L}{L-1} \tag{4.36}$$

The case of two paths ($L = 2$) is analysed in more detail because it is assumed to be important in reality. Let a be the ratio of the powers of the two paths. Then the PDF of X is:

$$f_X(x) = \frac{a+1}{a-1} \cdot [e^{-x(1+1/a)} - e^{-x(1+a)}] \tag{4.37}$$

and the average transmit power rise is:

$$E(Y) = \frac{a+1}{a-1} \cdot \ln(a) \tag{4.38}$$

For two multi-paths and antenna diversity with uncorrelated antennas, the result is effectively four paths, and the corresponding PDF and the average transmit power rise are:

$$f_X(x) = 4 \cdot \left(\frac{a+1}{a-1}\right)^2 \cdot \left[e^{-2(1+a)x} \cdot \left(x - \frac{a}{1-a^2}\right) + e^{-2(1+1/a)x} \cdot \left(x + \frac{a}{1-a^2}\right)\right] \tag{4.39}$$

$$E(Y) = 2 \cdot \left(\frac{a+1}{a-1}\right)^2 - 4 \cdot \frac{a+a^2}{(a-1)^3} \cdot \ln(a) \tag{4.40}$$

In Figure 4.24 the theoretical average transmit power rise from Equations (4.38) and (4.40) is plotted as a function of the average power difference of the two propagation paths.

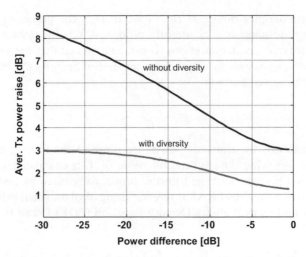

Figure 4.24 Theoretical average transmit power rise as a function of the power difference between the paths in a two Rayleigh path propagation channel.

Reproduced by permission of IEEE.

Realistic Power Control

In the uplink closed-loop PC of the reference WCDMA, the SIR is measured at the Node B and compared with an SIR threshold (SIR_{th}). If the measured SIR is below the SIR_{th} an 'up' command is sent to the UE, otherwise a 'down' command is sent. If the UE receives an 'up' command, it increases its transmit power by Δ dB, otherwise it decreases its transmit power by Δ dB, within the dynamic range of the UE. The closed-loop PC works at 1.6 kHz frequency, thus the TPC commands are given at 0.625 ms time intervals. The PC step size Δ has been 1 dB in the simulations of this study. In reality the closed PC is not ideal for at least the following reasons:

- power not adjusted continuously;
- power-adjusting step size limited, often constant;
- delay between the measurement and adjusting power accordingly;
- inaccurate SIR estimate;
- TPC commands sent in feedback channel are misinterpreted;
- PC range finite.

The effects of realistic fast closed-loop PC were studied with the help of simulations. In the simulator the 32 ksps uplink channel was implemented with a realistic channel and SIR estimation. The user data rate was 8 kbps and the interleaving interval was 10 ms. The propagation channel consisted of two uncorrelated Rayleigh paths with an average level difference of 12.5 dB. This is the Pedestrian A channel converted to the bandwidth of the reference system. Uncorrelated space diversity was assumed, meaning 2 + 2 Rayleigh paths for the RAKE receiver. AWGN was added to the signal after the propagation channel. In the RAKE receiver model, ideal finger allocation was assumed. More about RAKE receivers can be found in [24].

Table 4.11 Average E_b/I_0 required for BER $= 10^{-3}$ with and without fast power control and the average transmit power rise. Channel: Pedestrian A, antenna diversity assumed.

Maximum Doppler frequency	TPC off	TPC on	
	Average received E_b/I_0 [dB]	Average received E_b/I_0 [dB]	Average transmit power rise [dB]
5 Hz	13.1	4.9	2.1
20 Hz	11.5	5.7	2.0
40 Hz	9.7	6.0	1.6
100 Hz	7.9	6.0	0.8
240 Hz	6.5	6.3	0.2

Reproduced by permission of IEEE.

The simulations were performed at different UE speeds without and with fast closed-loop PC. The simulation length was 10000 frames for the pedestrian UE speed (max. Doppler frequency 5 Hz) and 3000 frames for other speeds. In the simulations the received and transmitted powers were collected slot by slot. The required received average E_b/I_0 was estimated to achieve a BER of 10^{-3}. The average power rise was calculated as the average difference between transmitted and received powers. Table 4.11 gives the numerical results.

By comparing Figure 4.24 and Table 4.11 it can be seen that although there are many sources for non-ideality of PC the average power rise with low UE speed is close to the theoretical model. Also it can be seen that in these simulations the average transmit power in every case is lower with the fast PC than without it, directly indicating higher capacity.

Estimation of Average Interference and its Effect on Cell Capacity

The average power rise caused by the fast closed-loop PC compensating multi-path fading should be taken into account in network level calculations when estimating interference and capacity. By following the logic presented in [25, ch. 6] one can conclude that the average power rise raises the average interference experienced at a Node B. It does not raise the average interference from the UEs connected to this particular cell, but it does raise the interference from the UEs connected to surrounding cells as in the case of shadow fading when modelled by a log-normal distribution.

The net effect of the reduced received E_b/I_0 and the average power rise due to fast PC can be illustrated by the following example. Given that processing gain is G, the required E_b/I_0 is ρ, the effective service activity is ν, the allowed loading is η_0, the other-to-own-cell-interference ratio is i, the number of connections at the nominal loading η_0 can be approximated by:

$$M_{\eta_0} = \eta_0 \cdot \frac{G}{\rho \cdot \nu \cdot (1 + i)} \qquad (4.41)$$

Table 4.12   Example of estimated cell capacity (number of connections) with fast power control off and on.

Maximum Doppler frequency	Capacity at 75% load (number of connections)	
	TPC off	TPC on
5 Hz	18	98
20 Hz	26	82
40 Hz	40	80
100 Hz	60	87
240 Hz	83	85

Reproduced by permission of IEEE.

Assuming $\eta_0 = 0.75$, $G = 4.096 \times 10^6/8000$ (8 kbps speech), $\nu = 0.67$, control channel overhead added to 0.5 voice activity, $i = 0.55$ in the case of fast PC off, $i = 0.55 \times$ (average transmit power rise from Table 4.11) in the case of fast PC on and $\rho =$ average received E_b/I_0 from Table 4.11, one gets the capacity numbers given in Table 4.12. In reality the capacity is affected by many factors not modelled here – e.g., SHO. The effect of SHO on average power rise is studied in the next section.

Impact of Fast Power Control on Cell Range

In network dimensioning, the average received E_b/I_0 requirement, ρ, is usually the basic number used in calculating the uplink cell range – i.e., an estimate is made of the maximum path loss that can be subtracted from the maximum UE transmit power to achieve ρ. With fast PC a fast-fading margin, or in other words TPC headroom, should be taken into account in addition to a shadow-fading margin to get the correct results for the cell range. From the single link point of view, the fast PC does not increase the cell range. This can be understood by the fact that the furthest point from a Node B where a UE can move is when it is transmitting constantly with maximum power. From a capacity point of view this is, however, not desirable.

When a UE approaches the cell edge and the transmit power is near its peak, the quality will deteriorate, and as a result the outer-loop PC should start to raise the target, after which the connection will be maintained for a while. The cell edge effect is studied here briefly with the help of the simulation results made by limiting the PC range above the E_b/I_0 setpoint. Only a one Rayleigh path propagation channel was simulated, with a maximum Doppler frequency of 20 Hz. The results are presented in Figure 4.25. The x-axis of Figure 4.25 is the target E_b/I_0 towards which the PC tries to target the received E_b/I_0. The y-axis is the required headroom for the PC, so that BER $= 10^{-3}$ performance was achieved. Thus moving along the x-axis from left to right emulates approaching the cell edge. It can be seen that by adding just a few decibels to the target E_b/I_0 (≈ 4.8 dB with infinite dynamic range) the required headroom decreases significantly. As soon as the target E_b/I_0 is above 7 dB the sum of the target E_b/I_0 and the headroom is approximately constant and equal to the required E_b/I_0 without PC (Figure 4.26). In practice this means that the cell edge has been reached and any outer-loop action cannot help the situation.

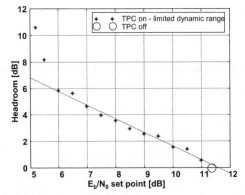

Figure 4.25 Link-level simulation results with limited power control headroom. Propagation channel: one Rayleigh path.

Reproduced by permission of IEEE.

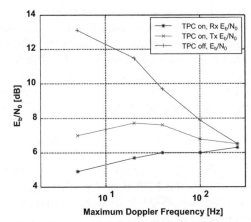

Figure 4.26 Link-level simulation results to demonstrate the effect of UE speed on the efficiency of fast power control.

Reproduced by permission of IEEE.

Definition of TPC Headroom

Although the previous example is theoretical because of a special propagation channel, it is helpful in understanding what happens near the cell edge and how TPC headroom should be defined. Based on this, it is proposed that for the fast closed-loop PC:

$$TPC\ headroom = Average\ required\ received\ E_b/I_0\ without\ fast\ PC$$

$$- Average\ required\ received\ E_b/I_0\ with\ fast\ PC \quad (4.42)$$

As an example one can take the numbers from the first two columns of Table 4.11 and estimate TPC headrooms of 8.2, 5.8, 3.7, 1.9 and 0.2 dB corresponding to maximum Doppler frequencies 5, 20, 40, 100 and 250 Hz, respectively, for the Pedestrian A channel. These numbers are, however, only for a single isolated cell because SHO is not taken into account. The effect of SHO is studied further in the next section.

4.7.1.2 Impact of Soft Handover on Transmit Power Control Headroom and Transmit Power Increase

The analysis in Section 4.7.1.1 was done for the single-link case only. The motivation of this section is to extend the approach to multiple links by estimating the gains in average received and transmitted power and also in the required TPC headroom due to SHO. The gains in SHO are achieved in the first place because, of all the cells in the active set, the best received frame can be selected on a frame-by-frame basis, and second because the fast PC no longer has to compensate for the deepest fades. The results presented here are based on simulations made with an uplink link-level simulator. The simulator model included SHO with two Node Bs. Simulations were made for two multi-path propagation channel profiles. The received and transmitted power statistics were collected as a function of the average power-level difference between the Node Bs in the active set. The simulations were repeated for different UE speeds.

The SHO gains presented in this section should not be confused with the so-called SHO gains against shadow fading, estimated, for example, in [25]. In this section only the gains of the rapid frame selection and less peaky PC due to SHO are calculated. This models directly the benefits of having several simultaneous radio links in uplink. The studied cases consisted of the Pedestrian A and Vehicular A propagation channels both simulated with maximum Doppler frequencies of 5, 20, 40, 100 and 250 Hz corresponding to UE speeds of 3, 11, 22, 54 and 135 km/h, respectively. The same multi-path channel was assumed for both SHO branches. Each study case was repeated by setting the average level difference of the SHO branches to 0, 3, 6 and 10 dB. To emulate also the single-link case for comparison, a level difference of 40 dB was simulated. In all simulations the BER $= 10^{-3}$ performance was searched. For the 5 Hz maximum Doppler frequency 5000 frames were simulated; in all other cases 3000 frames were considered already enough.

The gains in received and transmitted powers are presented in Tables 4.13 and 4.14 for the Pedestrian A channel. Vehicular A channel results can be found in Tables 4.15 and 4.16. Received power was always measured from the stronger link. The numbers presented correspond to BER $= 10^{-3}$ performance. In the last column (single-link) of Tables 4.14 and 4.16, 'Transmitted E_b/I_0' means the average transmitted E_b over received I_0. This differs from the 'Received E_b/I_0' in Tables 4.13 and 4.15 because of the average transmit power rise caused by TPC following multi-path fading.

The SHO gains are larger for the Pedestrian A channel than for the Vehicular A channel. This is natural, because the Pedestrian A channel has less multi-path diversity. For both channels the SHO gains are largest at a maximum Doppler frequency of 20 Hz. The 20 Hz case shows the worst performance in the single-link case when measured from the transmitted power. For the Pedestrian A channel there is almost no gain at all when the level difference between the SHO links is 10 dB. In the case of the Vehicular A channel this already happens at a level difference of 6 dB. TPC bit errors were not generated in the simulator and thus observed selection-combining gains might be a little too optimistic.

Table 4.13 Soft handover gains in received power for the Pedestrian A channel.

Maximum Doppler frequency	Level difference between the SHO links				Single link
	0 dB	3 dB	6 dB	10 dB	Received E_b/I_0 [dB]
	SHO gain in received power [dB]				
5 Hz	1.6	0.7	0.3	0.1	4.9
20 Hz	1.6	1.0	0.5	0.0	5.7
40 Hz	1.7	0.8	0.3	0.0	6.0
100 Hz	1.4	0.5	0.2	0.0	6.0
250 Hz	1.3	0.1	0.1	0.0	6.3

Reproduced by permission of IEEE.

Table 4.14 Soft handover gains in transmitted power for the Pedestrian A channel.

Maximum Doppler frequency	Level difference between the SHO links				Single link
	0 dB	3 dB	6 dB	10 dB	Transmitted E_b/I_0 [dB]
	SHO gain in transmitted power [dB]				
5 Hz	2.7	1.4	0.6	0.1	7.0
20 Hz	2.7	1.7	1.0	0.1	7.7
40 Hz	2.4	1.2	0.5	0.1	7.5
100 Hz	1.7	0.7	0.2	0.0	6.8
250 Hz	1.3	0.1	0.1	0.0	6.5

Reproduced by permission of IEEE.

Table 4.15 Soft handover gains in received power for the Vehicular A channel.

Maximum Doppler frequency	Level difference between the SHO links				Single link
	0 dB	3 dB	6 dB	10 dB	Received E_b/I_0 [dB]
	SHO gain in received power [dB]				
5 Hz	1.1	0.3	0.1	0.0	6.0
20 Hz	1.2	0.4	0.2	0.0	6.3
40 Hz	0.7	0.2	0.1	0.0	6.1
100 Hz	0.8	0.1	0.0	0.0	6.2
250 Hz	1.1	0.1	0.1	0.1	6.6

Reproduced by permission of IEEE.

Table 4.16 Soft handover gains in transmitted power for the Vehicular A channel.

Maximum Doppler frequency	Level difference between the SHO links				Single link
	0 dB	3 dB	6 dB	10 dB	Transmitted E_b/I_0 [dB]
	SHO gain in transmitted power [dB]				
5 Hz	1.3	0.4	0.1	0.0	6.4
20 Hz	1.6	0.6	0.2	0.1	7.0
40 Hz	1.0	0.3	0.1	0.1	6.6
100 Hz	1.2	0.4	0.0	0.0	6.7
250 Hz	1.2	0.1	0.2	0.1	6.7

Reproduced by permission of IEEE.

Soft Handover Gain in Transmit Power Control Headroom

In this section the SHO gain in the required TPC headroom is estimated in the case of a two-way SHO.

The Pedestrian A channel was simulated with maximum Doppler frequencies of 5, 20, 40, 100 and 250 Hz. As before, the level difference of the two SHO links was adjusted stepwise, but now TPC was not used, corresponding to the furthest location of the UE from the BS site. At this location the UE is transmitting constantly with maximum power. The average received E_b/I_0 (which in this case is the same as the average transmitted E_b/I_0) required for BER $= 10^{-3}$ was measured. The results are presented in Table 4.18.

The basic number in link budget calculations is typically the average received E_b/I_0 with TPC on in the single-link case. Thus the required TPC headroom needed in the link budget can be calculated in the SHO situation by subtracting the 'TPC on' column of Table 4.17 from the columns of Table 4.18 corresponding to different level differences of the two best radio links in the active set. In conclusion, the SHO gains for different Doppler frequencies in TPC headroom, or equivalently in the cell range

Table 4.17 Transmit power control headroom for the Pedestrian A channel in the single-link (no SHO) case.

Maximum Doppler frequency	TPC off	TPC on	TPC headroom [dB]
	Average received E_b/I_0 [dB]	Average received E_b/I_0 [dB]	
5 Hz	13.1	4.9	8.2
20 Hz	11.5	5.7	5.8
40 Hz	9.7	6.0	3.7
100 Hz	7.9	6.0	1.9
250 Hz	6.5	6.3	0.2

Reproduced by permission of IEEE.

Table 4.18 Required average received E_b/I_0 in soft handover without transmit power control for the Pedestrian A channel.

Maximum Doppler frequency	Level difference between the SHO links			
	0 dB	3 dB	6 dB	10 dB
	Received E_b/I_0 with TPC off [dB]			
5 Hz	8.5	10.2	11.4	12.4
20 Hz	7.0	8.5	10.4	11.1
40 Hz	6.4	8.0	9.1	9.7
100 Hz	5.4	7.0	7.8	7.9
250 Hz	5.2	6.0	6.4	6.5

Reproduced by permission of IEEE.

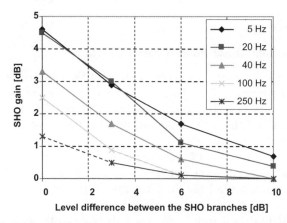

Figure 4.27 Soft handover gain in transmit power control headroom for different user equipment speeds as a function of level difference between the SHO links. Multi-path channel: Pedestrian A.

Reproduced by permission of IEEE.

under the constant loading assumption, are presented in Figure 4.27. It can be noticed that the gains are largest for the low-speed UE, which is quite natural, since TPC is able to follow fast fading.

In this section the uplink gains of SHO in a WCDMA system have been analysed with the help of link level simulations. In the simulation model, a two-way SHO connection was implemented using frame-based selection combining and the special form of fast closed-loop PC applied in SHO, in which the transmit power of the UE is always controlled by the strongest link in the active set. The SHO simulation results were compared with the single-link case and gains were estimated. The gains were measured for the average transmitted and received power and for the required PC headroom as a function of the average level difference between the SHO links and the UE speed. The gains estimated here are different from the 'traditional' SHO gains which are against shadow fading. The analysis made here only deals with SHO gains due to frame-by-frame selection combining within the active set and due to more stable UE transmit powers.

The SHO gains estimated from the link level simulations can be applied in radio network dimensioning and planning. The SHO gains in the average transmit and receive powers are mapped to reduced interference and thus to increased uplink capacity. The gain in the received E_b/I_0 should be applied when estimating the interference caused by the UEs connected to the same cell with the desired cell. The gain in the transmitted E_b/I_0 should be applied when estimating interference from the UEs connected to the surrounding cells.

To estimate SHO gains for cell range calculations, the required maximum transmit powers in the single-link and SHO connection cases must be compared. In this study significant gains were found from the simulations, especially for slowly moving UEs and in poor multi-path diversity conditions. SHO clearly reduces the required PC headroom above the average transmit power, which is required to maintain quality near the cell edge.

4.7.1.3 Handover Control and Power Control Conclusion

There are a number of interactions between HC and PC as these two RRM functions both affect the radio bearers. Their bonding is due mainly to the fact that SHO gain is dependent on PC efficiency. To summarise their relationship, these interactions are:

- SHO gains depend on the type of channel and the degree of PC imperfection. It is usually higher with imperfect PC.
- SHO diversity can reduce the PC headroom, thus improving the coverage.
- The transmit and receive power differences as a result of SHO measurement errors and SHO windows can affect the PC error rate in the uplink, reducing uplink SHO gains.
- In the uplink, SHO gain is translated into a decrease in the outer-loop PC's E_b/N_0 target.

4.7.1.4 Impact of Compressed Mode on Network Performance

CM, used for generating gaps in which a one-receiver mobile can make inter-frequency and/or inter-system measurements, naturally influences the performance of a RAN, depending on the method used as described in Section 4.3.6. The loss comes partly from the reduced performance of fast PC during CM, and partly from the reduced power of interleaving. Both lead to a higher E_b/N_0 requirement during the CM and therefore to higher transmit powers. This issue is less critical close to the BS where enough power for a single link is available to cope with the increased average transmit power and the headroom needed for fast TPC. At the cell edge, however, CM can have significant impact. Usually the systems are either capacity limited in the downlink or coverage limited in the uplink. [2] presents an example of a 2 dB higher E_b/N_0 requirement. In the downlink this translates into a decrease in capacity of roughly 20% if all users are in CM and 2% if only 10% of the users are in CM. In the uplink the same 2 dB higher E_b/N_0 requirement results in 2.4 dB reduced coverage. Studies in [26] for the downlink and [27] for the uplink also indicate that link performance is significantly affected only at the cell edge. It is thus inevitable that if CM is used in a network, inter-frequency and inter-system measurements are started early enough inside the cell so that appropriate HO can be performed before a significant loss occurs.

4.7.2 Radio Resource Management Optimisation Examples

The performance of WCDMA can be further tuned by finding appropriate parameters settings to achieve a high level of harmony between various RRM functions. The sensitive parameters are those that affect the performance directly at the link level. When the link budget has enough margins to maintain a reliable link, the setting of the parameters at the cell level becomes more important. This is true when the network is reaching congestion.

Three factors involved in the parameter sensitivity evaluation are important:
- environmental factors such as noise floor characteristics;
- dimensioning assumptions, especially the link budget;
- setting of RRM functions.

4.7.2.1 Soft Handover Optimisation

As discussed in Section 4.7.2.4 and Chapter 9, the setting of congestion control parameters is based on the expected load of the cell, so that the designed QoS can be achieved. It takes into consideration the amount of traffic as a result of SHO. This amount of traffic depends on SHO gains. If the SHO gain is negative (meaning a loss), then the expected increase in the load is quite significant. This is more likely to happen in the downlink. In the uplink, SHO always results in capacity gain. The effect of downlink SHO on the capacity of the cell is mainly due to two factors: (1) the power increase due to the increase in the number of links with poor SHO gains; and (2) heavy fluctuation in SHO probability. As a result there is a need to decrease the offered load for the same *PtxTarget*. Table 4.19 shows the difference in the power level at the Node B with respect to the SHO overhead. It also shows the uplink capacity improvement with higher SHO probability.

Based on the relationship between SHO probability and downlink Node B power, some capacity optimisation can be done. First, if the downlink power is high and the SHO probability is high, we can set the SHO windows smaller to reduce the downlink load. Alternatively, we can lower the load by setting the *PtxTarget* lower. In this case the downlink will not accept more SHO branches. In any case, tight coordination in the

Table 4.19 Impact of soft handover on capacity (speech at 37 Erlangs/cell, Vehicular A channel).

SHO probability (1, 2, 3-way) [%]	Average total downlink power	Uplink noise rise
67, 22, 11	4.37 W	1.77 dB
56, 25, 19	6.32 W	1.68 dB
45, 27, 28	8.56 W	1.58 dB

setting of SHO windows and the *PtxTarget* is essential to support the target capacity of
the cell. The other aspect of looking at the relationship between SHO and LC is the
increased margin in the link budget as a result of diversity. The diversity in effect can
tolerate higher outage probability in each link in the downlink. This will allow having a
higher *PtxTarget* or *PrxTarget* in the uplink, increasing the load of the cell.

In these example cases the HO algorithm (see Section 4.3) has been parameterised for
simplicity so that the active set weighting coefficient, *W*, has been set to zero.

Addition Window Optimisation

Figures 4.28 and 4.29 show the impact of the addition window and the drop window.
The addition window determines the relative difference of the cells at the UE end that
are to be included in the active set. It is essential that the addition window is optimised
so that only the relevant cells are in the active set. An addition window that is either too
large or too small will result in reduced capacity.

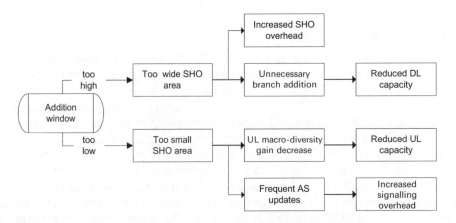

Figure 4.28   Impact of the addition window.

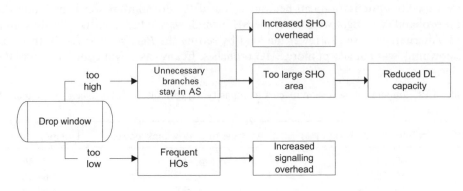

Figure 4.29   Impact of the drop window.

Drop Window Optimisation

Typically, the drop window is set relative to the addition window. The hysteresis is a couple of decibels, meaning that the drop window is slightly larger than the addition window. If the drop window is too large, the wrong cells stay in the active set, resulting in reduced uplink and downlink capacity. If the drop window is too small, frequent and thus delayed (relevant cells ping-pong in the active set) HOs will also degrade the capacity, and in addition the HO-related signalling is increased.

Replacement Window Optimisation

The replacement window is utilised in the case of branch replacement (see Figure 4.8). It determines the relative threshold that is used by the UE to trigger reporting event 1C. A cell can replace the weakest cell in a full active set if the difference between the P-CPICH of the cells is equal to or greater than the threshold. The impact of the replacement window is shown in Figure 4.30. In case of a too large threshold, branch replacement will happen too slowly, which causes the active set to be non-optimal. A non-optimal active set will result in increased transmit powers and thus in reduced capacity and quality in the uplink and downlink directions. In case of a too rapid replacement the ping-pong effect will be present and the signalling overhead related to SHOs will increase.

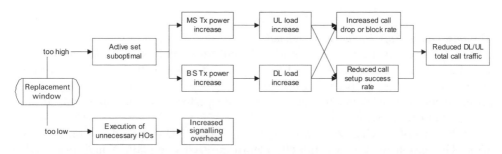

Figure 4.30 Impact of the replacement window in the case of branch replacement.

The Impact of Maximum Allowed Active Set Size

The impact of active set size on HO performance, in the case of a large maximum active set size, is secondary. If the SHO parameters controlling the candidates entering the active set are set correctly, the actual parameter controlling the maximum active set size is not significant. Therefore, it is recommended to start the optimisation of SHO performance with the window parameters. Should it happen that the maximum active set size is set too large and the other HO control parameters are also set wrongly, the impact of a too large active set is in reduced capacity in both the uplink and downlink (Figure 4.31).

The relevance of maximum active set size would be different if it had been set too low. This will result in frequent branch replacements, with delayed handovers degrading the performance in both the uplink and downlink. The reduced BLER performance will

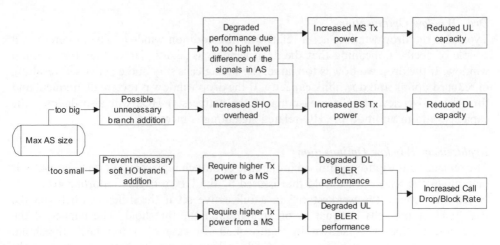

Figure 4.31 Impact of the maximum active set size.

cause the transmit powers to go up. Increased transmit powers in an interference limited environment mean reduced capacity.

Optimisation of Soft Handover Overhead

The target level for the SHO overhead (RT and NRT total) is 20–30%. This target is a value for a mature network. The target value is partly based on hardware requirements – e.g., to prevent running out of codes. In addition to the hardware requirements it is essential to optimise the HO performance for high and effective utilisation of radio resources.

At the start of 2G CDMA network operations, the SHO overhead was generally too high. The reason for this was the HO algorithm itself (see [28]). A high overhead is generally not such a big problem for the uplink. In the uplink direction the power from the UE can be slightly increased owing to non-optimal HO conditions. At the Node B the overhead increases processing load. For the downlink, however, reaching an optimal overhead value is more important. At this optimum, the HO gain is maximal. This means that all Node Bs in the active set contribute positively to the received signal at the UE. By following the call drop rate, call success rate and Node B transmit power, it is possible to get an indication of the gains.

The most important parameters for optimising the SHO overhead are addition window and drop window. These are tuned first together. Changing the active set size will also have a considerable impact. Maximum active set size tuning has to be done very carefully and only on a cell basis. The value of drop timer has only a small effect on the overhead and should be set according to the environment. Finally, one can tune P-CPICH transmit power to change the SHO overhead. This is not recommended, however, as it will also affect many other aspects of the system. This parameter has to be set to an initial value in the radio network planning phase (Figure 4.32).

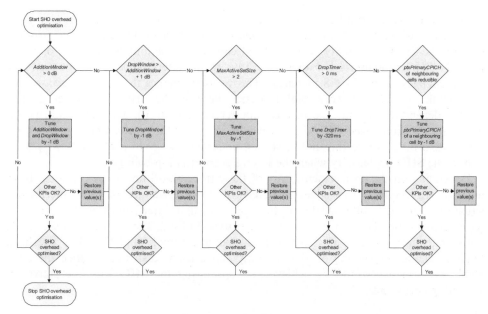

Figure 4.32   Optimisation of soft handover overhead [29].

The Impact of the Active Set Weighting Coefficient

The examples above were presented without the impact of the active set weighting coefficient. In this section the relevance of this coefficient is discussed. It is used to compare the currently received energy in the active set with the improvement if a branch were added. This prevents active set updates in the case when the contribution of a cell to SHO performance is small and thus the cell is not worth adding to the active set. The conclusions of this section are based upon [28], in which the basic cdmaOne algorithm, the algorithm with active set weighting coefficient values set to 0 (relative threshold), and the algorithm with the value set to 1 (slope algorithm) were compared. The results are shown in Table 4.20 in terms of Active Set Update (ASU) period and Total SHO (total diversity HO) probability, including all SHO types. OCNS refers to Other Cell Noise Source level.

Table 4.20   Collected simulation results.

	Basic cdmaOne		Relative threshold		Slope algorithm	
OCNS	ASU period	Total SHO	ASU period	Total SHO	ASU period	Total SHO
5 W	18 s	48%	17 s	50%	24 s	40%
10 W	22 s	28%	29 s	36%	44 s	33%
15 W	24 s	23%	38 s	27%	62 s	31%
20 W	28 s	5%	70 s	15%	87 s	30%
Max. Δ	10 s	43%	53 s	35%	63 s	10%

Reproduced by permission of IEEE.

From Table 4.20 it can be seen that the ASU period of the basic cdmaOne algorithm is quite insensitive to the interference level, but the increase in interference strongly reduces the SHO probability. The modified cdmaOne functions in quite the opposite way: the SHO probability is rather insensitive to the interference level, but the ASU period changes as the OCNS increases. This property makes the optimising task of a real cellular CDMA network slightly easier, because the SHO boundaries do not change depending on the time of day (interference situation). The algorithm, which makes the decision based upon the relative differences of the received pilot E_c/I_0 levels, rather than on absolute thresholds, shows moderate change in both the ASU period and in the SHO probability when the interference situation changes.

Depending on the requirements, any of these three algorithms could be used. If an algorithm is required that gives relatively stable performance in each interference situation, can also offer SHO in the case of high interference and is easy to parameterise, the choice would be the algorithm utilising the relative thresholds. If the interference situation in a network is changing rapidly and the SHO overhead tends to change depending on the currently active traffic distribution, then the most stable performance would be achieved by optimising the SHO with the active set weighting coefficient.

4.7.2.2 Power Control Optimisation

The PC loops as described in Section 4.2 depend very little on optimisation-related issues. The settings for the P-CPICH and other downlink control channel powers (see Table 4.1) are another matter. It is recommended to set the P-CPICH power to about 5–10% of the value of the total cell transmit power in order to guarantee adequate P-CPICH coverage and quality. The actual optimisation of this power setting would then aim for optimal utilisation of Node B resources in terms of power consumption. In Section 9.3.6.1 the autotuning feature for this is introduced. Furthermore, the same steps and methodology can be utilised in manual tuning. It is worth noting that the other control channel powers can be set relative to the P-CPICH power setting; thus the optimisation can first concentrate on P-CPICH performance optimisation. Another possibility for optimisation in PC could be in selecting the appropriate PC Algorithm 1 or 2 depending on the environment and the average speed of the UEs in the cell.

4.7.2.3 P-CPICH Transmit Power Optimisation

The P-CPICH is one of the most important common channels in the network. Not only is it the phase reference for most other common channels and needed in the initial access to a cell, but also the powers of the other common channels are derived from its power (see Section 4.2.2). If set too high, the sum of all common channels' power can easily eat up in the order of 50% of the cell's power resources, leaving not enough power for the traffic channels (dedicated as well as shared ones), thus creating a shortage in cell capacity. The power of the P-CPICH more or less also defines the cell coverage area and, if set too small, coverage holes will be the result. Furthermore, the power of the P-CPICH is one of the main determining factors in mobility management, as its reception (signal level and/or quality) is responsible where and to

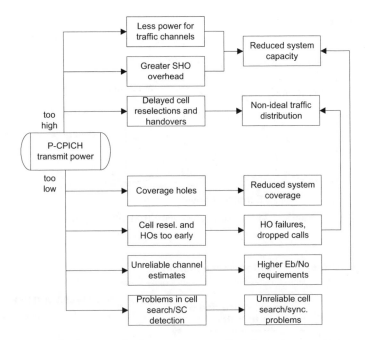

Figure 4.33   Impact of primary common pilot channel transmit power.

which cells HOs (soft or hard) are performed. Figure 4.33 gives an overview on the consequences of too high or too low P-CPICH power, from which some guidelines on its settings can be derived. Section 9.3.6.1 presents a method of autotuning the P-CPICH transmit power.

4.7.2.4   Congestion Control Target Level Optimisation

For most economical dimensioning, there is always a tradeoff between coverage and capacity. Only in very few cases, such as pico-cells, is coverage not an issue at all. In optimisation one needs to balance the noise rise and the total path loss so that one can guarantee that the maximum transmit power is not exceeded within the tolerable outage probability. When the network is already in place, the parameters that can be manipulated are those that control the fluctuation of the receiver noise. These are mainly in LC and PS such as *PrxTarget* and *PrxOffset*. Correspondingly, the downlink total Node B power needs to be controlled. This is done by setting the parameters *PtxTarget* and *PtxOffset*. Figure 4.34 helps to visualise the tradeoff.

Figure 4.34 shows that larger cells can allow a smaller noise rise due to their smaller interference margin, whereas smaller cells can tolerate a higher noise rise. In order to determine how much load can be accommodated in the cell, it is also important to see how the noise level (either noise rise in the uplink or total Node B power in the downlink) fluctuates. In the uplink, the total noise fluctuation at the Node B depends mainly on the traffic load and the behaviour of the noise floor. Then the LC tries to limit the fluctuation of the noise by setting the upper limits. The setting of the upper

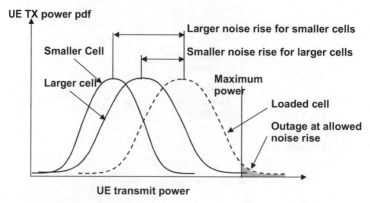

Figure 4.34 Coverage capacity tradeoff in uplink.

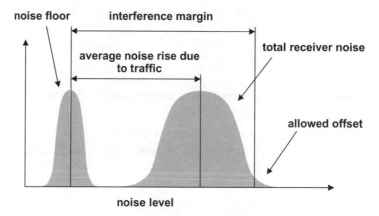

Figure 4.35 Noise rise and interference margin in uplink.

limits must correspond to the required outage probability of the given cell. To illustrate these points, Figure 4.35 shows the noise dynamics.

By looking at Figure 4.35, the challenge is how to set the cell's LC parameters so that the required QoS can be met. If the cell were loaded over the allowable noise rise, the result would be higher call dropout or lower quality rating than expected. On the other hand, if the cell is only allowed to accept less traffic, a high blocking probability for speech or a high queuing probability for packets will result. Therefore, the setting of the LC parameters must enable the required QoS defined by the network planner to be achieved. But prior to the discussion on parameter setting for optimisation purposes, the planner needs to understand the behaviour of the noise rise and that of downlink total Node B power.

In Section 9.3.5 the autotuning of the uplink and downlink operating points and relevant Key Performance Indicators (KPIs) related to them are introduced. Similar processes can be applied to optimisation of congestion control.

4.7.3 Call Setup Delay

Call setup delay has a significant impact upon end-user experience and the perception of system performance. GSM system performance defines initial expectations and provides a benchmark from which the WCDMA system must improve. The overall delay associated with establishing an active end-to-end service connection can be divided into a set of distinct phases. The precise set of phases depends upon the type of service connection being established. It also depends upon the initial state of the UE and whether or not it already has an RRC connection. Call establishment may also include mobility management procedures – e.g., a location area update. All connection establishment procedures from idle mode start with a random access phase during which the UE establishes an RRC connection using a three-way handshake on the PRACH, S-CCPCH and DPCH physical channels. Each of these physical channels must be capable of providing reliable signalling. Some messages are relatively large and do not use L2 retransmissions. In this case single transport block errors necessitate retransmission of the entire message from L3 and thus introduce relatively significant delays. Once an RRC connection has been established the main delays are associated with call control signalling and the RAB establishment procedure.

Call setup delay can be affected by the radio network plan, by the RNC databuild and by the system design. If the radio network plan includes areas of poor coverage then there is an increased probability of requiring L2 and L3 retransmissions. An example of a parameter from the RNC databuild which can have a significant impact upon call setup delay is T300. This is a 3GPP-specified parameter which defines the delay between successive retransmissions of the RRC Connection Request message. If this parameter is configured with a low value then the delay between successive attempts is reduced but the correlation of the radio conditions between those attempts is increased. This means that if one attempt has failed then there is an increased probability that the next attempt will also fail. System design defines characteristics such as the signalling radio bearer bit rate. Assigning a high bit rate to the signalling radio bearer increases the rate at which the signalling procedures can be completed but also places a greater requirement upon coverage and Iub transmission resources.

A detailed analysis of call setup delay requires an understanding of the signalling completed by each protocol. In the case of a UE which has registered with the network but is in RRC idle mode, the first phase of a mobile-originated speech call is to establish an RRC connection with the RNC. From the UE perspective, RRC connection establishment involves three RRC messages and air interface synchronisation. From the network perspective, RRC establishment involves RRC, NBAP, Access Link Control Application Part (ALCAP) and frame protocol signalling as well as air interface and Iub synchronisation.

The most common approach to measuring call setup delay is to record the time difference between the first uplink RRC: Connection Request message and the downlink Alerting message. The delay should be measured from UE log files rather than from network log files although the network allows a more detailed analysis. Table 4.21 presents an example of the speech service call setup delay recorded by a UE which used a 3.7 kbps signalling bearer.

Table 4.21   Example speech service call setup delay.

Message	Incremental delay [ms]	Cumulative delay [ms]
RRC Connection Request	0	0
RRC Connection Setup	401	401
RRC Connection Setup Complete	40	441
CM Service Request	460	901
Security Mode Command	361	1262
Security Mode Complete	1	1263
Setup	249	1512
Call Proceeding	311	1823
Radio Bearer Setup	481	2304
Radio Bearer Setup Complete	250	2554
Measurement Control	495	3049
Alerting	96	3145

A relatively simple change that can be made to reduce the call setup delay is to change the signalling bearer bit rate from 3.7 kbps to 14.8 kbps. The main drawback associated with increasing the signalling bearer bit rate to 14.8 kbps is an increase in Iub transmission resources. A reduced call setup delay can also be achieved by improving coverage and reducing the requirement for L2 retransmissions. In addition, the RNC databuild could be refined using, for example, the parameters T300, N300, N312. In the case that synchronised radio link reconfiguration is used then the time offset for the CFN during which the radio bearer becomes active plays an important role. The RNC sends an NBAP: Radio Link Reconfiguration Prepare message to Node B. This message informs Node B of how its existing radio link will need to be reconfigured to be capable of supporting the new service. Node B responds with the NBAP: Radio Link Reconfiguration Ready message but without yet applying the reconfiguration. Once transmission resources have been reserved the RNC sends the NBAP: Radio Link Reconfiguration Commit message to Node B. This message informs Node B of the Connection Frame Number (CFN) during which it must start to apply the radio link configuration which includes the new radio bearer. The CFN must be defined such that the UE is able to apply the new configuration at the same time as Node B. The UE is informed of the CFN within the subsequent Radio Bearer Setup message. This means that the CFN must be defined such that it occurs after the UE has received the Radio Bearer Setup message. If it is defined to be significantly after the Radio Bearer Setup message then it has a corresponding impact upon call setup delay.

. Call setup delay has a significant dependence upon the design of the CN as well as the design of the RAN. The CN can be designed to run certain procedures in parallel rather than sequentially. For example, in the case of a mobile-to-mobile call, the CN could be designed to page the terminating mobile after the radio bearer has been established with the originating UE. Alternatively the CN could be designed to page the terminating UE in parallel to establishing the radio bearer with the originating UE. This example illustrates the importance of considering the signalling to the terminating end of the connection as well as to the originating end. The signalling used depends upon whether the connection is mobile-to-mobile or mobile-to-fixed.

References

[1] 3GPP, Technical Specification 25.214, Physical Layer Procedures (FDD), v5.10.0, January 2005.

[2] Holma, H. and Toskala, A. (eds), *WCDMA for UMTS* (3rd edn), John Wiley & Sons, 2004.

[3] 3GPP, Technical Specification 25.435, UTRAN Iub Interface UP Protocols for Common TrCH Data Streams, v5.7.0, April 2004.

[4] Baker, M.P.J. and Moulsley, J.T., *Power Control in UMTS Release '99*, 3G Mobile Telecommunication Technologies, Conf. Publ. N.471, IEE, May 2000, pp. 36–40.

[5] Holma, H., Soldani, D. and Sipilä, K., Simulated and measured WCDMA uplink performance. *Proc. VTC 2001 Fall Conf., Atlantic City, Seattle, October 2001*, pp. 1148–1152.

[6] 3GPP, Technical Specification 25.331, Radio Resource Control (RRC) Protocol Specification, v5.11.0, December 2004.

[7] 3GPP, Technical Specification 25.101, UE Radio Transmission and Reception (FDD), v5.13.0, January 2005.

[8] 3GPP, Technical Specification 25.433, UTRAN Iub Interface NBAP Signalling, v5.11.0, January 2005.

[9] 3GPP, Technical Specification 25.133, Requirements for Support of Radio Resource Management (FDD), v5.13.0, January 2005.

[10] 3GPP, Technical Specification 25.215, Physical Layer Measurements (FDD), v5.5.0, September 2003.

[11] Kourtis, S., Code planning strategies for UMTS FDD networks. *Proc. VTC 2000 Spring Conf., Tokyo, Japan, May 2000*, pp. 815–819.

[12] Östberg, C., Wang, Y.E. and Janecke, F., Performance and complexity of techniques for achieving fast sector identification in an asynchronous CDMA system. *Proc. of Personal Multimedia Communications Conf., Yokosuka, November 1998*.

[13] European Radiocommunications Committee (ERC), Draft ERC Recommendation (00)07, Border Coordination of UMTS/IMT-2000 Systems, Annex 4, p. 6.

[14] Kolding, T.E., Pedersen, K.I., Wigard, J., Frederiksen, F. and Mogensen, P.E., High-speed downlink packet access: WCDMA evolution. *IEEE Vehicular Technology Society (VTS) News*, 50(1), February 2003, pp. 4–10.

[15] 3GPP, Technical Specification 25.211, Physical Channels and Mapping of Transport Channels onto Physical Channels (FDD), v5.6.0, September 2004.

[16] 3GPP, Technical Specification 25.308, UTRA High-Speed Downlink Packet Access (HSDPA); Overall Description; Stage 2, v.5.7.0, December 2004.

[17] 3GPP, Technical Specification 25.213, Spreading and modulation (FDD), v5.5.0, January 2004.

[18] Holtzman, J.M., CDMA forward link waterfilling power control. *Proc. VTC 2000 Fall Conf., Boston, Massachusetts, September 2000*, pp. 1663–1667.

[19] Holtzman, J.M., Asymptotic analysis of proportional fair algorithm. *Proc. PIMRC 2001, San Diego, California, September/October 2001*, pp. F33–F37.

[20] Sipilä, K., Laiho-Steffens, J., Wacker, A. and Jäsberg, M., Modelling the impact of the fast power control on the WCDMA uplink. *Proc. VTC 1999 Spring Conf., Houston, Texas, May 1999*, pp. 1266–1270.

[21] Sipilä, K., Jäsberg, M., Laiho-Steffens, J. and Wacker, A., SHO gains in a fast power controlled WCDMA uplink. *Proc. VTC 1999 Spring Conf., Houston, Texas, May 1999*, pp. 1594–1598.

[22] Japan's Proposal for Candidate Radio Transmission Technology on IMT-2000: W-CDMA, June 98, available online at ⟨*http://www.itu.int/imt*⟩

[23] Proakis, J.G., *Digital Communications* (3rd edn), McGraw-Hill, 1995.

[24] Ojanperä, T. and Prasad, R., *Wideband CDMA for Third Generation Mobile Communications*, Artech House, 1998.

[25] Viterbi, A.J., *CDMA: Principles of Spread Spectrum Communication*, Addison-Wesley, 1995.

[26] Gustafsson, M., Jamal, K. and Dahlman, E., Compressed mode techniques for inter-frequency measurements in a wideband DS-CDMA system. *Proc. PIMRC 1997, Helsinki, Finland, September 1997*, pp. 231–235.

[27] Toskala, A., Lehtinen, O. and Kinnunen P., UTRA GSM handover from physical layer perspective. *Proc. ACTS Summit, Sorrento, Italy, June 1999*.

[28] Laiho-Steffens, J., Jäsberg, M., Sipilä, K., Wacker, A. and Kangas, A., Comparison of three diversity handover algorithms by using measured propagation data. *Proc. VTC 1999 Fall Conf., Houston, Texas, May 1999*, pp. 1370–1374.

[29] Buot, T., Zhu, H., Schreuder, H., Moon, S. and Song, B., Soft handover optimisation for WCDMA. *4th International Symposium on Wireless Personal Multimedia Communications, Aalborg, Denmark, September 2001*.

5

WCDMA–GSM Co-planning Issues

Kari Heiska, Tomáš Novosad, Pauli Aikio, Chris Johnson and Josef Fuhl

5.1 Radio Frequency Issues

The frequency allocations for different radio systems have a major effect on possible interference scenarios. If another system is operating next to the Wideband Code Division Multiple Access (WCDMA) system, there is the possibility of signal power leakage due to imperfect transmitter filtering of the other system and receiver filtering of the WCDMA system. Frequency usage around 2 GHz in the different parts of the world is shown in Figure 5.1. In the case of larger frequency separation, intermodulation can be an issue. When co-siting WCDMA with GSM900 (Global System for Mobile communication at the 900 MHz band), special attention should be paid to the second harmonics that can fall into the WCDMA uplink band. Third-order intermodulation products can be problematic when GSM1800 (GSM at the 1800 MHz band) and WCDMA networks operate in the same area.

In the USA 2G and 3G (second and third generation) networks operate in the same band. This implies very good receiver and transmitter filtering to avoid interference coupling between these systems. To achieve tight enough filtering, especially in mobiles, is certainly a challenge.

5.1.1 Thermal Noise

In a practical system there is always some minimum detectable signal level determined by the noise floor. The noise floor depends mainly on the thermal noise power and the excess noise generated inside the system. Thermal noise is caused by the random movement of atoms in material and its spectrum is white.

The root mean square (r.m.s.) voltage v_n over a resistor R can be expressed by the formula:

$$v_n = \sqrt{\frac{4hfBR}{e^{hf/kT} - 1}} \tag{5.1}$$

Radio Network Planning and Optimisation for UMTS Second Edition
Edited by J. Laiho, A. Wacker and T. Novosad   © 2006 John Wiley & Sons, Ltd

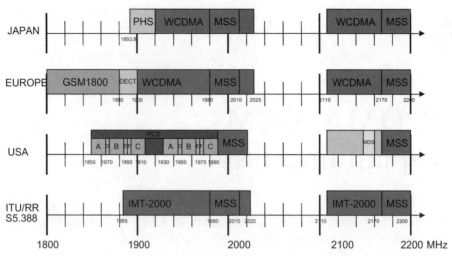

Figure 5.1 Frequency usage around 2 GHz in different parts of the world.

where h is the Planck constant $(6.626 \times 10^{-34} \, \mathrm{J\,s})$; f is the frequency in Hz; B is the bandwidth in Hz; R is the resistance in ohms; k is the Boltzmann constant $(1.380 \times 10^{-23} \, \mathrm{J\,K^{-1}})$; and T is the absolute temperature in Kelvin.

In WCDMA frequencies $hf \ll kT$; using the two first terms of a Taylor series expansion:

$$e^x = 1 + x + \frac{x^2}{2!} + \frac{x^3}{3!} + \cdots \tag{5.2}$$

we can approximate the denominator in the following way:

$$e^{hf/kT} - 1 \approx \frac{hf}{kT} \tag{5.3}$$

Now the equation for v_n can be simplified to:

$$v_n = \sqrt{4kTBR} \tag{5.4}$$

The noise power in the resistor R can be expressed as:

$$P_n = \left(\frac{v_n}{2R}\right)^2 R = \frac{v_n^2}{4R} = kTB \tag{5.5}$$

From this equation it can be seen that the thermal noise power P_n is directly proportional to the temperature T and the bandwidth B. This equation is valid for those cases where $hf \ll kT$ and this is true for the frequencies used in mobile telecommunications.

The WCDMA radio network is limited by interference, as opposed to the GSM, which is limited mainly by spectrum availability. The background noise level has a considerable effect on the coverage and capacity of a WCDMA radio network. The thermal noise power P_n within a bandwidth of 3.84 MHz can be calculated by Equation (5.5) and is:

$$P_n = kTB = -108.1 \, \mathrm{dBm} \tag{5.6}$$

This sets the ultimate limit to WCDMA receiver sensitivity, excluding the Noise Figure (NF) and its processing gain. Assuming the NF $NF = 3.0$ dB, speech service with a data rate $R = 12.2$ kbps, chip rate $W = 3.84$ MHz, $E_b/N_0 = 4.0$ dB and noise rise $NR = 1.5$ dB, the sensitivity of the receiver would be:

$$P_{Rx,min} = P_n + NF + E_b/N_0 + NR - 10 \cdot \log_{10}(W/R) = -124.6 \, \text{dBm} \qquad (5.7)$$

5.1.2 Man-made Noise

Man-made noise can be classified as either intentional or unintentional. Intentional emissions come mainly from radio transmitters, such as radars, mobile networks and broadcast systems. Unintentional sources include, for example, motor traffic, industrial equipment, consumer products and lighting systems.

The main sources that contribute to background noise are traffic, other radio systems, industrial machinery, power lines and lightning. Spread spectrum systems such as WCDMA are less sensitive to narrowband interfering signals, owing to their processing gain. If the narrowband signal level is so high that it exceeds the input power of the Low-noise Amplifier (LNA), it can be saturated and the wanted signal shadowed and prevented from proper detection and decoding.

There are some reports quantifying noise level increase caused by human activities ([11] and [12]). According to these reports, noise rise is around 4–8 dB at frequencies of 1.8–2.0 GHz. It is greater at lower frequencies and there are models to describe the noise characteristics at these frequencies [13]. In practice, noise levels should be measured, because the level depends on the environment and the time of day.

5.1.3 Interference Scenarios

Interference can originate in a system itself or it can come from external sources, mainly from other radio systems. At a WCDMA Base Transceiver Station (BTS) all the received power other than the desired mobile power can be treated as interference power. Besides thermal noise power, this interference power consists of adjacent channel power, leaked transmitted power through the duplexer, transmissions from other mobiles and radio systems, and unintentional external noise sources such as radio emissions from traffic.

If systems are not operating in adjacent frequency bands, the most probable interference sources are system non-linearities. Harmonic distortion can be a problem if the victim system's frequency is a multiple of the offender system's; that is:

$$f_{HD} = mf_1 \qquad (5.8)$$

where f_{HD} is the harmonic distortion frequency; m is a positive integer; and f_1 is the fundamental frequency. The order of the harmonic distortion is determined by m.

The same kind of equation can be expressed for intermodulation:

$$f_{IMD} = nf_1 \pm mf_2 \qquad (5.9)$$

where f_{IMD} is the intermodulation frequency; f_1 and f_2 are fundamental frequencies; and n and m are integers. The order of intermodulation distortion is defined to be $|m| + |n|$. The most harmful intermodulation distortions are considered to be third-order

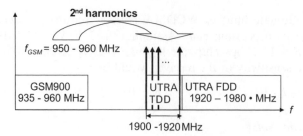

Figure 5.2 Second order harmonic distortion from the GSM900 band falling into the WCDMA band.

products, owing to their relatively high power compared with products of higher order. The intermodulation products generated inside a transmitter can be filtered down by transmit filters, whereas those generated inside a receiver cannot be filtered out. In practice, the most probable place where intermodulation products can be generated inside a receiver is the LNA and the mixer. Passive intermodulation distortion can be generated in places where high power is used, such as transmitter antenna lines, connectors, filters and antennas. The levels of intermodulation distortion generated in passive components are much lower than those generated in active components. Intermodulation distortion can be a problem in multi-system networks, such as some indoor solutions with shared feeders, filters and repeaters.

The situation in the case of GSM900 and WCDMA is shown in Figure 5.2. GSM900 third order intermodulation products do not fall within the WCDMA Frequency Division Duplex (FDD) uplink receive band. However, GSM1800 third order intermodulation products may be generated within the WCDMA uplink receive band. For example, if the GSM1800 frequency 1805.2 MHz is used then any frequency between 1862.6 MHz and 1880.0 MHz will generate a third order intermodulation product within the WCDMA uplink band – i.e., between 1920 and 1980 MHz. So, it is recommended that GSM1800 frequency planning is completed in such a way that third order intermodulation products do not fall within the WCDMA uplink band. This reduces the isolation requirement to that defined by spurious emissions and receiver blocking. On the other hand WCDMA third order intermodulation products do not fall within the GSM900 nor GSM1800 frequency bands. This means that GSM900 and GSM1800 inter-system isolation requirements are not affected by intermodulation.

If the different systems are operating next to each other in the frequency domain, the main problems arise from filtering non-idealities and are considered in Section 5.4.

5.1.4   Interference Reduction Methods

The most common causes of intersystem interference are spurious emissions (non-ideal transmit filtering), receiver blocking (non-ideal receiver filtering) and intermodulation (non-linear mixing). A WCDMA Node B may experience interference from a GSM Base Station (BS) if there is not sufficient isolation between the GSM transmitter and the WCDMA receiver. [14] specifies GSM BS spurious emissions requirements. These requirements are presented in Table 5.1.

Table 5.1 GSM900 and GSM1800 base station spurious emission performance requirements.

Conditions	Frequency band	Maximum power	Measurement bandwidth
Co-located base stations	1920–1980 MHz	−96 dBm	100 kHz

A maximum power of −96 dBm measured in 100 kHz is equivalent to −80 dBm measured in 3.84 MHz. If it is assumed that the WCDMA Node B has an NF of 3 dB then the thermal noise floor is −105 dBm. This means that if there is 25 dB of isolation between the WCDMA Node B and the GSM BS then the noise floor of Node B would be increased by 3 dB. The GSM BS is likely to be transmitting more than a single RF carrier. If the GSM BS is transmitting with four RF carriers then spurious emissions would be increased by 6 dB. In practice GSM BSs are likely to perform better than the specifications. The figures within the specifications are usually adopted as a set of worst case assumptions. Figure 5.3 illustrates the increase in the interference floor which is generated at the WCDMA Node B receiver for a range of isolation figures.

An isolation of 40 dB results in a relatively small interference floor increase for up to eight GSM RF carriers. The Third Generation Partnership Project (3GPP) [15] specifies the WCDMA Node B receiver blocking requirements. These requirements are presented in Table 5.2.

A GSM BS typically radiates a power of between 37 dBm and 45 dBm. Assuming a GSM BS transmit power of 45 dBm means that 29 dB of isolation is required such that the GSM signal is attenuated to 16 dBm prior to reaching the WCDMA Node B. This isolation requirement is less than that associated with GSM spurious emissions.

A GSM BS may experience interference from a WCDMA Node B if there is not sufficient isolation between the WCDMA transmitter and the GSM receiver. 3GPP [15] specifies the WCDMA Node B spurious emission requirements. These requirements are presented in Table 5.3.

A maximum power of −98 dBm measured in 100 kHz is equivalent to −95 dBm measured in 200 kHz. If it is assumed that the GSM BS has an NF of 2 dB then the

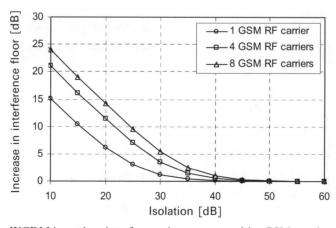

Figure 5.3 WCDMA receiver interference increase caused by GSM spurious emissions.

Table 5.2   WCDMA Node B receiver blocking performance requirements.

System	Frequency band	Interfering signal	Wanted signal	Interferer
GSM900	921–960 MHz	16 dBm	−115 dBm	CW carrier
GSM1800	1805–1880 MHz	16 dBm	−115 dBm	CW carrier

Table 5.3   WCDMA Node B spurious emission performance requirements.

System	Conditions	Frequency band	Maximum power	Measurement bandwidth
GSM900	Co-located base stations	876–915 MHz	−98 dBm	100 kHz
GSM1800	Co-located base stations	1710–1785 MHz	−98 dBm	100 kHz

thermal noise floor is −119 dBm. This means that if there is 24 dB of isolation between the WCDMA Node B and the GSM BS then the noise floor of the BS would be increased by 3 dB. The WCDMA Node B may be transmitting more than a single RF carrier. If the WCDMA Node B is transmitting with two RF carriers then the spurious emissions will be increased by 3 dB. In practice WCDMA Node Bs are likely to perform better than the specifications. The figures within the specifications are usually adopted as a set of worst case assumptions. Figure 5.4 illustrates the increase in interference floor which is generated at the GSM BS receiver for a range of isolation figures.

Similar to the scenario of a WCDMA Node B receiver, an isolation of 40 dB results in a relatively small interference floor increase. 3GPP [14] specifies the GSM900 and GSM1800 receiver blocking requirements. These requirements are presented in Table 5.4.

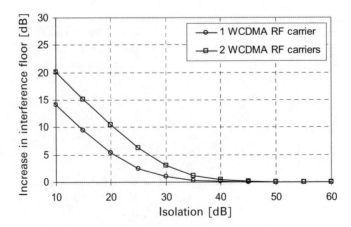

Figure 5.4   GSM receiver interference increase caused by WCDMA spurious emissions.

Table 5.4 GSM900 and GSM1800 base station receiver blocking performance requirements.

System	Frequency band	Interfering signal	Wanted signal	Interferer
GSM900	925–12750 MHz	8 dBm	Sensitivity + 3 dB	CW carrier
GSM1800	1805–12750 MHz	0 dBm	Sensitivity + 3 dB	CW carrier

Assuming a 43 dBm WCDMA Node B transmit power means that 35 dB of isolation is required such that the WCDMA signal is attenuated to 8 dBm prior to reaching the GSM BS. A GSM1800 BS must be capable of receiving 0 dBm of interference from the WCDMA band while maintaining its quality requirements with a speech signal which is 3 dB above reference sensitivity. Assuming a 43 dBm WCDMA Node B transmit power means that 43 dB of isolation is required such that the WCDMA signal is attenuated to 0 dBm prior to reaching the GSM BS.

Both transmitter and receiver filtering can be considered as a Radio Frequency (RF) method, and advanced interference cancellation receivers are covered by baseband methods. Antenna placement comprises both RF and installation methods.

In general, filtering in a transmitter plays an important role, especially in out-of-band and spurious emissions. Filtering in a receiver has the same role but just attenuates incoming unwanted signals below the desired level.

Most intermodulation distortion problems can be avoided by proper frequency planning in 2G networks. This can be done so that frequencies that can produce third or higher order intermodulation products are not used in the same sector or site if WCDMA and 2G BSs are co-sited.

5.2 Noise Measurements

Having a clear enough spectrum is a prerequisite for proper WCDMA operation. Background noise measurements provide a baseline that can be used to plan and optimise the parameters that influence WCDMA system performance.

5.2.1 Acceptable Radio Frequency Environment

In a Code Division Multiple Access (CDMA) system, background noise levels above the thermal noise limit mean:
- Reduced cell sizes for coverage (i.e., uplink) limited parts of the network and therefore more BSs. Figure 5.5 shows the relative reduction in uplink cell range and cell size with increasing background noise level. As a general approximation, for background noise levels close to the theoretical limit a 1 dB increase in noise level reduces the cell size by about 6% and consecutively the cell area by some 11%. This in turn means that 11% more BSs are required to cover the same area for an increase of only 1 dB in background noise level.

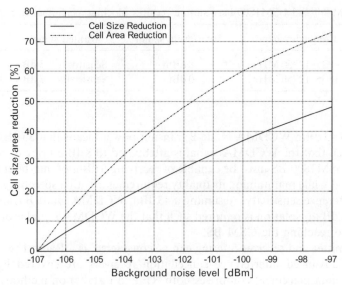

Figure 5.5   Relative reduction of cell size and cell area due to increasing background noise (assuming the Okumura–Hata propagation model).

- Reduced capacity for services where cell sizes are not uplink limited. The number of users K for one service type for a given BS output power P is as follows:

$$K = \frac{P}{\dfrac{E_b/N_0}{W/R} \cdot \nu \cdot [P_{noise} \cdot L_{av} + P \cdot (1 - \alpha + i)]} \qquad (5.10)$$

where E_b/N_0 is the ratio of required energy-per-bit-to-spectral-noise density; W is the chip rate; R is the L2 user data rate; ν is service activity; P_{noise} is the power of the background noise; L_{av} is the average downlink path loss in the cell; α is the orthogonality; and i is the other-to-own-cell-interference ratio in the downlink.

The relative capacity reductions for different values of L_{av}/L_{max} are shown in Figure 5.6. The capacity reduction due to increasing background noise depends heavily on the value of the average path loss a cell is designed for in relation to the maximum allowed path loss – i.e., on the network topology. The closer the ratio is to 1, the stronger the influence of increasing background noise.

The maximum acceptable increase in background noise level is defined as the value where:
- for coverage uplink limited scenario: cell range is reduced by 10% (area by 20%);
- for capacity downlink limited scenario: relative capacity is reduced by 10%.

This yields the maximum allowed noise increases as shown in Table 5.5. This table proves that the nearer mobiles are to the BS (compared with the cell size defined by the uplink limit) the smaller the influence of background noise. In addition, the uplink is the far more critical link concerning the influence of background noise.

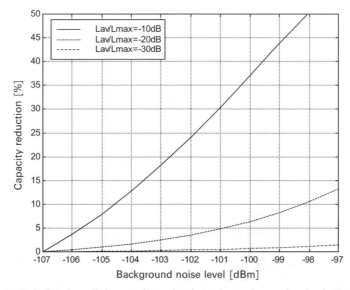

Figure 5.6 Relative downlink capacity reductions due to increasing background noise.

Table 5.5 Maximum allowed noise increase for uplink and downlink limited environments.

Limiting link	Criterion	Ratio of actual average pathloss to maximal average path loss (L_{av}/L_{max}) [dB]	Maximum noise increase over thermal background level [dB]
Uplink	Cell size reduction of 10%	Not applicable	1.6
Downlink	Relative capacity reduction of 10%	−10	2.4
Downlink	Relative capacity reduction of 10%	−20	8.7
Downlink	Relative capacity reduction of 10%	−30	18.1

5.2.2 Conducting Measurements in a Real Environment

A practical measurement setup capable of doing onsite background noise measurements is shown in Figure 5.7. The equipment needed for these measurements is:

- laptop for data collection (with customised software);
- General Purpose Interface Bus (GPIB);
- Spectrum Analyser (SA);
- cables to connect the antenna to the SA;
- adapters and connectors;
- LNA, power source for the LNA;
- filter;
- receiving antenna.

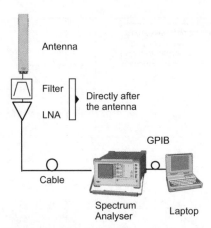

Figure 5.7   Setup for noise measurements.

As BS antenna, a standard Universal Mobile Telecommunications System (UMTS) BS antenna is used. The antenna is followed by a filter prior to the LNA to ensure that no intermodulations are caused by the LNA due to excessive GSM1800 or GSM900 signals, since LNAs used for such measurements usually have an amplification range covering, as well as the UMTS frequency band, some bands where high-power signals (e.g., GSM900/GSM1800 signals) might be present.

It must be ensured that the SA has a sufficiently high dynamic range to satisfy the requirements for the noise floor measurements. The comparatively high NFs of SAs are negligible due to the high LNA gain.

The total NF of the measurement system is thus given only by the bandpass filter, the cables, connectors and the LNA itself, and follows from the general Friis formula as denoted in Equation (6.1):

$$F_{System} = NF_{Filter} + A_{Filter} \cdot (F_{LNA} - 1) + \frac{A_{Filter} \cdot (NF_{Cable} - 1)}{G_{LNA}} \qquad (5.11)$$

where F_{System} denotes the NF of the measurement system; NF_{Filter} is the NF of the filter; A_{Filter} is the attenuation of the filter in the measurement band; F_{LNA} is the NF of the LNA; G_{LNA} the gain of the LNA; and NF_{Cable} is the NF of the cable used between the LNA and the SA (given by the cable attenuation, A_{Cable}). Inserting the values for the equipment actually used, we obtain $F_{System} = 2.2\,dB$.

The NF determines the noise levels that can be measured. The lower bound N_{Limit} is given by $N_{Limit} = N_{Thermal} + F_{System} = -104.8\,dBm$, where $N_{Thermal}$ denotes thermal noise (being $-107\,dBm$ for a bandwidth of 5 MHz and a temperature of 290 K).

Before doing the first actual measurements a standard process ensuring correct operation of the measurement setup should be invoked as follows:

1. Check the LNA for overload state. To ensure a meaningful measurement, the LNA output signal level (=input signal to the SA) should be at least 10 dB below the 1 dB compression point of the LNA – i.e., $P_{SA} \leq G_{LNA,1\,dB} - 10\,dB$, where P_{SA} denotes the input power to the SA and $G_{LNA,1\,dB}$ the 1 dB compression point of the LNA. To check for this condition, either of the following two methods may be applied:

o check for the power of the main peak within the amplification band of the LNA and ensure that the above condition holds. In Europe the main peaks are typically the transmit signals of the GSM900 and GSM1800 systems;

o check for the power of the main peak within the amplification band of the LNA and ensure that the above condition holds. Additionally, insert attenuators before the LNA; the output power measured on the SA should decrease perfectly linearly by the attenuation value of the inserted attenuator.

2. Check the SA's amplification. Ensure that the condition $P_{SA} \le P_{SA,1dB}$ – where $P_{SA,1dB}$ is the 1 dB compression point of the SA – holds. If this condition does not hold, insert the appropriate RF attenuation at the input of the SA. However, bear in mind that insertion of the attenuator decreases the sensitivity of the measurement setup accordingly.

3. Check the measured values by applying post-processing: arrange the measurement setup without connecting the antenna output to the filter input. Instead connect a standard 50 Ω load to the filter input. The measured and calculated mean value and the median value should agree to be:

$$E(n) = N_{Thermal} + G_{LNA} - (A_{Cable} + A_{Filter}) \qquad (5.12)$$

where $E(n)$ is the expectation (mean value) of the noise process.

Typically, one has to correct the values read from the SA display by adding 2.5 dB. This is due to two facts [17]. First, the SA uses a peak rather than an r.m.s. detector (correction by 1 dB). The second reason is that since video filtering is applied after the logarithmic amplifier, the mean of the logarithm is calculated, not – as it should be – the logarithm of the mean (correction by 1.5 dB).

5.2.3   Measurement Results

As shown in Section 5.2.1 the uplink is the more critical link. To validate a probable location of a UMTS BS antenna, background noise measurements at the BS antenna location could be one task in the process.

In order to get a general impression of the European situation, a measurement campaign has been performed in Austria for probable locations for UMTS BS antennas during the peak traffic hours of a working day. In most of these locations a GSM900 and/or GSM1800 antenna is already present. The noise levels were determined by pointing the UMTS receive antenna at one location in four typical directions (conducting measurements for one hour in each direction) to cover the whole angular spectrum.

When conducting these background noise measurements the main aims are:

- to determine the background noise level in the UMTS uplink band;
- to search for and – if present – identify interference and its sources;
- to look for the time behaviour of the noise and identify typical patterns;
- to assess the suitability of the location for carrying a UMTS antenna from the noise level point of view.

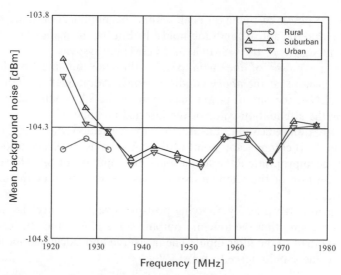

Figure 5.8 Comparison of example mean noise values for urban, suburban and rural areas.

Figure 5.8 shows an example of mean background noise levels for urban, suburban and rural rooftop measurement locations.

The observed noise levels in channels 4–12 (1935–1980 MHz) were around 104.5 dBm (with smooth behaviour), which is 0.3 dB above the noise limit of the measurement setup. Therefore, we conclude that no measurable noise rise caused by the environment is found at rooftop locations. Virtually no difference in the noise levels between suburban and urban areas was observed. For rural areas only the noise values from 1920 to 1935 MHz are considered valid, since in the remaining UMTS band strong micro-wave signals are present. If the trend of the three measurement values in rural areas is extrapolated to the whole UMTS band, about the same mean noise values as for urban and suburban areas are observed. However, for urban and suburban areas slightly higher noise values are being observed for the first two to three frequency channels (1920–1930 (1935) MHz) when compared with the other channels.

As one can see in Figure 5.8, the ripple in the noise values for different bands is attributed to imperfections when equalising the measurement setup during post-processing; it can be considered as virtually flat.

Comparing the noise levels of rooftop and street level measurements in suburban areas shows that street level noise values are 0.5 dB above the values found on rooftop locations (see Figure 5.9). This is caused by the presence of more noise sources of human origin in close vicinity to the receive antenna when it is located at street level.

Conducting the same analysis for dense urban areas shows that the environmental noise levels at street level are about 1.9 dB above the levels found on rooftop locations (see Figure 5.10).

To further characterise background noise, the changes in mean noise power within 0.5 s in each 5 MHz band are analysed, with the results shown in Figure 5.11. These graphs show that the changes in mean noise values measured in the real environment do

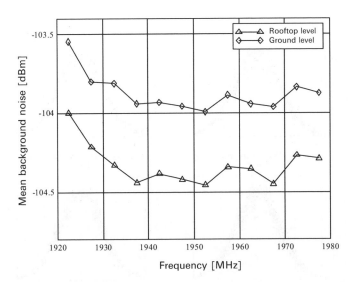

Figure 5.9 Mean environmental noise levels for suburban street- and rooftop-level measurements.

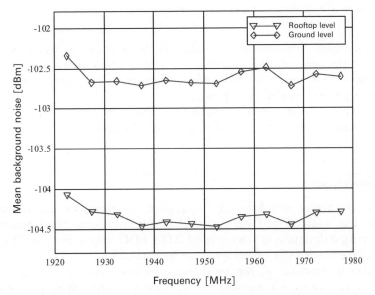

Figure 5.10 Mean environmental noise levels for urban street- and rooftop-level measurements.

not differ for different morphological types and match almost perfectly the changes predicted by applying the theoretical Gaussian noise model. If this analysis is repeated for longer timelags, the same behaviour is observed.

The results shown so far are generally applicable to many measurement positions, though for specific locations behaviour differing from these results is observed:

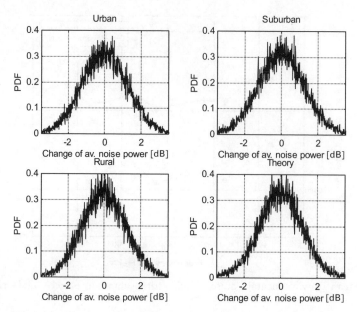

Figure 5.11   Changes in average noise power within 0.5 s for a 5 MHz UMTS channel for urban, suburban and rural environment, and the theoretical curve.

- For measurement locations close to the border of countries which have not yet cleared the frequency band in question for UMTS applications, weak (interfering) signals have been observed. This effect, however, will vanish as soon as these countries have also cleared their bands.
- Measurements close to military/air control stations also showed some interference for small time periods, which needs further investigation.

5.2.4   Conclusions

The main parameters found for the individual environments are summarised in Table 5.6. The minimal and maximal values for the mean noise power (μ), the average standard deviation (σ) in the full UMTS FDD uplink band and the maximal variation of the mean noise power level can be obtained.

The maximal variation of the mean noise power level, μ, over the band (per measurement direction) is a measure of the equality of the 5 MHz portions in terms of background noise power. It is always smaller than 0.5 dB, with lower values occurring at the upper end of the band.

The mean background noise power for all measurements lies in the range from −105.7 dBm up to −102.7 dBm. The standard deviation is about 2.9 dB. For all measurements the noise power is lowest for the higher frequency bands. These results show the following:

- The average mean background noise floor power is about −104.5 dBm, therefore close to the theoretical measurement limit of the measurement setup used.

Table 5.6 Summaries of the main results for individual environments.

Environment	Min μ Max μ [dBm]	Average σ [dB]	Maximal variation of μ per direction in band [dB]
Urban rooftop level	−105.3 −104.3	2.9	0.3
Urban street level	−103.3 −102.7	2.9	0.4
Suburban industrial	−105.7 −104.1	2.9	0.5
Rural	−105.1 −104.5	2.9	0.5

- There is no significant difference in the background noise floor between urban and several suburban environments.
- Antenna locations at street level in the vicinity of heavy motor traffic increase the noise floor by about 1.9 dB for urban and 0.5 dB for suburban environments.
- The FDD background noise floor seen at the BS shows smooth behaviour over the full band, with the lower bands showing dependence on slightly higher values.
- In general, site locations in countries using the same systems as in central Europe can be assumed to be not too sensitive to background noise values, the exception being UMTS BS locations close to countries that have not yet cleared the frequency band in question for UMTS applications. Locations near military/air traffic control should be evaluated further by onsite measurements.
- UMTS BSs have very tough restrictions on their out-of-band emissions. From the GSM BTS point of view, no restriction for co-siting of GSM and UMTS was found.
- No interfering signals directly attributable to GSM900/GSM1800 systems have been observed. From the UMTS point of view, no restrictions on co-siting of GSM and UMTS with separated antennas are observed, provided the specified isolation is kept.

5.3 Radio Network Planning Issues

The aim of radio network planning is to design networks that enable effective use of the existing spectrum and equipment at a reasonable cost. For most current network cases, a design with constraints is characteristic. The constraints follow from the fact that the UMTS network must overlay the existing network. The overlay network uses different technology along with similar or different frequencies. The scenario usually brings several challenges into the design, especially:

- coverage–performance challenges;
- site solutions and co-siting;
- transmission methods and planning;

- technology comparison and perception by the end-user;
- tight usage of the frequency spectrum by different technologies in some cases.

At first, coverage should be planned according to the set targets. These may be different for different services and area types. The easiest way to improve coverage is to add more sites, but this is probably not the cheapest way. Other means of improving coverage could be, for example, different kinds of diversity methods, higher transmit powers or better antenna selection. For multi-system networks coverage can be improved by inter-system handovers if there are multi-mode terminals available.

Capacity becomes an issue in mature networks. It can be increased by adding more hardware into existing sites, adding more sites and switching to hierarchical cell structures where the micro- and pico-cells are put into hotspot areas (see Chapter 6).

The coverage–performance relationship might be generally challenging for the areas where there is requirement for deploying 1.9 or 2 GHz UMTS technology and usage of existing sites of the network designed for 850 or 900 MHz. To keep an acceptable service coverage, coverage supporting solutions are needed along with intersystem interworking features to allow those services possibly to work in both 3G and 2/2.5G technologies with seamless coverage and user experience.

The quality of the network is a combination of service availability and quality experienced by users. All of these demands should be met at a reasonable price. Cost plays a very important role in an operational network, and end-user costs should be included in any assessment of overall quality.

5.3.1 Co-planning Process

The typical case would be an operator who already has a GSM network and has received a licence for a WCDMA network. To reduce the site acquisition effort, using existing GSM sites and co-locating WCDMA sites is the preferred solution from an operator's point of view.

The quality of the existing 2G network puts some limits on the quality of the WCDMA network in the case of one-to-one site reuse. If the existing 2G network has interference problems, the same would also be confronted in a WCDMA network. In some cases it would be even preferable not to reuse all of the existing sites. For that reason the problem areas of the existing network should be identified and taken into account during the planning process.

The simplest case is a one-to-one reuse of existing GSM sites. Figure 5.12 shows a schematic view of the radio network planning process for this case.

In the first stage, the quality of the 2G network should be evaluated, paying special attention to dominance areas, since WCDMA is an interference limited system and does not have a frequency plan as in 2G systems.

As a next step, the availability of physical space for the extra equipment at sites should be checked. It might well be that some sites cannot be reused owing to lack of space in the equipment room or on the rooftop.

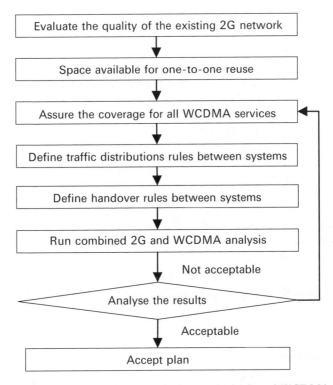

Figure 5.12   Radio network planning process for co-sited 2G and WCDMA networks.

Once site positions have been decided, the coverage areas required for the services to be offered should be checked and the plan should be changed in order to meet coverage requirements with a pre-defined load.

After the traffic distribution and handover rules are applied, a combined 2G and WCDMA analysis can be performed and the results analysed to check whether the planned network fulfils the targets set by the operator.

5.3.1.1 Site Reuse

Site reuse would be beneficial for an operator who already has some other radio network up and running. By reusing existing sites, site acquisition costs can be cut and some savings can be achieved from the reduction in the amount of civil engineering work. By co-locating different radio system BSs, near–far effects can be avoided because the path loss difference between serving and interfering systems is low.

Antenna sharing reduces the number of physical antennas needed and saves space in the antenna masts or in the other antenna installation places like rooftops or building walls. In some places, such as church towers, there may be no space for additional antennas. If single-band antennas with spatial diversity are in use, they can be changed to multi- or wideband antennas with polarisation diversity, enabling higher order diversity if needed. Antenna sharing does not allow independent mechanical adjustment of antenna bearing and tilting and therefore is not always an optimal solution from this point of view.

Isolation between co-located radio systems can be achieved by filtering or by antenna installation. For standardisation, the isolation between antennas is assumed to be at least 30 dB [16]. If there is not enough space in the cable conduits, antenna lines can be shared and di- or triplexers should be used to combine and split more than one signal into and from the single antenna line.

In practice, reusability of a site will be determined by several non-RF factors, such as:
- lease contract agreements;
- building permit conditions;
- floorspace availability;
- antenna space availability;
- transmission connectivity;
- mechanical constraints (windloads, cabling, etc.);
- power consumption;
- radiation constraints.

The above restrictions might make it hard to reach 100% one-to-one site reuse in practice.

5.3.1.2 Site Solutions

In many cases existing GSM operators will additionally acquire a 3G licence and expand their existing GSM network towards UMTS. For reasons of cost, rollout speed and simplicity, they will prefer to reuse their GSM sites also for UMTS. Apart from the RF reasoning discussed in previous sections, there are a number of possible mechanical restrictions in doing so.

Co-located Sites
If the existing rental contract does not allow installing more hardware elements (BSs, cables, antennas, etc.), it will need to be renegotiated or the existing equipment will need to be shared between the GSM and UMTS infrastructure. This should be verified on a site-by-site basis before new site configurations are planned. This may restrict the usage of, for example, six-sector cells, MHA or similar. If antennas need to be shared between both systems, then the antenna bearing, gains, tilts, etc. will be identical (being mechanically coupled) for both systems. This severely restricts the planner's freedom in optimising site configuration.

Separate feeder networks and antennas for both systems are preferable, although they will not be feasible in a number of cases. Shared feeders and antennas need additional diplexers to separate the individual signals (GSM transmitter, GSM receiver, UMTS transmitter, UMTS receiver). This introduces additional losses into the link budget (and hence slightly reduces cell capacities for UMTS) and increases the system NF.

Installing six-sector sites has a clear theoretical advantage concerning supported capacity and coverage. This may be outweighed by the number of additional antennas, MHAs and feeder networks to be installed. In many networks sites will also be shared by at least two operators.

Co-siting

The term 'co-locating' shall be used when BSs are installed at the same site.

When sites are co-located *and* share feeders and antennas this shall be called 'co-siting'. Basically the same isolation requirements are still valid as in co-located sites but the means to achieve this could be different.

Different kinds of sharing situations may be distinguished, such as antenna and/or feeder sharing, or there could even be multi-mode BSs that share the same cabinets, site support equipment, transmission, feeders and antennas.

In the basic situation, there would be an existing BS with the required site support equipment, feeders and antennas, and the operator installs a WCDMA BS on the same site. If the existing system is GSM1800, the attenuation of the feeder would be of the same order as in WCDMA, but in the case of GSM900 the attenuation of the feeder should be checked and changed if needed. Single-mode antennas can be replaced by multi-mode antennas. One example is shown in Figure 5.13.

Based upon the preceding discussion and the assumed NFs in Section 5.1.4, the isolation between a WCDMA BS cabinet and a GSM900 BS cabinet should be at least 40 dB whereas the isolation between a WCDMA BS cabinet and a GSM1800 BS cabinet should be at least 45 dB. If the receiver NF is greater than that assumed – e.g., for an active distributed antenna system indoor solution – then the isolation requirement can be reduced.

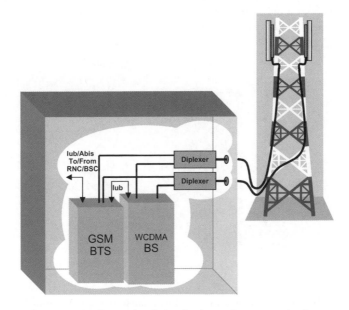

Figure 5.13   Example of site, feeder and antenna sharing.

The way in which this isolation requirement is achieved depends upon the detailed site design. If a diplexer is being used to combine the WCDMA and GSM signals such that they can share the same feeders then the diplexor provides the majority of the isolation requirement. A diplexer typically offers 40 dB of isolation between the GSM and WCDMA systems.

By changing the single-mode BS to dual- or triple-mode, the space required could be smaller due to the single-site support package.

5.3.1.3   Antenna Configurations

Interference between other systems and the WCDMA band depends heavily on the antenna configurations used for both systems. The main problem has been identified with the GSM1800 band; all other systems pose little or no risk of blocking and/or intermodulation with the WCDMA band. Therefore, only the GSM1800 case is further investigated.

If antennas for both GSM and UMTS systems have to be mounted on a single carrier pole, the usual 120° three-sector configuration with vertical stacking of GSM and UMTS antennas seems to be a suitable solution, providing isolation values of approximately 30 dB between sectors and systems.

If diversity reception is needed, the diversity branches of both systems can be handled by a single physical antenna (assuming dual-band antennas). This is beneficial when the diversity antenna is as far as possible from the (possibly interfering) GSM transmit antenna. Such a triple-stack antenna requires tall poles and may not be feasible in many locations (Figure 5.14).

On large flat roofs, isolation between antenna positions can be improved by setting GSM and UMTS antennas physically apart, so that no direct Line-of-Sight (LOS) connection between them exists. One way to do this is by lowering one set of

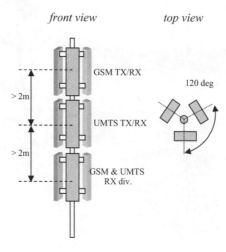

Figure 5.14   Possible antenna configuration for dual-system GSM and UMTS site (with diversity).

antennas down over the edge of the rooftop, if that position is available and suitable from a radio propagation perspective.

5.3.1.4   Traffic and Service Distribution between Systems

Traffic between systems could be separated according to the type of service – e.g., voice and low-speed data traffic could be directed mainly into the 2G network, whereas higher speed data traffic can be directed into the WCDMA.

Traffic sharing between layers can be implemented so that the high-speed data traffic is concentrated in pico- and micro-cells and the low-speed data and voice traffic in macro-cells. This is reasonable, because in WCDMA the coverage is tightly bound to the data speeds through processing gain, the higher data rate implying smaller coverage. The services can be handed over as a function of the loading – e.g., speech services can be handed over from WCDMA to 2G if loading is higher than 10% – which in practice directs speech services into the 2G network.

Subscribers could be classified into different groups that have different rights depending on their subscription, and accordingly redirected to the relevant systems. Subscribers with lower priority could be redirected to the 2G network, which has lower maximum data rates for different services. Packet data users who might suffer from excessive delays could be handed over to whichever network has the most extra capacity available.

5.3.1.5   Coverage and Capacity

At the beginning of WCDMA deployment, coverage will not be continuous, but it could be extended by selective handover to the 2G network. In areas where WCDMA coverage is continuous, dual-mode or multi-mode mobiles could be set to start their calls in the WCDMA network by proper setting of idle mode parameters. By doing this, the loading between 2G and WCDMA networks can be balanced and in some cases reduce the traffic in overloaded 2G networks: see Figure 5.15.

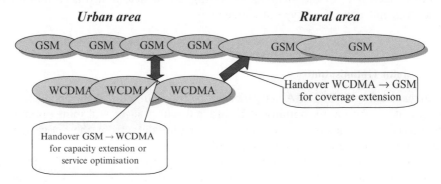

Figure 5.15   Schematic view of handovers between GSM and WCDMA networks for load and coverage reasons.

Table 5.7 Trunking gain in the case of load sharing between EDGE and WCDMA. The blocking probability used was 2% and the capacity of EDGE is the same as that of WCDMA.

	Number of channels	WCDMA or EDGE [Erl]	WCDMA + EDGE [Erl]	Combined capacity [Erl]	Trunking gain [%]
Speech	60	49.6	99.3	107.4	8.2
64 kbps	10	5.1	10.2	13.2	29.7
144 kbps	5	1.7	3.3	5.1	53.4
384 kbps	2	0.2	0.4	1.1	145.2

Such handover for coverage reasons should be initiated sufficiently early, because during compressed mode measurements higher power is needed if spreading factor splitting is used. If the mobile is located at the cell edge and is already transmitting with full power, it cannot increase its transmit power further and the connection might be lost if handover is not started early enough. To avoid this kind of problem, handover statistics can be used to determine the sites where inter-system handovers happen most frequently and trigger compressed mode measurements early enough.

Instead of switching to compressed mode, blind handover can be performed if both systems are located at the same site, since path loss remains the same. Blind handover is especially useful for Non-Real Time (NRT) users, as synchronisation for Real Time (RT) users might take too long, leading to deterioration of connection quality below acceptable limits.

Load sharing between 2G and WCDMA networks can be exploited to make full use of their capacity and to achieve some trunking gain, as their resources are in the same pool (see Table 5.7). It is seen that the trunking gain increases as the used data rate increases.

Load sharing operation is closely related to how traffic and services are distributed between systems. Speech users can be kept in the 2G network as long as the loading of that network is below the pre-defined threshold, whereas high-speed data users can always be handed over to the 3G network if it is available. The order of the mobiles that are handed over to the other radio system can be determined according to their service, transmit power and type of connection.

5.3.1.6 Joint Optimisation

Resources in 2G and WCDMA networks can be fully utilised if their management and deployment can be jointly optimised. In order to effect successful joint optimisation, there should be a means of gathering performance data from the active network, analyse it and change the parameters accordingly.

Handover parameters can be adjusted to balance the load between different systems and to take full advantage of the common resource pool to achieve trunking gains from it. By adjusting idle mode parameters, the initial camping of the mobile can be directed to the desired radio system and unnecessary handovers can be avoided.

5.3.2 Transmission Planning

The aim of transmission planning is to connect BSs to BSCs or Radio Network Controllers (RNCs). Transmission media can be copper wire, coaxial cable, microwave links or fibre-optic line. Microwave links are flexible and can easily be located at the same places as BSs, whereas the copper wire solution will need more civil engineering work. Fibre-optic lines are deployed if there is a need for high-capacity links.

The main difference between radio network and transmission planning is that in the latter case the network should be planned to fulfil the capacity demands throughout the network's lifespan. The topology of the transmission network determines its capacity, protection and expandability, therefore topology changes should be avoided if possible.

5.3.2.1 Transmission Topologies

Co-siting of WCDMA and GSM BSs means that the whole network will be affected, both access and core. Together with capacity growth, the content of the carried signal moves from circuit switched speech to packet data, both RT and NRT.

Upgrading means important modifications in three areas. There could be topological changes, site configuration changes, and media upgrading and changes.

The topologies used can be divided into five structures: chain, star, tree, loop and mesh. Chain topology can be used, for example, along highways but gives poor protection against faults. Loop and mesh topologies can provide good protection but they are quite expensive solutions.

In any case a major upgrade of the transmission backbone for 3G systems is needed, compared with a GSM network. While a standard $4 + 4 + 4$ GSM site can be fitted to a single E1 trunk, a single WCDMA TRX (transmit and receive unit, or transceiver) delivers up to 1.5 Mbps of data on the Iub interface.

In a typical urban European network, macro-cells with one carrier have been simulated to have an average throughput of 700–1000 kbps. Including 30% soft handover overhead, various protocol overheads, and so on, this adds up to a total of typically 1.5 Mbps per TRX, meaning a typical WCDMA $1 + 1 + 1$ site will need a transmission capacity of approximately 5 Mbps for 3G traffic. This is additional to existing GSM traffic. Note that GPRS does not contribute extra traffic, since it is handled via the GSM air interface, which has a direct mapping to the Abis interface (non-blocking).

On the access network (Abis, Iub) the existing chain and loop topologies must be investigated and modified to accommodate the additional 3G traffic. This is likely to cause redesign of transmission topologies, or at least of traffic routing. In any case the issue leads to additional capacity needs. A factor of approximately 4 in additional capacity is needed.

5.3.2.2 Transmission Methods

The transmission method defines the structure of the data and control stream in a transmission medium. In 2G networks the data and control streams were structured according to E1 or T1 trunks; the method was based on Time Division Multiplex

(TDM). In 3G transmission networks the new method will be Asynchronous Transfer Mode (ATM) and, in the future, Internet Protocol Radio Access Network (IP RAN).

The main difference between 2G and 3G traffic is assumed to be the burstiness of 3G services, as the packet data share will increase more than circuit switched data. The variety of services in 3G networks will also benefit from the statistical multiplexing gain achieved in ATM networks. The delay characteristics of ATM networks are looser than those of TDM networks where in practice the delay is constant. In all-IP networks the delay characteristics will be specified. All-IP deployment enables the combining of different services and technologies under the same protocol, which will reduce system building and operating costs.

5.3.2.3 Transmission Sharing between Systems

Sharing of the transmission systems between GSM and WCDMA would be useful in order to make full use of the existing hardware and to prevent the building of a totally new transmission network. By sharing hardware resources, some trunking gain can be achieved, and statistical multiplexing gain can also be obtained for 2G network services if the ATM or all-IP transmission network is deployed. In most cases there would be no strict necessity to change geographical topology and therefore sharing can be done by just adding or changing low-capacity devices to higher capacity ones.

5.3.3 Perception of Different Technologies by the End-user

Technology comparison is a natural issue if the UMTS technology layer is included in an existing 2–2.5G network. As the network needs to meet customer expectations from the end-user point of view, new technology must meet very good interworking standards from the beginning. This means especially inter-system handovers and cell reselection functionality along with similar or better Call Drop Rate (CDR) experience by the end-user. The property is important for services like speech where behaviour and quality is known from GSM and other cellular systems. Thus, the CDR must be the same or better. Let us consider the behaviour of speech quality in the situation when the mobile is moving out of the coverage area and it is not feasible to make a handover to a better cell either in the same or another technology. In this situation the desired behaviour of the mobile is to drop the connection after a similar period of bad quality as would have happened in the already used technology. Such behaviour represents an optimum between customer churn on one side and effective usage of technology on the other side. This should be adjusted by parameters and it is a natural optimisation target from the beginning.

5.3.4 Tight Usage of Frequency Spectrum by Different Technologies

In some cases the frequency regulator issues a technology-independent licence. Thus, the operator can handle the spectrum owned quite freely. One possibility is thus that the spectrum is tightly used by different cellular technologies. Tight frequency use of the spectrum by different technologies brings challenges to the additional filtration solution for 2G BSs. The minimum coupling requirements between different BSs and different cellular systems are specified in [15]. Concrete solution of such cases depends on the spectrum situation of the specific case. 3GPP technical specifications for BS radio transmission and reception in FDD mode are in [15] and the Mobile Station (MS) is specified in [1]. The impact of narrowband technologies with tight frequency separation from the UMTS band is discussed in the next section.

5.4 Narrowband and WCDMA System Operation in Adjacent Frequency Bands

Utilisation of WCDMA outside the 2 GHz UMTS core band – e.g., in the GSM1800 band or in the US Personal Communication System (PCS) 1900 MHz band – is now discussed. When the adjacent system to WCDMA is some narrowband mobile telecommunication system, such as GSM/EDGE, TDMA or narrowband CDMA, the evolution of mobile network systems from 2G to 3G requires flexible utilisation of available frequency bands. Operation of the WCDMA system when there are adjacent narrowband systems working in the same geographical area is, however, different from operation with the basic frequency allocation because of increased interference between the narrowband system and the WCDMA system.

In the 3GPP specifications the coexistence of WCDMA with the spectrally adjacent narrowband system has been taken into account. 3GPP Release 5 specifies both the characteristics for the WCDMA BS and the User Equipment (UE) respective MS when operating at the same band with the narrowband system – the PCS system in this case ([1] and [15]). The most essential requirements covered by the specifications are blocking for the BS and out-of-band emission levels, as well as requirements for the narrowband blocking and intermodulation characteristics of the MS.

New 3G multimedia services and enhanced capacity require more user bandwidth, which in turn causes decreased tolerance to interference from systems operating at adjacent frequency bands. This is due to the more demanding design of the wideband, linear components and also because a wideband receiver is more exposed to various interference sources. Also, the new frequency allocation schemes set additional requirements for the components. For example, the narrower duplex gap in the case of the PCS band sets more stringent requirements for duplex filters at the MS.

In interference limited systems such as WCDMA, the increased interference causes a need for additional power in order to maintain the link quality, which in turn effects additional capacity and coverage degradation. In the adjacent channel operation of WCDMA and narrowband systems, several possible interference sources or interference mechanisms are present. The relative importance of various interference mechanisms is dependent on implementation of different network elements, locations of interfered and interfering sites with respect to each other, and the type and size of the cells. Performance degradation can be decreased by introducing guardbands around the WCDMA carrier, by frequency planning, by careful site and power planning or by co-siting with the interfering system. The general frequency allocation scenario showing

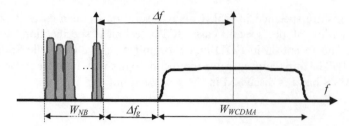

Figure 5.16 Frequency allocation with narrow and wideband systems including a guardband.

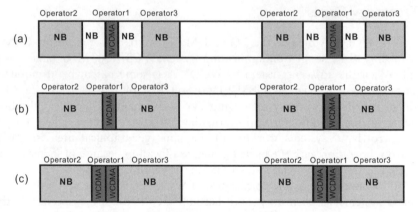

Figure 5.17 Different frequency scenarios: (a) embedded scenario; (b) 5 MHz operation scenario; (c) 10 MHz operation scenario. The upper frequency allocations are for downlink and the lower ones are for uplink transmission directions.

the WCDMA band W_{WCDMA}, the band allocated for the narrowband system W_{NB} and the guardband Δf_g are shown in Figure 5.16.

By co-siting, it is possible to avoid the near–far effect between WCDMA and narrowband systems. The near–far effect here means, for example, that when the narrowband mobile is close to the WCDMA site and far away from its own site there will be uplink interference from the narrowband mobile to the WCDMA BS, and also that when the WCDMA MS is close to the narrowband BS there will be a large downlink interference component from the narrowband system to the WCDMA system. These same interference mechanisms also occur from the WCDMA system to the narrowband system, but the effect is smaller. Figure 5.17 shows some of the principal frequency allocation schemes associated with the WCDMA narrowband co-operation case. The upper scheme shows the situation where operator 1 has one WCDMA carrier and several narrowband carriers and the other operators have only narrowband carriers. The middle scheme shows the situations where operator 1 has only one WCDMA carriers and adjacent to that there are narrowband carriers of other operators. In the lower scheme operator 1 has two adjacent WCDMA carriers.

In the first scheme, operator 1 can coordinate the usage of WCDMA and its own narrowband systems by co-siting them. By doing this the uncoordinated narrowband system is spectrally far away from the WCDMA system, decreasing the interference levels considerably. In the second case, operator 1 has only one WCDMA carrier just next to adjacent operators' bands. In this case the interference is high, since the sites of different operators are usually not co-located. There is a possibility that the WCDMA and narrowband systems interfere each other, and such interference has to be taken into account in radio network planning and dimensioning. Interference between narrowband and CDMA systems has also been studied in [8]–[10]. In the last frequency scenario, operator 1 has two adjacent WCDMA carriers. In this case the performance degradation of the WCDMA system due to additional interference can be avoided with inter-frequency handover between WCDMA carriers.

5.4.1 Interference Mechanisms

Figure 5.18 shows the main interference mechanisms between WCDMA and narrowband systems. In the following sections these interference mechanisms will be discussed. More detailed information about different interference mechanisms can be found, for example, from [4].

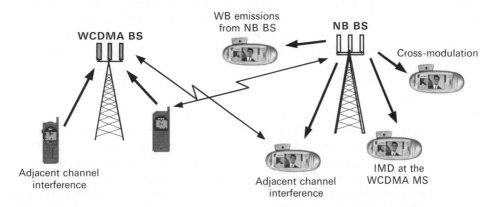

Figure 5.18 Main interference mechanisms between the narrowband system and the WCDMA system.

5.4.1.1 Adjacent Channel Interference

Adjacent Channel Interference (ACI) results from non-ideal receiver filtering outside the band of interest. Even with an ideal transmitter emission mask, there is interference coming from adjacent channels because of ACI. Adjacent channel filtering and therefore ACI depend on the implementation of analogue and digital filtering at the MS in the downlink and at the BS in the uplink. Additionally, ACI is dependent on the power of the interfering system as well as the frequency offset between the interferer and the interfered systems. Usually, ACI is most severe when the channel separation between the own band and the interfering band is low. The effect of ACI decreases rapidly outside the receive band, so ACI can be eliminated with an adequate guardband beside the WCDMA band.

5.4.1.2 Wideband Noise

Wideband noise refers to all out-of-band emission components coming from the transmitter outside the wanted channel of the interfering system. It includes unwanted wideband emissions, thermal noise, phase noise and spurious emissions as well as transmitter intermodulation. These interference mechanisms usually appear at frequencies which are far away from the band of interest and therefore these mechanisms can be considered as wideband. The allowed upper limit of wideband noise is usually described in the specifications of the narrowband system.

5.4.1.3 Intermodulation Distortion at the Receiver

Intermodulation Distortion (IMD) is caused by non-linearities in the RF components of the receiver or transmitter. Intermodulation takes place in the non-linear component when two or more signal components reach it and the signal level is high enough for the operating point to be in the non-linear part of the component. When two or more signals are added together in the non-linear element, the resulting outcome from the element includes, in addition to the desired signal frequency, higher order frequencies caused by the higher order non-linearities. Third-order IMD is particularly problematic, because it is typically strongest and falls close to the band of interest. In the case of two interfering signals on frequencies f_1 and f_2, in the proximity of the desired signal, third-order IMD products are those falling on frequencies $2f_1 - f_2$ and $2f_2 - f_1$ (Figure 5.19). Higher order IMD products exist but are usually less strong.

Usually, the receiver IMD is the most relevant source of intermodulation, since the active components in the receiver are less linear than those in the transmitter; therefore, only the receiver IMD is considered here. Furthermore, we can focus on the downlink, since the active components in the BS are more linear than those in the MS. This is because, when increasing the linearity of the receiver, the power consumption increases as well, which is usually more critical in the design of the MS.

The IMD in the downlink is caused by the mixing of products of the narrowband BS with carrier frequencies f_1 and f_2. Assuming that these frequencies have equal powers, so that $P_{f1} = P_{f2}$, the third-order intermodulation power reduced to the input of the nonlinear element is given by:

$$P_{IMD}^{in} = 3 \cdot P_i - 2 \cdot IIP_3 \tag{5.13}$$

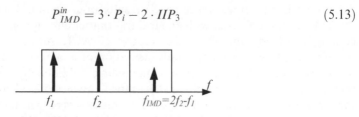

Figure 5.19   Third-order intermodulation distortion.

where P_i [dBm] is the power at the input of the non-linear component; and IIP_3 [dBm] is the third-order input intercept point of the same. So, the strength of this mechanism depends on the output power of the interfering BS as well as the receiver linearity. The strength of the IMD is proportional to the third power of P_i so that it is large when the receiver is close to the interfereing source but decreases rapidly as distance and therefore path loss increase.

5.4.1.4 Transmission Intermodulation Distortion

In CDMA systems, mobile transmission and reception occur simultaneously and a portion of the transmitted signal leaks into the receiver due to non-idealities of the duplex filter. Therefore, another IMD mechanism results from the interaction of a single strong interferer and the leaking transmission signal. Figure 5.20 illustrates this phenomenon, here referred to as Transmission Inter Modulation Distortion (TxIMD). If the interfering frequency, f_I, is below the mobile transmission frequency, f_{Tx}, so that $f_{TxIMD} = 2f_{Tx} - f_1$, the intermodulation power at the input of the LNA is given as:

$$P_{TxIMD} = P_i + 2 \cdot P_{MS,leak} - 2 \cdot IIP_3, \qquad \text{if } f_{Tx} > f_I \qquad (5.14)$$

where P_i is the interferer power at the receiver input; and $P_{MS,leak}$ is the leakage power from the mobile transmission. If the interfering frequency, f_I, is above the mobile transmission frequency, f_{Tx}, so that $f_{TxIMD} = 2f_I - f_{Tx}$, the intermodulation power at the input of the receiver is given as:

$$P_{TxIMD} = 2 \cdot P_i + P_{MS,leak} - 2 \cdot IIP_3, \qquad \text{if } f_{Tx} < f_I \qquad (5.15)$$

The severity of the TxIMD depends on the particular frequency scenario. In mobile telecommunication applications the component given in Equation (5.15) is usually more relevant, because it corresponds to the case where the frequency of the interferer is located within the receive frequency band and has no attenuation due to band-selective filtering. It should be noted that TxIMD is proportional to the strength of the leakage power, which is subsequently dependent on the isolation properties of duplex filtering. If the isolation of the duplex filter is large enough, the TxIMD has quite a minor effect on system performance.

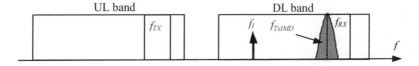

Figure 5.20 Transmission intermodulation.

5.4.1.5 Cross-modulation Distortion

Cross-modulation Distortion (XMD) is also caused by third-order receiver non-linearities. It is produced in the non-linear receiver element when an amplitude-modulated mobile transmission signal is mixed with a narrowband interferer located

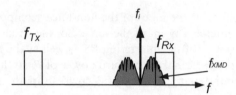

Figure 5.21   Cross-modulation spectrum modulated around the outside interferer.

close to the mobile reception frequency. The power of the cross-modulation at the input of the receiver can be written as:

$$P_{XMD} = P_i + 2 \cdot P_{MS,leak} - 2 \cdot IIP_3 - C_{XMD} - L_{XMD}(\Delta f_c) \qquad (5.16)$$

where $P_{MS,leak}$ represents transmission leakage; P_i is the narrowband interferer power; and C_{XMD} is a factor that depends on the transmit signal modulation index. Reduction of cross-modulation power as a function of channel separation, due to the partial overlap with the wanted signal, is given by:

$$L_{XMD}(\Delta f) = \begin{cases} -10 \cdot \log_{10}\left[2\dfrac{\Delta f}{B}\left(1 - \dfrac{(\Delta f - B/2)}{B}\right)^2\right], & \text{if } \Delta f \le 3B/2 \\ \infty, & \text{if } \Delta f > 3B/2 \end{cases} \qquad (5.17)$$

For further description of cross-modulation see [4]–[6], for example. Figure 5.21 shows the cross-modulation spectrum around the narrowband signal which flows into the reception band.

5.4.2   Worst Case Analysis

In this section a simple comparison between different interference components is carried out. The target is to show the interfering power of the most important interference mechanisms in different cases in a situation where the interfering BS or MS is very close to the interfered BS/MS. Table 5.8 shows the basic parameter values used for this analysis.

Table 5.8   Parameters used for worst case analysis.

Bandwidth	5	MHz
WCDMA MS transmission power	21	dBm
WCDMA BS transmission power	43	dBm
Narrowband MS transmission power	30	dBm
Narrowband BS transmission power	43	dBm
Rejection loss of duplexer in MS	40	dB
IIP_3 at the MS	-10	dBm
C_{XMD}	5	dB
XMD attenuation at 1 MHz channel separation	0.5	dB
Minimum coupling loss between MS and BS (MCL)	75	dB
Time averaging of MS power (one over eight slots $= -10 \cdot \log_{10}(1/8)$)	9	dB

Table 5.9 shows the assumptions of the out-of-band noise and adjacent channel filtering of both the MS and BS of the narrowband system and WCDMA.

Table 5.10 shows the worst case analysis based on these parameter, out-of-band emissions and filtering values. From these results we can see that the worst direction for interference is from the BS of the narrowband system to the WCDMA mobile.

The uplink interference from the narrowband MS to the WCDMA BS is also considerable, since the increase of interference level at the BS influences the coverage area of the whole cell (Figure 5.22). Assuming that the interference limit for the narrowband

Table 5.9 Assumed out-of-band emission and adjacent channel filtering values for worst case analysis.

		WCDMA MS	WCDMA BS	Narrow-band MS	Narrow-band BS
Out-of-band emissions, [1], [2] and [15]	No guardband	−26.8 dBc at 200 kHz	−48.8 dBc at 200 kHz	−11.5 dBc at 5 MHz	−11.5 dBc at 5 MHz
	1 MHz guardband	−42.0 dBc at 200 kHz	−60.8 dBc at 200 kHz	−53.1 dBc at 5 MHz	−64.5 dBc at 5 MHz
Channel filtering	No guardband	10 dB	20 dB	41 dB over 5 MHz	45 dB over 5 MHz
	1 MHz guardband	30 dB	45 dB	67 dB over 5 MHz	70 dB over 5 MHz

Table 5.10 Interference levels with minimum coupling loss.

	Uplink	Downlink
Channel separation 200 kHz		
WCDMA → NB	$21 - 26.8 - 75 = -80.8$ dBm (WB)	$43 - 48.8 - 75 = -80.8$ dBm (WB)
	$21 - 75 - 45 = -99$ dBm (ACI)	$43 - 75 - 41 = -73$ dBm (ACI)
NB → WCDMA	$30 - 11.5 - 75 - 9 = -65.5$ dBm (WB)	$43 - 11.5 - 75 = -43.5$ dBm (WB)
	$30 - 75 - 20 - 9 = -74$ dBm (ACI)	$43 - 75 - 10 = -42$ dBm (ACI)
		$43 - 75 + 2 \times (21 - 35) - 2 \times (-10) - 5 = -45$ dBm (XMD)
		$2 \times (43 - 75) + (21 - 35) - 2 \times (-10) = -58$ dBm (TxIMD)
Channel separation 1 MHz		
WCDMA → NB	$21 - 42.1 - 75 = -96.1$ dBm (WB)	$43 - 60.8 - 75 = -92.8$ dBm (WB)
	$21 - 75 - 55 = -109$ dBm (ACI)	$43 - 75 - 67 = -99$ dBm (ACI)
NB → WCDMA	$30 - 53.1 - 75 - 9 = -107.1$ dBm (WB)	$43 - 64.5 - 75 = -96.5$ dBm (WB)
	$30 - 75 - 45 - 9 = -99$ dBm (ACI)	$43 - 75 - 30 = -62$ dBm (ACI)
		$43 - 75 + 2 \times (21 - 35) - 2 \times (-10) - 5 - 0.5 = -45.5$ dBm (XMD)
		$2 \times (43 - 75) + (21 - 35) - 2 \times (-10) = -58$ dBm (TxIMD)

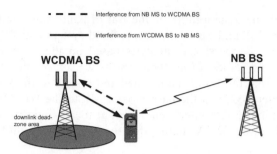

Figure 5.22   Uplink interference when a narrowband mobile station is close to the WCDMA base station.

system is 90 dBm in the downlink, when the WCDMA interference reaches this limit the coupling loss is $43 - 41 + 90 = 92$ dB, assuming 41 dB filtering at the GSM MS over the 5 MHz band. This means that when the coupling loss is below 92 dB the GSM MS is in outage, due to interference from the WCDMA BS. With 92 dB path loss the interference at the WCDMA BS is $30 - 9 - 92 - 11.5 = -82.5$ dBm. When the coverage target is 95 dBm this means that there might be a coverage reduction of 12.5 dB. This happens only in those cases where the narrowband mobile is allocated to the next carrier and is very close to the WCDMA cell.

5.4.3   Simulation Case Study with a Static Simulator

This section shows two simulation studies of narrowband and WCDMA operators working in the same geographical area. In this scenario, the figures shown in colour can be found at the weblink (*www.wiley.com/go/laiho*). In these studies there is no guardband between operators. In the first case study, both the WCDMA and the narrowband operator utilise   macro-cells. The sites of these two operators are assumed to be independently located. In the second case study, one operator uses macro-cells and the other uses micro-cells. Only downlink interference effects have been studied here because of their relative importance as concluded in Section 5.4.2. Table 5.11 shows the main input values for the simulation case.

The main interference-related parameters are collected in Table 5.12. Narrowband BS powers have been assumed to be constant in the simulations. Out-of-band noise emissions have been considered as constant over the spectrum.

In the presence of narrowband interference the Signal-to-Interference Ratio (SIR) in the downlink is given by:

$$\left(\frac{S}{I}\right)_i = \frac{p_i/L_{p,i}}{(1 - \alpha_i) \cdot P + I_{oth,i} + I_{WB,i} + I_{ACI,i} + L_1 \cdot I_{XMD,i} + N} \tag{5.18}$$

where P is the transmitting power of the BS; α_i is the orthogonality of the propagation channel; $I_{oth,i}$ is the interference coming to mobile i from other BSs of the WCDMA systems; $I_{WB,i}$, $I_{ACI,i}$ and $I_{XMD,i}$ are the narrowband interference powers of wideband emissions, ACI and cross-modulation, respectively; L_1 is the attenuation between the antenna and the LNA; N is thermal noise; and $L_{p,i}$ is the link loss between the

Table 5.11   Main system parameters.

Chip rate	3.84 Mcps
Base station maximum transmit power	43 dBm (macro-cells)
	37 dBm (micro-cells)
Mobile station maximum transmit power	21 dBm
Mobile station minimum transmit power	−50 dBm
Shadow fading correlation between base stations/sectors	50 %/80 %
Channel profile	ITU vehicular
Service	12.2 kbps, speech
Mobile station speed	3 km/h
MS/BS noise figures	11 dB/5 dB
CPICH power	30 dBm
Combined power for other common control channels	23 dBm
Orthogonality	50 %
BS antennas	65°, 17.5 dBi (macro-cells)
	Omni (micro-cells)
MS antennas	Omni, 0 dBi
Cable losses	0 dB
Propagation model	Okumura–Hata (macro-cells)
	Ray-tracing model (micro-cells)

Table 5.12   Main interference-related parameters.

Transmission power of narrowband macrocell	43 dBm
Transmission power of narrowband microcell	35 dBm
Rejection loss of duplex filter	40 dBm
Noise emission level from narrowband BS	−13 dBm
Narrowband bandwidth	3 MHz
MS IIP_3	−10 dBm
C_{XMD}	5 dBm
Guardband	0 MHz
Assumed slope function for MS filter (f = separation between WCDMA and narrowband carriers in MHz)	$-21 \cdot (f - 2.5) - 11\,\mathrm{dB}$

WCDMA BS and MS i, including antenna gains. The contribution of the IMD and the TxIMD has been neglected in these simulations.

The transmission power for each link is allocated to meet the SIR requirement of the service carried on that link. Increased downlink interference also increases the needed transmission power at the BS, and the link will be blocked if the maximum allowed transmission power is exceeded. This could happen especially when the mobile is very close to the interfering BS. This might cause deadzones around narrowband BSs and thus mainly limits the coverage. The capacity of the WCDMA cell is limited by the maximum allowed transmission power of the BS and, therefore, increased interference also decreases the capacity of the WCDMA system. This effect is visualised in Figure 5.23.

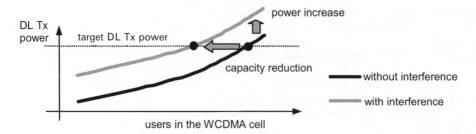

Figure 5.23   The effect of narrowband interference on WCDMA downlink capacity.

When a large number of mobiles in a WCDMA cell are affected by narrowband interference and the power needed for each link increases, the total transmission power needed increases as well. In the Radio Resource Management (RRM) functionalities of the WCDMA, such as admission control, packet scheduling or load control, the total transmission power is measured to detect the downlink load. The load of the system is adjusted to the target level and RRM actions take place when the load exceeds the target level. Thus, the increased average interference reduces the maximum number of users that the system is able to support.

If the link-specific power is above the maximum allowed link power, the mobile is unable to get the required service. Also, if the target BS power is exceeded, the system has to limit the number of users. These two limitations have also been taken into account in system simulations. When link-specific maximum powers are limiting, those users who need large power levels – e.g., for increased narrowband interference – will be dropped from the calculations. This, in fact, most probably increases the overall maximum capacity of the system since – after dropping the large power allocated to one user – it is now available to support the remaining users (interference is large only in limited areas around narrowband BSs).

Because of different propagation conditions in micro-cells and macro-cells, interference also depends on cell type. In micro-cells the radiowave propagates through street canyons and the shadowing effects of individual buildings can be significant. One example is the street corner effect, where the signal strength drops by 10–30 dB when the mobile moves from LOS to Non LOS (NLOS). In particular, the worst case scenario might be when the interfering narrowband BS is micro-cellular and the WCDMA cell is macro-cellular, so that the mobile can be in LOS to the interfering BS and in NLOS to its own BS. In the basic interference case both the interfering and interfered cells are in macro-cellular networks; this means that the antennas are above the rooftops, the proportion of LOS area is negligible and the effect of interference is rather small.

Generally speaking the effect of the external interference depends on the relative cell sizes of the interfering and the interfered cells. When the average size of the interfering cell is small compared with the interfered cell, the effect of interference is large compared with the case where the interfering network is sparse and the interfered network is denser. This means that, for example, in the rollout phase of the new system, the macro-cells that are usually preferred are those that are, however, more exposed to interference problems.

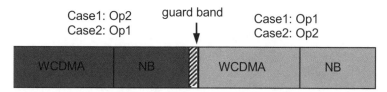

Figure 5.24 Frequency scenario for operators 1 and 2 used in the simulation case study. Operator 1 has 12 cells and operator 2 has 13 cells.

In the first case study, the effect of macro-cell narrowband interference on WCDMA macro-cell downlink capacity and coverage was studied. It was assumed that operators do not share sites but plan their networks independently. In the simulation study, two operators were assumed to have both their WCDMA networks and narrowband networks co-sited. The frequency scenario used is shown in Figure 5.24. In the first sub-case the lower frequencies were allocated to operator 2 and the upper frequencies to operator 1, and in the second sub-case vice versa. Interference between the WCDMA carriers of different operators was neglected and only the capacity effects of the WCDMA systems are considered here. ACI, wideband noise and XMD have been taken into account in these simulations. The main reason for the increased interference in this example is the ACI, because no guardband between operators was used.

5.4.3.1 Macro-cell–Macro-cell Scenario

The network scenario of the macro-cell case, as well as the initial user distribution of operator 1, are shown in Figure 5.25. In these simulations the channel spacing of the narrowband system is 200 kHz. In the macro-cell case each narrowband cell has been allocated to its own frequency channel so that cell #1 has the channel closest to the WCDMA carrier of the adjacent operator with 2.6 MHz channel separation, cell #2 is allocated to the next carrier with 2.8 MHz channel separation and so on. The overall narrowband spectrum used by operator 1 is $12 \times 0.2 = 2.4$ MHz and by operator 2 is $13 \times 0.2 = 2.6$ MHz.

The service probability in the downlink for 12.2 kbps users was simulated first without and then with narrowband interference. The respective coverage results with no interference are shown in Figures 5.26–5.28. The initial number of users requesting service in the downlink was 400 for both WCDMA macro-cell operators. Two downlink power allocation schemes were also tested, with and without the link-specific power limitations. With no link-specific power limitations, the numbers of satisfied users were 382 and 366 for operators 1 and 2, respectively, giving a service probability of 95% and 92% for the two operators. If the downlink link powers of 12.2 kbps users were limited to 30.4 dBm, the numbers of satisfied users would be 391 and 393 for operators 1 and 2, respectively, now giving a service probability of about 98%. When the link-specific downlink powers were limited, the service probability increased slightly, because those users who need more power also reduce the power resources for other users. Figure 5.26 shows the downlink coverage analysis for operator 1 in the case without narrowband interference. Each pixel in the coverage

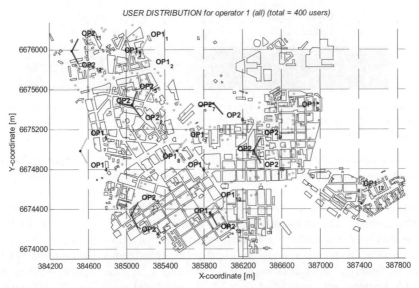

Figure 5.25 Macro-cell–macro-cell network scenario and the initial user distribution: 12 macro-cells of operator 1 (green sites) and 13 macro-cells of operator 2 (red sites). See *www.wiley.com/go/ laiho* for colour images.

map represents the needed link-specific transmit power from the best server cell for that point in current interference conditions. It can be seen that the needed power is usually above 15 dBm, and at the cell borders about 30 dBm. The calculated coverage percentage, however, is very high at 99.9%. Figure 5.28 shows the cumulative distribution of the needed powers from the best server cell. According to these results, about 25–27 dBm power per link would be needed in order to reach 95% coverage probability.

When narrowband interference was present the number of users without downlink power limitations were 340 and 366 for operators 1 and 2, respectively, and the service probabilities were 85% and 92% with respect to simulation cases 1 and 2 in Figure 5.24. When using the link-specific power limitation of 30.4 dBm the number of served users were 378 and 391, respectively. Thus, the narrowband interference decreases the capacity of operator 1 from 391 to 378 and that of operator 2 from 393 to 391. The overall capacity reduction was then 11% and 3.3% without and with the link-specific DL power limitations for operator 1, and 0% and 0.5% for operator 2, respectively. The final results of this capacity analysis are summarised in Table 5.13. Figures 5.29–5.31 show the area coverage in the case when narrowband interference is present. The coverage probability is 97.5% for operator 1 and 97.4% for operator 2. There are, however, very large deadzone areas around the other operator's BS. For operator 1 the problems are concentrated on areas where there are cells working at the closest carriers, which are sites 1–3 of operator 2 using carriers with 2.6, 2.8 and 3.0 MHz separation from the WCDMA carrier. In other locations there are only minor deadzone areas. In the case of operator 2, there are also deadzones around the closest carriers (sites 1–3 of operator 1), but even larger deadzone areas exist on the left-hand side of the figure at carriers 4 and 5, which are 3.2 and 3.4 MHz away from the WCDMA centre frequency.

Table 5.13 Capacity simulation results in the macro-cellular case. The initial number of users was 400 in each of the two operators' networks.

	Without narrowband interference		With narrowband interference	
	Without power limitation	With power limitation	Without power limitation	With power limitation
Operator 1	382	391	340	378
Operator 2	366	393	366	391

This is because of the more open propagation environment around sites 4 and 5, so that the path loss difference between the own and the interfering site is much larger in this case. It has to be noted that narrowband interference only includes the contribution from the adjacent operator, and the effect of the own narrowband network has not been taken into account.

Figure 5.31 shows the cumulative distribution of the required link-specific transmission powers with the narrowband interference. The needed power in this case in order to achieve 95% coverage probability was 28–29 dBm, which is 2–3 dB higher than in the case without narrowband interference.

Macro-cell–Macro-cell Case without Interference (Figures 5.26–5.28)

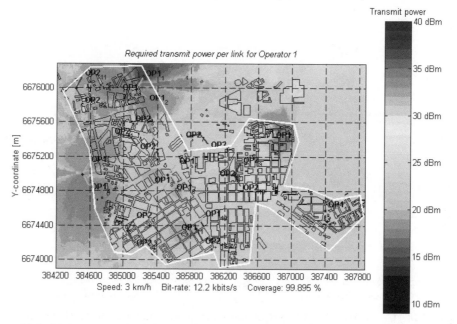

Figure 5.26 Link powers needed in WCDMA macro-cells (operator 1) when narrowband interference is not present. Downlink coverage of 12.2-kbps service is 99.9%.

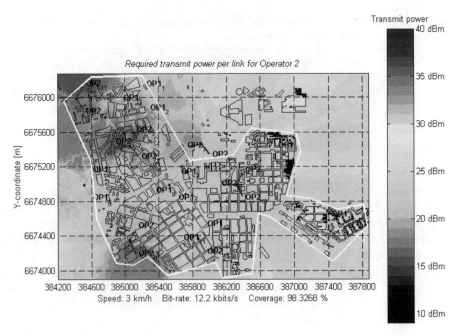

Figure 5.27 Link powers needed in WCDMA macro-cells (operator 2) when narrowband interference is not present. Downlink coverage of 12.2 kbps service is 98.3%.

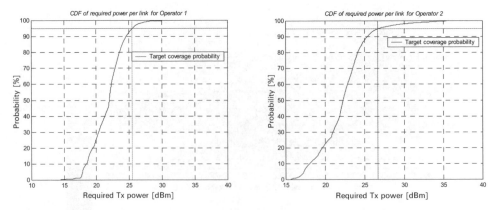

Figure 5.28 Cumulative distributions of downlink powers needed for 12.2 kbps service for operator 1 (left) and operator 2 (right) with no narrowband interference from the adjacent operator.

Macro-cell–Macro-cell Case with Interference (Figures 5.29–5.31)

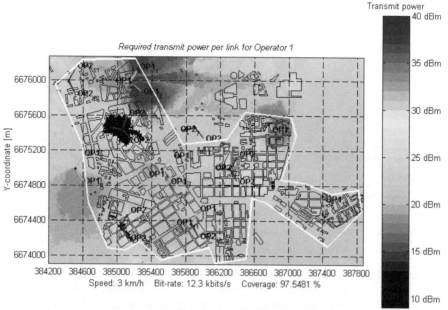

Figure 5.29 Link powers needed in WCDMA macro-cells (operator 1) when narrowband interference is present. Downlink coverage of 12.2 kbps service is 97.5%.

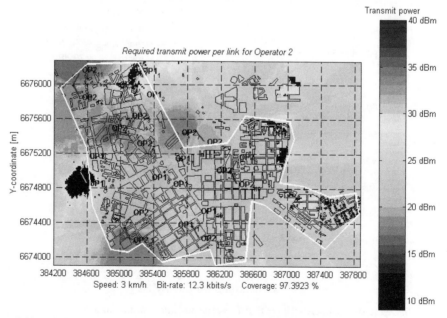

Figure 5.30 Link powers needed in WCDMA macro-cells (operator 2) when narrowband interference is present. Downlink coverage of 12.2 kbps service is 97.4%.

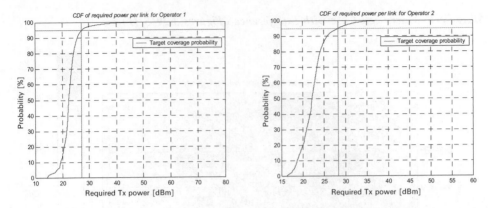

Figure 5.31 Cumulative distributions of downlink powers needed for 12.2 kbps service for operator 1 (left) and operator 2 (right), assuming narrowband interference from the adjacent operator.

5.4.3.2 Micro-cell–Macro-cell Scenario

The micro-cell simulation case was basically similar to the macro-cell case but with operator 2 now using micro-cells. With this scenario it is possible to study the extent to which the narrowband micro-cellular network interferes with WCDMA macro-cells and also the extent to which the narrowband macro-cell system interferes with the

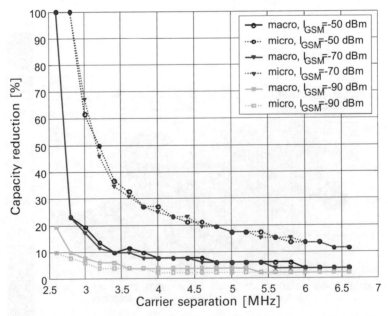

Figure 5.32 Macro-cell–micro-cell network scenario and initial user distribution for the macrocell: 13 macro-cells of operator 1 (green sites) and 35 micro-cells operator 2 (red sites). See *www.wiley.com/go/laiho* for colour images.

WCDMA micro-cellular system. The used scenario consisted of 12 macro-cells (for operator 1) and 35 micro-cells (for operator 2) and is shown in Figure 5.32. The micro-cell antennas were 10 m high and were horizontally omni-directional. Propagation data were computed with the ray-tracing propagation model.

The micro-cell frequencies were allocated among 15 carriers, thus requiring the 3-MHz frequency band. The channels were allocated as follows (cell number|channel number): 1|5, 2|4, 3|3, 4|2, 5|1, 6|15, 7|14, 8|13, 9|12, 10|11, 11|10, 12|9, 13|8, 14|7, 15|6, 16|5, 17|4, 18|3, 19|2, 20|1, 21|15, 22|14, 23|13, 24|12, 25|11, 26|10, 27|9, 28|8, 29|7, 30|6, 31|5, 32|4, 33|3, 34|2 and 35|1. Channel 1 corresponds to the carrier that is closest to the WCDMA carrier, and channel 15 is the farthest, with 5.4 MHz channel separation from the WCDMA carrier.

The capacity without the downlink link-specific power limitation when narrowband interference is not present was 446 for the macro-rcell network and 776 for the micro-cell network. Initially, there were 500 and 800 mobiles for the macro-cell and micro-cell networks, respectively. If the link-specific powers were limited, the capacities were 452 and 778, respectively. The service probability was then 89.2% and 97% without the power limitation, and 90.4% and 97.2% with the power limitation. So the power limitation increases the overall service probability of the system. In the case of power limitation, the maximum allowed power per link for the 12.2 kbps service was 30.4 dBm. Figures 5.33–5.35 show the downlink coverage of the macro-cells of operator 1 and the micro-cells of operator 2 with no interference from the adjacent operator. The respective coverage percentage is 99.9% in both cases. Figure 5.35 shows the cumulative distribution of the transmission power required from the best server of each pixel, which is about 25 dBm in the macro-cell case and 18 dBm in the micro-cell case.

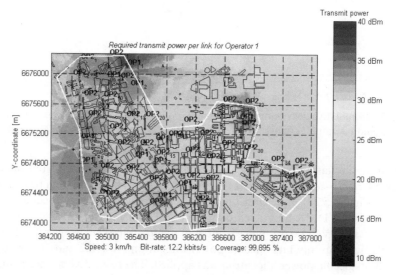

Figure 5.33 Link powers needed in WCDMA macro-cells (operator 1) when narrowband interference is not present. Downlink coverage of 12.2 kbps service is 99.9%.

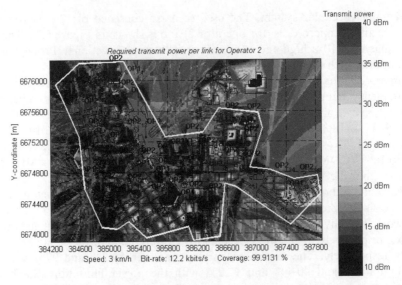

Figure 5.34 Link powers needed in WCDMA micro-cells (operator 2) when narrowband interference is not present. Downlink coverage of 12.2 kbps service is 99.9%.

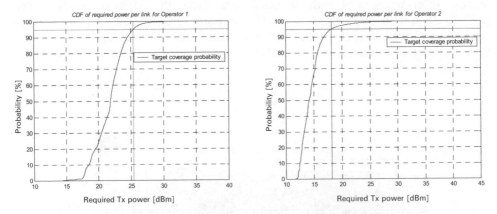

Figure 5.35 Cumulative distributions of downlink powers needed for 12.2-kbps service for operator 1 (left) and operator 2 (right) with no narrowband interference from the adjacent operator.

In the next simulation case the effect of narrowband interference from operator 2's narrowband micro-cellular network to operator 1's WCDMA macro-cell network (Case 1 in Figure 5.24) and from operator 1's narrowband macro-cellular network to operator 2's WCDMA micro-cell network (Case 2 in Figure 5.24) were studied. The capacities of the WCDMA macro-cell and micro-cell systems were 351 and 761, respectively, without any link-specific power allocation, and 422 and 763, respectively, when the link-specific power allocation was applied. The respective service probabilities in this case were 70.2% and 95.1%.

When the link power limitation of 30.4 dBm is utilised, the respective service

Table 5.14 Capacity simulation results in the micro-cellular case. The initial number of users was 500 in operator 1's network and 800 in operator 2's network.

	Without narrowband interference		With narrowband interference	
	Without power limitation	With power limitation	Without power limitation	With power limitation
Operator 1	446	452	351	422
Operator 2	776	778	761	763

probabilities were 84.4% and 95.4%. So the service probability in macro-cells dropped by 6% when narrowband interference was introduced when downlink powers were limited (Table 5.14).

Figures 5.36–5.38 show the coverage of the micro–macro case study results when the cells are fully loaded – i.e., the very high interference situation. From Figure 5.36 we can see that there will be large coverage holes around narrowband micro-cell BSs. The size of the deadzone is mainly dependent on three factors: (1) the extent of the LOS around the narrowband BS, (2) the distance between the narrowband BS and the own WCDMA BS, and (3) the carrier separation between the own WCDMA carrier and the narrowband carrier. Also, the antenna pattern of the narrowband system as well as its

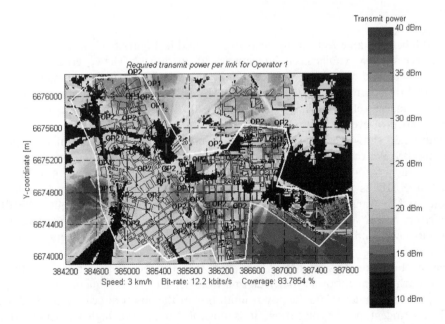

Figure 5.36 Link powers needed in WCDMA macro-cells (operator 1) when narrowband interference is present. Downlink coverage of 12.2 kbps service is 83.8%. Black areas indicate locations that cannot be served.

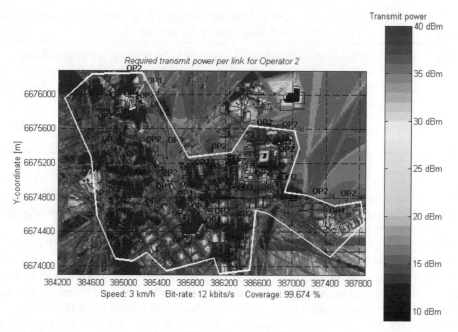

Figure 5.37 Link powers needed in WCDMA micro-cells (operator 2) when narrowband interference is present. Downlink coverage of 12.2 kbps service is 99.7%.

vertical lobe have an effect on the deadzone around it. Figure 5.37 shows the coverage area of the WCDMA micro-cellular network of operator 2. The coverage in this case does not change much (from 99.9% to 99.6%), since the narrowband macro-cells of operator 1 do not interfere with micro-cells. This is because the minimum coupling loss from the macro-cell is typically well above 70 dB, whereas in the micro-cell case it is around 60 dB or even below.

Figure 5.38 shows the cumulative distribution of the required transmission power from the BS. We can see that, when introducing narrowband interference, the needed power to achieve 95% coverage probability increases from 26 dBm to 41 dBm. The effect of XMD is very low in this case, since the mobile powers are low. The maximum MS transmit power is about −14 dBm, which leads to about −109 dBm interfering power according to Equation (5.16), assuming 35 dBm interfering BS transmit power and 60 dB coupling loss. The reason why the MS powers are so low is that we have only considered the 12.2 kbps speech service.

We conclude from this simulation case study that a particular performance reduction is very much dependent on the particular network scenario, the frequency scenarios, the cell types (micro/macro) and the power limitation or the power allocation method used. With a stringent power allocation, it is possible to achieve high capacities in the interference situation, since the most interfering mobiles will be dropped from the network. On the other hand, a tight power allocation increases the number of deadzone areas around BSs and thus decreases the quality of the network.

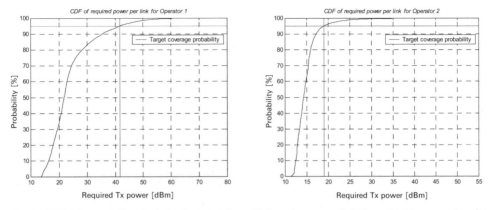

Figure 5.38 Cumulative distributions of downlink powers needed for 12.2 kbps service for operator 1 (left) and operator 2 (right), assuming narrowband interference from the adjacent operator.

5.4.4 Capacity Reduction

5.4.4.1 Downlink Effects

Section 5.4.3 described the detailed planning exercise carried out with a static system simulator. In this section the analytical model [7] has been used in order to analyse the effect of a guardband between the WCDMA system and the narrowband system. The model determines the average transmit power at the WCDMA BS needed to support M users when narrowband interference is present. The average transmit power of the WCDMA can be computed as:

$$\bar{P}_{Tx} = \frac{\dfrac{M \cdot \rho \cdot R}{W} \cdot E[L_{p,i} \cdot (N + I_{NB,i})]}{1 - \dfrac{M \cdot \rho \cdot R}{W} \cdot [(1 - \alpha) + i_{DL}]} \qquad (5.19)$$

where $E[\cdot]$ is the expectation operator; $I_{NB,i}$ is the narrowband interference falling to the own band because of various interference mechanisms; α is the average orthogonality factor of the radio channel; $L_{p,i}$ is the path loss from the own BS to mobile i; $N = FN_0$ is the thermal noise power at the mobile; F is the NF of the mobile receiver; and i_{DL} is the other-to-own-cell-interference ratio. The thermal noise N_0 over the 5 MHz band is -107.5 dBm. By using this formula we can compute the total capacity per cell and the capacity reduction due to the narrowband interference of the BS, assuming a constant target level for WCDMA BS transmit power. Figure 5.39 shows the computed capacity reduction in the case of macro–macro (left) and macro–micro (right). From these figures we can see that ACI is the main interfering mechanism when own-cell size is small.

The simulation results from Section 5.4.3 showed that the number of users being served was reduced by 11% for the macro-cellular case and 21% for the micro-cellular case, which corresponds quite well with analytical results when the channel separation was 2.6 MHz. The out-of-band emissions from the narrowband BS (referred to as

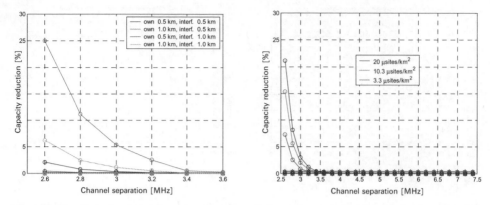

Figure 5.39 Capacity reduction as a function of channel separation Δf_g in the macro-cellular case (left) and micro-cellular case (right).

wideband noise) have been considered here as independent of the frequency. The wideband noise values from the system specification have been used for the GSM BS (-71 dBc at 5 MHz [1]) and for the Time Division Multiple Access (TDMA) BS (-6 dBm at 5 MHz [3]).

These results show that the effect of narrowband interference is dependent on the cell deployment of both the WCDMA system and the interfering system. If the interfering network is much denser than the WCDMA cell network, the effect of interference is large, which causes considerable capacity reduction as shown in Figure 5.40. The relative importance of various interference mechanisms is also dependent on site density. The cross-modulation component is remarkable when the own-cell size is large, because in that case the MS power is also large, which in turn increases the power leakage through the duplexer, as described in Section 5.4.1.5.

5.4.4.2 Uplink Effects

Uplink coverage in WCDMA depends on the total interference level at the BS. The uplink capacity of the system can therefore be defined as the maximum number of users for which the total interference level is below a certain threshold. The uplink interference level at the WCDMA cell can be written as:

$$I_W = I_{ownW} + I_{othW} + I_{NB} + N_W \tag{5.20}$$

I_W can be written more specifically with:

$$I_W = \frac{K_W \cdot R \cdot \rho}{W} \cdot I_W \cdot (1 + i_{UL}) + \sum_{l=1}^{M_{NB}} \sum_{j=1}^{K_{NB}} \frac{P_{NB,j,l}}{L_{p,j,l} \cdot L_{ACIR,j,l}(\Delta f)} + N_W \tag{5.21}$$

where K_W is the number of users in our own system; a narrowband mobile is assumed to transmit with power $P_{NB,j,l}$; $L_{p,i,l}$ is the path loss from the mobile to the WCDMA BS; adjacent channel attenuation, taking into account the transmit power spectrum and the receiver filter, is $L_{ACIR,j,l}(\Delta f)$; and thermal noise is N_W. Now we can solve the uplink

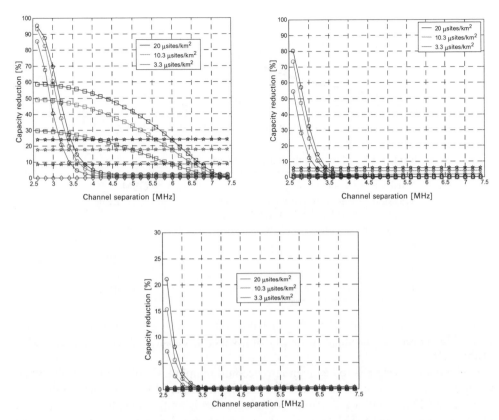

Figure 5.40 Capacity reduction as a function of channel separation Δf_g when the interfering network is micro-cellular. Three different own-cell ranges (maximum ranges) were considered here: 1.5 km (upper left figure), 1.0 km (upper right figure) and 0.5 km (lower figure). Symbols: square = XMD, circle = ACI, star = TDMA WB noise, diamond = GSM WB noise.

noise level as:

$$
I_W = \frac{\displaystyle\sum_{l=2}^{M_1}\sum_{j=1}^{K_1} \frac{P_{NB,j,l}}{L_{p,j,l} \cdot L_{ACIR,j,l}(\Delta f)} + N_W}{1 - \dfrac{K_W \cdot R \cdot \rho}{W}(1 + i_{UL})}
\tag{5.22}
$$

The definition of uplink capacity reduction due to narrowband interference is depicted in Figure 5.41. This definition guarantees that interference is always below a certain target level, and that uplink coverage exceeds a certain target.

The uplink capacity reduction due to GSM being located in an adjacent band was investigated with Monte Carlo simulations. Keeping the number of GSM users per WCDMA cell as constant, the link losses between the WCDMA cell and each GSM mobile $L_{p,j,l}$ were randomly generated from the path loss distribution of the WCDMA cell, which is either micro-cell or macro-cell. Also, the values for $L_{ACIR,j,l}(\Delta f)$ were generated by assuming that the mobile could be located with equal probability at any

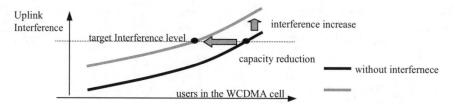

Figure 5.41 Definition of capacity reduction in uplink.

Table 5.15 Main simulation parameters.

Narrowband system	GSM
Transmit power of narrowband mobiles	21 dBm
Number of GSM users in WCDMA cell	7
Operator bandwidth	3 MHz
f	0.6
ρ	5 dB
R	12.2 kbps
Transmit power of the WCDMA cell	40 dBm

carrier within the narrowband operator's bandwidth, which was 2 MHz in these simulations. MS power was assumed to be 21 dBm, which is considered as an average value (Table 5.15).

Downlink blocking has to be taken into account in uplink simulations, as well. When the mobile is very close to the WCDMA cell, it might get blocked before it interferes with the WCDMA cell in the uplink. Blocking depends on the downlink ACIR as well as the strength of the GSM carrier. Figure 5.42 shows the capacity reduction in the

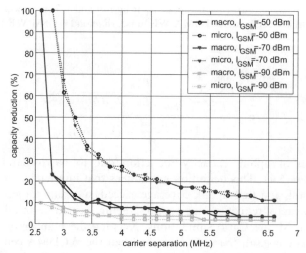

Figure 5.42 Capacity reduction in WCDMA uplink in the case of GSM interference. The WCDMA cell is either micro-cell or macro-cell. The blocking criterion for the GSM downlink is either −50 dBm, −70 dBm or −90 dBm.

uplink in the case of micro-cells and macro-cells. The maximum allowed interference that the GSM mobile can tolerate in in the downlink has been taken into account as well. It can be seen that in micro-cells the effect of uplink interference is high because of low coupling losses. In macro-cells the minimum coupling losses are higher, so they can tolerate interference much better. The micro-cell minimum coupling losses were around 55 dB, whereas in macro-cells the minimum coupling loss was around 70 dB because of higher antenna masts and the narrower vertical antenna pattern. The ray-tracing propagation model was used.

5.4.5   Summary and Radio Network Planning Guidelines

This concluding section describes some radio network planning aspects associated with the co-existence of a narrowband system and an adjacent WCDMA system in the same geographical area. The main difference between radio network planning in the core UMTS band and in the band allocated to the narrowband system is the interference between these two systems. Because narrowband power can be concentrated close to the edge of the WCDMA carrier, there are much more severe impacts on the WCDMA system than vice versa. Also, there is a high probability that WCDMA reception will be interfered with as a result of intermodulation products, because of the large bandwidth. Section 5.4.1 described the main downlink interference mechanisms as well as the respective calculation formulas for them. According to simple worst case calculations in Section 5.4.2, the predominant main interference mechanism is the interference betweem the narrowband system and the WCDMA system. In the downlink there are three main interference components: ACI, wideband noise and cross-modulation (XMD). In addition, there is intermodulation distortion (IMD or TxIMD) when two or more carriers cause mixing of products in the non-linear receiver element falling into the reception band. Downlink interference problems overcome the uplink, mainly because it is not possible to achieve such a high receiver performance at the mobile as at the BS. This is due to more stringent size and power consumption limitations at the MS, which cause lower linearity, lower adjacent channel filtering and lower duplex isolation at the MS.

ACI is determined by the adjacent channel selectivity characteristics of the filter chain in the MS. This depends on the design of the receiver filter which is specific to the mobile vendor and is not specified in the system specifications. ACI is very sensitive to channel separation between the carriers, so it can be reduced by using a guardband between the carriers. Without a guardband some capacity and coverage degradation might exist, as indicated in Section 5.4.3, and more careful network planning is required. The most significant factors affecting capacity and coverage are the spatial and spectral distance between the own and the adjacent carrier, the cell type (macro/micro) and the power levels used. Typically, in macro-cells the effect of ACI is not significant, since the coupling loss from the interfering BS is large. However, in some cases there might be large LOS areas around the interfering sites where the interfering power can be high. In such cases the location of the own site in the site planning phase of the own network is important.

Cross-modulation is determined by the non-linearity of the MS receiver, the duplexer isolation and the transmitting power of the mobile. It is not very sensitive to channel separation, so that even when the guardband is smaller than about 5 MHz, it has no significant effect on XMD levels. However, XMD is sensitive to cell sizes, since when the cell size increases the required MS transmission power increases, causing a relatively high increase in interference. This is because XMD is proportional to the square of transmitting power, so it is very sensitive to the transmission power of the MS. XMD starts to have a significant impact when the own-cell size is above 1.5–2 km. It occurs when the channel separation between the narrowband interferer and the WCDMA carrier is smaller than 7.5 MHz, so with larger channel separations XMD does not exist. XMD caused by narrowband outdoor BSs can be avoided by planning the network so that mobiles use low power when outdoors. MS powers are typically very high indoors when connected to outdoors, which causes problems when the adjacent operator has indoor solutions. In such cases, XMD problems are very likely. Another aspect is that when the uplink bit rate increases the power required also increases and so does the subsequent XMD. Interference caused by XMD can be reduced significantly by increasing the isolation in duplex filters in the MSs. This is, however, a difficult optimisation problem for the manufacturer, since larger isolation leads to less sensitive MS reception and less effective MS transmission. Also, the size of the duplexer is a limiting factor.

In the case of interfering micro-cells, the problems are more severe, since the path loss from the interfering site is relatively low and thus the interfering power is high in large areas around the interfering site. One possibility is to utilise several carriers so that the WCDMA system is able to affect inter-frequency handover between carriers for mobiles affected by interference. The worst case scenario would be for an operator to have only 5 MHz for the WCDMA, with micro-cell sites from two operators around the WCDMA carrier (Figure 5.17(b)). In that case WCDMA has to plan the network mainly with dense micro-cells in order to avoid large interference effects and to locate the sites close to interfering sites, mainly on the same streets.

In the uplink the capacity reduction due to adjacent narrowband interferer is large in micro-cells. This is mainly because the coupling loss from the interfering mobile to the WCDMA cell is lower. The capacity reduction is dependent largely on the blocking criteria in the downlink of the narrowband link. In the case of GSM, when the mobile is far away from the GSM BS, it might be dropped if the interference from the WCDMA is above the thermal noise level. If the mobile is closer to the GSM cell, it might tolerate more interference and avoid blocking. In such cases there might be a significant possibility of interference to the WCDMA cell. On the other hand, when the mobile is close to the GSM cell, its power is lower due to uplink power control in GSM. Therefore, interference is dependent on the power control of the narrowband system. When the available continuous spectrum is larger than 5 MHz, one possibility for avoiding interference problems is to allocate the adjacent channels to the own narrowband system (GSM/EDGE) and to co-site them with the WCDMA – i.e., by using the embedded type of scenario shown in Figure 5.17(a).

When co-siting interfering and interfered sites, the near–far effect vanishes, since when interference is large close to the interfering BS the path loss to the own BS is low, enabling the power control to reach the required SIR. Co-siting, however, requires that the original site planning for the existing system has been done properly to avoid any intrasystem interference in WCDMA. Also, Broadcast Control Channel (BCCH) planning has to be considered properly to avoid interference from high-power BCCH channels. In partial co-siting, only a proportion of the sites are co-sited, so that the near–far effects remain in some sites. In this case the frequencies closest to WCDMA should be allocated to the co-sited BS, if possible.

References

[1] 3GPP, Technical Specification, UE Radio Transmission and Reception (FDD), TS 25.101, v5.13.0.

[2] European Telecommunications Standards Institute (ETSI), GSM 05.05. Available online at ⟨http://www.etsi.org⟩.

[3] National Archives and Records Administration, Code of Federal Regulations, 24.238 CFR, 47, Parts 20–39. Available online at ⟨http://www.access.gpo.gov/nara/cfr⟩.

[4] Perez, R., *Wireless Communication Design Handbook, Vol. 2: Terrestrial and Mobile Interference*. Academic Press, 1998.

[5] Ko, B.-K., Cheon, D.-B., Kim, S.-W., Ko, J.-S., Kim J.-K. and Park, B.-H., A nightmare for CDMA RF receiver: The cross modulation. *AP-ASIC'99*, pp. 400–402.

[6] Aparin, V., Butler, B. and Draxler P., Cross-modulation distortion in CDMA receivers. *Proc. IEEE MTTS*, pp. 1953–1956.

[7] Heiska, K., Posti, H., Muszynski, P., Aikio, P., Numminen, J. and Hämäläinen, M., Capacity reduction of WCDMA downlink in the presence of interference from adjacent narrowband system. *IEEE Transactions on Vehicular Technology*, **VT-51**(1), January 2002, pp. 37–51.

[8] Hamied, K. and Labedz, G., AMPS cell transmitter interference to CDMA mobile receiver. *Proc. Vehicular Technology Conf., 1996*, Vol. 3, pp. 1467–1471.

[9] Schilling, D.L., Garodnick, J. and Grieco, D., Impact on capacity to AMPS jamming CDMA/CDMA jamming AMPS in adjacent cells. *Proc. Vehicular Technology Conf., 1993*, pp. 547–549.

[10] Kwon, D.S., Hong, H.J. and Kang, S.G., CDMA mobile station intermodulation interference induced by AMPS competitor base station. *Proc. IEEE 4th International Symp. on Spread Spectrum Techniques and Applications, 1996*, Vol. 1, pp. 380–384.

[11] Lu, Y.E. and Lee, W.C.Y., Ambient noise in cellular and PCS bands and its impact on the CDMA system capacity and coverage. *Proc. ICC 1995, Seattle*, Vol. 2, pp. 708–712.

[12] Prasad, R., Kegel, A. and de Vos, A., Performance of microcellular mobile radio in a co-channel interference, natural, and man-made noise environment. *IEEE Transactions on Vehicular Technology*, **VT-42**, February 1993, pp. 33–40.

[13] ITU, Recommendation ITU-R PI.372-6, Radio Noise.

[14] 3GPP, Technical Specification, Digital Cellular Telecommunications System (Phase 2+); Radio Transmission and Reception, TS 05.05.

[15] 3GPP, Technical Specification, BTS Radio Transmission and Reception (FDD), TS 25.104, v5.9.0.

[16] 3GPP, Technical Specification, RF System Scenarios, TR 25.942, v5.3.0.

[17] Engelson, M., *Modern Spectrum Analyser Theory and Applications*. Artech House, 1984.

6

Coverage and Capacity Enhancement Methods

Chris Johnson, Achim Wacker, Juha Ylitalo and Jyri Hämäläinen

6.1 Introduction

An operator's site density and the associated site configurations are primarily determined by the service coverage and system capacity objectives. The site density defined for initial system deployment should account for both the present and future coverage and capacity requirements. In terms of capacity, site density should be sufficient to permit capacity upgrades without the requirement for interleaving new sites – i.e., by including additional carriers, downlink transmit diversity or additional sectorisation. In doing so the original radio network plan remains relatively unchanged as the system capacity is increased to support the maturing levels of Wideband Code Division Multiple Access (WCDMA) traffic. Achieving this requires the radio planner to have a strong understanding of both WCDMA traffic expectations and the techniques available for enhancing service coverage and system capacity.

Service coverage and system capacity may be studied analytically using link budgets in combination with the load equations derived in Section 3.1.1.1 for the uplink and Section 3.1.1.2 for the downlink. The process requires a feedback loop to ensure coherence between the two – i.e., the uplink cell loading used in the link budget is dependent upon the output of the load equation, and the quantity of traffic used in the load equation is dependent upon the output of the link budget. Analytical methods are generally used to provide an analysis of average system behaviour. A WCDMA radio network planning tool as introduced in Section 3.2 is able to provide a more detailed, scenario-specific analysis of coverage and capacity performance. Many of the assumptions that form inputs to the link budgets and load equations are generated as outputs by the radio network planning tool – e.g., the level of inter-cell interference.

It is often assumed that service coverage is uplink limited although it is relatively simple to identify specific scenarios where it is downlink limited. System capacity may be either uplink or downlink limited dependent upon site configuration, mobile terminal performance and traffic profile. A cell is uplink capacity limited when it reaches its maximum permissible level of uplink load. A cell is downlink capacity limited when it reaches its maximum transmit power. An uplink capacity limited scenario may occur in a rural environment where the network has been planned with a relatively low uplink cell load for the benefit of a lower interference margin. A downlink capacity limited scenario is more likely in an urban environment where the network has been planned with a higher uplink load to increase system capacity. Understanding which link is limiting the network capacity is fundamental in being able to define a strategy for increasing capacity.

This chapter opens by describing WCDMA link budgets, system load equations and uplink and downlink capacity limited systems. These sections provide the basis for understanding the subsequent solutions for enhancing service coverage and system capacity. The chapter closes with an overview of a typical capacity upgrade process and a summary of the various site configurations.

6.2   Techniques for Improving Coverage

Most of the existing literature makes the assumption that service coverage is uplink limited. This assumption should always be checked for a specific site, service and Radio Network Controller (RNC) databuild configuration. For example, a site which uses Mast Head Amplifiers (MHAs) for a service whose data rate is asymmetric and an RNC databuild which limits the maximum downlink transmit power is more likely to be downlink service coverage limited.

The simplest method for studying service coverage performance is using a link budget. Link budgets often form the basis of system dimensioning exercises to provide an initial indication of cell range and thus site count requirement. A link budget is also useful to identify parameters that need to be improved to enhance service coverage performance.

The main techniques for improving service coverage are active antennas, MHAs, higher order receive diversity, increased sectorisation and repeaters. Some of these techniques improve coverage at the cost of capacity, while others improve both coverage and capacity. This is dependent upon which link budget parameters are affected.

6.2.1   Uplink and Downlink Coverage Limited Scenarios

WCDMA link budgets follow the same basic principles as those for the Global System for Mobile communications (GSM). The main differences are the inclusion of processing gain, E_b/N_0 requirement, soft handover gain, target uplink cell loading and a headroom to accommodate the inner-loop power control. The processing gain and E_b/N_0 requirement combine to generate the more familiar Carrier-to-Interference ratio (C/I) target. Target loading is the main capacity-related parameter appearing within the link budget. A low target load figure corresponds to a larger cell range but a lower cell capacity.

Table 6.1 presents a typical link budget for a data service supporting 384 kbps on the downlink and 64 kbps on the uplink. The service coverage is uplink limited as indicated by the lower allowed propagation loss figure. In this example the allowed propagation

Table 6.1 An example link budget for an asymmetric data service.

Parameter	Uplink	Downlink	
Uplink bit rate	64	384	kbps
Maximum transmit power	21.0	40.0[1]	dBm
Antenna gain	0.0	18.5	dBi
Body/Cable loss	0.0[2]	2.0	dB
Transmit EIRP	21.0	56.5	dBm
Processing gain	17.8	10.0	dB
Required E_b/N_0	2.0	4.5	dB
MDC gain	0.0	1.2	dB
Target loading	50	80[3]	%
Rise over thermal noise	3.0	7.0	dB
Thermal noise density	−174.0	−174.0	dBm/Hz
Receiver noise figure	3.0	8.0	dB
Interference floor	−168.0	−159.0	dBm/Hz
Receiver sensitivity	−117.9	−99.9	dBm
Receiver antenna gain	18.5	0.0	dBi
Cable/Body loss	2.0	0.0[2]	dB
Fast fading margin	3.0	0.0	dB
Soft handover gain	2.0	2.0	dB
Isotropic power required	−133.4	−101.9	dBm
Allowed propagation loss	154.4	158.4	dB

[1] 40 dBm is a typical limit placed upon a downlink traffic channel for a 43 dBm power amplifier module to prevent an excessive share of base station power being allocated to a single user.
[2] It has been assumed that data services do not incur a body loss.
[3] The downlink target loading is a function of the traffic mix loading the cell. 80% is a typical figure.

loss for the uplink is 4.0 dB less than that for the downlink. To illustrate an example scenario where service coverage is downlink limited, consider a Base Station (BS) configured with 37 dBm power amplifiers. A maximum of half of the total transmit power is generally allocated to a single traffic channel – i.e., 34 dBm (similar to the original example where 40 dBm represents half of the total transmit power capability of a 43 dBm power amplifier). This decreases downlink allowed propagation loss by 6 dB and results in a downlink limited service coverage. Including MHAs or making the bit rates asymmetric to a greater extent would further increase downlink coverage limitation. Downlink link budgets should always be checked to ensure that service coverage is uplink limited as generally assumed. As will be illustrated later, downlink link budgets are also essential when evaluating BS transmit power requirements.

6.2.2 Link Budget Analysis

Table 6.2 presents a series of typical uplink link budgets for a range of service data rates. The differences in the allowed propagation loss figures may be used to estimate the difference in site count requirements for various service coverage objectives. For example, based upon a path loss gradient of 3.5, providing coverage for the 128 kbps

Table 6.2 Example uplink link budgets illustrating the impact of service data rate.

	Service type			
	Speech	Data	Data	Data	
Uplink bit rate	12.2	64	128	384	kbps
Maximum transmit power	21.0	21.0	21.0	21.0	dBm
Antenna gain	0.0	0.0	2.0^1	2.0^1	dBi
Body loss	3.0	0.0^2	0.0^2	0.0^2	dB
Transmit EIRP	18.0	21.0	23.0	23.0	dBm
Processing gain	25.0	17.8	14.8	10.0	dB
Required E_b/N_0	4.0	2.0	1.5	1.0	dB
Target loading	50	50	50	50	%
Rise over thermal noise	3.0	3.0	3.0	3.0	dB
Thermal noise density	−174.0	−174.0	−174.0	−174.0	dBm/Hz
Receiver noise figure	3.0	3.0	3.0	3.0	dB
Interference floor	−168.0	−168.0	−168.0	−168.0	dBm/Hz
Receiver sensitivity	−123.1	−117.9	−115.4	−111.1	dBm
Receiver antenna gain	18.5	18.5	18.5	18.5	dBi
Cable loss	2.0	2.0	2.0	2.0	dB
Fast fading margin	3.0	3.0	3.0	3.0	dB
Soft handover gain	2.0	2.0	2.0	2.0	dB
Isotropic power required	−138.6	−133.4	−130.9	−126.6	dBm
Allowed propagation loss	156.6	154.4	153.9	149.6	dB

[1] It has been assumed that user equipment supporting higher data rates is superior in terms of antenna configuration.
[2] It has been assumed that data services do not incur a body loss.

service requires 39% fewer sites than the 384 kbps service when planned for the same level of coverage probability. There is less difference between the 12.2 kbps, 64 kbps and 128 kbps services. This highlights the importance of the decision as to whether or not a specific area should be planned for, say, 384 kbps or 128 kbps service coverage.

Table 6.2 illustrates that the highest data rate service defines the cell range in terms of allowed propagation loss. Planning the network for 384 kbps service coverage will be sufficient to ensure acceptable coverage performance for lower data rate services. In some cases an operator will define different coverage probability objectives for each service. In this case one of the lower data rate services may become the limiting service. For example, specifying a coverage probability objective of 85% for the 384 kbps data service and 95% for the 128 kbps data service leads to a difference in slow fading margins of 6.1 dB (based upon a path loss gradient of 3.5 dB and a slow fading standard deviation of 9 dB). In this case the radio network should be planned according to the coverage requirements of the 128 kbps data service.

Improving any of the parameters in the link budget will lead to an improvement in service coverage performance. Improving service coverage leads to a greater average BS transmit power requirement per downlink connection. If the system capacity is uplink limited, then this is of no consequence other than the system moves closer to becoming downlink limited. However, if the system capacity is downlink limited then improving service coverage will lead to a loss in system capacity. If BS E_b/N_0 performance is improved then it is possible to simultaneously enhance both service coverage and system capacity. This is a result of the E_b/N_0 requirement appearing in both the link budget and load equation.

6.3 Techniques for Improving Capacity

The simplest and most effective way to increase system capacity is to add one or more carriers. It will be illustrated in Section 6.5.1 that upgrading a site from single carrier to dual carrier can generate more than double the capacity. This can be done without changing the radio plan or requiring a new antenna configuration. However, once all of the available carriers have been used, other techniques must be sought to increase capacity further. These techniques include High-speed Downlink Packet Access (HSDPA), transmit diversity, beamforming, additional scrambling codes, increased sectorisation and micro-cells.

The simplest method of studying system capacity performance is using the load equations combined with link budgets. Link budgets are required in downlink capacity limited scenarios to evaluate the BS transmit power requirement. Load equations and link budgets are also useful for identifying which parameters need to be improved to increase system capacity. WCDMA radio network planning tools may be used to provide a more accurate and detailed evaluation of system capacity.

6.3.1 Uplink and Downlink Capacity Limited Scenarios

An uplink capacity limited scenario occurs when the maximum uplink load is reached prior to the BS running out of transmit power. This means that no additional users can be supported without degrading the planned service coverage performance. This is likely to occur in environments where the capacity requirements are relatively low and the network has been planned with a low uplink cell load to maximise cell range and thus reduce the requirement for sites. The traffic associated with an uplink capacity limited scenario is generally relatively symmetric. Increasing system capacity for an uplink limited scenario requires the uplink load equation to be enhanced.

A downlink capacity limited scenario occurs when the BS runs out of transmit power. Additional users cannot be added without modifying the site configuration. Downlink capacity limited scenarios are likely to occur in suburban or urban environments where the network has been planned to a relatively high uplink cell loading. The traffic associated with a downlink capacity limited scenario is generally asymmetric with a greater amount of traffic on the downlink. Downlink capacity limited scenarios are also likely to occur where the network has been configured with low BS transmit power capability, which may have been done in some circumstances to reduce the requirement for power amplifier modules. Increasing system capacity for a downlink limited scenario requires either the downlink load equation or the downlink link budget to be enhanced. Table 6.3 summarises the characteristics of uplink and downlink capacity limited scenarios.

Table 6.3 Uplink and downlink capacity limited scenarios.

	Uplink limited	Downlink limited
Limiting factor	Uplink cell load	BS transmit power
Common reasons	Planned to a low uplink cell load High BS transmit power capability Relatively symmetric traffic	Planned to a high uplink cell load Low BS transmit power capability Greater traffic on the downlink
Indications	BS transmit power not at maximum Uplink cell load at maximum	BS transmit power at maximum Uplink cell load not at maximum
Solution	Improve uplink load equation	Improve downlink load equation Improve downlink link budget

6.3.2 Load Equation Analysis

Separate load equations are used to study the uplink and downlink directions. Both include E_b/N_0 requirement, processing gain, activity factor and inter-cell interference. The downlink load equation also includes an orthogonality factor and soft handover overhead. In the uplink an increase in the level of inter-cell interference – as a result of considering the power control in adjacent cells to be uncorrelated with the fading to the BS receiver – is included.

The uplink load equation is presented as Equation (3.8) in Section 3.1.1.1. It can be seen that the cell loading is directly proportional to both the E_b/N_0 requirement and bit rate. For any given bit rate and level of uplink load, a lower E_b/N_0 requirement allows the system to support a greater number of users. The relationship between E_b/N_0 and bit rate determines how the overall cell throughput can be maximised.

Table 6.4 illustrates the variation of cell throughput for various service mixes. The number of users presented in each column corresponds to an uplink cell load of 50%. The first column only includes speech users. The remaining columns include a mix of speech and data users. The approach taken has been to add as many data users as possible without exceeding the 50% limit. Any spare capacity has then been assigned to speech users.

The results indicate that overall cell throughput is maximised when there is a high proportion of high data rate users. Thus an operator may benefit from allocating high bit rates for short periods of time rather than low bit rates for longer periods of time. Doing so, however, places greater requirements on service coverage performance.

Table 6.4 The variation in uplink cell throughput with service mix.

	12.2 kbps speech	64 kbps data	128 kbps data	384 kbps data
Data users	—	10	5	2
Speech users	51	2	7	4
Total throughput	622 kbps	664 kbps	805 kbps	817 kbps

Table 6.5 A comparison of uplink and downlink cell loads.

	Speech service		Asymmetric data service	
	Uplink	Downlink	Uplink	Downlink
Number of users	51	51	5	5
Bit rate	12.2 kbps	12.2 kbps	64 kbps	128 kbps
E_b/N_0 requirement	4.0 dB	6.5 dB	2.0 dB	5.0 dB
MDC gain	0.0 dB	1.2 dB	0.0 dB	1.2 dB
Activity factor	0.67	0.58	0.1	1.0
Inter-cell interference	0.65	0.65	0.65	0.65
Increase in inter-cell interference	1 dB	—	1 dB	—
Soft handover overhead	—	0.40	—	0.40
Orthogonality	—	0.50	—	0.50
Cell loading	50%	63%	2%	64%

The downlink load equation is presented as Equation (3.9) in Section 3.1.1.2. In this case the load equation includes the overhead due to soft handover and a reduction in own-cell interference due to the orthogonality between channelisation codes. In general, for any given number of users the load on the downlink will be greater than that on the uplink. This is illustrated in Table 6.5 for both the 12.2 kbps speech service and the 64/128 kbps data service. In the case of the speech service the downlink load is greater but the difference is not large. This service would tend to generate an uplink capacity limited scenario. For the data service a 1 : 10 uplink-to-downlink activity ratio has been assumed. Combined with the asymmetric data rates, this leads to a significant difference in uplink cell load. This service would tend to generate a downlink capacity limited system. Demonstrating an uplink capacity limited system requires the uplink load equation alone. Demonstrating a downlink capacity limited system requires the downlink load equation combined with the uplink and downlink link budgets. The uplink link budget is used to define the cell range that in turn forms an input to the downlink link budget.

6.3.3 Identifying the Limiting Link

Table 6.5 illustrated the uplink and downlink loads for the 12.2 kbps speech and 64/128 kbps data services. These appeared to indicate that for a 50% loaded system the speech service would be uplink capacity limited and the data service would be downlink capacity limited. Confirmation of this requires evaluation of the BS transmit power requirement based upon the downlink link budget [1]. This is done in Table 6.6.

It has been assumed that 20 W of BS transmit power is available. 20% of this has been assigned to the Common Pilot Channel (CPICH) and Common Control Channels (CCCHs). The allowed propagation loss figures have been taken from the uplink link budgets presented in Table 6.2. Note that the figure corresponding to the 64 kbps service has been used in both cases. It is unlikely that a WCDMA network would be planned based only on the speech service. In the case of the speech service it is

Table 6.6   Downlink link budget used to evaluate the base station transmit power requirement.

	Speech service		Asymmetric data service	
Number of users	51		5	
Soft handover overhead	0.40		0.40	
Pilot and common channel power	36	dBm	36	dBm
Allowed propagation loss	154.4	dBm	154.4	dBm
Allowed propagation loss correction	6	dB	0	dB
Downlink E_b/N_0 requirement	6.5	dB	5.0	dB
MDC gain	1.2	dB	1.2	dB
Noise rise	4.3	dB	4.5	dB
BS antenna gain	18.5	dBi	18.5	dBi
BS cable loss	2	dB	2	dB
MS antenna gain	0	dBi	0	dBi
MS body loss	3	dB	0	dB
Transmit power required/TCH	19.4	dBm	31.3	dBm
Total transmit power required	41.4	dBm	41.3	dBm

assumed that there are sufficient users to safely assume a uniform distribution of users across the cell. The subsequent calculation is then based upon the average allowed propagation loss, which is 6 dB less than that at the cell edge [2]. In the case of the data service all users are assumed to be at the cell edge and the worst case allowed propagation loss is used. The noise rise shown corresponds to the number of users and the downlink cell loading presented in Table 6.5.

The results indicate that a cell loaded with 51 speech users requires a BS transmit power of 41.4 dBm. This confirms that the scenario is uplink capacity limited; adding another speech user would cause the 50% uplink load limit to be exceeded without consuming the entire 43 dBm of BS transmit power. The results also indicate that the 64/128 kbps data service leads to a downlink capacity limited scenario. The four data users generate an uplink load of only 2% while the BS transmit power requirement is 42.4 dBm. Adding another user would require more power than the 43 dBm available.

6.4   Uplink Cell Load and Base Station Transmit Power

Uplink cell load and BS transmit power form the limiting factors for uplink and downlink capacity limited systems. The appropriate level of uplink cell load should be evaluated during initial radio network planning. The level chosen should be sufficient to support both initial and medium- to long-term traffic expectations. Once the network has been deployed, it is relatively difficult to increase the cell loading limit without having to interleave additional sites to maintain service coverage performance.

The BS transmit power requirement also needs to be planned during initial radio network planning. However, it is relatively easy to upgrade BS transmit power without changing the layout of the radio network plan. In this case the initial configuration can be based upon short-term traffic expectations alone. This allows the operator to delay expenditure on BS hardware. This must be balanced against the cost of subsequent site upgrades.

6.4.1   Impact of Uplink Cell Load

The maximum permissible level of uplink cell load determines the interference margin that appears in any link budget calculation. The greater the cell loading, the greater the number of sites as well as the greater potential capacity per site.

For an uplink capacity limited system, the capacity per site is directly proportional to the maximum permissible level of uplink cell load. Each user that establishes a connection and has the same E_b/N_0 requirement and activity factor increases the cell load by the same amount. Doubling the maximum cell load results in double the cell capacity for an uplink capacity limited system. The impact upon cell range is dependent upon the absolute levels of cell load. The relationship between the cell load and maximum allowed propagation loss is exponential. Equation (3.11) from Section 3.1.1.2 describes the relationship between the cell load and the resulting increase in the receiver interference floor. As the cell load approaches 100%, the receiver inter-ference floor increases without limit. This condition never occurs in practice because the population of mobile terminals have a finite transmit power capability. The minimum recommended uplink cell load to which a network should be planned is 30%, whereas the maximum is 70%. Planning for a 30% uplink cell load leads to 1.5 dB of inter-ference margin, whereas planning for a 70% cell load leads to 5.2 dB of margin. Assuming a propagation path loss gradient of 3.5, this corresponds to 61% more sites than when the network is planned for 30%. In addition, if the system capacity is uplink limited then the capacity per cell would increase by a factor of 2.3. For a network based upon three-sector sites, this results in a system capacity 3.7 times greater than when the network is planned for a 30% cell load. When the system capacity is downlink limited, planning to a higher uplink cell load will also result in a higher capacity per cell. This is due to the maximum allowed propagation loss being less and the BS having a lower average transmit power requirement per user. Figure 6.1 illustrates the relationship between capacity and the allowed propagation loss for a macro-cellular environment.

If, for example, a cell configured with a 43 dBm power amplifier has an allowed propagation loss of 160 dB when planned for 30% loading, the capacity would increase from 37 to 52 speech users if the planned cell load was increased to 70%. The relative increase in capacity is dependent upon the absolute values of the allowed propagation loss. If the original path loss figure had been 150 dB, the increase would have been much less. Table 6.7 summarises the impact of increasing the planned level of cell loading for uplink and downlink capacity limited systems.

6.4.2   Impact of Base Station Transmit Power

The transmit power assigned to a WCDMA cell must be shared amongst all active users belonging to that cell, including those connected by soft handover. A lower average transmit power requirement results in a higher cell capacity. In general, approximately

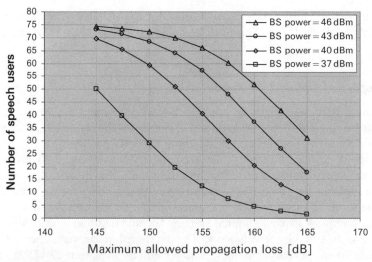

Figure 6.1 Relationship between the maximum allowed propagation loss and macro-cell capacity, with base station transmit power as a parameter.

Table 6.7 Impact of increasing the planned level of uplink cell load.

	Uplink capacity limited	Downlink capacity limited
System capacity	Increased number of sites leads to increased capacity	Increased number of sites leads to increased capacity
Per-cell capacity	Increased cell loading leads to a proportional increase in capacity	Capacity increases as a result of the lower average base station transmit power requirement

20% of the cell power is assigned to the CPICH and the CCCHs. The remaining 80% is available to support Traffic Channel (TCH) capacity. A range of typical WCDMA BS transmit power configurations is presented in Table 6.8.

The capacity offered by each transmit power configuration is a function of the traffic profile as well as the maximum allowed propagation loss defining the cell range. The greater the propagation loss, the greater the average transmit power requirement and the lower the cell capacity. This section considers a population of speech users uniformly distributed in a cell.

Figure 6.1 provides an illustration of how capacity varies with the allowed propagation loss and cell transmit power capability. Returning to the link budgets shown in Table 6.2, a cell planned for 384 kbps has an associated maximum allowed propagation loss of 149.6 dB.

According to Figure 6.1 this corresponds to a capacity of 68 speech users for a 20 W transmit power capability. Doubling the power to 40 W increases the capacity by 7%. In this case the relatively small cell range allows the level of downlink loading to

Table 6.8 Typical base station transmit power configurations.

Base station transmit power per cell per carrier	Application
37 dBm (5 W)	Provides low macro-cell capacity for scenarios where the main objective is providing coverage. Also appropriate for indoor and micro-cell solutions
40 dBm (10 W)	Provides medium macro-cell capacity per carrier. Two carriers each configured with 10 W provide greater capacity than a single carrier configured with 20 W
43 dBm (20 W)	Provides high macro-cell capacity per carrier
46 dBm (40 W)	Provides increased macro-cell capacity per carrier when the maximum allowed propagation loss is relatively high

become high prior to the cell running out of transmit power. The result of the high downlink cell loading is that each additional user generates a relatively large increase in mobile terminal interference floor and any additional transmit power capability is rapidly consumed. The overall cell capacity would be greater if the additional 20 W of transmit power had been assigned to a separate carrier. A cell configured with two 20 W carriers has a capacity at least twice that of a single 20 W carrier.

A cell planned for the 128 kbps service has a higher maximum allowed propagation loss and in this case the difference between the various capacity figures is greater. This is a result of the lower transmit power configurations running out of power before the cell becomes heavily loaded. Any additional users do not then generate large rises in the mobile terminal interference floor. Nevertheless the overall cell capacity remains greatest when the BS transmit power is shared across the available carriers. Considering a cell configured with 40 W and a maximum allowed propagation loss of 155 dB, a single-carrier configuration results in a capacity of 66 speech users. If the same power is shared across two carriers then the capacity is greater than 114 users.

The impact of high downlink cell loading is illustrated in Figure 6.2. This presents the BS transmit power required as a function of cell loading, with the allowed propagation loss as a parameter. A transmit power of 36 dBm has been assigned to the CPICH and the CCCHs (this represents 20% of 20 W). As the cell loading increases, the incremental transmit power requirement for each additional user increases. This is a result of the exponential increase in receiver interference floor with cell loading and is the reason for the inefficiency in assigning large quantities of transmit power to a single carrier.

The impact of BS transmit power capability upon service coverage performance is dependent upon the uplink and downlink link budgets. Under normal circumstances service coverage is uplink limited and BS transmit power has no effect upon service coverage. However, in some cases, such as a 5 W BS with asymmetric data services and the use of MHAs, service coverage can become downlink limited. Here, service coverage is directly related to the maximum power assigned to a single TCH.

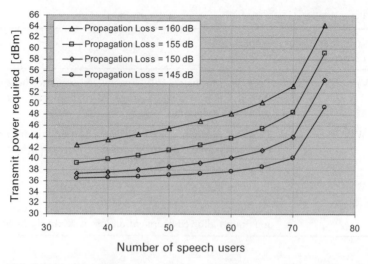

Figure 6.2 Relationship between the base station transmit power requirement and cell loading with the allowed propagation loss as a parameter. 20% of the base station power has been reserved for the downlink common channels.

6.5 Additional Carriers and Scrambling Codes

Additional carriers form the simplest and most effective way of increasing system capacity. Service coverage performance is generally improved as a result of reducing the uplink load on each carrier. This improvement is reduced as the traffic increases and the cell load becomes as high on each carrier as it was originally. When a BS whose capacity is downlink limited has limited transmit power capability, system capacity is maximised by sharing the power across the available carriers. For example, the capacity of two carriers each configured with 10 W can be significantly greater than that of a single carrier configured with 20 W.

Additional scrambling codes are used when the system capacity is hard limited by the number of available channelisation codes. In general, for a macro-cell environment the capacity is limited by either uplink cell loading or BS transmit power capability. Channelisation codes become the limiting factor under relatively high throughput scenarios. This is more likely to occur in either micro-cell or indoor scenarios where the cell range is limited and code orthogonality is high.

6.5.1 Impact of Additional Carriers

The majority of WCDMA operators have more than a single carrier. In general two or three carriers are licensed per operator by the local regulatory authority. Operators must then define a strategy for distributing their carriers across the network hierarchy. Carriers can be either dedicated or shared between layers. Sharing carriers can provide the highest overall system capacity and spectrum efficiency but requires very careful radio network planning to ensure adequate isolation between macro- and micro-layers. For cell deployment strategies see also Chapter 3.

The impact of assigning multiple carriers to a single cell depends upon whether the system capacity is uplink or downlink limited, upon the transmit power assigned to each carrier and upon whether the RRM function supports load balancing between carriers. Although maximising the number of carriers at a cell provides the most efficient way of utilising the resources in terms of air interface loading and transmit power requirements, operators will generally add carriers only as and when required by the growth in demand for system capacity. This allows the operator to delay any capital expenditure on BS hardware.

In the simplest case adding a second carrier with the same configuration as the first will always at least double the cell capacity. If the system supports load balancing between carriers, there will be an additional trunking gain and the capacity will be more than double. The impact upon service coverage depends largely upon the original level of uplink cell load. Assuming that traffic is evenly distributed across the carriers, which represents the most efficient way of utilising the resources in terms of air interface loading and transmit power requirements, the cell load will be halved. If the original cell load was low, halving it will not have a significant impact upon coverage performance. If the cell load was relatively high, however, the coverage performance will improve. This improvement will diminish as the levels of network traffic increase over time and the cell load on each carrier approaches that originally on the first carrier.

BS cabinets generally accommodate a fixed number of power amplifier and transceiver modules. Often cabinets can be cascaded, but this is not an attractive solution in terms of cost and cabinet space requirements. For scenarios where the BS runs out of space for additional power amplifier modules, the power of the existing modules can be shared as additional carriers are added. As an example, consider a BS cabinet that supports 12 transceiver modules and six power amplifier modules. The power amplifier modules installed could have any practical transmit power capability. Assume for this example that the operator has planned the network for 384 kbps coverage and has thus purchased 20 W power amplifier modules, since 40 W modules bring little gain when the allowed propagation loss is relatively low. Once the operator has reached a $2 + 2 + 2$ configuration, there is no additional room for the power amplifier modules required to upgrade to $3 + 3 + 3$. Half of the amplifier modules could be swapped out for 40 W modules, in which case all of the carriers could be assigned 20 W. Alternatively the six existing 20 W modules could be used to provide one carrier with 20 W and the remaining two carriers with 10 W each. The capacity loss in doing so is only 10–20% relative to having 20 W per carrier.

Table 6.9 presents the capacity of various multiple carrier configurations. The analysis simplifies matters by assuming that only speech users are loading the cell. This permits the Erlang B equation – Equation (10.1) from Section 10.1.6.1 – to be used to evaluate trunking gain.

Comparing the second and third rows of Table 6.9 illustrates that – when doubling the number of carriers while maintaining an equal number of power amplifier modules – leads to only 15% less capacity than if the number of power amplifier modules were doubled, too. In the case of migrating to three carriers, the capacity loss when maintaining six 20 W power amplifier modules relative to upgrading three of them to 40 W is 9%. This demonstrates that system capacity is significantly more sensitive to the number of carriers than the power with which those carriers are configured.

Table 6.9 Typical capacity figures for a downlink capacity limited cell with various carrier and transmit power configurations.[a]

Site configuration	Power amplifier modules	Erlang B capacity per cell per carrier	Erlang B capacity per cell
1 + 1 + 1	3 × 20 W	57 users	57 users
2 + 2 + 2	3 × 20 W	53 users	106 users
2 + 2 + 2	6 × 20 W	62 users	124 users
3 + 3 + 3	6 × 20 W	57 users	171 users
3 + 3 + 3	3 × 20 W + 3 × 40 W	63 users	189 users

[a] Based upon 12.2 kbps speech users with 2% blocking and a maximum allowed propagation loss of 150 dB. Figures are only indicative. In practice the cell will be loaded by a combination of speech, circuit switched data and packet switched data users. Assumes radio resource management supports load balancing between carriers.

6.5.2 Impact of Additional Scrambling Codes

The capacity of a macro-cell is generally limited by the air interface – i.e., by either the uplink cell load or the cell's downlink transmit power capability. When the system is downlink capacity limited the number of users supported depends on downlink channelisation code orthogonality. Improvements in orthogonality generate proportional reductions in the levels of own-cell interference. In this case the cell capacity becomes more sensitive to other-cell interference. Orthogonality is primarily dependent upon the radio channel linking the users to the BS. High levels of multi-path degrade orthogonality because, unless the multi-path is accurately resolved, the received signal interferes with itself. Micro-cells are generally associated with radio channels that have strong line-of-sight components and relatively low delay spread. In this case orthogonality is significantly improved – typically 0.9 for a micro-cell compared with 0.5 for a macro-cell. Micro-cells are studied in greater detail in Section 6.15. Table 6.10 presents the air interface capacity for micro-cell scenarios both with and without transmit diversity when loaded with a range of service bit rates.

Transmit diversity (see Section 6.9) has no impact upon cell capacity for the 12.2 kbps speech and 64/64 kbps data services. This is because these services are uplink capacity limited and transmit diversity only improves the downlink. The impact of transmit diversity for these services may be observed as a reduction in BS transmit power requirement. In the case of the 64/128 kbps and 64/384 kbps data services, transmit diversity increases cell capacity by approximately 70%. These air interface capacity figures may be compared with the capacity limitations of the channelisation code tree.

Table 6.11 presents the downlink channelisation code limited capacities for both macro-cells and micro-cells based upon a single scrambling code.

The availability of channelisation codes becomes the limiting factor for higher data rate services when transmit diversity is being used. A second scrambling code may be used to introduce a second channelisation code tree. However, the two code trees will not be orthogonal to one another. Users assigned a channelisation code from the first code tree will incur a different downlink cell load and link budget relative to those who have been assigned a channelisation code from the second tree. As an example consider

Table 6.10 Air interface capacities for micro-cell scenarios with and without transmit diversity, based upon an allowed propagation loss of 144.7 dB (64 kbps uplink link budget with 70% loading).

	Service	Downlink capacity per cell [users]	Uplink load [%]	Base station transmit power requirement [dBm]
Micro-cell without transmit diversity	12.2 kbps speech	79	69.9	39.8
	64/64 kbps data	17	66.3	40.7
	64/128 kbps data[a]	12	4.7	42.4
	64/384 kbps data[a]	4	1.6	41.4
Micro-cell with transmit diversity	12.2 kbps speech	79	69.9	37.2
	64/64 kbps data	17	66.3	38.7
	64/128 kbps data[a]	18	7.0	42.7
	64/384 kbps data[a]	7	2.7	42.0

[a] Includes an activity factor ratio of 1 : 10 for uplink-to-downlink traffic channel activity.

Table 6.11 Traffic channel limitations of a single orthogonal variable spreading factor code tree.[a]

	Downlink bit rate [kbps]	Air interface bit rate [kbps]	Spreading factor	Number of possible traffic channels
Macro-cell[b]	12.2	60	128	89
	64	240	32	22
	128	480	16	10
	384	960	8	5
Micro-cell[c]	12.2	60	128	104
	64	240	32	25
	128	480	16	12
	384	960	8	5

[a] $C_{ch,256,0}$ used for CPICH; $C_{ch,256,1}$ used for P-CCPCH; $C_{ch,64,1}$ used for S-CCPCH; $C_{ch,256,2}$ used for AICH; and $C_{ch,256,3}$ used for PICH.
[b] Based upon a soft handover overhead of 40%.
[c] Based upon a soft handover overhead of 20%.

a micro-cell loaded by 15 users each with a downlink service bit rate of 128 kbps. The first 12 of these users are supported by the first code tree. The remaining three are assigned a channelisation code from the second code tree. In terms of air interface loading, each of the first 12 users sees 11 orthogonal users and three non-orthogonal users. Each of the second three users sees two orthogonal users and 12 non-orthogonal users. Thus the downlink link budget is more favourable for users assigned channelisation codes from the first code tree.

6.6   Mast Head Amplifiers and Active Antennas

MHAs and active antennas are used to reduce the composite Noise Figure (NF) of the BS receiver sub-system. In doing so the uplink link budget is improved and service coverage performance increases. The impact of MHAs and active antennas upon system capacity is dependent upon whether the capacity is uplink or downlink limited. If it is uplink limited then they are not likely to have an effect upon capacity. The network will, however, move closer to becoming downlink capacity limited. If the network capacity is downlink limited then the use of MHAs or active antennas will decrease system capacity. There are two reasons for this: an MHA introduces an insertion loss of typically 0.5 dB in the downlink direction, thus reducing the available Equivalent Isotropic Radiated Power (EIRP); additionally, the fact that the MHA has improved coverage means that users are now supported in locations which require greater BS transmit power. The same applies to active antennas.

The main difference between MHAs and active antennas is that MHAs are connected externally to the antenna casing. MHAs are connected between the antenna and feeder cable. Two short pieces of jumper cable are used for the connection on either side. Active antennas include the low-noise amplifier as an integrated part of the antenna itself.

Having active components as either part of or located adjacent to the antenna means that a power supply must be provided. This can generally be fed into the feeder cable without the requirement for an additional cable. This is done using a so-called 'bias-T' as connector.

6.6.1   Mathematical Background

The reduction in composite NF provided by MHAs and active antennas can be evaluated using Friis' equation ([3] and [4]). This models the behaviour of a cascaded line of active and/or passive components (Figure 6.3).

Friis' equation is presented in Equation (6.1).

$$Composite\ noise\ figure = NF_1 + \frac{(NF_2 - 1)}{G_1} + \frac{(NF_3 - 1)}{G_1 G_2} + \frac{(NF_4 - 1)}{G_1 G_2 G_3} + \cdots \quad (6.1)$$

The NF performance of the first component is the most important and the composite NF can never be less than this. The higher the gain of the first component, the closer the composite NF is to that of the first component. In this case the first component is the MHA or active antenna. The number of stages to be considered depends upon the architecture of the BS receiver sub-system. Figure 6.4 illustrates possible architectures

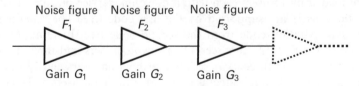

Figure 6.3   The concept of a cascaded receiver sub-system.

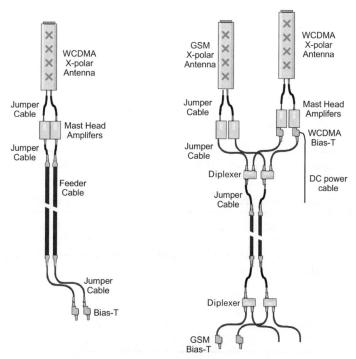

Figure 6.4 Typical base station receiver sub-systems: dedicated WCDMA and joint GSM/WCDMA with shared feeders.

for a dedicated WCDMA receiver sub-system and a joint GSM/WCDMA receiver sub-system with shared feeder lines.

If the WCDMA BS does not share feeders with a co-sited GSM BS, the stages are likely to be MHA, feeder cable, bias-T and the BS itself. Jumper cables are assumed to have very low loss and are ignored in this analysis. If the WCDMA BS shares feeders with a co-sited GSM BS, the cascade of stages will also include a pair of diplexers. In this example, the GSM system also uses MHAs and already has a pair of bias-Ts supplying the feeders with a voltage. The WCDMA system cannot always draw current from the same source and an additional pair of bias-Ts is required adjacent to the WCDMA MHAs. This requires rigging up an additional power cable.

To use Friis' equation, each component must have an associated NF and gain. In the case of passive components the NF is equal to the loss. Table 6.12 presents typical figures for each component.

6.6.2 Impact of Mast Head Amplifiers and Active Antennas

MHAs and active antennas are used to improve service coverage performance. Their effectiveness depends upon the receiver sub-system architecture and the associated parameter set. Table 6.13 presents the resultant NFs for the dedicated and shared feeder architectures as a function of feeder loss.

Table 6.12 A typical parameter set for a mast head amplifier receiver sub-system.

Element	Uplink gain [dB]	Noise figure [dB]	Downlink loss [dB]
MHA	12.0	2.0	0.5
Feeder	−2.0[a]	2.0[a]	2.0[a]
Diplexer	−0.3	0.3	0.3
Bias-T	−0.3	0.3	0.3
BS	—	3.0	—

[a] The feeder loss is dependent upon the site architecture and the associated feeder quality and length.

Table 6.13 The benefit of using mast head amplifiers as a function of feeder loss (in dB).[a]

Feeder loss	Dedicated feeders			Shared feeders		
	NF without MHA	NF with MHA	Benefit	NF without MHA	NF with MHA	Benefit
1.0	4.0	2.6	1.4	4.6	2.7	1.9
2.0	5.0	2.8	2.2	5.6	2.9	2.7
3.0	6.0	3.0	3.0	6.6	3.2	3.4
4.0	7.0	3.3	3.7	7.6	3.5	4.1

[a] Based upon the parameter set presented in Table 6.12 but with an MHA gain of 9 dB.

The benefit of using MHAs in terms of gain in the uplink link budget increases with feeder loss. The gain is greater for the shared feeder architecture as a result of the feeder loss being increased by the pair of diplexers. In the case of dedicated feeders, feeder loss figures less than or equal to 3 dB result in a gain that is always at least as great as the feeder loss itself.

Greater values of feeder loss result in a gain that is less than the feeder loss. This is an important point to note, since the impact of MHAs is often modelled by removing feeder loss from the uplink link budget. This is clearly pessimistic for feeder loss figures below 3 dB and optimistic for feeder loss figures above 3 dB. The precise crossover point changes with changes in MHA gain. A relatively conservative MHA gain has been assumed for this example. In the case of shared feeders, the crossover point occurs at a feeder loss greater than 4.0 dB.

The impact of using MHAs upon system capacity is dependent upon whether the capacity is uplink or downlink limited. If it is uplink limited then there will be no change in system capacity, although the BS will be using more of its transmit power capability. If the system capacity is downlink limited there will be a loss in capacity. This loss results from both the MHA insertion loss and increased allowed propagation loss. Table 6.14 quantifies this loss for both dedicated feeder and shared feeder architectures.

The loss in capacity increases with feeder loss as a result of the increased coverage gain from using the MHAs. Capacity loss is greater for the shared feeder architecture since this scenario benefits from a greater MHA coverage gain.

Table 6.14 Capacity loss when using mast head amplifiers for a downlink capacity limited system.

Feeder loss [dB]	Capacity loss[a]	
	Dedicated feeders [%]	Shared feeders [%]
1.0	6.1	7.9
2.0	9.1	11.0
3.0	12.3	14.3
4.0	15.6	17.7

[a] Includes the capacity loss due to both the MHA insertion loss and the increased allowed propagation loss.

The improved receiver sensitivity provided by MHAs makes the system more susceptible to external sources of interference – i.e., interference from another system that was originally safely below the interference floor may begin to impact system performance. The external source of interference would prevent the system from taking full advantage of the reduced composite NF. Instead the unloaded system interference floor would remain relatively high and the corresponding uplink link budget would be less than optimal.

6.6.3 Practical Considerations

Practical considerations applicable to the use of MHAs include mast space and loading as well as the requirement to have power provided at the remote end of the feeder cable. The weight and wind loading of the MHA must be added to that of the antenna, bearing in mind that two MHAs will be required per crosspolar antenna. Active antennas have the advantage of integrating the amplifier unit within the antenna casing, thus removing the requirement to rig up an additional unit. Figure 6.4 illustrates how a bias-T may be used to supply power to the feeder cable to prevent the requirement for any additional cable runs to the antenna. When feeder cables are being shared with a GSM system which is already configured with MHAs or active antennas additional power cables may be required.

6.7 Remote RF Head Amplifiers

Remote RF head amplifiers allow the physical separation of a BS's RF and baseband modules. The baseband modules remain within the BS cabinet, whereas the RF modules can be located remotely adjacent to the antenna sub-system. In this context, the RF modules include transmit power amplifiers, receiver front-ends and RF filtering. The RF and baseband modules are connected via an optical link. This means that cells can be located at site locations that would normally require prohibitively long runs of feeder cable. For example, a building where it is beneficial to make use of cabinet space

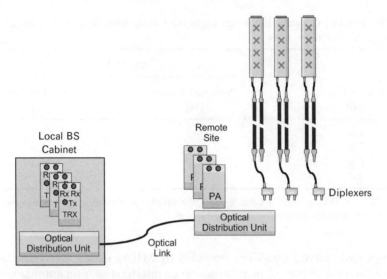

Figure 6.5 Architecture of a remote RF head amplifier base station.

in the basement can make use of rooftop antennas without having to incur large feeder losses between the basement and roof – an optical link connects the remote RF modules located on the roof. Figure 6.5 illustrates the BS architecture when using remote RF head amplifiers. The optical link can be as much as 2 km in length. The concept has some similarities to the use of repeaters – i.e., service is provided at a location remote to the BS cabinet. There are, however, some important differences.

A comparison of remote RF head amplifiers and repeaters is presented in Table 6.15. Repeaters are studied in greater detail in Section 6.14. The most important difference is that a remote RF head amplifier generates its own logical cell whereas a repeater provides an extension of an existing logical cell. In practical terms, a repeater is easier to deploy since the forward and return links to the BS are often via a radio link.

Table 6.15 A comparison of remote RF head amplifiers and repeaters.

	Remote RF head amplifier	Repeater
Application	Locating the entire logical cell at a location normally requiring a long feeder run	Extending the coverage of an existing logical cell
Hardware at remote location	Transmit power amplifiers and receiver front-ends	Complete receiver and transmitter chain for both uplink and downlink directions
Connection to BS	Optical link	Usually a radio link
Function	Normal RF functions of the BS	Non-intelligent retransmission

6.7.1 Mathematical Background

A simple analysis of remote RF head amplifiers does not require any new modelling techniques beyond those already presented. It can be assumed that the optical link is transparent to the BS's performance, so that remote RF head amplifiers may be studied using the same link budget and load equation analysis as a standard BS.

6.7.2 Impact of Remote RF Head Amplifiers

It is of interest to compare the cell performance of a site equipped with remote RF head amplifiers with that associated with a standard site configuration based upon the same cabinet and antenna locations. This corresponds to comparing a site with high feeder losses with another with low feeder losses.

In terms of service coverage performance, the gain is simply the difference between the feeder loss when using remote RF head amplifiers and the feeder loss when using a standard site configuration. For example, if the feeder loss when using a standard site configuration is 6 dB and the feeder loss when using remote RF heads is 0.5 dB, then the coverage gain is 5.5 dB. When the feeder loss associated with a standard site configuration is relatively small there is less requirement for the use of remote RF head amplifiers. As the feeder loss increases then so, too, does the benefit of using remote RF head amplifiers.

If remote RF head amplifiers are used to improve service coverage performance, cell capacity remains unchanged. This is because the RF head amplifiers have increased the cell's maximum allowed propagation loss but at the same time have increased the available EIRP. This is in contrast to MHAs, which can improve service coverage performance but at the cost of some loss in cell capacity when the system is downlink capacity limited.

Alternatively, remote RF head amplifiers may be used to maintain the same service coverage performance while increasing cell capacity – i.e., the available EIRP increases while the allowed propagation loss remains fixed. Table 6.16 presents a set of typical capacity gains.

Table 6.16 An evaluation of the capacity gain when using remote RF head amplifiers.

Feeder loss without RF head [dB]	Feeder loss with RF head[a] [dB]	EIRP without RF head[b] [dBm]	EIRP with RF head[b] [dBm]	Capacity gain [%]
2	0.5	59.5	61	3
4	0.5	57.5	61	8
6	0.5	55.5	61	17
8	0.5	53.5	61	31

[a] Typical figure representing a relatively short feeder run between remote RF head and antenna.
[b] Based upon a 43 dBm power amplifier and an 18.5 dBi antenna.

Capacity gain becomes significant as the feeder loss increases. Nevertheless this gain is only negating the loss in capacity caused by higher feeder losses – i.e., cell capacity when using RF head amplifiers is the same as that of a standard cell configuration, which has an equally small feeder loss.

6.7.3 Practical Considerations

The use of remote RF head amplifiers allows antenna positioning at locations that would normally require prohibitively long runs of feeder cable. Nevertheless, there is a requirement to run an optical link between the BS cabinet and the remote antenna location. The size, weight, wind loading and power consumption of remote RF head amplifiers must also be accounted for.

6.8 Higher Order Receive Diversity

Receive diversity provides an effective technique for both overcoming the impact of fading across the radio channel and increasing the resulting Signal-to-Interference Ratio (SIR). The former is achieved by ensuring uncorrelated fading between antenna branches – i.e., not all antennas experience fades at the same time. The latter is achieved by ensuring uncorrelated interference – i.e., coherently combining two branches of the desired signal results in a 6 dB increase in power, whereas combining two branches of uncorrelated interference results in a 3 dB increase in power. In general, the standard configuration for a WCDMA BS includes two-branch receive diversity achieved with a single crosspolar antenna (polarisation diversity) or two vertically polarised antennas (space diversity) [5].

Higher order receive diversity implies more than two receive branches. The optimal number of receive branches depends upon the particular radio environment [6]. In the case of WCDMA, the wideband signal results in a high delay spread resolution, permitting potentially large gains from multi-path diversity. Multi-path diversity has a significant impact upon uplink performance and the relative incremental gain that can be achieved from higher order receive diversity. The impact upon the downlink is less, owing to the limitations in baseband processing and the number of fingers in the RAKE receiver.

In some radio environments – e.g., rural or micro-cellular – the level of multi-path diversity may be small. A lack of multi-path diversity can be partially compensated for by the time diversity achieved by channel coding and interleaving. Time diversity is most effective for high-mobility users who experience frequent but narrow fades. Multi-path and time diversity are relatively poor solutions for low-mobility mobile terminals experiencing low levels of delay spread. For these users, system performance can be significantly improved by taking advantage of higher order receive diversity. Receive diversity improves both the uplink fast fading margin as well as the uplink E_b/N_0 requirement. Improving the latter also results in an improvement of both service coverage and uplink system capacity.

6.8.1 Impact of Higher Order Receive Diversity

Table 6.17 presents a set of simulation results comparing four-branch and eight-branch receive diversity with two-branch receive diversity in a macro-cellular radio environment.

In each case Maximal Ratio Combining (MRC) has been used as a combining algorithm in the receiver. The modified ITU Vehicular A channel is characterised by six delay spread components with relative tap powers of $0, -1.9, -7.3, -10.4, -10.9$ and $-17.3\,$dB. The ITU Pedestrian A channel is characterised by two delay spread components with relative tap powers of 0 and $-12.9\,$dB. The angular spread of the BS received signal power has been modelled as a Laplacian distribution according to field trial experience [7]. The simulations were based upon a $5°$ angular spread. The current 3GPP specification [8] was used to set the operating point to a level corresponding to a 1% BLER.

The simulation results indicate that when receive branches are completely uncorrelated the reduction in E_b/N_0 requirement is greatest for low-speed mobile terminals. These are the scenarios in which the time diversity provided by channel coding and interleaving is relatively poor. The relatively large reduction in E_b/N_0 requirement when increasing the number of receive branches from four to eight is partly due to the eight-branch receiver benefiting from 16 RAKE fingers, whereas the four-branch receiver has only eight. The simulation results for the Pedestrian A channel illustrate the potentially high impact of four-branch and eight-branch receive diversity. In this scenario, time and multi-path diversity gains are relatively low resulting in high incremental gain from higher order receive diversity.

The impact of higher order receive diversity upon service coverage is twofold. In the first place the reduction in E_b/N_0 requirement translates to a direct gain in the uplink link budget. In addition, for any given quantity of traffic, uplink load and thus interference margin will also be reduced. This is a result of uplink load being directly proportional to the E_b/N_0 requirement. This means that when the E_b/N_0 reduction provided by higher order receive diversity equals the reduction in composite noise figure provided by MHAs, higher order receive diversity will provide a greater increase in service coverage performance.

The impact upon system capacity is dependent upon whether the capacity is uplink or downlink limited. If it is uplink limited then higher order receive diversity will increase

Table 6.17 Reduction in speech service E_b/N_0 requirement associated with higher order receive diversity relative to that of two-branch receive diversity.

Antenna configuration	Modified Vehicular A			Pedestrian A
	3 km/h [dB]	50 km/h [dB]	120 km/h [dB]	3 km/h [dB]
Four uncorrelated antennas	3.0	2.5	2.3	5.9
Four partially correlated antennas	1.5	2.2	2.0	4.2
Eight uncorrelated antennas	6.9	5.4	5.0	10.3

Table 6.18 A comparison of uplink capacity between a four-branch higher order receive diversity cell and a two-branch receive diversity cell (uplink loading limit is 30%).

		12.2 kbps speech	64 kbps data	128 kbps data	384 kbps data
Two-branch receive diversity	Data users	—	6	3	1
	Speech users	30	1	1	7
	Total throughput	366 kbps	396 kbps	444 kbps	469 kbps
Four-branch receive diversity (2.5 dB gain)	Data users	—	11	5	2
	Speech users	54	0	5	7
	Total throughput	659 kbps	704 kbps	781 kbps	853 kbps

Table 6.19 A comparison of the downlink capacity losses associated with mast head amplifiers and four-branch receive diversity when the maximum allowed propagation loss is increased by 3 dB.

	Capacity loss
MHAs[a]	12.3%
Four-branch receive diversity[b]	10.2%

[a] Based upon dedicated WCDMA feeders with 3 dB feeder loss – i.e., 3 dB reduction in receiver noise figure.
[b] Based upon a 2.5 dB reduction in E_b/N_0 and a 0.5 dB reduction in interference margin.

system capacity. Table 6.18 illustrates the capacity gain for a cell planned for 30% uplink loading.

Capacity gain is about 80% for each service mix. The relative gain would be less if the cell had a 50% uplink load limit, because the cell would rapidly become downlink capacity limited and BS transmit power would become the limiting factor.

If system capacity is downlink limited there will be a loss in capacity when higher order receive diversity is used. For equal gains in service coverage performance, higher order receive diversity results in capacity loss lower than that caused by MHAs. This is because higher order receive diversity does not generate downlink insertion loss and the BS's EIRP remains at its maximum. Table 6.19 compares the downlink capacity loss associated with MHAs and four-branch receive diversity when the maximum allowed propagation loss is increased by 3 dB.

6.8.2 Practical Considerations

The relatively large antenna configurations associated with higher order receive diversity form a significant practical consideration. Large antenna configurations raise issues regarding antenna space, weight, wind loading and feeder rigging. Environmental aspects are increasing in importance as the number of operators and

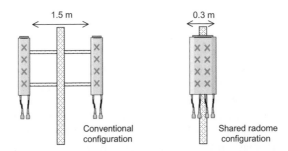

Figure 6.6 Typical antenna dimensions associated with four-branch receive diversity.

cellular systems grow. In addition, even if practical circumstances prove acceptable, the site owner is likely to charge higher rent for a large antenna configuration. The use of crosspolar antennas helps to decrease the number of antenna units required. In some environments this may result in a small polarisation loss. Figure 6.6 illustrates typical antenna dimensions for four-branch receive diversity with cross-polar antennas.

The first configuration corresponds to the uncorrelated antenna results presented in Table 6.17. In this case the antennas are horizontally separated by ten wavelengths. The second configuration corresponds to the partially correlated scenario. The two sets of crosspolar elements are housed within the same radome; this decreases the diversity gain but improves the practicalities of deploying four-branch receive diversity.

Practical considerations mean that it is not generally feasible to deploy higher order receive diversity with more than four branches. Instead, beamforming techniques (Section 6.10) may be used to improve system performance further.

Although this section has focused upon receive diversity at the BS, it is also feasible to benefit from diversity at the mobile terminal. In this case the manufacturer has to consider the implications upon size, power consumption and cost. WCDMA services are likely to be supported by a range of mobile terminal types. For example, a laptop computer may be used to provide wireless Internet applications. In this case it is possible to adopt similar receive diversity techniques as for the BS. An advantage the mobile terminal has over the BS is that the local environment is typically rich in scatterers. Antenna separations can be as small as half a wavelength [9]. In a traditional mobile terminal handset two-branch or three-branch receive diversity may be feasible using a dual-polarised patch antenna or a combination of a monopole and patch antenna.

6.9 Transmit Diversity

3GPP does not define receive diversity as mandatory at the mobile terminal. If an operator wishes to improve downlink performance, a transmit diversity scheme from two BS antennas can be adopted. 3GPP defines downlink transmit diversity as mandatory for the mobile terminal. The greatest challenge in achieving high performance is having accurate knowledge of the downlink radio channel at the BS transmitter. The frequency division nature of UTRA FDD means that uplink channel estimations are not applicable to the downlink. Without knowledge of the downlink radio channel, it is not possible to calculate the optimal complex weights for each of the two transmit antennas. One possibility is for the mobile terminal to measure the downlink channel from each antenna and feed the information back to the BS. To obtain optimal weights this technique would require an excessive overhead [10].

The WCDMA air interface specification [8] defines two approaches for accomplishing transmit diversity: closed-loop and open-loop. In the case of closed-loop mode, the mobile terminal provides a relatively coarse degree of feedback to the BS regarding the state of the radio channel from each of the two transmit antennas. In the case of open-loop mode, a simple space–time coding scheme is used. This helps maintain mobile terminal complexity at a reasonable level. The benefit of open-loop mode is its robustness and ease of implementation.

Taking advantage of transmit diversity requires the mobile terminal to separate the signals from the two BS antennas. The WCDMA system facilitates this by transmitting the Primary Common Pilot Channel (P-CPICH) from each of the two antennas with the same channelisation and scrambling code but with different symbol sequences. The pilot is sent from one antenna such that the complex symbol sequence appears as $\{A, A, A, A, A, A, \ldots\}$, and from the second antenna as $\{-A, A, A, -A, -A, A, \ldots\}$, where A denotes a complex symbol $1 + j$. The first sequence is the same as that used by a standard non-transmit diversity cell. These orthogonal symbol sequences allow the mobile terminal to evaluate independent channel impulse responses from each antenna.

Closed-loop Mode

Two closed-loop modes of transmit diversity are defined by the 3GPP WCDMA air interface specification [8]. In each case the principle is for the mobile terminal to measure the channel impulse response from each of the two transmit antennas, then using the feedback channel to return the relative phase shift between the two antennas. The BS is then able to generate complex weights for each transmit antenna such that the two signals received by the mobile terminal are as coherent as possible. In closed-loop mode 1 the phase of one antenna is adjusted relative to the other using 1 bit accuracy per slot. In closed-loop mode 2 both the relative phase and amplitude are adjusted. The relative phase is adjusted with 3 bit accuracy and the amplitude with 1 bit accuracy. Both modes provide feedback at a rate of a single bit per slot – i.e., 1.5 kHz. Figure 6.7 illustrates the principle behind the closed-loop modes.

Closed-loop modes offer better performance than the open-loop mode but at the cost of additional complexity. For example, improved closed-loop performance requires

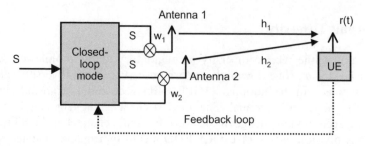

Figure 6.7   Principle used to apply WCDMA closed-loop transmit diversity.

antenna verification at the mobile terminal. This helps to ensure that the weight applied by the BS is actually the same as that commanded by the mobile terminal and that any potential errors in the feedback channel are detected.

Open-loop Mode
The WCDMA open-loop mode of transmit diversity is based upon Space–Time Transmit Diversity (STTD). Space–time processing techniques exploit diversity in both the spatial and temporal domains in an open-loop fashion ([11] and [12]). Space–time coding includes both space–time block codes and space–time trellis codes. Figure 6.8 illustrates the principle of how STTD encodes two Quaternary Phase Shift Keying (QPSK) symbols across two antennas during two symbol periods.

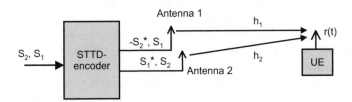

Figure 6.8   Principle behind the WCDMA space–time transmit diversity coding scheme.

The encoding technique makes the two transmitted signals orthogonal to one another allowing relatively simple detection at the mobile terminal receiver. Using the notation of Figure 6.8 the STTD coded signals $r(t)$ and $r(t + T)$ received at the mobile terminal in consecutive time intervals t and $t + T$ can be expressed according to Equation (6.2):

$$\left. \begin{array}{l} r(t) = r_1 = S_1 \cdot h_1 + S_2 \cdot h_2 + n_1 \\ r(t + T) = r_2 = -S_2^* \cdot h_1 + S_1^* \cdot h_2 + n_2 \end{array} \right\} \tag{6.2}$$

where h_1 and h_2 are the channel impulse responses associated with the two transmit antennas; and n_1 and n_2 represent the composite noise plus interference received during the two time intervals. The mobile terminal estimates the symbols by linear combination according to:

$$\left. \begin{array}{l} \hat{S}_1 = \hat{h}_1^* \cdot r_1 + \hat{h}_2 \cdot r_2^* \\ \hat{S}_2 = \hat{h}_2^* \cdot r_1 - \hat{h}_1 \cdot r_2^* \end{array} \right\} \tag{6.3}$$

where ^ denotes an estimated value. The space–time combining rule generates symbols that are proportional to the sum of the channel powers from each of the antennas.

6.9.1   Impact of Transmit Diversity

Downlink orthogonality has a large influence upon the performance of transmit diversity schemes for WCDMA. The additional multi-path generated by transmit diversity may result in a loss of downlink channelisation code orthogonality. The main benefit of using transmit diversity is a reduction in the downlink E_b/N_0 requirement. Simulations used to quantify the performance benefits must include a population of co-channel mobile terminals sharing the same channelisation code tree. The Geometry parameter, or G-parameter, is often used to define the ratio of partially orthogonal intra-cell interference to non-orthogonal inter-cell interference. Inter-cell interference is assumed to be Gaussian. The G-parameter is defined by Equation (6.4):

$$G = \frac{\hat{I}_{or}}{I_{oc}} \tag{6.4}$$

where $\hat{I}_{or}$ is the intra-cell interference power spectral density; and I_{oc} is the inter-cell interference power spectral density. Large values of the G-parameter correspond to being close to the BS where intra-cell interference dominates. Small values correspond to being close to the cell edge. Typical values range from $-3\,\text{dB}$ to $12\,\text{dB}$ [13]. A useful measure of downlink performance is the ratio of transmit power requirement per radio link connection to total BS transmit power. This is evaluated at the operating point providing the specified QoS. The ratio can be denoted by transmit $I_c/\hat{I}_{or}$, where I_c is the power spectral density of the transmit power requirement per radio link. A low ratio corresponds to high performance. A value of $-20\,\text{dB}$ indicates that only 1% of the total BS transmit power is required for the single radio link.

Figure 6.9 illustrates an example set of simulation results comparing the performance of single antenna transmission with open-loop STTD and closed-loop transmit diversity mode 1. A 4% BER has been assumed for both the inner-loop power control commands and the feedback information bits of closed-loop mode.

A significant challenge associated with specifying the performance of downlink transmit diversity is identifying a representative mobile terminal performance. The algorithms used by the mobile terminals have a significant impact upon how well the transmit diversity scheme performs in practice. For example, a mobile terminal must be able to make an accurate estimation of the channel impulse response and SIR.

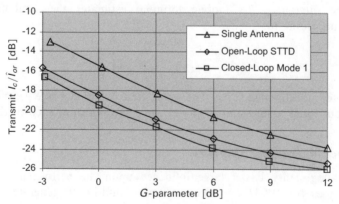

Figure 6.9 A comparison of single antenna transmission with open loop space–time transmit diversity and closed-loop mode 1 transmit diversity as a function of the Geometry parameter for an ITU Pedestrian A channel.

Table 6.20 Reduction in E_b/N_0 requirement provided by open-loop mode and closed-loop mode 1 transmit diversity relative to single-antenna transmission (average gains for the 12.2 kbps to 128 kbps services, BLER 1%).

Diversity mode	Modified Vehicular A			Pedestrian A
	3 km/h [dB]	50 km/h [dB]	120 km/h [dB]	3 km/h [dB]
Open-loop mode	1.0	0.5	0.5	3.0
Closed-loop mode 1	1.5	1.0	0.0	3.5

Table 6.21 Typical capacity increases when using open-loop and closed-loop transmit diversity in macro-cell and micro-cell scenarios. Based upon a downlink capacity limited scenario.

Diversity mode	Macro-cell capacity gain [%]	Micro-cell capacity gain [%]
Open-loop mode	25	50
Closed-loop mode 1	35	70

Table 6.20 presents the transmit diversity performance gains for both the modified ITU Vehicular A and Pedestrian A channel responses. A G-parameter of 3 dB has been assumed. In a similar fashion to the gains achieved with higher order receive diversity, the impact of transmit diversity is greatest when the performance of time and multipath diversity is relatively poor – i.e., a low-mobility user in an environment with little multi-path. High mobile terminal speeds in a Vehicular A environment result in negligible performance gains. The high levels of multi-path associated with a Vehicular A channel response reduce downlink orthogonality, which subsequently impacts upon transmit diversity performance.

These improvements in E_b/N_0 requirement impact upon both downlink system capacity and downlink service coverage. The gain in service coverage is of greatest importance when the coverage is downlink limited. This is most likely to occur in micro-cell scenarios when BS transmit power is relatively limited. Table 6.21 presents a set of typical capacity gains for both macro-cell and micro-cell scenarios. These gains are relative to that of a downlink capacity limited scenario with a single-element transmitter. The figures indicate that the capacity gain is greatest for the micro-cell scenario.

6.9.2 Practical Considerations

According to the 3GPP WCDMA air interface specification [8], the modes of transmit diversity described above are mandatory features for terminals but optional for BSs. This means that operators have the choice of whether or not to apply transmit diversity. Transmit diversity schemes are particularly appropriate for micro-cell scenarios where capacity gains are large and other techniques such as beamforming are less appropriate due to the large angular spread [14]. Transmit diversity provides a relatively simple capacity upgrade solution in terms of configuring additional hardware. Two antenna elements are required on the downlink: either a single crosspolar antenna or two vertically polarised antennas. In terms of power amplifier modules, the operator may be able to share the existing power. This is possible when the site is configured with multiple carriers and multi-carrier power amplifiers. For example, a $2 + 2 + 2$ site configured with six 20 W power amplifiers can be upgraded to include downlink transmit diversity with 10 W per transmit element without increasing the number of amplifier modules. This is not possible for single-carrier scenarios and supplementary power amplifiers must be included.

6.10   Multiple Input Multiple Output in UTRA FDD

The important property of diversity systems, relying on multiple transmit and/or receive antennas, is the ability to mitigate the detrimental effects of multi-path fading. This property was successfully employed in UTRA where receive diversity can be applied independently from the standards. Transmit diversity modes when using two antennas are specified in [19].

Letting both the number N of transmit antennas and the number M of receive antennas be more than 1 leads to a Multiple Input Multiple Output (MIMO) system. According to present 3GPP standards a $2 \times M$ diversity MIMO that uses two-antenna transmit diversity and MRC combining over M receive antennas can be applied. Although impressive performance gains can be achieved through the diversity systems, it is known from recent studies ([20]–[22]) that information MIMO approaches, where up to $\min\{M, N\}$ independent data streams are transmitted, provide means to achieve high data rates in a bandwidth-efficient manner. In 3GPP the standardisation process for MIMO algorithms are discussed under TSG RAN WG1 (Technical Specification Group Radio Access Network Work Group 1). At the moment some technical reports are available ([23] and [24]), but final specifications are not ready. The following sections first recall the theoretical background of MIMO, discuss its impact upon UTRA FDD and go through the practical problems arising. Then briefly the 3GPP MIMO candidate algorithms are presented and, lastly, uplink-specific challenges are considered.

6.10.1   Mathematical Background

In the field of academic research MIMO has been a huge success. This is due to the fact that MIMO promises an improvement of magnitudes in the performance of wireless links without any extra spectrum demand. In this section only the simplest fundamental results are introduced in order to illustrate the expected benefits of MIMO; further details can be found in [25].

Let $\mathbf{H}$ be an $M \times N$ matrix that consists of normalised complex channel coefficients $h_{m,n}$. Then the bound for the information rate of a memoryless diversity MIMO system with Additive White Gaussian Noise (AWGN) is given by:

$$C_D = \log_2\left(1 + \frac{\rho}{N}\sum_{m=1}^{M}\sum_{n=1}^{N}|h_{m,n}|^2\right) \quad \text{bits/s/Hz} \tag{6.5}$$

where ρ is the average Signal-to-Noise Ratio (SNR) and it is divided by N since the total transmission power is shared equally between N antennas. For the information MIMO system the bound for the information rate is of the form:

$$C_I = \sum_{k=1}^{K}\log_2\left(1 + \frac{\rho}{N}\lambda_k\right) \quad \text{bits/s/Hz}; \qquad K = \min\{M, N\} \tag{6.6}$$

where λ_k is an eigenvalue of $\mathbf{H}\mathbf{H}^*$ (or eigenvalue of $\mathbf{H}^*\mathbf{H}$ if $N < M$). Equations (6.5) and (6.6) are valid if transmission power in the system is fixed and there is no Channel State Information (CSI) in the transmitter. Equation (6.6) defines the link capacity of the MIMO system while Equation (6.5) provides the link capacity for Single Input Multiple Output (SIMO) and Multiple Input Single Output (MISO) if $N = 1$ or $M = 1$, respectively.

Figure 6.10 depicts the expected capacity/information rate in case of *i.i.d.* flat Rayleigh fading channels when one to two receive antennas and one, two or four transmit antennas are applied.

While the gain from transmit diversity is small, receive diversity provides noticeable gain, and information MIMO (dashed curves) using two or four transmit antennas clearly indicates the best performance. Hence, according to basic theory, it seems that information MIMO is superior to diversity methods. Furthermore, if it is

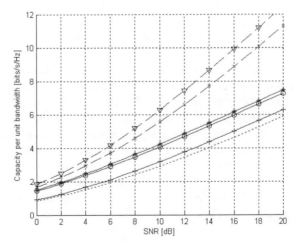

Figure 6.10 Link capacity/information rate versus average signal-to-noise ratio when Rayleigh fading is assumed and $(M, N) = (1, 1)$ (dotted curve), $(M, N) = (1, 2)$ (+), $(M, N) = (2, 1)$ ($\bigcirc$), $(M, N) = (2, 2)$ ($\otimes$) and $(M, N) = (2, 4)$ (∇). Dashed curves refer to multiple input multiple output information.

assumed that there is CSI in the transmitter, then even more impressive gains can be obtained by information MIMO. However, in practical systems such as UTRA theoretical capacities do not necessarily reflect directly on system performance.

6.10.2 Impact of MIMO

At the moment MIMO discussion in 3GPP has concentrated on HSDPA, because the capacity demand imposed by projected data services (e.g., Web browsing) burdens more heavily the downlink and, on the other hand, the best gains from MIMO processing are achieved when fast power control is not applied – as is the case in the High-speed Downlink Shared Channel (HS-DSCH) where BS transmission power is fixed [23].

Evaluation of the various MIMO schemes is a very challenging task and intensive discussions are carried out in TSG RAN WG1. At the present only draft specifications exist and both link-level and system-level evaluations are being made by different standardisation parties. Since there is not yet a mature MIMO technique in UTRA FDD final conclusions on the impact of MIMO cannot be drawn. Therefore, only the widely accepted basic impacts of MIMO on UTRA FDD are briefly explained. Section 6.10.3 discusses the so-called 'wish list' that shows the practical requirements and expected impacts of MIMO on UTRA FDD, forming the basis for the MIMO standardisation process.

MIMO algorithms can roughly be divided into two groups according to the number of data streams – namely, algorithms that apply multiple antennas at both ends but support only one data stream represent the *diversity MIMO* approach, while algorithms that support multiple data streams represent the *information MIMO* approach.

The main benefit from using diversity MIMO in the HS-DSCH is basically similar to the case of transmit diversity where improved performance leads to a reduced E_b/N_0 requirement. This gain has a positive impact on both downlink capacity and coverage, the service coverage gain being the main advantage. The drawback to diversity MIMO is that it does not solve the problem related to shortage of orthogonal channelisation codes.

In hotspot cells with several active high data rate users the shortage of channelisation codes may limit system capacity. Then information MIMO can provide a good solution to increase individual data rates without a need for additional channelisation codes. This is especially true when high data rate users are near to the BS.

In the uplink the benefit of MIMO is limited since accurate transmit power control is employed to avoid the near–far effect. Diversity MIMO reduces the mobile transmission power leading to better service coverage and reduced inter-cell interference but system capacity gain is usually not noticeable. By information MIMO individual data rates can be increased without heavy code puncturing – not the case in SIMO – leading to some system capacity gain [29].

6.10.3 Practical Considerations

Let us recall from [23] some of the most important requirements that should be taken into account while evaluating different candidate MIMO techniques (the 'wish list').

In the following each paragraph begins with a direct reference to requirements given therein.

MIMO proposals shall be comprehensive to include techniques for one, two and four antennas at both the base station and UE. This requirement is motivated by the fact that deploying multiple antennas in the mobile terminal or BS to support MIMO techniques is not straightforward due to concerns of cost, complexity and visual impact. This is especially true of today's mobile terminals, where basic products with large production volumes may have at most two antennas. Multi-mode terminals supporting, for example, WCDMA, GSM and GPS may already require several antennas even without applying MIMO processing. Macro-BSs typically employ two or four antennas, and it is expected that two-antenna BSs will dominate in number in the near future. Thus, in practice, mobile terminals and data modems may have four antennas at the maximum, while two antennas represent the most likely solution.

For each proposal, the transmission techniques for the range of data rates from low to high SIR shall be evaluated. This is a trivial but important requirement since the gain from information MIMO greatly depends on the SIR/SNR as is seen from Figure 6.10. Especially in macro-cell environments the operating SIR/SNR in HSDPA is most of the time less than 10 dB and the practical performance differences between various diversity MIMO and information MIMO techniques need not to be as large as Figure 6.10 hints.

Operation of MIMO technique shall be specified under a range of realistic conditions. The conclusion drawn from this requirement is that there should be realistic channel models for simulations. This topic has been considered in [24]. Moreover, to imitate realistic conditions implementation non-idealities should also be taken into account.

The MIMO technique shall have no significant negative impact on features available in earlier releases. Let us give an example of a serious backward compatibility problem that may arise when introducing MIMO. According to present standards there are at maximum two P-CPICHs applied in UTRA FDD downlink to aid channel estimation in the mobile terminal. To support four-antenna MIMO a straightforward solution would be to define two additional P-CPICHs. However, since total transmission power in the BS cannot be increased due to network interference and capacity reasons, the transmission power per antenna needs to be halved when doubling the number of transmit antennas in the BS. But then UEs that are made according to earlier standard releases and can identify only two common pilot signals would receive in a four-antenna cell only half of the pilot power when compared with the pilot power that they would receive in a two-antenna cell. This would lead to serious performance losses.

MIMO techniques shall demonstrate significant incremental gain over the best performing systems supported in the current release with reasonable complexity. Although the capacity curves of Figure 6.10 suggest that information MIMO would give remarkable gains over various diversity systems, it is found that – especially when the number of antennas is only two at both ends – the practical gains from information MIMO can be small in some cases [26]. Not only does increasing the number of antennas increase the gain of information MIMO, but the implementation complexity also grows rapidly and backward compatibility issues – such as the above-mentioned pilot design problem – need to be faced.

6.10.4 Candidate MIMO Algorithms in 3GPP Standardisation

The standardisation of MIMO is still ongoing and there are many candidate algorithms that are proposed by different parties. In the following sections the proposed algorithms are briefly summarised. A more detailed description and performance analysis can be found in [23] and corresponding standardisation contributions.

6.10.4.1 Per-Antenna Rate Control

According to information theory results ([27] and [28]) the capacity limit for an open-loop MIMO link can be achieved by transmitting separately encoded data streams from different antennas with equal power but possibly with different data rates. This idea provides a background for the basic Per-Antenna Rate Control (PARC) architecture that is given in Figure 6.11 in case of $N = 2$.

PARC shows how the HS-DSCH data stream is demultiplexed into two low-rate streams. Both streams are turbo-encoded, interleaved and mapped onto either QPSK or 16 State Quadrature Amplitude Modulation (16QAM) symbols. Code rates and symbol mappings can vary between low-rate streams, and therefore the number of information bits assigned to each stream can be different. Symbols are further demultiplexed into a maximum of K sub-streams, where K is the maximum number of High-speed Physical Downlink Shared Channels (HS-PDSCHs) defined by the mobile terminal capability. After spreading these sub-streams – employing distinct Orthogonal Variable Spreading Factor (OVSF) channelisation codes denoted by OC_1–OC_K in Figure 6.11 – they are summed and modulated by a scrambling code. The resulting antenna-specific WCDMA signal is transmitted from the associated antenna.

The data rates for different antennas are selected in the BS based on antenna-specific Signal-to-Interference-and-Noise Ratio (SINR) feedback. If the SINR for a particular transmit antenna is too low to support even the lowest data rate, then transmission

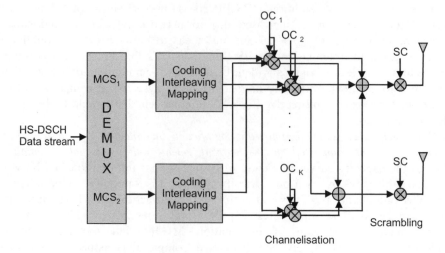

Figure 6.11 Transmitter structure for per-antenna rate control.

through that antenna is suspended. For this purpose the mobile terminal estimates the CSI for all antennas and sends the required information to the BS through a feedback channel. Since the Modulation and Coding Scheme (MCS) for each antenna is selected using SINR feedback, the design of feedback quantisation is an important task. In fact, quantised CSI defines a mapping onto the table giving the modulation, coding and number of spreading codes used for each transmit antenna. Since the total number of possible transport format combinations is large, a suitable subset of combinations should be designed in order to avoid large signalling overhead.

6.10.4.2 Double STTD with Sub-group Rate Control

Double STTD with Sub-group Rate Control (DSTTD-SGRC) is designed for a system with $2N$ transmit and at least N receive antennas. The basic idea is to divide antennas into N sub-groups each containing two antennas and apply adaptive modulation and coding along with STTD-based transmission by each group to transmit data. Within the sub-group both antennas apply the same MCS but the data rates of separate groups can be adjusted independently or jointly by selection of suitable MCSs. In the framework given by present 3GPP standardisation the maximum number of transmit antennas is expected to be four and thus, at maximum, two independent data streams can be transmitted.

DSTTD-SGRC can be viewed as an extension to conventional STTD supported by Release '99 standards – STTD was introduced in Section 6.9. While conventional STTD employs two transmit antennas and a single data stream, DSTTD-SGRC doubles the number of transmit antennas and data streams, provided that the mobile terminal is equipped with at least two antennas. From this viewpoint it can be expected that DSTTD-SGRC attains good backward compatibility with previous standard releases.

Figure 6.12 shows the structure of the DSTTD-SGRC transmitter when four antennas are being used. The incoming HS-DSCH data is divided into two streams by the demux module and transmitted by the first and second sub-groups. The applied

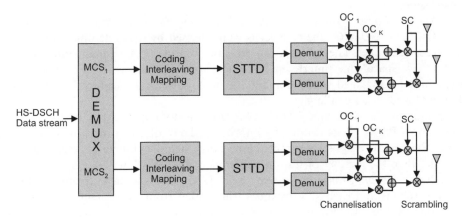

Figure 6.12 Transmitter structure for double space–time transmit diversity with sub-group rate control.

MCS and the number of spreading codes define the number of information bits allocated to each stream. For both streams information bits are coded, interleaved and modulated according to the selected MCS. The two symbol streams obtained after STTD encoding are then split into K parallel streams corresponding to K spreading codes. In the last stage the streams are combined, scrambled and transmitted.

6.10.4.3 Other proposed MIMO algorithms

Besides PARC and DSTTD-SGRC six other MIMO algorithms are proposed in [23]. Since most of these schemes are not as well-documented as PARC and DSTTD-SGRC they are introduced here only very briefly.

In *Rate-Control Multi-Paths Diversity (RC-MPD)* each data stream is transmitted from at least two antennas and the number of data streams is equal to the number of transmit antennas. Furthermore, a pair of data streams that share the same two antennas apply the same MCS. The basic idea is to transmit another copy of the signal after a 1 chip delay by using STTD encoding. Hence, if there are two antennas, two data streams and the corresponding symbols are s_1 and s_2, then the transmitted signal consists of symbols s_1 and s_2 at time T and symbols $-s_2^*$ and s_1^* at time $T + T_C$ where T_C is the chip interval. The aim in the method is to achieve multi-path diversity that is orthogonalised through STTD encoding.

The *single-stream closed-loop MIMO* is a four-antenna extension of the two-antenna closed-loop mode 1 that is supported by Release '99 standards – it was introduced in Section 6.9. There are two basic problems with this method. First, only a single data stream is supported limiting achievable peak data rates. Second, for the phase reference four common pilots instead of two are needed. This leads to backward incompatibility with previous standard releases.

Per-User Unitary Rate Control (PU²RC) is based on the singular value decomposition of MIMO channels. In this method transmit weights are computed based on the unitary matrix that is a combination of the selected unitary basis vector from all mobile terminals. The aim is to utilise multi-user diversity on top of MIMO transmission.

In *Transmit Power Ratio Control for Code Domain Successive Interference Cancellation (TPRC for CD-SIC)* the receiver is characterised by the code domain successive interference canceller. The goal is to suppress the impact of code domain interference in addition to space–time interference. System performance is further boosted by employing the so-called 'code domain transmit power ratio control' that requires additional feedback signalling.

The aim of the *Selective PARC (S-PARC)* is to improve the performance of conventional PARC. This is done by improving the feedback format of conventional PARC. Performance gains are expected especially when the number of receive antennas is smaller than the number of transmit antennas or SNR is low.

Finally, in *Double Transmit Antenna Array (D-TxAA)* the data stream is split into two sub-streams and each sub-stream is transmitted from two antennas by applying either one of the closed-loop methods according to Release '99. Hence, the total number of transmit antennas is four. Again the same common pilot problem as in the case of single-stream closed-loop MIMO is faced.

Various performance results for the above-mentioned candidate algorithms have been presented during the 3GPP standardisation process. However, since there is no wide agreement concerning the mutual ranking of the candidate algorithms and even simulation assumptions are under consideration, no performance results are shown here.

6.10.5 MIMO in UTRA FDD Uplink

So far, MIMO discussions in 3GPP have focused on HSDPA. However, when new services such as videophones become more popular, it is extremely important to reach high spectral efficiency in the uplink direction as well. Furthermore, if multi-antenna mobiles are deployed for HSDPA, it is important to study the gain of multiple transmit antennas in the uplink.

In the UTRA framework, the feasibility of different MIMO methods varies between the uplink and downlink. While intra-cell users in the downlink are separated by different orthogonal channelisation codes, and the capacity is limited by the shortage of channelisation codes, in the uplink, different users are separated by long scrambling codes, and a single user may use the entire family of orthogonal channelisation codes.

Transmit power control is an inherent characteristic of the asynchronous WCDMA uplink. Due to non-orthogonality of the users' channelisation codes multi-user interference cannot be avoided. Accurate transmit power control is indispensable to uplink performance and should be taken into account when designing MIMO algorithms.

In [29] simple diversity and information MIMO approaches were studied assuming the UTRA FDD framework. Results show that the uplink coverage and capacity of the UTRA FDD mode are significantly increased by SIMO and MIMO. While the performance increase from additional BS antennas reflects to coverage and capacity results straightforwardly, the transmit diversity gain from additional antennas at the mobile end is relatively small. This is due to the fact that link-level power control converts the increased diversity to a decrease in required transmission power. On the contrary, if user bit rates higher than 2 Mbps are needed, the gain from information MIMO is large, because heavy code puncturing can be avoided. Thus, multiple transmit antennas should be used in the mobile terminal for spatial multiplexing rather than for transmit diversity. Furthermore, the simplest information MIMO algorithms only require minor changes to the present UTRA FDD specifications.

6.11 Beamforming

Whereas higher order receive diversity improves uplink performance and transmit diversity improves downlink performance, beamforming improves both uplink and downlink performance. If the antenna array has between two and eight elements, uplink receive diversity provides approximately the same uplink gains as beamforming. However, antenna arrays with more than two elements can provide greater downlink gains than those provided by transmit diversity. This is a result of spatial filtering, which confines downlink interference to a limited angular spread. The choice of whether to use beamforming or higher order receive diversity combined with transmit diversity is dependent upon the specific radio environment as well as the maturity of each technology.

6.11.1 Mathematical Background

Directing a beam in a particular direction can be achieved using a phased array antenna. A common solution is the uniform linear array, which adjusts the phase shift for each antenna element such that the desired signal sums coherently at a specific Direction of Arrival (DoA). Figure 6.13 illustrates the phase difference between two adjacent antennas of a four-element array for a DoA θ. The phase shift relative to the reference element increases linearly from element to element. Compensating for the phase shifts corresponding to a specific DoA results in coherent summation.

The phase shift at element m is a function of the inter-element spacing d, DoA θ and carrier wavelength λ. Equation (6.7) expresses the relationship:

$$\varphi_m = \frac{2 \cdot \pi}{\lambda} \cdot \Delta l_m = \frac{2 \cdot \pi}{\lambda} \cdot (m-1) \cdot d \cdot \sin\theta, \qquad m = 1, \ldots, M \qquad (6.7)$$

The response vector $\mathbf{a}$ of an antenna array with M elements describes the complex antenna weights for the beam directed towards DoA θ:

$$\mathbf{a} = [1, \exp(j \cdot \varphi_1), \ldots, \exp(j \cdot \varphi_M)] \qquad (6.8)$$

There are two fundamental approaches to beamforming: either multiple fixed beams or user-specific beams. Orthogonal fixed beams can be generated using the Butler matrix, which defines the parallel sets of phase shifts associated with each beam. Table 6.22 presents the phase shifts of a four-element array used to generate four orthogonal beams.

Figure 6.14 illustrates the corresponding beam patterns with respect to a hexagonal cell footprint. This figure takes account of the beam pattern of each individual antenna element.

The fixed beam approach can be implemented in a relatively simple manner by integrating analogue phase shift components into the antenna panel. In this case multiple users are assigned to each beam. The user-specific approach to beamforming

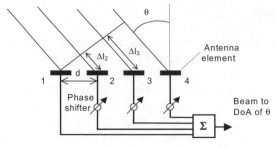

Figure 6.13 Geometry of a uniform linear array for a planewave in the direction of arrival θ.

Table 6.22 Phase shifts φ_m for the 4×4 Butler matrix.

Beam #	Antenna element #			
	1 [°]	2 [°]	3 [°]	4 [°]
1	0	−135	−270	−405
2	0	−45	−90	−135
3	0	45	90	135
4	0	135	270	405

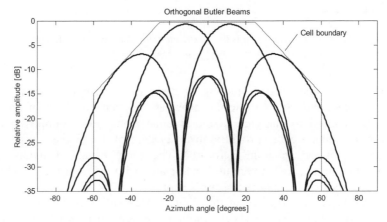

Figure 6.14 Beam pattern of a four-element array based upon the Butler matrix of Table 6.22.

is more complex and requires a separate response vector to be assigned to each mobile terminal.

6.11.2 Impact of Beamforming

Table 6.23 presents a set of link-level simulation results comparing the uplink performance gains for a range of antenna configurations. The beamforming results correspond to the fixed beam approach rather than the user-specific beam approach. The $4 + 4$ configuration implies two sets of four beams separated by polarisation diversity. The gain is presented in terms of a reduction in E_b/N_0 requirement relative to two-branch receive diversity. E_b/N_0 reductions improve both coverage and capacity in the uplink direction.

The gain is relatively insensitive to the DoA of the mobile terminal – i.e., whether it is towards the centre of a beam or between two beams. This is a result of the angular diversity gain being at a maximum between two beams while the beamforming gain is at a maximum in the direction of a beam. In the Pedestrian A environment which exhibits only two delay spread components, the fixed eight-beam approach performs no better than four-branch MRC.

Table 6.23   Reduction in uplink E_b/N_0 requirements provided by fixed beam beamforming and four-antenna MRC relative to the E_b/N_0 requirement of a two-branch receiver for a 12.2 kbps speech service with a BLER of 1%.

Antenna configuration	Modified Vehicular A			Pedestrian A
	3 km/h [dB]	50 km/h [dB]	120 km/h [dB]	3 km/h [dB]
4-antenna MRC[a]	3.0	2.5	2.3	5.9
8 beams[b]	4.9	5.2	5.1	5.9
8 beams[c]	4.4	4.9	4.8	5.8
4 + 4 beams[b]	5.5	5.7	5.9	7.0
4 + 4 beams[c]	4.4	4.3	4.5	6.0

[a] Uncorrelated antennas.
[b] Mobile terminal direction of arrival towards the maximum beam gain, eight RAKE fingers.
[c] Mobile terminal direction of arrival between two beams, eight RAKE fingers.

Beamforming provides spatial filtering of downlink transmit power towards the desired mobile terminal. Spatial filtering provides two benefits. First of all transmit power can be reduced by the gain of the antenna array. For example, in an ideal scenario a four-antenna array provides an array gain of 4 and the transmit powers can be reduced by a corresponding factor of 4. The second benefit of spatial filtering is the reduction in interference between users associated with different beams. This allows a significant increase in the number of users supported.

The physical layer performance of the WCDMA downlink is dependent upon the mobile terminal's ability to accurately estimate the channel impulse response and measure the received SIR. In the case of single transmit antenna configurations, the 3GPP specifications define a reliable phase reference in terms of the P-CPICH. When an operator deploys fixed beam beamforming Secondary CPICHs (S-CPICHs) are used to provide a separate and reliable phase reference for each beam. It is possible to evaluate the downlink beamforming gains based upon the mobile terminal's reception of CPICHs [15].

Table 6.24 presents a set of simulation results for a macro-cell environment as a function of the BS antenna configuration and the angular spread of the radio environment. The angular spread at the BS antenna array has been modelled as a Laplacian distribution. The gains have been evaluated by averaging over all azimuths. The results indicate that beamforming provides an effective technique for improving downlink performance, especially in environments with low angular spread.

6.11.3   Practical Considerations

The requirements of beamforming techniques have been taken into account throughout the standardisation of WCDMA. The fixed beam approach is more mature than the user-specific beam approach. Fixed beams are usually generated by analogue phase shifters. In the case of user-specific beamforming, a different beam points in the

Table 6.24 Reduction in downlink E_b/N_0 requirement associated with fixed beam beamforming relative to a cell configured with a single transmit element.

Antenna configuration	Angular spread			
	2° [dB]	6° [dB]	10° [dB]	20° [dB]
Two-beam	2.2	2.2	2.1	1.8
Four-beam	5.1	5.0	4.5	3.7
Six-beam	6.9	6.3	5.8	4.5
Eight-beam	8.8	8.0	7.0	5.2

direction of each mobile terminal. User-specific beamforming necessitates the use of the pilot sequence within the Dedicated Physical Control Channel (DPCCH), which reduces link performance by 2–3 dB relative to when using the P-CPICH. The power of the DPCCH can be varied, but excessive powers lead to inefficient use of downlink transmit power and a corresponding loss in capacity. User-specific beamforming can be implemented either fully digitally or as a hybrid analogue/digital solution.

The WCDMA specification favours adoption of the fixed beam approach. Reasons include the following:

- Mobile terminal functions are well-specified. Beam-specific S-CPICHs can be exploited allowing standard channel impulse response and SIR estimation algorithms to be used.
- Primary and secondary scrambling codes can be assigned across the beams belonging to a cell. This helps alleviate the issue of limitations in the channelisation code tree.
- One or more downlink shared channels can be assigned to each beam to help improve packet scheduling for shared channels. This can lead to improved trunking efficiency.
- The impact upon RRM functionality is minimal.

The fixed beam approach is also attractive because of its strong physical layer performance and reasonable mobile terminal complexity requirement. The largest drawback with the user-specific approach is the increase in complexity and the requirement for non-standard functionality. In addition, the specification for user-specific beamforming does not support transmit diversity and there is a relatively large impact upon RRM functions. Finally, the fact that user-specific beamforming does not provide significant performance gains over the fixed beam approach means that the fixed beam approach is likely to be the preferred technique for WCDMA.

A significant advantage of beamforming is that the antenna array can be constructed within a single antenna radome. The relatively high gain of the array means that the vertical dimensions of the antenna panel can be reduced while maintaining service coverage and system capacity performance.

6.11.4 Impact of Fixed Beam Approach upon Radio Resource Management Algorithms

The spatial filtering that is characteristic of beamforming means that the loading per beam varies as a function of the azimuth distribution of the traffic and multiple access interference. Mobile terminals using high data rate services tend to generate a non-uniform spatial traffic and interference distribution. The admission control and load control schemes should recognise when cell loading is non-uniformly distributed and react accordingly.

The conventional power-based admission control algorithms used with standard sectorised sites can be modified to cope with the fixed beam configuration ([16]–[18]). Power-based admission control algorithms monitor received interference power as well as BS transmit power. Users are granted access to the system if both the receiver interference floor and the BS transmit power are below certain pre-defined thresholds. In the case of power-based admission control with fixed beam beamforming a new user is granted access if the angular power distribution remains satisfactory – i.e., the total BS power and interference level thresholds in each fixed beam are not exceeded. The power increase in each beam depends upon the angular spread and the DoA of the mobile terminal as well as the beam patterns themselves. Figure 6.15 illustrates a fixed beam antenna configuration with a new user attempting to access the system.

If the new user is granted access to beam $P(\theta_4)$ then not only will the load of this beam increase but also those of beams $P(\theta_1)$, $P(\theta_2)$ and $P(\theta_3)$. This is caused by the side lobes of each beam leaking and receiving power across the entire coverage area of the cell. Figure 6.14 shows the side lobes from a four-beam antenna array. The capacity provided by this form of admission control is greatest for uniform traffic and interference loading the cell.

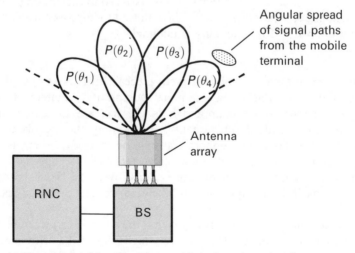

Figure 6.15 An illustration of the effective transmit and receive azimuth power spectrum from a base station configured with a fixed four-beam beamforming antenna array.

6.12 Rollout Optimised Configuration

Rollout Optimised Configuration (ROC) is based upon sharing power amplifiers between cells. Section 6.5.1 described how BS power amplifier modules can be shared between carriers. Doing so generally reduces site capacity but also reduces the requirement for power amplifiers and therefore the capital expenditure associated with the BS. For some uplink capacity limited scenarios the use of ROC may not affect system capacity. This is dependent upon the BS transmit power requirement.

The uplink of an ROC BS appears identical to that of a standard BS – i.e., there are separate transceiver modules for each cell. The downlink is characterised by a splitter dividing the total downlink power between sectors. The downlink appears as a single logical cell configured with a single scrambling code. This is a result of the same signal being transmitted from all three sectors. The downlink antenna gain patterns effectively combine and it is possible to receive multi-path signals from multiple antennas. The combination of the three downlink antenna patterns needs careful consideration, since nulls are likely to appear. Figure 6.16 illustrates the architecture of an ROC BS.

The downlink may be configured with one or two power amplifier modules to share between sectors. Adding a third means that the splitter can be removed and the BS evolves to a standard configuration. In addition, an ROC BS can be configured with multiple carriers. Following the arguments presented in Section 6.4, BS capacity will be greater if power amplifiers are assigned a carrier each rather than being shared across the same carrier.

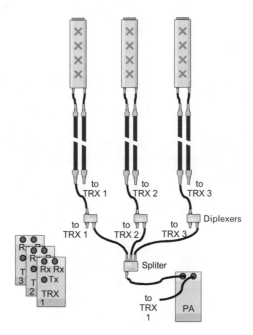

Figure 6.16 Architecture of a rollout optimised configuration base station.

6.12.1   Impact of Rollout Optimised Configuration

If service coverage is uplink limited then the ROC configuration has the same service coverage performance as a standard three-sector site – i.e., the uplink link budget does not change and the cell range remains similar to that of a standard site configuration. If service coverage is downlink limited then the ROC configuration is likely to have a lower coverage performance. This is because there is less downlink transmit power available from each sector. In interference limited scenarios this has little impact because the level of interference is also lower for a population of ROC sites but in a thermal noise limited scenario the service coverage is reduced.

The impact upon system capacity is dependent upon whether the system is uplink or downlink capacity limited. For downlink capacity limited scenarios, the use of an ROC will reduce capacity as a result of the lower BS transmit power capability, although the downlink inter-cell interference ratio is also reduced to a level comparable with that of an omni-directional site configuration. The extent of the loss is dependent upon the allowed propagation loss. A site planned for the 64 kbps data service and having a relatively large allowed propagation loss will incur a greater loss in capacity than a site planned for the 384 kbps data service having a smaller allowed propagation loss.

Consider an ROC BS configured with a single 20 W power amplifier. The 20 W are shared between the three sectors. This means that a maximum of 6.7 W are transmitted to each sector. Typically, 0.5 W of this 6.7 W must be assigned to the P-CPICH and a further 1 W to the Primary and Secondary Common Control Physical Channels (P-CCPCH and S-CCPCH). This results in 5.2 W being available for TCHs. However, not all of the entire 5.2 W are useful power. The ROC configuration leads to a significant transmission power overhead as a result of the same signal being transmitted to all three sectors, as illustrated in Figure 6.17.

User 1 resides within a single cell and is not in softer handover. The downlink transmit power is non-intelligently split between sectors, with no discriminating based on the location of the user. This generates a 200% overhead. In fact only one-third of the 5.2 W is useful TCH power. The remaining two-thirds comprises signal power intended for users in the other two sectors.

Tables 6.25 and 6.26 compare the capacity of a conventional $1 + 1 + 1$ BS configuration with that of a $1 + 1 + 1$ ROC. Table 6.25 is based upon an allowed propagation loss corresponding to a cell planned for the 64 kbps data service.

Table 6.26 is based upon a larger allowed propagation loss, corresponding to a cell planned for the 12.2 kbps speech service.

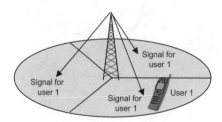

Figure 6.17   Rollout optimised configuration's inherent downlink transmit power overhead.

Table 6.25 A comparison of the capacity associated with a conventional base station configuration and a rollout optimised base station configuration, based upon an allowed propagation loss of **154.4 dB**.

Base station transmit power	Service	Downlink capacity per site [users]	Uplink load [%]	Downlink load [%]
Conventional 1 + 1 + 1	12.2 kbps speech	233	75.5	78.1
20 W per sector	64/64 kbps data	31	50.2	75.5
(12 W total assigned to	64/128 kbps data[a]	17	2.8	74.7
CPICH and CCCHs)	64/384 kbps data[a]	7	1.1	75.5
ROC 1 + 1 + 1	12.2 kbps speech	84	27.3	25.8
20 W shared between sectors	64/64 kbps data	11	16.8	23.1
(4.5 W assigned to	64/128 kbps data[a]	6	0.9	22.3
CPICH and CCCHs)	64/384 kbps data[a]	2	0.4	23.1
ROC 1 + 1 + 1	12.2 kbps speech	134	43.4	41.0
40 W shared between sectors	64/64 kbps data	17	27.3	37.5
(9 W assigned to	64/128 kbps data[a]	10	1.5	37.6
CPICH and CCCHs)	64/384 kbps data[a]	4	0.6	37.5

[a] Includes an activity factor ratio of 1 : 10 for uplink-to-downlink traffic channel activity.

Table 6.26 Comparison of the capacity associated with a conventional base station configuration and a rollout optimised base station configuration, based upon an allowed propagation loss of **156.6 dB**.

Base station transmit power	Service	Downlink capacity per site [users]	Uplink load [%]	Downlink load [%]
Conventional 1 + 1 + 1	12.2 kbps speech	202	65.5	67.8
20 W per sector	64/64 kbps data	27	42.9	64.5
(12 W total assigned to	64/128 kbps data[a]	15	2.4	64.4
CPICH and CCCHs)	64/384 kbps data[a]	6	0.9	64.5
ROC 1 + 1 + 1	12.2 kbps speech	56	18.0	17.0
20 W shared between sectors	64/64 kbps data	7	11.0	15.0
(4.5 W assigned to	64/128 kbps data[a]	4	0.6	14.1
CPICH and CCCHs)	64/384 kbps data[a]	1	0.2	15.0
ROC 1 + 1 + 1	12.2 kbps speech	95	30.7	29.0
40 W shared between sectors	64/64 kbps data	12	19.0	26.1
(9 W assigned to	64/128 kbps data[a]	7	1.1	25.9
CPICH and CCCHs)	64/384 kbps data[a]	2	0.4	26.1

[a] Includes an activity factor ratio of 1 : 10 for uplink-to-downlink traffic channel activity.

These tables demonstrate the principles described in Section 6.4 – i.e., that as the allowed propagation loss increases the downlink capacity becomes dominated by the BS transmit power capability rather than the level of downlink loading. This means the BS runs out of power before reaching the 'elbow' in the exponential rise in interference floor. When the cell range is small the elbow in the exponential is reached before the BS runs out of power and the subsequent sharp increase in interference floor means that the BS runs out of power relatively independently of its transmit power capability. Table 6.25 illustrates the fact that when planning for 64 kbps uplink coverage the 20 W ROC configuration's capacity is ~35% of the conventional configuration.

Table 6.26 illustrates that when the cell range is increased the capacity becomes more sensitive to the BS transmit power capability and the 20 W ROC configuration has a capacity of approximately 25% of the conventional configuration.

The results for the 40 W ROC configuration demonstrate that larger allowed propagation loss figures lead to greater relative increases in capacity as the transmit power is increased. It is evident that the cell capacity of an ROC BS is almost always downlink capacity limited. The only result that indicates the possibility of an uplink capacity limited system is the speech row for the 40 W ROC configuration. In this case the uplink loading figures are 43.4% and 30.7%. This means that if the radio network has been planned for 30% loading and the traffic is dominated by speech users then the cell capacity will be uplink limited. In this case, there is no loss in capacity by using the ROC configuration compared with the conventional configuration. This makes the ROC configuration particularly applicable to rural scenarios where the network has been planned for a relatively low uplink cell load.

6.12.2 Practical Considerations

The antenna sub-system and cabinet requirements for an ROC BS are similar to those of a standard BS with the addition of a splitter to divide the downlink power between sectors. This chapter focused upon describing a three-sector ROC configuration. The same principle may be applied to any number of sectors. Two-sector ROC sites are often appropriate providing coverage along roads. The reduced cost of ROC BSs must be balanced against the relatively low capacity and the need for future upgrades.

6.13 Sectorisation

The term 'sectorisation' refers to increasing the number of sectors belonging to a site. Sectorisation is used primarily as a technique to increase system capacity, although service coverage is generally improved at the same time. This is a result of the increased antenna gain associated with more directional antennas. Antenna selection is a critical part of planning for increased sectorisation. Levels of inter-cell interference and soft handover overhead must be carefully controlled. For example, upgrading a three-sector site to a six-sector site does not involve simply rigging an additional three antennas but also changing the original three. For this reason it is useful to plan the requirement for high sectorisation during initial system rollout. It may be advantageous to deploy

Table 6.27 The application of various levels of sectorisation.

Level	Application
1 sector	Micro-cell or low capacity macro-cell
2 sectors	Sectored micro-cell or macro-cell providing roadside coverage
3 sectors	Standard macro-cell configuration providing medium capacity
4 or 5 sectors	Not commonly used but may be chosen to support a specific traffic scenario
6 sectors	High capacity macro-cell configuration

highly sectorised configurations during initial rollout to reduce the requirement for subsequent upgrades.

Increasing the number of sectors at a BS places a greater requirement upon the quantity of hardware required within the BS cabinet. In general, doubling the number of sectors will require twice as many transceiver modules, twice as many power amplifier modules and twice as much baseband processing capability. If the site uses multiple carriers and multi-carrier power amplifiers, the existing transmit power may be shared across carriers. For example, a $2 + 2 + 2$ site configured with dedicated 20 W multi-carrier power amplifiers for each carrier of each cell can be upgraded to a $2 + 2 + 2 + 2 + 2 + 2$ configuration without increasing the requirement for power amplifier modules. The existing six power amplifiers may be shared across the carriers belonging to each cell, such that 10 W are available to each carrier in each cell. The configurations associated with various degrees of sectorisation are presented in Table 6.27.

6.13.1 Impact of Sectorisation

The most important factor influencing the system performance of a sectorised site is the choice of antenna. To a large extent this determines the levels of inter-cell interference, soft handover overhead and any changes in the maximum allowed propagation loss. System capacity is directly affected by all three. Service coverage is affected by changes in the maximum allowed propagation loss. Table 6.28 presents a set of typical figures for the sectorisation of both macro-cells and micro-cells.

Micro-cell sectorisation does not normally exceed two sectors. Antennas must be placed with extreme care to ensure adequate isolation between cells. The nature of micro-cellular radio propagation means that simply pointing antennas in different directions is not sufficient to ensure clearly defined dominance areas with adequate inter-cell isolation.

In the case of macro-cells, it is common to consider up to six sectors per site. As the level of sectorisation increases then so too does the associated antenna gain and level of inter-cell interference. Antenna side lobes are also likely to be greater for more directional antennas. The soft handover overhead should be maintained at approximately 30% with the help of the relevant RRM parameters – e.g., defining the active set size and soft handover window.

Table 6.28 Typical antenna, inter-cell interference and soft handover overhead assumptions for various levels of sectorisation.

Cell type	Level of sectorisation [sectors]	Typical antenna beamwidth and gain [°/dBi]	Typical inter-cell interference ratio	Typical soft handover overhead
Micro-cell	1	65/12.0	25%	20%
	2	65/12.0	Scenario-dependent	Scenario-dependent
Macro-cell	1	360/6.0	55%	30%
	2	90/16.5	60%	40%
	3	65/18.5	65%	40%
	4 or 5	65/18.5	75%	40%
	6	33/21.0	85%	40%

Tables 6.29 and 6.30 present typical downlink capacity figures per site. Uplink load is also presented to illustrate which scenarios are more likely to be uplink capacity limited. The level of downlink load is provided to indicate whether the BS is running out of transmit power due to high levels of system load (>80%) or simply as a result of the number of users combined with the allowed propagation loss. In the latter case, capacity may be increased by increasing BS transmit power capability.

Table 6.29 Impact of sectorisation upon site capacity, based on an allowed propagation loss of **154.4 dB** corresponding to the **64 kbps** uplink data service for the $1 + 1 + 1$ configuration.

Base station transmit power	Service	Downlink capacity per site [users]	Uplink load [%]	Downlink load [%]
Omni 20 W	12.2 kbps speech	83	75.4	76.5
	64/64 kbps data	11	50.0	73.8
	64/128 kbps data[a]	6	2.8	74.1
	64/384 kbps data[a]	2	1.0	73.8
$1 + 1 + 1$ 20 W per cell	12.2 kbps speech	233	75.5	78.1
	64/64 kbps data	31	50.2	75.5
	64/128 kbps data[a]	17	2.8	74.7
	64/384 kbps data[a]	7	1.1	75.5
$1 + 1 + 1 + 1 + 1 + 1$ 20 W per cell	12.2 kbps speech	410	75.7	80.7
	64/64 kbps data	55	50.5	78.4
	64/128 kbps data[a]	31	2.8	78.6
	64/384 kbps data[a]	12	1.1	78.4

[a] Includes an activity factor ratio of 1 : 10 for uplink-to-downlink traffic channel activity.

Table 6.30 Impact of sectorisation upon site capacity, based on an allowed propagation loss of **149.6 dB** corresponding to the **384 kbps** uplink data service for the $1 + 1 + 1$ configuration.

Base station transmit power	Service	Downlink capacity per site [users]	Uplink load [%]	Downlink load [%]
Omni 20 W	12.2 kbps speech	99	89.5	90.8
	64/64 kbps data	14	60.7	89.5
	64/128 kbps data[a]	8	3.4	89.3
	64/384 kbps data[a]	3	1.3	89.5
$1 + 1 + 1$ 20 W per cell	12.2 kbps speech	273	88.3	91.3
	64/64 kbps data	37	59.9	90.1
	64/128 kbps data[a]	21	3.4	90.1
	64/384 kbps data[a]	8	1.3	90.1
$1 + 1 + 1 + 1 + 1 + 1$ 20 W per cell	12.2 kbps speech	471	86.9	92.7
	64/64 kbps data	65	59.1	91.6
	64/128 kbps data[a]	36	3.3	90.7
	64/384 kbps data[a]	14	1.2	91.6

[a] Includes an activity factor ratio of $1 : 10$ for uplink-to-downlink traffic channel activity.

In each case, increasing the sectorisation from a single sector to three sectors leads to a capacity increase in the order of 2.8. Similarly, increasing the sectorisation from three sectors to six sectors leads to a capacity gain of approximately 1.8. Decreasing the cell's maximum allowed propagation loss means that more users can be supported before the BS runs out of transmit power. This is due to relatively low levels of downlink load as shown in Table 6.29. Table 6.30 indicates higher levels of downlink load. In this case, further reducing the allowed propagation loss or increasing the BS transmit power will not increase site capacity. Here, capacity can only be increased by enhancing some parameters within the downlink load equation – i.e., reducing the E_b/N_0 requirement or reducing inter-cell interference. The uplink load column illustrates the fact that when the traffic profile is dominated by speech or symmetric data services, there is a high likelihood of site capacity being uplink limited.

6.13.2 Practical Considerations

Deploying highly sectorised sites requires a correspondingly high quantity of hardware in terms of both the antenna sub-system and modules to be fitted within the BS cabinet. A single-carrier 6-sector site taking advantage of dual-branch receive diversity requires 6 crosspolar antennas, 12 runs of feeder cable, potentially 12 MHAs, 6 transceiver modules, 6 power amplifier modules and a significant quantity of baseband processing capability. Configuring an additional carrier at the site would require another 6 transceiver modules, potentially another 6 power amplifiers and twice as much baseband processing. If the power amplifiers are multi-carrier then it is feasible to share the original 6 modules between the 2 carriers with some loss in capacity. In some cases the additional transceivers and power amplifiers may require a second BS cabinet. Alternatively, standard transceiver modules can be upgraded to double-transceiver modules and 20 W power amplifier modules can be upgraded to 40 W modules.

6.14   Repeaters

Repeaters may be used to enhance or extend an area of existing macro-cell coverage. The repeater coverage area may be either an outdoor or indoor location. Repeaters are generally connected to their donor cell via a directional radio link. Using a directional radio link helps to provide favourable performance in terms of maximising antenna gain and minimising any interference and multi-path effects. In some cases an optical link may be used to connect the repeater to the donor cell. Repeaters are transparent to their donor cell, which is able to operate without needing to know whether or not a repeater is present. Inner-, outer- and open-loop power control algorithms are able to function transparently through the repeater. The main benefits of a repeater solution are the low cost and ease of installation. An important consideration when deploying a repeater for macro-cell coverage is configuring uplink and downlink repeater gains. The majority of repeaters allow configuring uplink and downlink gains independently. Downlink gain is typically configured relatively high to maximise the downlink coverage of the repeater. If uplink gain is also configured high then the donor cell may be desensitised by the thermal noise floor of the repeater. A repeater's uplink gain should usually be about 10 dB less than the link loss between the repeater and the donor cell. If the difference between uplink and downlink gains becomes too great then there is likely to be an impact upon soft handover performance. There is thus a requirement to balance the tradeoff between repeater coverage, donor cell desensitisation and soft handover performance. Multiple repeaters can be daisy-chained to extend areas of coverage beyond that feasible using a single repeater, but the inserted delays put a practical upper limit on the number of repeaters in a chain. Figure 6.18 illustrates the concept of using a repeater.

In general, digital repeaters have the advantage of allowing the received signal to be cleaned before retransmission by making hard decisions on the bit stream. In the case of

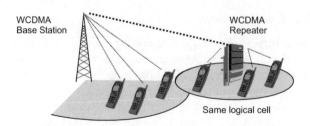

Figure 6.18   The concept of using a repeater.

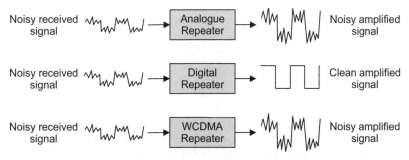

Figure 6.19 A comparison of analogue, digital and WCDMA repeaters.

WCDMA repeaters, the repeater cannot clean the bit stream unless it first applies scrambling and channelisation codes. The repeater has no knowledge of either of these and is forced to simply amplify the received signal plus noise in the same way as an analogue repeater. A comparison of the various types of repeater is illustrated in Figure 6.19.

Passing the WCDMA signal through two receiver sub-systems plus an additional transmitter degrades signal quality. This impacts directly upon the receiver E_b/N_0 requirement and indirectly upon system capacity and service coverage performance. If the system capacity is uplink limited then the capacity will be degraded by the repeater. If the system capacity is downlink limited then the impact upon capacity will depend upon the link budget between the donor cell and the repeater, the transmit power capability of the repeater, the allowed propagation loss between the mobile terminal and the repeater and the distribution of the traffic between the donor cell and the repeater. The majority of WCDMA BSs have dual-branch receive diversity whereas many repeaters do not have this functionality. This results in an increased fast fading margin and a greater uplink E_b/N_0 requirement. This further impacts upon the link budget for the coverage area of the repeater as well as the uplink capacity of the donor cell.

Soft handover does not occur between the donor cell and the repeater. This is because both belong to the same logical cell and transmit the same downlink signal with the same scrambling code. Mobile terminals located within the boundary area between the donor cell and the repeater may incur high levels of multi-path generated by the two sources of downlink transmission power and a corresponding loss in channelisation code orthogonality. Table 6.31 presents a typical specification for a WCDMA repeater.

Similar to the donor cell, the downlink transmit power must be sufficient to support the capacity requirements of the TCHs while reserving an allocation for the CPICH and

Table 6.31 Typical specification for a WCDMA repeater.

Downlink transmit power	Uplink transmit power	Delay	Uplink noise figure	Size	Weight
5.00 W	0.25 W	5 μs	3 dB	50 cm × 40 cm × 30 cm	25 kg

CCCHs. Repeaters introduce a delay in both uplink and downlink directions in the order of 5 μs. This delay is small enough – relative to the period of a slot (667 μs) – to be transparent to the performance of the inner-loop power control.

6.14.1 Impact of Repeaters

Repeaters are used primarily for extending the coverage area of an existing cell. The link budget performance of the donor cell remains unchanged. A second set of link budgets must be completed for the coverage area of the repeater. These link budgets are likely to be quite different from that of the donor cell. The parameters most likely to differ include E_b/N_0 requirement, receiver NF, antenna gain, cable loss and fast fading margin. Table 6.32 describes how these parameters may differ between the donor cell and the repeater. In addition, the difference between repeater gain and repeater-to-donor cell link loss should be accounted for within the link budgets. The combined effect of these parameters is likely to result in a lower maximum allowed propagation loss for the repeater when compared with the donor cell.

The impact of a repeater upon system capacity depends upon whether capacity is uplink or downlink limited. If it is uplink limited, there will be a loss of capacity by using a repeater. This is a direct result of the increased uplink E_b/N_0 requirement for those users linking to the donor cell via the repeater. The increased requirement depends largely upon whether or not the repeater benefits from receive diversity. Table 6.33 illustrates a typical loss in capacity when introducing a repeater to an uplink capacity limited cell.

In the case that system capacity is downlink limited, both the downlink load equation and downlink link budgets must be considered. The downlink link budgets include that of the donor cell as well as that of the repeater and the directional radio link between donor cell and repeater. The users linked to the donor cell via the repeater will have an increased E_b/N_0 requirement. This will increase the downlink loading of both the repeater and the donor cell. The increase in downlink cell loading will tend to decrease system capacity. In addition, the users located at the boundary area between the donor cell and repeater are likely to incur high levels of multi-path and

Table 6.32 Differences between link budgets of donor cell and repeater.

Factor	Difference
Uplink E_b/N_0 requirement	Repeater requires increased E_b/N_0, especially if it does not benefit from receive diversity
Receiver noise figure	Depends upon the repeater's receiver design
Receiver antenna gain	Depends upon scenario. Repeaters used to extend coverage along a road may use directional antennas
Feeder loss	Depends upon scenario
Fast fading margin	Repeater requires increased margin, especially if it does not benefit from receive diversity

Table 6.33 Impact upon uplink capacity in terms of speech users when a repeater is added to a cell planned for 30% uplink loading.

Service	E_b/N_0 requirement for users connected to donor cell [dB]	E_b/N_0 requirement for users connected via the repeater [dB]	Uplink capacity per cell[a] [users]
Three-sector site without repeater	4	—	30
Three-sector site with repeater benefiting from receive diversity	4	5	28
Three-sector site with repeater not benefiting from receive diversity	4	6	24

[a] Assuming an equal share of traffic between repeater and donor cell and no change in inter-cell interference when a repeater is included.

a corresponding loss of channelisation code orthogonality. This will also tend to increase downlink cell load and decrease system capacity. However, users linked to the donor cell via the repeater require a relatively low share of BS power as a result of the favourable link budget provided by the repeater gain and the directional radio link between donor cell and repeater.

6.14.2 Practical Considerations

Repeaters are often chosen for their low cost and ease of installation, requiring a minimum of configuration. They don't require any additional transmission links towards the controlling RNC. Their only requirement is a power supply. Repeaters are most applicable in scenarios where there is sufficient power to amplify and where there is relatively clear cell dominance.

6.15 Micro-cell Deployment

The coverage and capacity requirements within urban and dense urban environments lead directly to high site densities. Micro-cells become an attractive solution in terms of relative ease of site acquisition, increased air interface capacity and more efficient indoor penetration. Micro-cells may be realised by one of two generic BS solutions – either a dedicated micro-cell product or a macro-cell product with micro-cellular antenna placement. The dedicated micro-cell product provides the benefits of relative ease of installation and low cost. The macro-cell product provides the benefits of increased transmit power and baseband processing capability. Both solutions can support multiple carriers and multiple cells, although micro-cellular sectorisation is significantly more difficult than that for macro-cells. Both solutions are generally able to support dual-branch uplink receive diversity. Table 6.34 provides a comparison of the two solutions.

Table 6.34 A comparison of micro-cell solutions.

	Dedicated micro-cell product	Macro-cell product with below rooftop antennas
Cabinet	Compact, wall-mounted cabinet	Full-sized base station cabinet
Transmit power	Typically 8 W	Typically 10 W, 20 W or 40 W
Hardware limitations	Moderate processing capability	High processing capability
Cost	Low cost	Relatively high cost

6.15.1 Impact of Micro-cells

The propagation channel associated with a micro-cellular radio environment has a significant impact upon the air interface performance of a micro-cell solution. Micro-cellular propagation usually has a strong line-of-sight component with relatively weak multi-path, leading to high downlink orthogonality and correspondingly reduced intra-cell interference. The low intra-cell interference means that loading is more sensitive to inter-cell interference. However, the typical below-rooftop positioning of micro-cells leads to good inter-site isolation, and inter-cell interference is generally less than that for macro-cells. Good inter-site isolation also helps to manage the soft handover overhead. Table 6.35 presents the main differences between macro-cell and micro-cell capacity-related parameters.

Both the uplink and downlink micro-cell E_b/N_0 requirements are greater than those for a macro-cell. This tends to decrease uplink and downlink air interface capacities. The increased E_b/N_0 requirement is primarily a result of increased fading across the radio channel. This also impacts upon the coverage-related fast fading margin on the uplink. The increase in E_b/N_0 requirement is relatively large on the downlink as a result of the downlink figure including a contribution from the fast fading margin. The uplink increase in inter-cell interference is also greater for micro-cells. This figure combines with the inter-cell interference ratio in the uplink load equation to increase the level of inter-cell interference. For a macro-cell the resultant inter-cell interference is $0.65 + 1 \, dB = 0.82$ and for a micro-cell is $0.25 + 2 \, dB = 0.40$. The micro-cell's resultant uplink inter-cell interference remains significantly lower. The decrease in inter-cell

Table 6.35 Comparison of macro-cell and micro-cell capacity-related parameters.

	Macro-cell	Micro-cell
Uplink E_b/N_0 (12.2 kbps speech)[a]	4 dB	4.5 dB
Increase in inter-cell interference	1 dB	2 dB
Downlink E_b/N_0 (12.2 kbps speech)	6.5 dB	9.5 dB
Downlink orthogonality	0.5	0.9
Inter-cell interference ratio	0.65	0.25
Soft handover overhead	40%	20%

[a] Assumes dual-branch receive diversity for both macro-cell and micro-cell.

Table 6.36 A comparison of macro-cell and micro-cell capacities, based upon a macro-cell allowed propagation loss of 152.2 dB and a micro-cell allowed propagation loss of 144.7 dB (64 kbps uplink link budget with 70% loading) and 20 W assigned to both macro-cells and micro-cells.

	Service	Capacity per cell [users]	Uplink load [%]	Base station transmit power requirement [dBm]
Macro-cell without transmit diversity	12.2 kbps speech	72	70.0	40.4
	64/64 kbps data	11	52.8	42.1
	64/128 kbps data[a]	6	2.9	41.6
	64/384 kbps data[a]	2	1.0	40.4
Micro-cell without transmit diversity	12.2 kbps speech	79	69.9	39.8
	64/64 kbps data	17	66.3	40.7
	64/128 kbps data[a]	12	4.7	42.4
	64/384 kbps data[a]	4	1.6	41.4
Micro-cell with transmit diversity	12.2 kbps speech	79	69.9	37.2
	64/64 kbps data	17	66.3	38.7
	64/128 kbps data[a]	18	7.0	42.7
	64/384 kbps data[a]	7	2.7	42.0

[a] Includes an activity factor ratio of 1:10 for uplink-to-downlink traffic channel activity.

interference combined with the increase in downlink channelisation code orthogonality and decrease in soft handover overhead leads to a net increase in system capacity.

Table 6.36 provides a comparison of macro-cell and micro-cell capacity, assuming both are equipped with 20 W power amplifier modules. The speech service scenario is uplink capacity limited and the difference between the macro- and micro-cell capacities is relatively small – approximately 10%. The 64/64 kbps data service is downlink capacity limited for the macro-cell and uplink capacity limited for the micro-cell. This results in an intermediate capacity gain of approximately 55%. The remaining data services are downlink capacity limited for both the macro-cell and micro-cell scenarios and the capacity gain is 100%. Including downlink transmit diversity as part of the micro-cell solution further increases system capacity for the downlink capacity limited scenarios. The capacity increase is in the order of 70% beyond that of the micro-cell without transmit diversity and in the order of 350% beyond that of the macro-cell.

In practice it is common for micro-cells to have a lower transmit power. Table 6.37 presents the corresponding micro-cell capacities for a transmit power capability of 8 W.

Reducing the micro-cell transmit power to 8 W results in a loss in capacity. The loss is greatest for the downlink capacity limited scenarios.

Tables 6.36 and 6.37 present air interface capacities but take no account of the limitations of the downlink channelisation code tree. Table 6.38 presents these limitations for a micro-cellular environment.

Table 6.37   Micro-cell capacities when assigned 8 W of transmit power capability, based upon an allowed propagation loss of 144.7 dB (64 kbps uplink link budget with 70% loading).

	Service	Capacity per cell [users]	Uplink load [%]	Base station transmit power requirement [dBm]
Micro-cell without transmit diversity	12.2 kbps speech	79	69.9	38.8
	64/64 kbps data	15	58.5	39.0
	64/128 kbps data[a]	8	3.1	38.7
	64/384 kbps data[a]	3	1.2	38.7
Micro-cell with transmit diversity	12.2 kbps speech	79	69.9	35.1
	64/64 kbps data	17	66.3	37.3
	64/128 kbps data[a]	13	6.4	38.7
	64/384 kbps data[a]	5	1.9	38.9

[a] Includes an activity factor ratio of 1 : 10 for uplink-to-downlink traffic channel activity.

Table 6.38   Micro-cell traffic channel limitations of a single channelisation code tree.[a]

Downlink bit rate [kbps]	Air interface bit rate [kbps]	Spreading factor	Number of possible TCHs
12.2	60	128	104
64	240	32	25
128	480	16	12
384	960	8	5

[a] $C_{ch,256,0}$ used for the CPICH; $C_{ch,256,1}$ used for the P-CCPCH; $C_{ch,64,1}$ used for the S-CCPCH; $C_{ch,256,2}$ used for the AICH; and $C_{ch,256,3}$ used for the PICH. Based upon a soft handover overhead of 20%.

Comparing these figures with those presented in Tables 6.36 and 6.37 indicates that the availability of downlink channelisation codes may become a limitation for the 128 kbps and 384 kbps data services when the micro-cell is equipped with 20 W of transmit power and downlink transmit diversity. In these cases a second scrambling code may be introduced to provide a second channelisation code tree. However, this code tree will not be orthogonal to the first and its users will generate relatively large increments in downlink cell loading.

Micro-cell capacity can be increased by adding carriers or sectors in a similar fashion to macro-cells. The performance of sectorisation is, however, significantly more sensitive than that for macro-cells. If the sectors are not well-planned they are not likely to have clearly defined dominance areas and will incur high levels of inter-cell interference.

In terms of service coverage performance, micro-cells provide an effective solution for achieving a high degree of indoor penetration. Cell ranges tend to be smaller as a result of the below-rooftop antenna location and the relatively high gradient of the associated path loss characteristic. Table 6.39 presents the main differences between macro-cell and micro-cell coverage-related link- and system-level parameters.

Table 6.39 Comparison of macro-cell and micro-cell coverage-related parameters.

	Macro-cell [dB]	Micro-cell [dB]
Uplink E_b/N_0 (12.2 kbps speech)	4	4.5
Uplink fast fading margin	3	5
Downlink E_b/N_0 (12.2 kbps speech)	6.5	9.5

The uplink link budget of a micro-cell is characterised by an increased E_b/N_0 requirement and an increased fast fading margin. This results in a lower maximum allowed propagation loss. The downlink link budget is characterised by an increased E_b/N_0 requirement. Micro-cells configured with 8 W of transmit power capability and supporting asymmetric data services are likely to be downlink coverage limited.

Adjacent channel performance must also be considered when planning the deployment of micro-cells. The possibility of a low minimum coupling loss between the micro-cell antenna and users on the adjacent channel results in potentially harsh near–far effects. When the adjacent channel is being used by a second operator, near–far effects are significantly reduced if the second operator also uses that channel to deploy micro-cells.

6.16 Capacity Upgrade Process

There is a requirement for operators to have a process which allows them to identify when a capacity upgrade is necessary. This process should ensure that upgrades are completed prior to the network experiencing increased levels of connection blocking. However, the process should not be triggered too early otherwise it will result in operators increasing their capital expenditure sooner than necessary. Capacity upgrades, which involve changes to the network hardware are generally relatively expensive and should only be completed when necessary. It may be possible to increase system capacity and avoid a capacity upgrade by completing optimisation of the existing resources. Optimisation should always be completed prior to completing a capacity upgrade. Figure 6.20 illustrates an example capacity upgrade process.

RNC counters and Key Performance Indicators (KPIs) are typically used to trigger the capacity upgrade process. Operators should collect and monitor these data on a regular basis. For example, the data could be studied at the end of every week. The data should be recorded with a relatively high time resolution to avoid averaging peaks in traffic demand. If the time resolution becomes too high then the quantity of data becomes unmanageable. It is typical to use a time resolution of either 15 minutes or 1 hour. This time resolution may be greater than that used for other counters and KPIs recorded from the network. The KPIs should allow operators to evaluate whether or not system capacity limits are being approached. KPIs should be defined to quantify all aspects of system capacity. Example aspects of system capacity are uplink DPCH

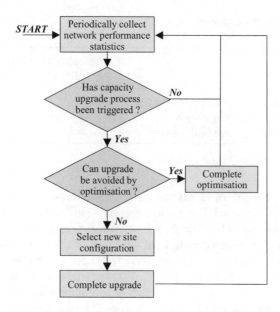

Figure 6.20   Example high-level capacity upgrade process.

capacity, downlink DPCH capacity, PRACH capacity, S-CCPCH capacity, channelisation code capacity, Node B baseband processing capacity and Iub capacity. Any of these could be responsible for triggering the capacity upgrade process. Some may only require changes to the RNC databuild rather than changes to the hardware configuration. For example, a second channelisation code tree could be introduced by allowing the use of a secondary scrambling code, or a second S-CCPCH could be configured. Some aspects of system capacity will be more critical than others. This means that it is likely that there will be a focus upon the most important aspects. Nevertheless, each aspect of system capacity should have its own set of KPIs and its own thresholds for triggering the upgrade process.

RNC counters and KPIs should be studied on a per-cell basis. If a specific cell triggers the capacity upgrade process then the existing performance of that cell should be studied in greater detail. For example, if downlink DPCH capacity has triggered the upgrade process then the dominance and soft handover overhead of the cell should be studied. It may be possible to improve the dominance of the cell and thus allow it to operate either more efficiently or across a smaller geographic area. If the soft handover overhead for the site is relatively high then it may be possible to reduce the number of soft handover connections and thus release some capacity. In both cases it may then be possible to postpone the capacity upgrade. If the BS baseband processing capacity becomes exhausted then it is less likely that optimisation can be used to avoid an upgrade. It may be possible to adjust dominance areas such that less traffic is loading onto the BS whose baseband processing capability has become exhausted. The RNC counters and KPIs should be studied subsequent to any optimisation activity to verify whether the optimisation has been sufficiently effective or an upgrade is in fact necessary.

If it is not feasible for optimisation activities to improve system capacity sufficiently then there is a true requirement for a capacity upgrade. Operators should have a well-defined site configuration upgrade path. For example, single RF carrier ROC sites could be upgraded to single RF carrier conventional sites, and then single RF carrier conventional sites could be upgraded to dual RF carrier conventional sites. Each site configuration should have an associated baseband processing and Iub configuration. A specific site configuration should be selected and the upgrade completed. Similar to the situation following any optimisation activity, the RNC counters and KPIs should be studied to verify that the upgrade has been effective.

Capacity upgrades are likely to have an impact upon the RNC databuild as well as the hardware configuration. For example, if the capacity upgrade is for the addition of a second RF carrier then the RNC databuild needs to be configured to ensure that mobile terminals establish connections on both RF carriers and that the load is relatively evenly distributed. The use of a second RF carrier also introduces the requirement for a new layer of scrambling codes and a new set of neighbour lists. Both intra-frequency and inter-frequency neighbours are required for both RF carriers. In addition, the impact of inter-frequency hard handovers should be evaluated. If ROC sites are being upgraded to conventional sites then the length of inter-system neighbour lists is likely to decrease whereas the length of intra-frequency neighbour lists is likely to increase.

6.17 Summary of Coverage and Capacity Enhancement Methods

Understanding the mechanisms for limitations in service coverage and system capacity forms an essential part of being able to enhance them. Coverage is generally uplink limited, although a low BS transmit power capability combined with asymmetric data services may lead to a downlink coverage limited scenario. Capacity may be either uplink or downlink limited dependent upon the planned level of uplink loading, BS transmit power capability, the traffic loading the network and the performance of the BS and mobile terminals.

Link budgets and load equations are effective at demonstrating the fundamental trends and principles prior to commencing detailed planning. Link budgets are associated with studying service coverage. Capacity analysis requires a combination of link budgets and load equations. Sophisticated WCDMA radio network planning tools are based upon the same type of link budgets and load equations as those used within this chapter.

The site density defined for initial system deployment should account for both present and future coverage and capacity requirements. In terms of capacity, site density should be sufficient to permit capacity upgrades without the requirement for interleaving new sites. This places great importance upon the initial definition of the planned uplink cell load, network traffic assumptions and choice of site configuration. Once the network has been deployed, it is relatively difficult to increase the planned uplink load without having to interleave additional sites to maintain service coverage performance.

The BS transmit power requirement also needs to be planned during initial system dimensioning, although it is relatively easy to upgrade BS transmit power without changing the layout of the radio plan. In general, a BS transmit power of 20 W is appropriate. The impact of exceeding 20 W is dependent upon the cell's maximum allowed propagation loss and the level of downlink load. If downlink load has reached the 'elbow' of its exponential characteristic there is little to be gained from increasing transmit power capability.

Additional carriers form the simplest and most effective way of increasing system capacity. When a BS whose capacity is downlink limited has limited transmit power capability, system capacity is maximised by sharing power across the maximum number of carriers. A trunking gain can also be achieved if RRM supports inter-carrier load control.

Additional scrambling codes become applicable when the system capacity becomes limited by the number of downlink channelisation codes. This is most likely to occur in micro-cell scenarios where the air interface capacity is relatively high. Users allocated channelisation codes under the second scrambling code are not orthogonal to those under the first scrambling code and therefore generate relatively large increases in downlink load.

MHAs and active antennas improve uplink coverage performance by reducing the composite NF of the BS receiver sub-system. Coverage gain is dependent upon the receiver sub-system architecture and the associated feeder loss. The benefit is greatest when the feeders are shared with the GSM. If system capacity is downlink limited then MHAs or active antennas will decrease system capacity. The loss in capacity is typically between 6% and 10%.

Remote RF head amplifiers allow the physical separation of a BS's RF and baseband modules, allowing cells to be located at locations which would otherwise require prohibitively long runs of feeder. Both the uplink and downlink link budgets are improved, meaning that coverage performance increases without a loss in capacity – i.e., the maximum allowed propagation loss increases but so too does the BS EIRP. This is in contrast to the MHA solution, which increases the maximum allowed propagation loss but reduces the BS EIRP as a result of insertion loss.

Higher order uplink receive diversity reduces the BS's E_b/N_0 requirement. The E_b/N_0 requirement appears in both the link budget and load equation, meaning that uplink coverage and capacity are simultaneously improved. Coverage gain tends to be greater than that for MHAs because the uplink link budget benefits from a reduced E_b/N_0 requirement as well as a reduced uplink load and a corresponding decrease in inter-ference margin. If system capacity is downlink limited then capacity will be reduced by the inclusion of higher order receive diversity. The loss will be less than that for MHAs, since there is no insertion loss reducing the BS's EIRP.

Downlink transmit diversity impacts upon the mobile terminal's E_b/N_0 requirement, channelisation code orthogonality and MDC gain. The net result is an increase in downlink system capacity in the order of 35% for macro-cells and 70% for micro-cells. There is no impact upon the uplink link budget. If service coverage is downlink limited, as may be the case for micro-cells, transmit diversity also improves service coverage performance.

The standardisation of MIMO is still ongoing in 3GPP TSG RAN WG1 where discussions are focused on HSDPA. It has turned out that the evaluation of various MIMO schemes is a very challenging task, and only a draft 3GPP specification document exists at the moment. Since MIMO is not a mature technique in UTRA FDD detailed conclusions cannot yet be drawn. In the UTRA framework, the feasibility of different MIMO algorithms varies between uplink and downlink since downlink capacity is code limited while uplink capacity is interference limited.

Whereas higher order receive diversity improves uplink performance and transmit diversity improves downlink performance, beamforming improves both uplink and downlink performance. Beamforming solutions are able to provide an increase in system capacity by limiting the aperture of transmitted and received signals. Beamforming solutions exist for either fixed or user-specific beams. In the uplink direction the reduction in E_b/N_0 requirement can be more than 2.5 dB beyond that provided by four-branch receive diversity. This has implications upon both uplink coverage and capacity. Likewise in the downlink direction, the reduction in E_b/N_0 requirement can be significantly greater than that provided by dual-antenna transmit diversity.

The ROC allows a BS to share power amplifiers between cells. Doing so generally reduces site capacity but also reduces the requirement for power amplifiers and the associated capital expenditure. The uplink of an ROC BS appears identical to that of a standard BS. For some uplink capacity limited scenarios the use of ROC may not affect system capacity. This is dependent upon the level of cell loading and the maximum allowed propagation loss.

Sectorisation is used primarily as a technique to increase system capacity, but service coverage is generally improved at the same time. Antenna selection is a critical part of planning for increased sectorisation. Levels of inter-cell interference and soft handover must be carefully controlled. Increasing the sectorisation from three to six sectors leads to a capacity gain of approximately 1.8. Micro-cell sectorisation is more difficult in terms of being able to achieve good inter-cell isolation. Micro-cells do not normally have more than two sectors.

Repeaters transparently extend the coverage area of an existing cell. An important consideration when deploying a repeater for macro-cell coverage is configuring the uplink and downlink repeater gains. Downlink gain is typically configured relatively high to help maximise the downlink coverage of the repeater. If uplink gain is also configured high then the donor cell may be desensitised by the thermal noise floor of the repeater. If the difference between uplink and downlink gains becomes too great then there is likely to be an impact upon soft handover performance. A large number of repeaters do not take advantage of uplink receive diversity. This leads to an increase in the uplink E_b/N_0 requirement. If system capacity is uplink limited, the capacity will be degraded by the repeater. If it is downlink limited, the impact upon capacity will depend upon the link budget between the donor cell and the repeater, the repeater gain, the allowed propagation loss associated with the repeater's coverage area and the distribution of the traffic between the donor cell and the repeater.

Micro-cells provide a high-capacity solution particularly suitable for urban and dense urban environments where there is a requirement for high site densities and macro-cell site acquisition becomes difficult. Micro-cells are characterised by increased E_b/N_0 requirements and fast fading margins but also increased channelisation code orthogonality and reduced levels of inter-cell interference and soft handover overhead. Micro-cells typically have twice the air interface capacity of equivalent macro-cells when configured with an equal transmit power. More effective in-building penetration is achieved by having below-rooftop antennas.

Operators should have a process which allows them to identify when a capacity upgrade is necessary. This process should ensure that upgrades are completed prior to the network experiencing increased levels of connection blocking. It is common to trigger the process using RNC counters and KPIs. It may be possible to increase system capacity and avoid a capacity upgrade by completing optimisation of existing resources. Optimisation should always be completed prior to a capacity upgrade. If a capacity upgrade is required then a new site configuration should be selected from a pre-defined list and the RNC databuild updated appropriately.

References

[1] Sipilä, K., Honkasalo, Z., Laiho-Steffens, J. and Wacker A., Estimation of capacity and required transmission power of a WCDMA downlink based on a downlink pole equation. *Proc. VTC 2000 Spring Conf., Tokyo, Japan, May 2000*, pp. 1002–1005.

[2] Laiho, J., Wacker, A. and Sipilä K., Verification of 3G radio network dimensioning rules with static network simulations. *Proc. VTC 2000 Spring Conf., Tokyo, Japan, May 2000*, pp. 478–482.

[3] Friis, H.T., Noise figures of radio receivers. *Proc. IRE, July 1944*, pp. 419–422.

[4] Friis, H.T., Discussion on noise figures of radio receivers. *Proc. IRE, February 1945*, pp. 125–127.

[5] Lempiäinen, J. and Laiho-Steffens, J., The performance of polarisation diversity schemes at a base station in small/micro-cells at 1800 MHz. *IEEE Transactions on Vehicular Technology*, **VT-47**(3), August 1998, pp. 1087–1092.

[6] Ylitalo, J. and Tiirola, E., Performance evaluation of different antenna array approaches for 3G CDMA uplink. *Proc. VTC 2000 Spring, Tokyo, Japan, May 2000*, pp. 883–887.

[7] Pedersen, K.I., Antenna arrays in mobile communications. Ph.D. dissertation, Center for Person-Kommunikation, Aalborg University, Aalborg, Denmark, 2000.

[8] 3GPP, Technical Specifications 25.101 and 25.211–25.214, Release 5, June 2004–January 2005.

[9] Lee, W.C.Y., *Mobile Communications Engineering*, McGraw-Hill, 1982.

[10] Gerlach, D., Adaptive transmitting antenna arrays at the base station in mobile radio networks. Ph.D. dissertation, Information Systems Laboratory, Stanford University, Stanford, CA, June 1995.

[11] Alamouti, S., A simple transmit diversity technique for wireless communications. *IEEE Journal on Selected Areas in Communications*, **SAC-16**(8), October 1998, pp. 2305–2314.

[12] Tarokh, V., Seshadri, N. and Calderbank, A., Space-time codes for high data rate wireless communication: Performance criterion and code construction. *IEEE Transactions on Information Theory*, **IT-44**(2), March 1998, pp. 451–460.

[13] Holma, H. and Toskala, A. (eds), *WCDMA for UMTS* (3rd edn). John Wiley & Sons, 2004.

[14] Andersen, S., Hagerman, B., Dam, H., Forssen, U., Karlsson, J., Kronestedt, F., Mazur, S. and Molnar, K.J., Adaptive antennas for GSM and TDMA systems. *IEEE Personal Communications*, June 1999, pp. 74–86.

[15] Tiirola, E. and Ylitalo, J., Performance evaluation of fixed-beam beamforming in WCDMA downlink. *Proc. of VTC 2000 Spring Conf., Tokyo, Japan, May 2000*, pp. 700–704.

[16] Liu, Z. and Zarki, M., SIR based call admission control for DS-CDMA cellular systems. *IEEE Journal on Selected Areas in Communications*, **SAC-12**(4), May 1994, pp. 638–644.

[17] Dziong, Z., Jia, M. and Mermelstein, P., Adaptive traffic admission for integrated services in CDMA wireless access networks. *IEEE Journal on Selected Areas in Communications*, **SAC-14**(9), December 1996, pp. 1737–1747.

[18] Ramiro-Moreno, J., Pedersen, K. and Mogensen, P., Directional power based admission control for WCDMA systems using antenna arrays. *Proc. VTC 2001 Spring Conf., Rhodes, Greece, May 2001*, pp. 53–57.

[19] 3GPP, Technical Specification 25.211, Physical Channels and Mapping of Transport Channels Onto Physical Channels (FDD), v5.6.0, September 2004.

[20] Foschini, G.J. and Gans J.M., On limits of wireless communications in a fading environment when using multiple antennas. *Wireless Personal Commuinications*, **6**, March 1998, pp. 311–335.

[21] Foschini, G., Layered space-time architecture for wireless communication in a fading environment when using multi-element antennas. *Bell Labs Technical Journal*, Autumn 1996, pp. 41–59.

[22] Telatar, E., *Capacity of Multi-antenna Gaussian Channels*. AT&T Bell Laboratories, Technical Memo, June 1995.

[23] 3GPP, Technical Report 25.876, Multiple Input Multiple Output in UTRA, May 2004.

[24] 3GPP, Technical Report 25.996, Spatial Channel Model for Multiple Input Multiple Output (MIMO) Simulations, September 2003.

[25] Gesbert, D., Shafi, M., Shiu, D., Smith, P.J. and Naguib, A., From theory to practice: An overview of MIMO space-time coded wireless systems. *IEEE Journal on Selected Areas in Communications*, **SAC-21**(3), April 2003, pp. 281–302.

[26] Fonollosa, J.R., Gaspa, R., Mestre, X., Pages, A., Heikkilä, M., Kermoal, J.P., Schumacher, L., Pollard, A. and Ylitalo, J., The IST METRA project. *IEEE Communications Magazine*, July 2002, pp. 78–86.

[27] Varanasi, M.K. and Guess, T., Optimum decision feedback multi-user equalisation with successive decoding achieves the total capacity of the Gaussian multiple-access channel. *Asilomar Conf. on Signals, Systems and Computers, November 1997*, pp. 1405–1409.

[28] Chung, S.T., Lozano, A. and Huang, H., Approaching eigenmode BLAST channel capacity using V-BLAST with rate and power feedback. *Proc. VTC Fall 2001 Conf., Atlantic City, New Jersey, October 2001*, pp. 915–919.

[29] Hämäläinen, J., Pajukoski, K., Tiirola, E., Wichman, R. and Ylitalo, J., On the performance of multi-user MIMO in UTRA FDD uplink. *EURASIP Journal on Wireless Communications and Networking*, Special Issue on Multi-user MIMO Networks, **2**, 2004, pp. 297–308.

7

Radio Network Optimisation Process

Jaana Laiho, Markus Djupsund, Anneli Korteniemi, Jochen Grandell and Mikko Toivonen

7.1 Introduction to Radio Network Optimisation Requirements

The operator business landscape has experienced a change during the last years. Third Generation (3G) networks are already commercially launched and the transition from voice to data services is at hand. The operator focus is moving from long-term technology strategies to shorter term revenue generation opportunities. There is a strong need to utilise existing GPRS (General Packet Radio Service) networks effectively and at the same time tune 3G networks and 3G services towards value-generating machinery. This work is supported by realistic business plans in terms of both future service demand estimates and the requirement for investment in network infrastructure. These are supported by system dimensioning tools capable of assessing both the radio access and the core network components. Having found an attractive business opportunity, system deployment must be preceded by careful network planning. The planning tool must be capable of accurately modelling the system behaviour when loaded with the expected traffic profile. Further, effective measurement-based feedback loops are the core of efficient network operation. The rapid transition from prediction-based performance estimation to measured facts about the network and service performance are the essence of operational efficiency.

UMTS traffic classes and user priorities, as well as the Radio Access Technology (RAT) itself, form the two most significant challenges in deploying a WCDMA-based 3G system. For 3G networks, the operators' task is to find a feasible capacity–coverage tradeoff and still provide competitive services. Also, a Network Management System (NMS) should identify not only a lack of capacity in the current network but also the potential for introducing data services where they currently do not exist. In [1] some of the issues relevant to 3G planning and management are listed:

- introduction of multiple services;
- Quality of Service (QoS) requirements;
- modelling of traffic distributions (e.g., traffic hotspots);

Radio Network Planning and Optimisation for UMTS Second Edition
Edited by J. Laiho, A. Wacker and T. Novosad © 2006 John Wiley & Sons, Ltd

- mobility impact on planning;
- hierarchical cell structures, and other special cell types;
- site synthesis;
- increasingly important role of the NMS.

When provisioning 3G network and services the control for the access part can be divided into three levels. These control levels are depicted in Figure 7.1.

The highest control layer in Figure 7.1 is for *statistical Non-Real Time (NRT) optimisation* and radio network performance tuning based on measured data from network elements. Measurements are combined with a cost function, and furthermore the output of the cost function is optimised. Optimisation is realised by tuning the configuration parameter settings. Automated support is needed for the cost function optimisation process as well as in configuration and measurement data retrieval. This is done inside the NMS. This loop statistically controls the behaviour of the other lower-level control loops closer to Network Elements (NEs). The loop also enables an automated troubleshooting process when performance faults can be corrected fast by delivering the information of alarms or reports to the optimisation engine and Configuration Management (CM) between NMS modules. Faults and monitored performance data can be easily passed to 'Optimizer' (in Figure 7.2) for further analysis, verification

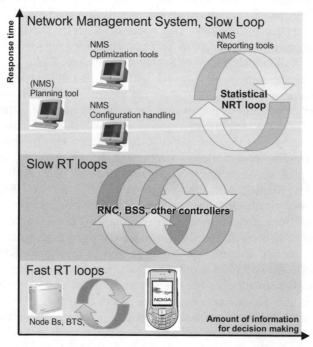

Figure 7.1 Hierarchy in the optimisation loops in a cellular network. *Note*: As much of the automation/optimisation as possible should happen at the low hierarchy layers. In this figure the pre-operational loop is combined with the network management system statistical optimisation loop.

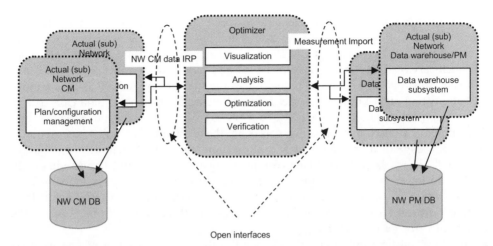

Figure 7.2 Interfaces required for network-wide optimisation (or automation) based on network statistics. In addition, an interface to alarm data can be provided. Alarms can be used as a pre-filter for optimisation.

and problem solving. Configuration data are available from 'Configurator' which also provides means for change implementation and provisioning. By adding centralised task management the whole process can be automated. However, the user keeps the control, defines the targets, approves results and schedules the implementation.

Optimisation can be targeted to improve the Radio Resource Utilisation (RRU) rate or to change the operating point on the capacity–coverage–cost tradeoff curve. Optimisation is also involved when the operator realises a need to enhance the network in terms of new sites or services, change in service provisioning, etc.

The role of optimisation is to provide automated or manual means to improve the performance of the network. The task of optimisation is to understand and translate the relationship between measured network performance and set QoS targets. The NMS level optimisation loop must interface to network configuration and measurement data as in Figure 7.2. 'Data warehouse' represents the interface to any measurement performed in the network in any network element. CM represents the database in which all the configuration parameters controlling the network are collected.

Additionally, a layer for the *pre-operational mode* can be modelled. This loop can be placed at the same level as the statistical network level optimisation loop. Pre-operational mode can be further split into two phases: initial planning (dimensioning) and detailed radio network planning. Pre-operational mode planning provides the first values for the performance iteration done with the statistical optimisation loop.

In a cellular system where all the air interface connections operate on the same carrier, the number of simultaneous users directly influences the receivers' noise floors. Therefore, in the case of UMTS the planning phases cannot be separated into coverage and capacity planning. For post-2G systems data services start to play an important role. The variety of services requires the whole optimisation process to overcome a set of modifications. One of these is related to Quality of Service (QoS) requirements. So far it has been adequate to specify the speech coverage and blocking probability only, but increasingly the indoor and in-car coverage probabilities have to be considered too. In the case of UMTS the problem is slightly more multi-dimensional. For each service, QoS targets have to be set and naturally also met. In practice this means that the tightest requirement determines the site density. As well as the coverage probability and other known measures for real time traffic, the packet data QoS criteria have to be assigned. The NRT traffic QoS is related to acceptable delays and throughput. Estimation of existing delays in the network planning phase requires good knowledge of user behaviour and understanding of the functions of the packet scheduler. Issues related to radio network planning are discussed in Chapter 3 in depth.

The two lower layers in Figure 7.1 consist of the real time feedback loops in Base Stations (BSs) and controllers – like the Radio Network Controller (RNC) and Base Station Controller (BSC). The main differentiator in these two real time loops is the time needed for decision making:

- The slow real time optimisation loop handles the dynamic control of system inter-working, self-regulation of radio network parameters (like load thresholds), etc. Depending on the functional split of network-controlling functions this loop can be placed into the actual NEs or it can be positioned in the NMS. The main benefit of utilisation of the NMS is the possibility to utilise statistical data covering the whole network area.
- Fast real time control loops are related to fast power control, fast congestion control, link adaptation and channel allocation. It is important to notice that this loop has an impact on the radio network planning process in terms of modelling power control and handover behaviour, etc.

Real time loops are also called RRM algorithms. RRM consists of a set of algorithms for admission control, power control, handover control, etc. and it is responsible for providing reasonable operation of the network. This is achieved by providing default parameter sets to control the network operating point in terms of capacity–coverage–cost – CAPEX or OPEX tradeoff. In short, this means that the operator needs to make business decisions related to the QoS. For example, does one offer high quality with reduced capacity, or does one aim for an expensive infrastructure that also has high coverage for high bit rate users, etc. Fast feedback loops in radio access network elements can be considered as adaptive RRM.

The statistical optimisation loop is needed to change the limits controlling RRM so that the network operating point is optimum in terms of capacity and quality. The capacity–quality tradeoff and interaction of optimisation and RRM is illustrated in Figure 7.3.

Real time tuning, which is related to RRM, was the subject of Chapter 4 and statistical tuning is the main scope of this chapter. Advanced optimisation methods in terms of effective visualisation means and automation are introduced in Chapter 9.

7.1.1   The Operations System's Role in the Optimisation Process

The high-level requirements for telecommunication management are introduced in [12]. Figure 7.4 shows the operation system's role in management interactions. The NMS is a

Figure 7.3 Capacity–quality tradeoff management. The task of the operator is to support the business strategy with correct weighting of the performance space. Radio resource management provides the upper bounds for the outer triangle; optimisation changes the shape of the inner triangle to support the operators' strategy.

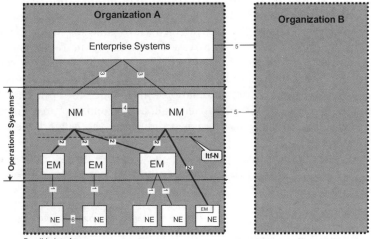

Possible interfaces:
1) Between the Network Elements (NEs) and the Element Manager (EM) of a single PLMN Organisation.
2) Between the EM and the Network Manager (NM) of a single PLMN Organisation.
NOTE: In certain cases the EM functionality may reside in the NE in which case this interface is directly from NE to NM. These management interfaces are given the reference name Itf-N and are the primary target for standardisation.
3) Between the NM and the Enterprise Systems (ES) of a single PLMN Organisation.
4) Between the NMs of a single PLMN Organisation.
5) Between ES and NMs PLMN Organisations.
6) Between NEs.

Figure 7.4 Management system interactions according to [12].

service that employs a variety of tools, applications and devices to assist human network managers in monitoring and maintaining operational networks. It can also be defined as the execution of the set of functions required for controlling, planning, allocating, deploying, coordinating and monitoring the resources of a telecommunications network. The process defined in the enhanced Telecom Operations Map (eTOM) is supported by NMSs.

The Element Management (EM) system is a prerequisite for the NMS. EM is perceived as lower-level management, targeted to specific NEs. It allows local and manual configuration changes and is responsible for collecting and transferring performance measurements and generated alarms/events to the NMSs. It should be noted that, depending on an operator's needs, measurement results may have to be transferred to EM alone, Network Management (NM) alone, or both. Depending on a vendor's implementation, measurement results may be transferred to NM directly from the NE or via EM. The distinction between EM and NM is implementation and thus vendor-specific, so no clear definition exists.

In the context of this book the NMS contains the configuration and performance data repositories, and allows mass modifications in NE configurations. Further, the NMS is capable of accessing alarm information, contains the analysis and optimisation logic to operate the network and provides support to provision and monitor mobile services. The logic to conclude Key Quality Indicators (KQIs) as described in [17] is located in the NMS.

From an optimisation perspective, the NMS has a key role in service-level optimisation and network tuning based on network-wide and/or statistical measurements. The granularity in how the measurements are collected from the network depends on the optimisation case. QoS-related monitoring in radio and Internet Protocol (IP) networks can occur at service level, at utilisation level (per cell or per router) and per Key Performance Indicator (KPI). Further, network diagnosis can be based on active measurements. More automation will be added for both statistical and real time optimisation. QoS control, capacity allocation and resource management policy will all be optimised using measurement data from the network elements and application servers.

The high-level description of the optimisation cycle in operational mode is depicted in Figure 7.5. The process starts with quality definition. The overall end-to-end quality

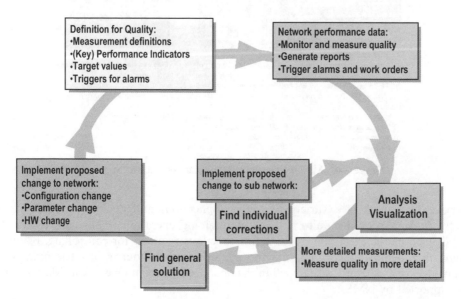

Figure 7.5 Optimisation process, network quality cycle. Actions from enhanced Telecom Operations Map assurance and fulfilment processes are required.

target is defined and for each service type the quality criteria are determined. The thresholds are then set for each related KPI. It is important to note that when setting the KPI targets, the operator provides the tools for capacity–quality tradeoff management. Network performance data can be gathered from NMSs, drive tests, protocol analysers and/or customer complaints. Network reporting tools provide statistical and pre-analysed information about the quality. Based on the network configuration and status of the network, quality in detail is analysed and individual corrections are done iteratively by tuning the individual parameters affecting the reported quality. Tuning of individual parameters or parameter sets is carried out in an iterative loop until the target quality is met. Finally, in addition to tuning single parameters, the general solution has to be found. After corrections have been implemented to the network, the quality cycle starts from the beginning.

The selection of the data for performance analysis consists of two aspects. First, the data are selected based on functional area (or a subset of that) – i.e., accessibility, reliability, traffic performance and distributions, to mention a few examples. All these functional areas are targeted to offer a picture of end-user-perceived quality. A good framework for this is provided by [6]. The other aspect is the purpose of the analysis. For getting an overall performance evaluation of the network the selection of the counters and other Performance Indicators (PIs) is different from those one would choose for optimisation or troubleshooting cases. The optimisation case is more focused, and thus more problem-specific indicators are required. In addition, uplink and downlink are often analysed separately. After optimisation has been performed and the changes implemented in the network, it is essential to check the function of the optimisation target, but it is equally important to derive the overall performance distribution and compare it with the pre-optimisation case. This is done to avoid the phenomenon that optimisation improves one subset of a functional area, but drastically decreases the performance of some of the others. If we invert the case: general performance information gives an indication that there is degraded operation in a functional area. To be able to derive the actual problem and find a solution for it, it is mandatory to change from the generic dataset to a more focused one.

In this section, the biggest challenges for the management of UMTS networks and the new requirements for UMTS compared with any existing network management system are listed. This section looks at the requirements mainly from an automation and NM point of view, since this is seen as an essential issue in the current, extremely competitive situation. Service driven management issues are discussed in Chapter 8. The tools and concepts introduced in this chapter are the basis for management of services.

7.1.1.1 Network-wide and Statistical Optimisation

The NMS has an important role in all the sub-processes of quality improvement. Because the main concern of the operator is the mobile service provided for the end-user, there has to be a tight mapping from the definition of service quality criteria to radio network configuration. The tools used in the process should be tightly integrated to each other in order to support an iterative optimisation process. In the future the automation requirements for the network will increase. Monitoring should

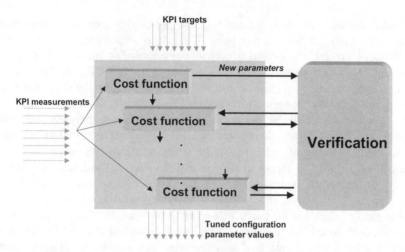

Figure 7.6 Optimisation can be simplified as the task of determining the key performance indicators to be optimised by the cost functions at hand. Before provisioning of the change the optimisation results need to be verified (see also Chapter 9).

automatically trigger the optimisation loop based on preset KPI or KQI thresholds. Also, the time-consuming detailed parameter analyses and consistency checking should be automated and more tools made available to experts for overall quality optimisation.

To guarantee the optimal performance of a cellular network, the operator must have means to visualise both the network and the services offered, and their quality. The tools should show the service distribution and related KPIs (see Section 7.3.3) overlaid on a map in order to easily locate the problem areas where QoS criteria are not met. Furthermore, the operator ought to have flexible means to set the QoS target based on the system KPIs and/or a cost function derived from those (see Figure 7.6). QoS targets could be set either for a cell cluster or on a cell-by-cell basis. QoS could be defined in terms of blocked calls due to hardware resources, 'soft' blocked calls (in interference limited networks), dropped calls, bad-quality calls, number of retransmissions and delay in the case of packet data, diversity handover probability, hard handover success rate, loading situation (uplink or downlink), ratio of packet switched data to circuit switched services, and so on.

Currently, RRM algorithms are parameterised separately: handover control, admission control, power control and several other parameter values are set independently, and in consequence one can identify cases where, for example, handover problems are due to wrong power control (CPICH or Common Pilot Channel) setting. Change in the admission control setting can result in a change in the quality of the packet data. Therefore, it is essential to have a functionality that monitors RRM as a whole, in terms of set QoS targets, rather than as individual functions.

In multi-radio environments (e.g., GSM–WCDMA) it is important to have the possibility of pooling the resources of both networks for optimised capacity, coverage and quality. This also requires an overall control functionality at a sufficiently high abstraction (KPI, cost function) level (Figure 7.6).

7.1.1.2 Interface Requirements for UMTS Network Management

A UMTS network will consist of many different types of components based on different types of technologies. Further, these components are often from different infrastructure vendors. There will be access, core, transmission and intelligent node networks, as well as several different access technologies like GSM/EDGE, WCDMA and GPRS. A complete management solution of a UMTS network will thereby consist of many Telecom Management Networks (TMNs) (see Figure 7.7) [13]. The Telecom Management (TM) architecture can vary greatly in scope and detail, because of its scale of operation and the fact that different organisations may take different roles in a UMTS. Some operators may prefer a high degree of centralisation, while others may have distributed management in regions [13].

The requirements and composition of TM for UMTS do not radically differ from those of 2G systems. Management processes will also be based on eTOM. However, there are four major requirements that are highly challenging for future management systems:

1. Management systems must support openness in order to provide long-term supportability, inter-operability, low life cycle cost, reasonable development times and good overall performance.
2. Management systems need to be more flexible to allow rapid deployment of services.
3. Support for managing equipment from different vendors will play a key role in the systems.
4. A change of focus from EM towards management of services is needed. Therefore, management systems should abstract the information to a higher level – i.e., hide the complexity of the network by hiding the individual parameters under more general goals (i.e., KPIs, QoS, etc.) that can be easily understood and verified.

Because UMTS networks are heterogeneous, inter-operability with other systems is essential. This means high modularity, scalability and well-defined open interfaces between the modules. Figure 7.7 illustrates the basic domains in a 3GPP (Third Generation Partnership Project) system, related management functional areas and introduces Interface-N (Itf-N).

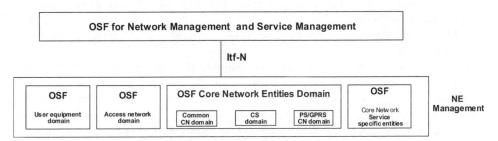

Figure 7.7 Overview of UMTS telecom management domains. Interface-N between network element operation system functions and network management/service management OSFs could be used by network management and service management systems to transfer management messages, notifications and service management requests via network element OSFs to network elements [13].

7.1.1.3   Network Resource Model

This section introduces the Network Resource Model (NRM) as defined in [17] and [18]. It defines an Integration Reference Point (IRP) through which an 'IRPAgent' (typically an element manager or network element) can communicate network-management-related information to one or several 'IRPManagers' (typically network managers). Reference [17] specifies a generic NRM (also referred to as a Management Information Model – MIM) with definitions of Managed Object Classes (MOCs).

The generic network resources' IRP here provides a base for all resource modelling. To summarise, the generic network resources' IRP main purpose is to define a generic NRM that constitutes a base from which other (more specialised) resource models can inherit or have associations with. This framework provides a reference against which vendor-specific resource models can be mapped. The generic, vendor-independent resource model is a prerequisite for multi-vendor-capable management systems. External objects represent the other vendor equipment in Figure 7.9.

Generally, an NRM is a model representing the actual TMN resources that a system provides through the subject IRP. An NRM describes MOCs, their associations, attributes and operations. The NRM describes an object structure, which is used in handling all network (element)-related configuration and monitoring data (CM, PM, FM) through a mutual management structure. For example, the desired power levels for each UtranCell are parameterised in configuration management and the power-level-related KPIs provided by performance management defined in [16] are all assigned to the same managed object – i.e., the UtranCell. This object indicates the network element level granularity the information is available with.

Reference [17] contains the following Information Object Class (IOC) definitions:

- *GenericIRP* IOC – represents the IRP capability associated with each IRPAgent. This IOC cannot be instantiated. It is defined for sub-classing purposes. At least one instance of a sub-class of GenericIRP shall be present for every IRPAgent instance.
- IOC *IRPAgent* – represents the functionality of an IRPAgent. It shall be present. For a definition of IRPAgent, see [14]. The IRPAgent will be contained under an IOC as follows (only one of the options shall be used):
  1. ManagementNode, if the configuration contains a ManagementNode;
  2. SubNetwork, if the configuration contains a SubNetwork and no ManagementNode;
  3. ManagedElement, if the configuration contains no ManagementNode or SubNetwork.
- *ManagedElement* IOC – represents telecommunications equipment or TMN entities within the telecommunications network that perform Managed Element (ME) functions – i.e., provide support and/or service to the subscriber. An ME communicates with a manager (directly or indirectly) over one or more interfaces for the purpose of being monitored and/or controlled. MEs may or may not additionally perform EM functionality. An ME contains equipment that may or may not be geographically distributed. An ME is often referred to as a 'network element'.
- *ManagedFunction* IOC – provides for sub-classing only. It provides attribute(s) that are common to functional IOCs. Note that an ME may contain several managed

functions. The ManagedFunction may be extended in the future if more common characteristics to functional objects are identified.

- *ManagementNode* IOC – represents a TM system (EM) within the TMN that contains functionality for managing a number of MEs. The management system communicates with the MEs directly or indirectly over one or more interfaces for the purpose of monitoring and/or controlling these MEs. This class has similar characteristics to the ME. The main difference between these two classes is that the ManagementNode IOC has a special association with the MEs that it is responsible for managing.

- *MeContext* IOC – introduced for naming purposes. It may support creation of unique Distinguished Names (DNs) in scenarios when some MEs have the same Relative DNs (RDNs) due to the fact that they have been pre-configured by the manufacturer. If some MEs have the same RDNs and they are contained in the same SubNetwork instance, some measure shall be taken in order to assure the global uniqueness of DNs for all IOC instances under those MEs.

- *SubNetwork* IOC – represents a set of managed entities as seen over the Itf-N. There may be zero or more instances of a SubNetwork. It shall be present if either a ManagementNode or multiple ManagedElements are present (i.e., ManagementNode and multiple ManagedElement instances shall have SubNetwork as their parents). The SubNetwork instance not contained in any other instance of SubNetwork is referred to as the 'root SubNetwork instance'.

- *Top* IOC – introduced for generalisation purposes. All IOCs defined in all technical specifications that claim to be conformant to [14] shall inherit from Top.

- The *VsDataContainer* managed object is a container for vendor-specific data. The number of instances of the VsDataContainer can differ from vendor to vendor. This IOC shall only be used by bulk CM IRPs for UTRAN, GERAN and CN NRMs. The usage of VsDataContainer is in Figure 7.10.

- *ManagementScope* – this association is used to represent relationships between one or more MEs and the ManagementNode that is responsible for managing the MEs. It has two roles, named 'Manager' and 'Subordinate'. Manager models the fact that a ManagementNode is responsible for managing zero or more MEs, and Subordinate models the fact that zero or one ManagementNode manages an ME. Each role is mapped in the IOC definition to a reference attribute with the same name.

Reference [18] utilises the concepts of [17] to introduce the actual network resource model for UTRAN (UMTS Terrestrial Radio Access Network). Similar models for the core network are also defined (see, for example, [20]). In Figures 7.8–7.10 the NRMs for UTRAN are presented. In order to provide better readability three different figures are provided. Figures are according to [18].

As said in [19] the number of instances of the VsDataContainer can differ from vendor to vendor. Vendor-specific parameters for handover control (like possible templates controlling the event-based measurement reporting for active set update) are mapped to VsDataContainer. In Figure 7.10 the usage of VsDataContainer in UtranRelation handling is depicted, as presented in [18].

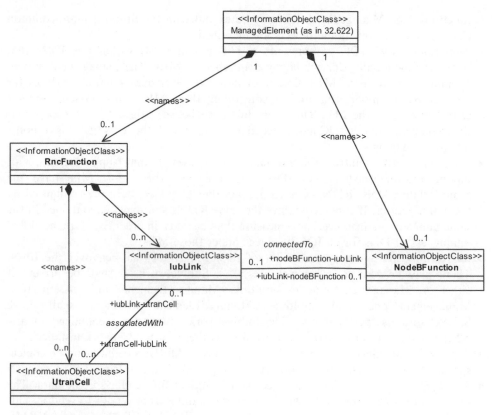

Figure 7.8 Transport view UTRAN containment/naming and association network resource model diagram from [18].

7.1.1.4 Automation Needs

During the 1990s there was a vast expansion in the area of cellular network deployment. Operators needed to introduce new services in mobile networks more quickly and with less manual intervention. Only a truly integrated performance management system can encompass all relevant technologies from the voice and data worlds. Network vendors have realised the need to hide the complex infrastructure of several radio access technologies. There will be more intelligence and real time automation in radio networks themselves, and WCDMA networks will be more automated than GSM networks. There is a clear analogy with the tendency in the process industry in the 1980s and early 1990s to install, for example, intelligent valves in process plants when Operational Expenditure (OPEX) became a major issue. Automation is needed not only for OPEX savings but also for faster service deployment. The key objectives are fast service creation, introduction and provisioning, and improved QoS at lower cost. These objectives can be achieved only through automation of customer care and operational support processes, and a strong automated linkage between the management of customer service offerings and the underlying network [2].

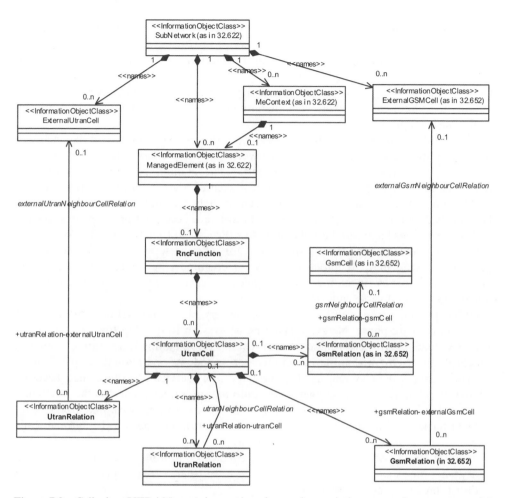

Figure 7.9 Cell view UTRAN containment/naming and association network resource model diagram from [18]. Relations for UtranCell are to support inter-system and intra-system handovers, including external GSM and UTRAN cells.

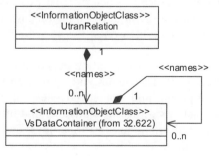

Figure 7.10 VsDataContainer containment/naming and association in UTRAN network resource model diagram from [18].

The level of automation and integration in current systems is lower than that needed by providers to remain competitive. Many service providers who face new competitive pressures or restructuring are now actively re-engineering their business processes to integrate and automate, thereby decreasing costs and improving customer-perceived value and performance. New entrants are developing operational processes based on automation of critical processes while planning to increase automation and integration as the business grows.

The problems associated with network expansion affect both day-to-day network operations management and strategic network growth planning. Each new network technology requires its own set of experts. The staffing requirements alone for managing large, heterogeneous networks have created a crisis for many organisations. At the same time, the resources needed to run the networks must be limited in order to keep costs down to a competitive level. Operators are becoming more 'process-aware' and organisations have to reflect the processes. Process automation is a key element in cost savings and a mandatory requirement for efficient service deployment. Service providers facing hard competition have recognised the value of well-defined automated processes (including what is typically called 'network capacity planning') integrated across diverse environments.

Offline planning tools will be used in the delivery phase, and network management system in the care phase. Network planning is foreseen to develop towards operative, network load- and performance-driven expansion planning and tuning. This is practical, short- to medium-term activity performed by operational people using a network management system as their primary tool. Service assurance then becomes primarily proactive – i.e., triggered by automation rather than by the customer. Automation is needed in at least four functionality areas in network management:

- Data retrieval: configuration, performance and alarm data need to be automatically collected and forwarded to the management system.
- Monitoring of network traffic and performance conditions is crucial as network degradation usually precedes network failure. Further, network performance has a direct impact on perceived service quality.
- Detection and forwarding of problems at an early stage can improve customer perception of service quality.
- Automation is also essential to the network planning and development processes, since it gives early warning of exhaustion of network capacity. KPIs can trigger troubleshooting tools automatically when a value exceeds or falls below a certain threshold value. The process is essential to support the network management, service management and customer care life cycles.

Performance management for next-generation networks consists of two components. The first is a set of functions that evaluates and reports on the behaviour of telecommunication systems' equipment and the effectiveness of the network or network elements. The second is a set of various sub-functions that includes gathering statistical information, maintaining and examining historical logs, determining system performance under natural and artificial conditions, and altering system modes of operation.

Network configuration management and automated parameter tuning can be triggered by the PM monitoring process. There are many parameters in the network that can be tuned automatically without any manual intervention. Ideally, as much as possible of this kind of tuning should happen in the network in real time. However, there are also processes that need more statistical data or overall knowledge of the network, and therefore the tuning functionality is needed in a network management system.

Fault management tools should detect and log network problems, and wherever possible fix them automatically in order to keep the network running effectively.

Network management needs to identify not only lack of capacity in the current network but also the potential to introduce data services where they currently do not exist.

7.1.1.5 Workflow Support

Service providers must be able to monitor and manage traffic levels and concurrent network congestion to optimise network performance. They must analyse the data to correlate end-to-end service performance, and take action based on a complete understanding of network behaviour. The ability to control network performance effectively can be achieved with an application that allows service providers to control the operation. The application must monitor network performance, analyse relevant data and direct network action accordingly. A well-supported workflow for optimisation and automated tuning is needed. The system has to be sufficiently modular that the operator or service provider can integrate the tools to support the organisation's workflow. Operators must be able to customise processes according to their needs. Optimisation is often based on strategic decisions. The service provider has to be able to define the KPIs and the criteria for optimisation. The quality experienced by the end-user and the network performance of competitors together define the targets for quality in certain areas of the network or for certain services.

The NMS should reflect the needs for the workflow support. Network management processes are distributed among several applications. High-level operator processes dictate application inter-operability requirements. The Graphical User Interfaces (GUIs) of different management applications should be integrated so that the most frequently used information is accessible from a single screen. Additionally, since the operator may use the same piece of information in several ways, the information needs to be easily transferred between applications. IRPs are introduced to ensure inter-operability between different applications. The three cornerstones of the IRP concept are [13]:

- Top-down, process-driven modelling approach. The purpose of each IRP is automation of one specific task, related to TMF TOM (TeleManagement Forum Telecom Operations Map). This allows taking a 'one step at a time' approach with a focus on the most important tasks.
- Technology-independent modelling. To create from the requirements an interface technology-independent model. This is specified in the IRP information service.
- Standards-based technology-dependent modelling. To create one or more interface technology-dependent models from the technology-independent model. This is specified in IRP solution set(s).

The complexity and heterogeneous nature of a 3G system calls for easy integration (plug&play) of hardware and software and IRPs are an integral part of that solution.

In practice, there is no quick way to automate optimisation of UMTS networks. Network operational complexity will be hidden gradually. At the beginning, when there is a shortage of UMTS competence, optimisation will present demanding challenges. Questions will arise as to which are the most critical KPIs to be monitored and tuned, how the KPIs are mapped into other vendors' networks, how different services are prioritised, etc.

The first phase will include the provision of mainly visualisation tools for better management of services and their distribution in the network. Geographical Information Systems (GISs) will have a key role in visualisation in order to show where and how the services are distributed in reality.

In the second phase, additional analysis tools for pre-processing the data will be provided. However, triggering of the tools will be manual or semi-manual, and it is the operator's choice how to use the tools. They can be used for knowledge acquisition and for a better understanding of network behaviour. When the analysis tools have been well-tested and competence with the system has grown, the major focus in improving service delivery will be on automated optimisation tools. Automation should be introduced as soon as the size of the UMTS network justifies it.

Although there will be much real time automation in radio networks themselves, that will not exclude the need for automation tools in network management systems.

7.2   Introduction to the Telecom Management Network Model

The Telecom Management Network (TMN) model ([3]–[5]) provides a widely accepted view about how the business of a service provider is managed. The TMN model consists of four layers, usually arranged in a triangle or pyramid, with business management at the top, service management the second layer, network management the third layer and element management at the bottom (see Figure 7.11). Management decisions at each

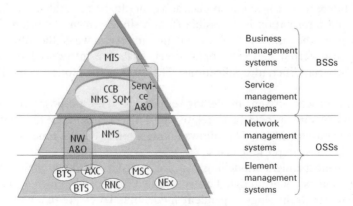

Figure 7.11   Telecom Management Network model. 'NW A&O' and 'Service A&O' refer to layers where network and service analysis and optimisation actions take place.

layer are different but related to each other. Working from the top down, each layer imposes requirements on the layer below. Working from the bottom up, each layer provides an important source of data to the layer above. The TMN of the TMF sets the guidelines for optimisation functionalities and processes. 3GPP ([12] and [13]) has adopted the same model. The scope of TMF is to:

- find a standardised way of defining service quality;
- set requirements for networks in terms of QoS measurements; and
- make it possible to have QoS reports between providers and systems that implement the service.

According to the TMN model, information from upper-level systems flows down, guaranteeing seamless operation and optimisation possibilities for the network. The TMN model is depicted in Figure 7.11. The information flow from the business management layers all the way down to the service management and network management layers is essential since business aspects have to be considered carefully in the optimisation and network development process. The TMN model demonstrates the change of the abstraction level in the operator's daily work. The efficiency of the business plan can be measured in terms of CAPEX, OPEX and revenue. The wanted business scenario is then translated to offered services, service priorities and service QoS requirements. On the lowest (network element) level of the TMN model business-related issues are converted into configuration parameter settings.

The following functions are supported by TMN's business management systems:
- creation of an investment plan;
- definition of the main QoS criteria for the proposed network and its services;
- creation of a technical development path (expansion plan) to ensure that the anticipated growth in subscriber numbers is provided for.

The functions supported by service management systems include:
- management of subscriber data;
- provisioning of services and subscribers;
- accounting and billing operations for services offered;
- creation, promotion and monitoring of services.

The following functions are supported by network management systems:
- planning the network;
- collecting information from underlying networks;
- pre-/post-processing of raw data;
- analysis and distribution of information;
- optimisation of network capacity and quality.

Element management systems can be considered as part of network element functionality with responsibility for:
- monitoring the functioning of the equipment;
- collecting raw data (performance indicators);
- providing local GUIs for site engineers;
- mediating with the network management system.

In addition to TMN, TMF also defines a TOM and an eTOM. The TOM was developed to drive a consensus around the processes, inputs, outputs and activities required for service provider operations management. Its focus and scope were operations and operations management. The TOM model continues to be the core of the eTOM Business Process Framework as it develops to deal with current issues, needs and trends, such as ebusiness integration. The augmentation of TOM is called 'eTOM'.

Telecom and data service providers must apply a customer-oriented service management approach using Business Process Management (BPM) methodologies to manage their businesses cost-effectively and to deliver the service and quality customers require. The TOM identifies a number of operations management processes covering customer care, service management and network management. It uses the layers of the TMN model as core business processes, but divides the service management layer into two parts: customer care, and service development and operations. Customer interface management is separately delineated, because it may be managed within the individual customer care sub-process or in combination across one or more of the customer care sub-processes. Figure 7.12 shows the high-level structure of network management processes and the supporting function set groups. The TOM links each of the high-level processes to a series of component functions (arranged in function set groups). It then identifies the relationships and information flows between them. In Figure 7.12 the TOM and its components are presented. The functionalities of the layers are the same as in Figure 7.11 to indicate the corresponding management layers. The modules contributing to optimisation and statistical QoS management activities are marked grey.

The TOM model demonstrates effectively the challenges of optimisation work. Optimisation functionalities are relevant in many of the TMN layers and thus fluent information flow from one layer to another is essential. This applies not only at the network element level but also to human interactions.

Figure 7.13 depicts eTOM level 1. In eTOM level 1, service management and service operations focus on knowledge of services and includes all functionalities necessary for the management and operations of communications and information services required by or proposed to customers. The focus is on service delivery and management, not on the underlying network and IT. Some of the functions involve short-term service capacity planning for service instance, service design to specific customers or improving service performance. These functions are closely connected with day-to-day end-user experience [5].

The operations process area contains the direct operations' vertical process groupings of Fulfilment, Assurance & Billing, together with the Operations Support & Readiness process grouping. Resource Management and Operations maintains knowledge of resources and is responsible for managing them. These operations are responsible for ensuring that the network and IT infrastructure supports the end-to-end delivery of the required service. Assurance and Fulfilment processes are an integral part of optimisation. Fulfilment in eTOM is responsible for providing customers with their requested products in a timely and correct manner. It translates the customer's needs into a solution which can be delivered using the specific products in the enterprise's portfolio. Assurance is a vertical end-to-end process that is responsible for reactive and proactive maintenance activities to ensure the services so that Service

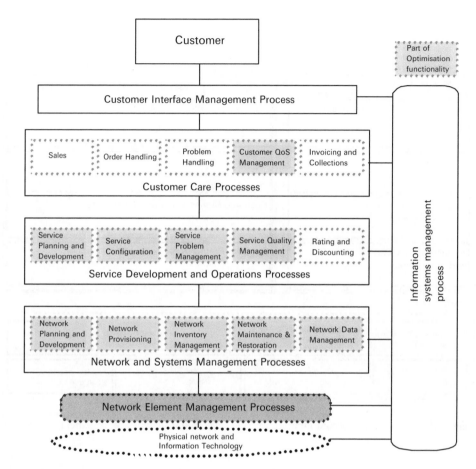

Figure 7.12 Telecom Operations Map, business process framework. Optimisation and network development position in TOM.

Level Agreements (SLAs) or QoS requirements are met. It performs continuous resource status and performance monitoring (of network and services) and detection of failures. It tries to identify problems and solve them before the customer notices it.

eTOM level 2 defines the process more accurately. In the 'Fulfillment' area there are optimisation related tasks like service configuration and activation and (network) resource provisioning. Provisioning functions as such are not in the focus of an analysis and optimisation tool, but it can help during the process to find, for example, optimal parameters to provision. In 'Assurance' there are items for:

- service problem management;
- service quality management;
- (network) resource trouble management;
- (network) resource performance management;
- (network) resource data collection.

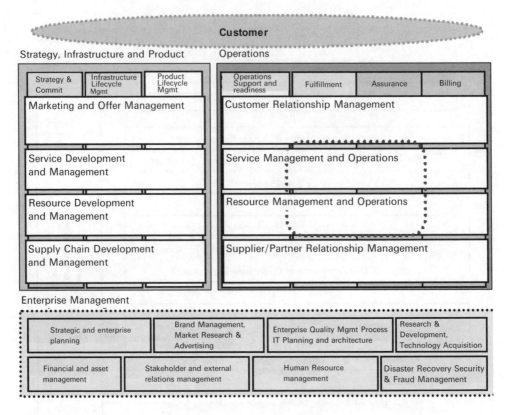

Figure 7.13 Enhanced Telecom Operations Map level 1 and network management and service management analysis and optimisation.

When the optimisation process itself is mapped to the TOM it can be seen that the network management and service management layers need to be tightly connected. The performance measures and targets are seen as service management functions, whereas the network management layer is responsible for deriving the actual solution for improving performance.

7.3   Tools in Optimisation

The following sections introduce the methods and tools provided by the planning, RNC, network management system and field measurement tool to support the network development and optimisation process.

The methods and tools developed for monitoring, measuring and visualising KPIs and PIs are described in Section 7.3.3. The network management system platform should be flexible in order to enable both 2G and 3G network management, optimisation and/or automation application development. Also, combined 2G/3G networks

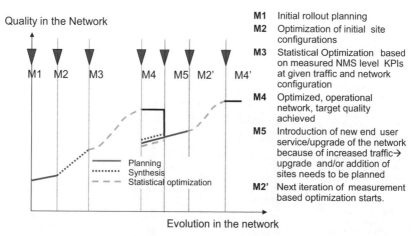

Quality in the Network

M1   Initial rollout planning
M2   Optimization of initial site
     configurations
M3   Statistical Optimization based
     on measured NMS level KPIs
     at given traffic and network
     configuration
M4   Optimized, operational
     network, target quality
     achieved
M5   Introduction of new end user
     service/upgrade of the network
     because of increased traffic→
     upgrade and/or addition of
     sites needs to be planned
M2' Next iteration of measurement
     based optimization starts.

M1 M2 M3   M4 M5 M2'   M4'

Planning
Synthesis
Statistical optimization

Evolution in the network

Figure 7.14   Optimisation is done in different phases of the network evolution.

have to be co-optimised (handovers, traffic balancing, coverage enhancement, etc.) to fully utilise the possible trunking gain.

In the following sections, optimisation in the planning phase and statistical optimisation and autotuning are discussed in greater depth. Figure 7.14 depicts the optimisation actions in different evolution phases.

First, planning tool level issues are handled. In the following section the role of the network management system in the statistical optimisation process is described, and some issues related to automation of the correction loop are presented. The monitoring and reporting functions and the measurement content are discussed in the subsequent sections. UMTS quality definition and QoS management issues are introduced in Chapter 8.

7.3.1   Planning Tool Level Optimisation

The process in the radio network planning phase is depicted in Figure 7.15. Initial planning (i.e., system dimensioning) provides the first and most rapid evaluation of the network size as well as the associated capacity of those elements. This includes both the radio access network as well as the core network. The target of the initial planning phase is to estimate the required site density and site configurations for the area of interest. Inputs needed for dimensioning are exactly the same as for other process phases: target values related to quality, capacity and coverage. In operator workflow the planning is performed after rough network dimensioning. During dimensioning the site density is estimated and then more detailed planning with the planning tool can proceed. Network planning includes the following tasks:

- dimensioning;
- KPI metrics definition;
- existing network assessment;
- creation of a nominal plan;

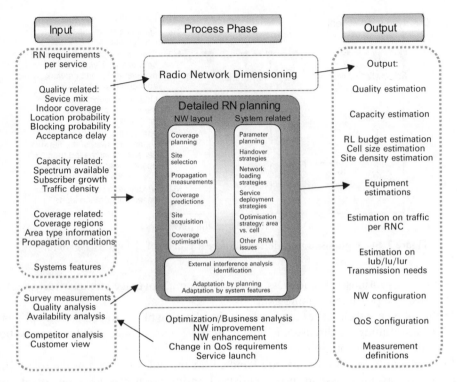

Figure 7.15 Radio network planning process including dimensioning and detailed planning items. Further, the feedback from business decisions to the process input is highlighted.

- capacity/coverage/quality extension plan;
- site selection;
- site-specific parameter setting;
- IP core planning;
- circuit switched core planning;
- pre-optimisation;
- KPI fulfilment.

The purpose of pre-launch optimisation is to assure a smooth and successful public network launch. In addition, it promotes efficient network resource utilisation by improving the overall network quality as perceived by mobile subscribers. Initial pre-optimisation covers a number of different activities and tasks, including the collection, post-processing and analysis of test data extracted from the network:

- existing network assessment;
- KPI metrics definition;
- drive test/network management system level monitoring;
- antenna related optimisation;
- site-specific parameter optimisation;
- KPI fulfilment and reporting.

In Figure 7.14 the phase from M2 to M3 is pre-optimisation. It consists of optimisation of coverage and handover performance with a planning tool and field testing. In the detailed planning phase, the dimensioned site density is transferred to a digital map, taking into account the physical limitations from, for example, site acquisition. The WCDMA analysis itself is an iterative process. Capacity requirements are taken into account as discrete User Equipment in the WCDMA simulation. In the detailed planning phase, a multiple analysis is performed to verify whether the set requirements are actually met. In the planning phase, the means of optimisation can be interference control in terms of proper antenna and site selection or antenna tilting. Furthermore, network performance can be brought closer to the required targets by using, for example, mast head amplifiers or rollout optimised configuration (see Chapter 6).

Statistical optimisation is based on a large number of performance samples from the network, thus the usability of this functionality is limited during the pre-optimisation phase, the exception being network expansions and possible vendor swap cases, since in such a case there is a solid customer base and the feedback from the network in terms of measurements is quickly available. This is depicted in Figure 7.14. In the M5 to M2' phases statistical optimisation is an effective means of performance and capacity utilisation improvements.

If the operator's business strategy changes, dimensioning and detailed planning can provide valuable information on network expansion. The measured traffic information can be imported to the planning tool and further used when verifying the capacity and coverage capabilities of the planned network.

The modelling and methodology within the planning tool (not visible to the user) are described in Chapter 3, which also provides practical examples related to plan optimisation – e.g., in terms of proper antenna selection. Issues related to capacity and coverage improvement can be found in Chapter 6.

7.3.1.1 Automated Cell Planning

Optimisation of the radio plan in order to reduce interference and thus improve capacity is a tedious task. Therefore, more advanced methods are proposed in [7]–[10]. A new trend in radio network planning research is plan synthesis, meaning automatic generation of BS site locations depending upon cost function output. In [7] the target is to utilise a cost function to minimise implementation costs, maximise the coverage, maximise the offered traffic and maximise the Signal-to-Interference Ratio (SIR) in the network. An additional challenge for this type of approach is to take antenna directions, number of sectors, and antenna bearing and tilting into account. A similar idea using neural networks can be found in [9]. Radio network planning synthesis using a genetic approach is introduced in [10]. Limitations for this type of approach arise from the fact that site locations in practice are limited. Further, antenna bearing changes are in practice costly and there are practical limitations for the tilt angles as well. Thus, the configuration pool from which the algorithm can choose the optimum locations and tilts is limited. The results of plan synthesis can be utilised to bring improvements to the radio network plan.

In Figure 7.16 a site synthesis example is presented. In the initial ('before') case the simulated service success was 87%, Soft handover overhead was 63% on average and

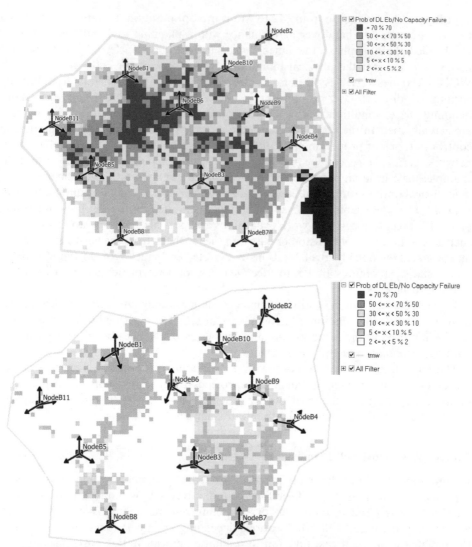

Figure 7.16 Interference minimisation using Genetic-algorithm-based site synthesis. See *www.wiley.com/go/laiho* for colour images.

more than 99% of the failure cases were because of downlink interference. After the synthesis, service success was increased to more than 95% by reducing downlink interference. Also, soft handover overhead was reduced to 56% by tuning the antenna tilts, bearings and pilot powers. Site locations were kept fixed.

Site synthesis (or automated planning) is a powerful tool for site configuration optimisation in terms of powers, site locations, antenna bearings and tilts. Site synthesis is still done using predicted data, so it is important to start utilising measured information as soon as the network is operational.

7.3.2   *Configuration Management in a Network Management System*

Configuration management, in general, provides the operator with the ability to assure correct and effective operation of the 3G network as it evolves. Configuration management actions have the objective of controlling and monitoring the actual configuration of network elements and network resources, and they may be initiated by the operator or by functions in operations systems or network elements.

Configuration management actions are needed as part of an implementation programme (like additions and deletions), as part of an optimisation programme in terms of modifications and to maintain the overall performance and/or QoS. Configuration management actions are initiated either as single actions on single network elements of the 3G network, or as part of a complex procedure involving actions on many resources/objects in one or several network elements.

The high-level requirements for network configuration management are listed in [13], which makes a distinction between three phases and states for a network and different degrees of stability. Once the first stage is over, the system will iterate between the second and third phases. This is known as the network life cycle and includes:

1. The 3G network is installed and put into service.
2. The 3G network reaches a certain stability and is only modified (dynamically) to satisfy short-term requirements – e.g., by (dynamic) reconfiguration of resources or parameter modification; this stable state of a 3G network cannot be regarded as the final one because each item of equipment or software modification will lead the 3G network progress to an unstable state and require optimisation actions again.
3. The 3G network is being adjusted to meet the long-term requirements of the network operator and the customer – e.g., with regard to performance, capacity and customer satisfaction – through the enhancement of the network or equipment upgrade.

During these phases, the operators will require adequate management functions to perform the necessary tasks. When a 3G network is installed and initialised for the first time, all network elements need to be introduced to the network management; the data for initialisation and software for proper functioning also need to be provided. All these actions are carried out to create network elements and to initialise them.

Whilst in service, the operator needs to react to short-term incidents such as traffic load requirements that are different from current network capabilities, network elements/network resources need to be reconfigured and parameters need to be adapted to follow these day-to-day requirements.

As the 3G network grows and matures new equipment is installed and understanding of system behaviour increases. Subscriber requirements/wishes may demand that operators modify their system. In addition manufacturers improve the infrastructure components and add features to their products, hence the operator will start modifying the 3G network to profit from these changes and to improve subscriber satisfaction. Additionally, the 3G network configuration will be modified (i.e., it will be updated or upgraded) to cope with a need for increasing or decreasing network capacity. These actions are carried out according to the long-term strategy of the operators to optimise the network. Whenever the 3G network needs to be improved for reasons of reducing failures, the system will be updated. In this case software or equipment will be replaced without adding new functionality or resources to the network.

System upgrade may affect all areas of 3G network activities and can be described as enhancements, whereby either new features or new facilities are implemented. This configuration management aspect also covers extensions, reductions or further replications of existing facilities. The configuration management functions employed are:

1. Creation of network elements and/or network resources.
2. Deletion of network elements and/or network resources.
3. Modification of network elements and/or network resources.

In configuration management principles, two types of configuration management functions are identified: passive and active. Passive configuration management (configuration overview) mainly provides to the network management information about the current configuration changes by means of notifications, and allows retrieval and synchronisation of configuration-related data on network management request. Whereas active configuration management offers an operator a capability to change the current network configuration. In [13] there are also two approaches to configuration management: basic CM and bulk CM. Basic CM is characterised by the use of singular operations to retrieve and activate configuration parameters over Itf-N from single network elements, or a collection of network elements. Bulk CM is meant for mass actions and is characterised by bulk (file-oriented) data retrieval and provisioning/ download of configuration parameters over Itf-N from single network elements, a collection of network elements or the whole network. Further, bulk CM allows network-wide activation of those parameters through a single operation – this is an active aspect of bulk CM.

The consistency between the actual local representation of physical and logical resources is a challenge in its own right and the standards do not take a stand on actual implementation. Peer-to-peer data consistency between NM–EM and EM–NE does not guarantee overall data consistency from a network point of view. It is however possible for the NM to maintain consistency on the network level, as far as the information in the Management Information Base (MIB) for the Itf-N is concerned, by comparing related information (MOIs and attributes) in all connected systems (element managers and network elements) in the managed network.

The bulk CM approach and the mechanisms for data consistency answers the challenges of operators. Each site in a mobile network contains in excess of a couple of hundred configuration parameter settings. Currently, many operators use offline spreadsheets or simplistic databases to record and manage network data. This is not only a difficult task, but it is also inherently risky. Each new cell added to the network can potentially break data consistency and seamless mobility. Mistakes in the definition of neighbour cells and handover control parameters (among others) can cause unpredictable network behaviour and will impact network performance.

Parameter changes currently involve highly skilled, experienced individuals making changes through element management systems. If such changes are logged at all, it is rare that an explanation for the modification or rationale is recorded. Simply put, there is little version control and understanding the network configuration remains the preserve of the individuals responsible. This is unsustainable as parameter changes can and do result in serious problems.

As discussed in Section 7.2 the eTOM fulfilment process is responsible for configuring the network and services. One of the challenges related to service configuration is that the settings in network elements must support the fulfilment and assurance of a service. With a network management system the bulk provisioning of the configuration parameter values for network elements can be performed as a mass operation from a central location. This diminishes the number of errors and is a fast method. Configuration management functionality provides operators with the basic functionality to interface planning tools, edit plans and provision plans into the network. Configuration management tools should support fast 2G and 3G radio network rollout and expansion and provides an efficient means for network configuration changes. Parts of the configuration management system are:

- Data repository, which contains the actual settings in network elements.
- Radio network planning management for provisioning/activating plans in the network.
- Support for rapid network rollout by automating the most common rollout-related parameter provisioning tasks like site creation and object reparenting.
- A rule system which comprises of a set of configurable standard rules (value-range-checking rules or consistency rules that define relations between network elements) that define the technology and vendor-specific consistency requirements for configurattion settings.
- Data exchange with external tools (requires, for example, an XML interface for the radio access planning data module).

For optimisation purposes the versioning of configuration settings should be supported. In cases where unexpected network behaviour results from the provisioning of a new configuration, 'rollback' functionality is needed in order to return to the original configuration.

In addition to the network configuration the service configuration needs to be controlled. A service configurator is used when configuring customer-facing services. While customer-facing services are deployed in the network, QoS/priority aspects are one of the so-called network-facing service components of the customer-facing service. This means that while deploying a new customer-facing service to the network, the service configurator will select what kind of treatment the corresponding traffic will experience while carried through the mobile network. The service configurator has to be aware of which QoS profiles/priorities are available and what the QoS attributes of each of those priority pipes are. Based on the needs of the newly created service the operator will select the most suitable pipe available in the network and assign the service to use that. The role of service configurator is, within the pre-defined priority settings, to take care of the correct configuration settings in packet core network elements. Service configuration settings need to be in accordance with the settings in the Home Location Register (HLR).

Generally, optimisation actions are perceived to utilise measured data. Optimisation, which utilises configuration data only, is related to mobility management in terms of adjacency definitions. Such optimisation cases are, for example, adjacency synchronisation between regions or vendors and distance-based adjacency optimisation for soft handovers, inter-frequency handovers and inter-system handovers. The latter require knowledge of site locations.

7.3.3 Performance Management in Network Elements and in the Operations System

In [6] the dependence of service performance and network performance is addressed. The scope of this section is to focus on the requirements of the network management layer in order to serve the service management layer.

In this section, measurement administration, definition of measurements and counters, collection of measurements, aggregation and summarisation of measurements, and the reporting tools in a network management system are addressed. Moreover, real time monitoring possibilities like online monitoring and subscriber trace are presented.

7.3.3.1 Measurement Administration

The ranges of measurements that will be available from network elements are expected to cover all of the requirements described in [6]. However, all of these measurements are not required all of the time, from every occurrence, of every relevant network element. Therefore, it is necessary to administer measurements in order to determine which measurement types, on which measured resources, at which times, are executed. With GSM or UMTS mobile telecommunication systems it is also necessary to collect the measured data to perform consistent analysis of the results and to evaluate interactions between network elements and different management domains.

Further, together with the introduction of data services a new challenge to measurement management is introduced. First, measurements should be collected per 3GPP-based priority class (see Chapter 8). Second, busy hour definition differs for each service. Thus the definition of 'busy hour' needs to be checked for each priority class separately, at least circuit switched–packet switched separation ought to be supported.

Measurement tasks or jobs – i.e., the processes that are executed in the network elements in order to accumulate measurement result data and assemble it for collection and/or inspection – contain the following actions:

1. Measurement job creation and deletion. This action implies the instantiation and respectively the deletion of a measurement collection process within the network.
2. Definition of measurement job scheduling. This action defines the period or periods during which the measurement job is configured to collect performance data.
3. Measurement job modification – i.e., changing the parameters (specifically the schedule) of a measurement job that has been previously created.
4. Specification of the measurement types to be contained in the job – e.g., 'number of GPRS attach attempts'. In UMTS, measurement jobs may be administered per individual measurement type or per measurement family, which comprises a collection of related measurement types, see [16].

5. Identification of the measured resources – i.e., the network elements or network element components (e.g., trunkgroups, radio channels, transceivers) to which the measurements relate.
6. Possibility to suspend and resume a measurement job. The 'suspend' action inhibits the collection of measurement result data by a measurement job, regardless of its schedule, without deleting it. The 'resume' action will re-enable measurement result data collection according to the measurement job schedule.
7. Setting up any necessary requirements for the reporting and routing of results to one or more Operations Systems (OSs) (element management and/or network management).
8. Possibility to view current measurement job definitions.

A measurement job is thus characterised by a set of measurement types and/or measurement families, which all pertain to the same set of measured resources and share the same schedule. Typically, a large number of measurement jobs will run simultaneously within the network elements comprising the Public Land Mobile Network (PLMN), and one or more element management systems or network management systems is involved in the administration of those measurement jobs. In order for the operator to manage this large number of measurement jobs effectively and efficiently, it is necessary for administration functions not only to deal with individual measurements on individual network elements, but also scope the execution environment across the measured resources, and apply an additional filter to the resources/ network elements selected by the measurement scope. Scope selection should support at least the following use cases. In other words, the administration system needs to:

- execute the same (set of) measurement type(s) *on a set of identical resources within a single network element* – e.g., to measure the average BER on all channels in a cell, or all channels of the cell that match the filter criterion;
- execute the same (set of) measurement type(s) *on a set of identical network elements or resources* according to the hierarchical structure of the network – e.g., to measure inter-cell handover success for all cells attached to the same RNC;
- execute the same (set of) measurement type(s) across all resources/network elements of the same type that belong to a specific administrative domain – e.g., to measure the call setup success ratio in all cells located in a certain geographical area or within the responsibility area of a system operator.

Figure 7.17 gives a glimpse of how the measurements are managed in an RNC.

Figure 7.18 gives an example view of how defined measurement jobs in a network management system are visualised. It is possible to define multiple measurement jobs in a network management system, select the scope for the job (like RNCs) and execute them in the network management system.

In Figure 7.19 the measurement plan/job creation is demonstrated. The measurement types, data collection time in terms of weekdays and intervals is defined. The plan can be stored either in the network management system or the network element. The benefit of using a network management system for measurement administration is the fact that measurement collection can be initiated for multiple network elements simultaneously. Further, the same plan can be restarted when

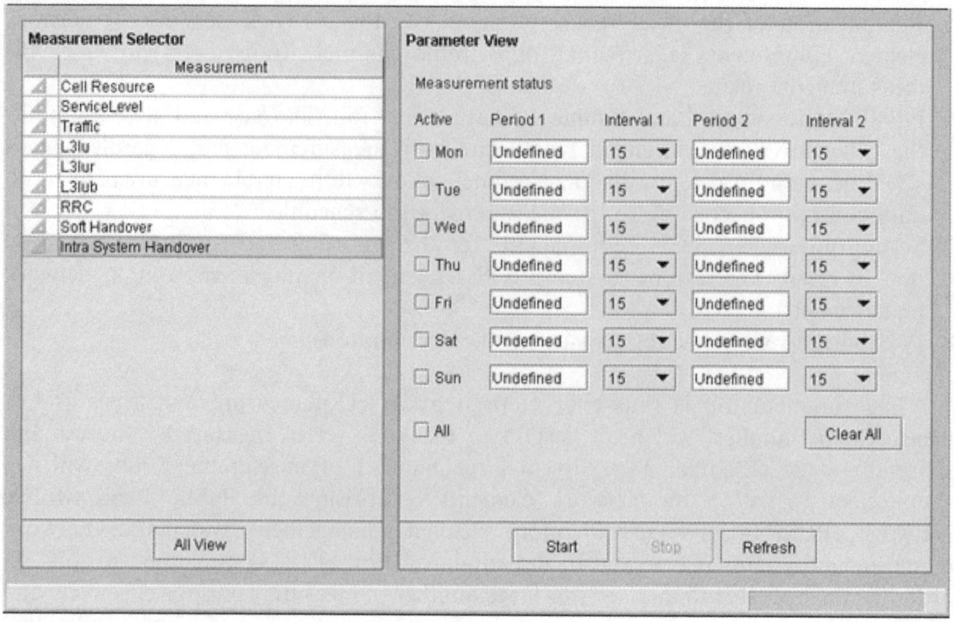

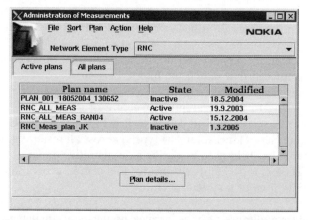

Figure 7.17   Element manager (radio network controller) view of measurement management.

Figure 7.18   Measurement administration in a network management system.

required. It is essential to do pre- and post-optimisation performance evaluation using exactly the same measurement definitions. The plan concept will ensure the consistency of the measurement reports in such a case.

In Figure 7.20 the GUI for starting the actual measurement is shown. The measurement plan/job is selected from the list, the scope (target RNCs) are defined and the actual start and stop time are given. The measurement definition (actual counters, gauges, etc.), the measurement interval and the weekdays to be measured are defined in the plan itself.

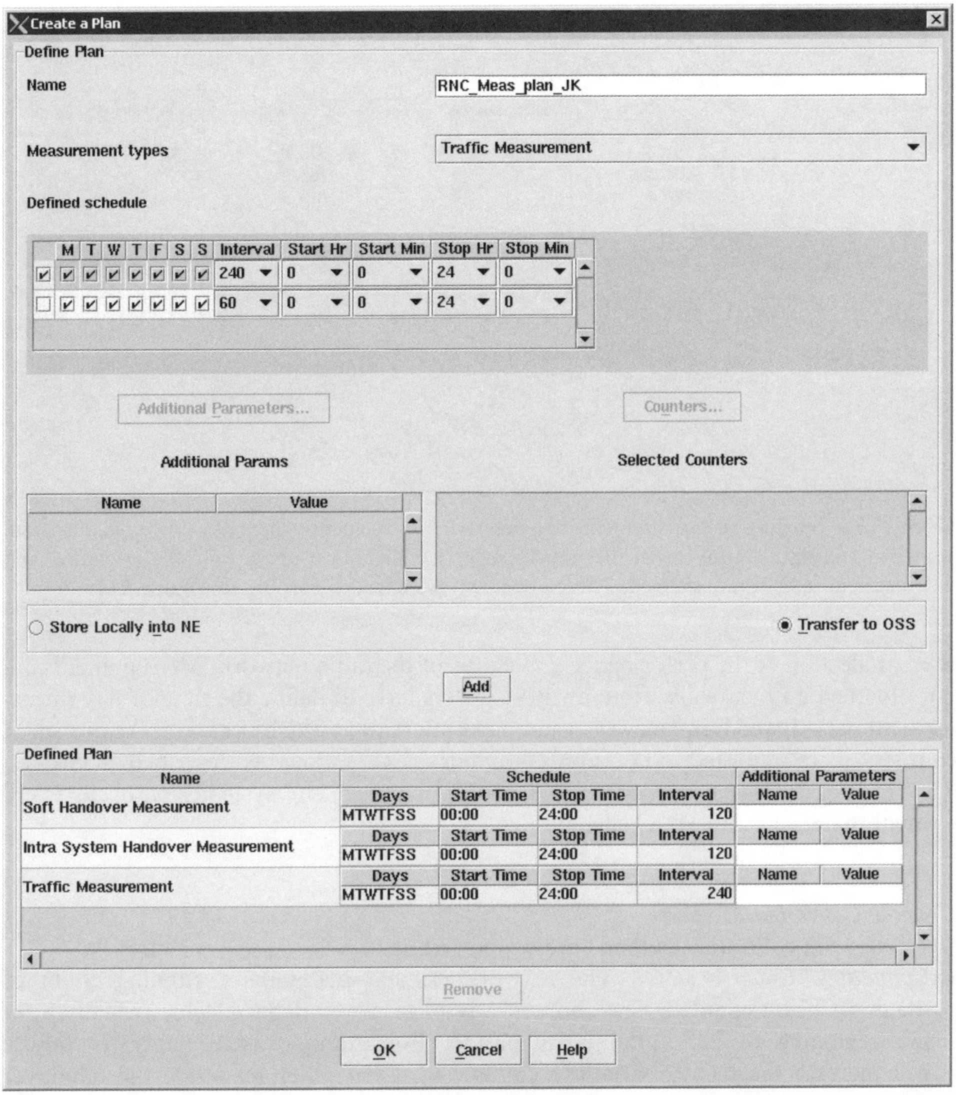

Figure 7.19 Example of measurement plan/job creation in the network management system.

7.3.3.2 Measurement Definition

As described in Section 7.1, the optimisation process requires not only configuration functions but also monitoring and reporting functions. Numerous measurements from the radio access network are required to support the optimisation feedback loop. From the network it is possible to collect an abundance of different kinds of information, such as data on RAB setup attempts. This is presented in the form of raw counters. Every counter belongs to a specific measurement area – e.g., soft handover measurement – according to the 'nature' of the counter. Through its measurements an RNC can give a

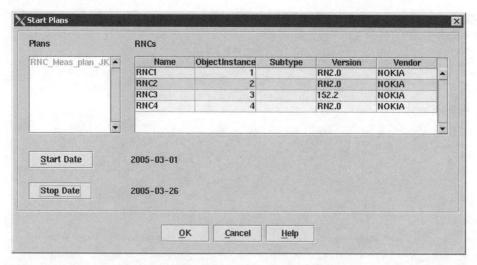

Figure 7.20   Starting of measurements in a network management system. At this stage the actual start and end date are given. In the measurement job definition phase only the weekdays and measurement interval are defined. The plan shown is defined according to Figure 7.19.

user a reflection of the performance and status of the radio network. Measurements are not obtained automatically from an RNC; users have to define the time of day (measurement period) and frequency (measurement interval) at which they wish the measured counters to be reported – i.e., apply administrative actions as described in Section 7.3.3.1. A user can be at either a network management system site or a 'local' element manager at an RNC site.

Counter Collection Methods
The following collection methods are supported:
- *Cumulative Counter (CC)*. The network element maintains a running count of the event being counted. The counter is reset to a well-defined value (usually 0) at the beginning of each granularity period. For example, each handover failure will increase the handover failure counter by 1. In order to derive the handover failure *ratio* we need another counter, which is updated every time handover is attempted.
- *GAUGE (dynamic variable)*, used when the data being measured can vary up or down during the period of measurement. Gauges represent dynamic variables that may change in either direction. Gauges can be integer- or real-valued. If a gauge is required to produce low and high tide marks for a granularity period (e.g., minimum and maximum call duration), then it shall be reinitialised at the beginning of each granularity period. An example of such a measurement would be a maximum hold for uplink interference measurements. If a gauge is required to produce a consecutive readout over multiple granularity periods then it shall only be reinitialised at the start of a recording interval. Recording the uplink interference level continuously is an example of the latter case.

- *Discrete Event Registration (DER)*, when data related to a particular event are captured and every *n*th event is registered, where *n* can be 1 or larger.
- *Status Inspection (SI)*. Network elements maintain internal counts for resource management purposes. These counts are read at a predetermined rate, the rate is usually based upon the expected rate of change of the count value. Status inspection measurements shall be reset at the beginning of the granularity period and will only have a valid result at the end of the granularity period.

Measurement Families

As described in [15] the requirements for measurement data originate from different needs and use cases. *Traffic* measurements provide the data from which, for example, the planning and operation of the network can be carried out. Traffic-related information includes among other things traffic load measurements on the radio or core network interfaces (signalling and user traffic); usage of resources within the network nodes; user activation information and use of supplementary services, etc. *Network configuration evaluation* is another application area. Once a network plan, or configuration changes to an existing plan, have been implemented it is important to be able to evaluate the effectiveness of these changes. Further, it is important to have a pre-defined set of measurements to support different functionality areas in the network. Typically, the measurements to support monitoring of the performance of admission control, handover control, packet scheduling and load control, not forgetting *Quality of Service* (QoS) related measurements. The user of a PLMN views the provided service from outside the network. That perception can be described in observed QoS terms. QoS can indicate the network performance experienced by the user. The QoS-related configuration parameters applied by the network to a specific service or users are also relevant to be monitored. It is important to verify the delivered QoS and determine the charges levied towards the user for the provision of those services. *Resource access and resource availability* set their own requirements on the measurements to be collected. For accurate evaluation of resource access, each measurement result would need to be produced for regular time intervals across the network, or for a comparable part of the network. The availability performance is dependent on the defined objectives – i.e., the availability performance activities carried out during the different phases of the life cycle of the system – and on the physical and administrative conditions.

The measurements are, as mentioned, divided between different needs. The same measurement can provide information for multiple cases. Nevertheless, it is essential that each measurement's relation to others is well-defined, otherwise a situation may arise where there is a gap in the wanted information – e.g., no measurement covers a needed area, or the same information is duplicated in two or more measurements. To avoid such situations, it is best to start by identifying the information required from the radio network so that the measurements can be related to each other at a desired level. The information required from the radio network is related to:

- QoS control and performance;
- control plane (C-plane);
- user plane (U-plane);

- power control performance;
- load control performance;
- admission control operation and performance;
- packet-scheduling performance;
- handover control performance.

   The information needs to be collected from radio and (packet) core network elements in order to fulfil usage case requirements. The above-mentioned list needs to be split into smaller more detailed measurement groups, or families, since the high-level definitions consist of many subfunctions – e.g., handover control performance contains soft handover, inter-frequency handover and inter-system handover performance figures.

   In [16] the measurement families are introduced. In total, 17 families are introduced. These include measurements related to GSM/UMTS inter-system changes, SRNS relocations, soft handover, hard handover and inter-radio access technology handover. In addition to the handovers, mobility management is its own entity containing measurements from Serving GPRS Support Node (SGSN). Measurements related to Radio Link Control (RLC), signalling and Radio Resource Control (RRC) also form their own measurement families. RAB, UMTS bearer service measurements and area measurements related to GPRS Tunnel Protocol (GTP) are also listed. These measurement families can be split into real measurements. In addition to the standardised measurement families there can be vendor-specific additions. These additions are often related to the monitoring of RRM algorithm performance and cell recourse utilisation. Figure 7.21 presents one subset of the measurement families in RAN, their relations to the Open Systems Interconnection (OSI) model, support for RRM functionalities/3GPP signalling protocol follow-up, and the QoS aspect.

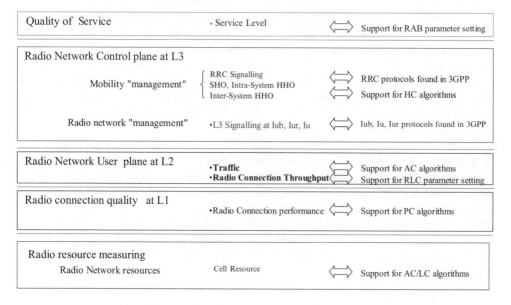

Figure 7.21   3G radio network measurement areas.

Counters and Network Resources

The defined physical measurements must always represent a collection of counters in relation to a network resource (Section 7.1.1.3). The most common example of a network resource is the UtranCell. Others are the RNC or a defined interface (Iub, Iur, Iu).

Before collecting the counters from the network and using them in making the right measurements, one characteristic of the radio access network must be noted and also shown in the measurements. This is the role of an RNC, which can act as an Serving RNC (SRNC), Drift RNC (DRNC) or Controlling RNC (CRNC). Some measurements are carried out in a CRNC and the counters are also separated for SRNC and DRNC. These measurements are currently:

- L3 signalling at Iub/Iur;
- soft handover/intra-system handover;
- traffic.

Other measurements are also carried out in a CRNC but counter separation into SRNC or DRNC is not needed. These measurements are currently:

- L3 signalling at Iu;
- RRC signalling;
- cell resource.

Some measurements are carried out in an SRNC. These are currently service level measurements. Figure 7.22 presents a view of the radio network measurements obtained from a radio access network.

Naming of the Counters

Every measurement must have a unique identifier – e.g., M1000C1. This means that only one object in the network can own the measurement in question and only one measurement type under that object can have that name. Other relevant information depends on the 'usage' of the measurement. Information common to all counters is the counter name, reason for update, and dependences on other counters. For fast and network-wide follow-up/optimisation, counter names should correspond to standardised names and terms – one naming convention is introduced in [16].

Reason for Counter Triggering

All measurement descriptions should include a sufficiently detailed description of the reason or *condition* for the update – e.g., signalling-related, failure, policy decision, together with dependences on other measurements, if any. For QoS and control- and signalling-related measurements the triggering point is essential information. This tells at what stage of a signalling sequence the counter was triggered. For example, for a mobile-terminated call a successful RRC connection setup can be triggered as in Figure 7.23.

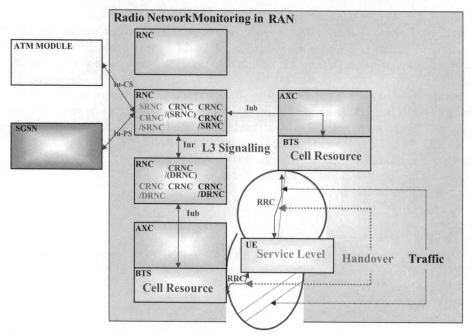

Figure 7.22   Radio network monitoring in a radio access network.

7.3.3.3   Measurement Storing and Post-processing

Storing of Raw Measurements

It is necessary for the network element to retain the measurement result data it has produced until they have been sent to, or retrieved by, the destination OSs. The storage capacity and the duration for which the data will be retained at the network element will be operator- and implementation-dependent. If the measurement result data are routed to a network manager via the element manager, then it is necessary for the element manager to retain the data at least until they have been successfully transferred to the network manager.

Typically the measurement results produced by network elements are transferred to an external network management system for storage, post-processing and presentation to the system operator for further evaluation. In a network with more than one network management system the data may be required by several OSs in order to handle the regional border areas. It is therefore necessary to support the possibility for multiple destinations for the transfer of measurement result data. From the network element to the element manager, the results of the measurement jobs can be forwarded in either of two standard ways: either as notification or as files. In the first case the scheduled result reports are sent to the element manager as soon as they are available. In the latter case the reports are stored in the network element (files) and transferred to or retrieved by the element manager when required.

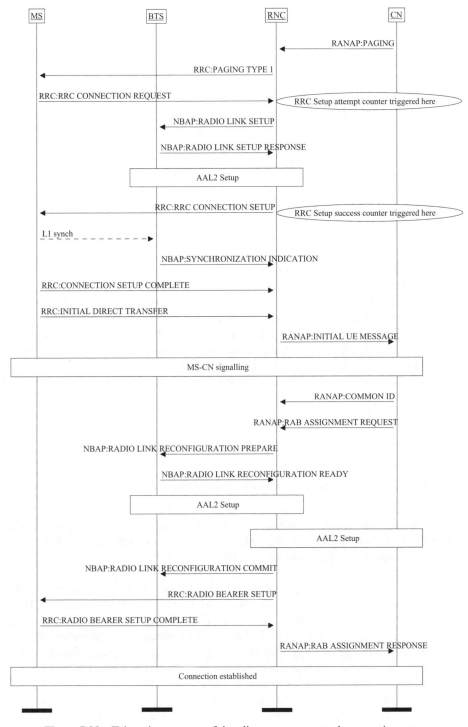

Figure 7.23 Triggering a successful radio resource control connection setup.

From the network to the network manager, measurement results can be forwarded via a bulk transfer (i.e., file-based) interface. It is an implementation option whether this interface to the network manager resides in the element manager or in the network elements.

There are several different users (e.g., network planners, operations engineers, business managers) for the measured data. Therefore, the collected data should be stored in databases where these users can access it easily. Because of the large number of users, data should be collected frequently and in the short term. For essential measurements, data collection and storage every hour is recommended. The amount of data collected can vary greatly, depending on the measurement activated. Usually a measurement includes between 50 and 300 counters. The sizes of the collected measurements together with their collecting frequencies should be taken into consideration when reserving storage capacity in a network element).

Post-processing of Raw Measurements
To obtain optimal benefit from the collected information, a network management system should have good post-processing facilities for the raw measurements. Usually it is very hard to notice problems in the network just by looking at the counter values; some graphical presentation is needed. With graphical views where a user can compare counters and other measurements with each other on a screen, problem solving is fast and more extensive. More about data processing and advanced visualisation and analysis methods for measured performance data is in Chapter 9.

7.3.3.4   Radio Access Network Key Performance Indicators

Operators have historically reported the performance of their networks against a set of KPIs. These KPIs are inherently network focused and they provide an indication of the end-to-end service delivery that the network supports. Nevertheless KPIs are an important measurement for network operations and will continue to be so for the foreseeable future.

By definition a PI is a measure that gives information about the performance of a network element, a process or a function in an NE or in a network subsystem. A KPI is considered an important performance measure to be followed. Later in this chapter, the term 'KPI' is used for both cases. In a radio network operator's organisation, different groups (management, marketing, operations) are interested in slightly different sets of KPIs in their reports – i.e., there is a need for several KPI sets to fulfil all parties' needs – and the definition of a 'KPI set' is always a subjective matter.

The move towards service-focused management leads to a requirement for a new 'breed' of indicators that are focused on service quality rather than network performance. These KQIs are discussed in more detail in Chapter 8.

Radio Access Network Measurements to Key Performance Indicators

Most KPIs are composed of several raw counters or other measurements collected from the network, mainly because a single raw measurement is at too detailed a level to be used as a KPI as such. However, there are exceptions where a collection of single raw measurement values can be used as a KPI. Also, the excessive number of raw counters would cause problems in understanding the real status of the network. Raw counters can be compiled to meaningful KPIs by using various equations – i.e., logical counters.

In a WCDMA radio access network there are tens of different KPIs to be followed, and there might be several different formulas to calculate a measure which may depend on object level, features used in network elements, etc. To define a formula for RAN calculations, a deep knowledge is required of the interest group's needs, the WCDMA system, vendor-specific counter implementation, basic mathematics, statistical methods and the statistics tools in use. 3GPP has done extensive work to define a uniform set of measurements that should be collected from the network elements of any vendor [16]. These KPIs include RAB management KPIs, like attempted RAB establishments for the CS/PS domain, successful RAB establishments without queuing for the CS/PS domain, failed RAB establishments without queuing for the CS/PS domain, successful RAB establishments with queuing for the CS/PS domain and failed RAB establishments with queuing for the CS/PS domain. In addition to RAB-related measurements numerous RRC-related measurements are defined. These include RRC connection establishment and release related-measurements. In the handover area numerous soft handover and hard handover measurements are proposed. Reference [16] also contains definitions for measurements in SGSN, GGSN, Iu and Gn interfaces. In addition to these standardised measurements vendor-specific KPIs are needed in order to monitor RRM performance and its impact on the network and service quality.

The first sample equation is presented in Equation (7.1): the RAB setup and access complete ratio for CS *voice* calls (i.e., a subset of CS domain measurements). The formula is based on cell-level measurements and related counters. The numerator is the sum of all successfully completed RAB setups (RAB setup and access phases included), and the denominator is the sum of all setup attempts. This formula could also be used, as such, for area levels covering several cells:

$$100 * \frac{\text{Sum(RABSetupAccCompforCSVoiceCall)}}{\text{Sum(RABSetupAttforCSVoiceCall)}} \% \qquad (7.1)$$

Equation (7.2) gives a sample formula for calculating a RAB setup complete ratio, including all traffic classes. When using a formula the study period may be different from case to case. One might be interested in busy hour (or busy period) averages from the previous week, while others might want to analyse 24 hr averages from the previous month. The formula itself does not change when the study period changes, but there might be a need to define several templates in the reporting tools used:

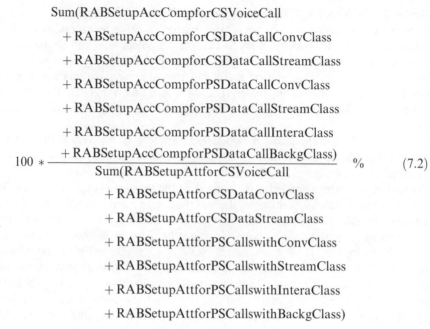

$$100 * \frac{\begin{aligned}&\text{Sum(RABSetupAccCompforCSVoiceCall}\\&+ \text{RABSetupAccCompforCSDataCallConvClass}\\&+ \text{RABSetupAccCompforCSDataCallStreamClass}\\&+ \text{RABSetupAccCompforPSDataCallConvClass}\\&+ \text{RABSetupAccCompforPSDataCallStreamClass}\\&+ \text{RABSetupAccCompforPSDataCallInteraClass}\\&+ \text{RABSetupAccCompforPSDataCallBackgClass)}\end{aligned}}{\begin{aligned}&\text{Sum(RABSetupAttforCSVoiceCall}\\&+ \text{RABSetupAttforCSDataConvClass}\\&+ \text{RABSetupAttforCSDataStreamClass}\\&+ \text{RABSetupAttforPSCallswithConvClass}\\&+ \text{RABSetupAttforPSCallswithStreamClass}\\&+ \text{RABSetupAttforPSCallswithInteraClass}\\&+ \text{RABSetupAttforPSCallswithBackgClass)}\end{aligned}} \% \quad (7.2)$$

KPI documentation plays a very important role, especially since terms such as Drop Call Ratio (DCR) are heavily overloaded – i.e., different parties use the same names when referring to slightly or even totally different things. Typically, each formula has a unique reference code, name, description and calculation formula. Alarms collected from the WCDMA RAN can also be used as KPIs. Some typical KPIs are listed in Table 7.1.

Table 7.1 Typical WCDMA key performance indicators. Note that some of the measurements are vendor-specific.

Object	Metrics
Capacity	Average uplink loading [dBm]
	Average downlink loading [dBm]
	Average random access channel throughput [kbps]
	Average forward access channel throughput [kbps]
	Average paging channel throughput [kbps]
	Uplink dedicated channel throughput [kbps]
	Downlink dedicated channel throughput [kbps]
Access	Radio resource control setup and access complete ratio [%]
	Radio access bearer setup and access complete ratio, for each traffic class [%]
Success	Radio resource control drop ratio [%]
	Radio access bearer drop ratio, for each traffic class [%]
Handover	Soft handover overhead [%]
	Hard handover failure ratio [%]

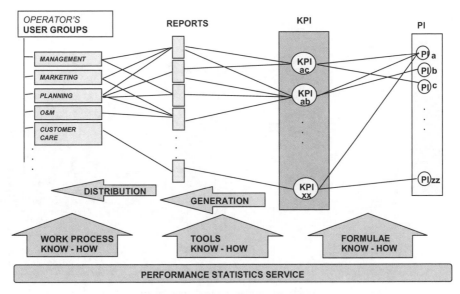

Figure 7.24   Reporting environment.

The KPI-reporting environment from raw measurement to KPIs is depicted in Figure 7.24. Before analysing the network's performance, special attention should be paid to how the KPI formula has been implemented in the reporting tool and how the tool has been used: the measurement period, the number of measurement samples during the study period and the averaging method used. In addition the user group plays an important role. Management needs are different from planning engineers' needs. At worst, one may end up with totally wrong conclusions when analysing the reported PIs if the items listed above are not considered and understood. Reference [16] contains an extensive list of KPIs as defined by 3GPP.

7.3.3.5   Radio Access Network Optimisation Process and Key Performance Indicators

The radio access network optimisation process can be defined in very many ways depending on the focus. In Figure 7.25 a simplified WCDMA radio access network optimisation process is described. An operator's master/business plan sets the framework for both short- and long-term performance criteria in terms of planned service area, call blocking and service mix used in dimensioning, soft handover overhead, etc. The process in Figure 7.25 can be divided into three different areas – namely, preparations, measurements and optimisation. During the process it is important to support the rollback functionality in order to return to the previous configuration if needed (see Section 7.3.2).

The starting point for the preparation phase is related to short- and long-term planning related to new services, capacity expansion needs, service area expansions and additional tasks related to new features in network elements or network element upgrades. During the preparations phase the performance targets are defined for

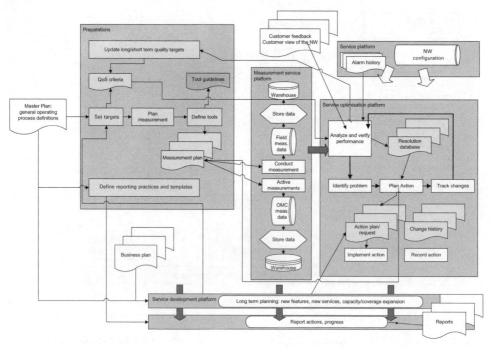

Figure 7.25   A simplified WCDMA radio access network optimisation process chart.

short- and long-term actions. KPI targets are defined for capacity-, coverage- and quality-related measures. Once those measures are defined the method to collect the information is chosen. Two main sources of measurements can be identified: statistics using the management system and the field measurement tool. For management system measurements the following items are defined:

- resources and their availability;
- list of KPIs to be followed;
- measurement schedules (start and stop time);
- summarisation level;
- scope (the objects to be followed);
- reporting.

In the case of the field measurement tool the measured indicators and their post-processing, measurement routes, services to be used (voice, CS/PS data) and number of calls to be generated are addressed.

During the measurement phase, measurements are collected either by storing the counters and other raw measurements in the performance management database or by collecting the information with the field measurement tool. In (signalling) trouble-shooting cases, measurements can be appended with information from a protocol analyser.

During the optimisation phase, measurements are post-processed and the acquired information is utilised in optimisation or troubleshooting activities. The alarm history and customer complaints can help the optimisation personnel to find and solve problems. For optimisation tasks, configuration information is also essential – as indicated in Figure 7.2.

During the optimisation phase, measurement data/reports are analysed together with the configuration information, alarm history and feedback from customer care. Possible problems are identified and an action plan is derived with the support of a resolution database. Two main types of resolutions can be introduced: hard ones and soft ones. Hard ones are associated with hardware-related changes, meaning antenna bearing, antenna type or antenna tilt change, site additions, network element hardware changes, etc. Soft changes are parameter changes that can be done centrally using management system capabilities. During the optimisation phase, cost and benefit considerations are made when deciding what type of change to make and in which order to perform the changes.

After the planned action has been implemented, the measurement results are analysed again to find out whether the taken action caused positive changes in the performance metrics. Service optimisation can be considered a never-ending task, aiming to maximise the amount of traffic without sacrificing quality.

7.3.4   Measurement Applications in Network Elements and in the Network Management System

In this section measurement support in the RNC and network-wide measurement support at the network management service level are introduced.

7.3.4.1   Real Time Monitoring in the Radio Network Controller

In addition to earlier described and presented measurements – which are processed with a delay and thus the results cannot be utilised immediately – there should also be real time tools providing monitoring possibilities for looking more deeply at the radio network. The one presented here is online monitoring, in which the user selects a specific monitored item – e.g., monitoring type – and according to parameters – e.g., cell-by-cell – tells from which part of the radio network the detailed information should be obtained. In Figure 7.26 the administration window is shown.

Two different screenshots are chosen as examples:
1. The monitoring of load behaviour in a cell.
2. The monitoring of handovers in a cell.

The result of (1) can be presented as in Figure 7.27 and (2) in Figure 7.28, respectively.

The first screenshot shows the load behaviour in the uplink of a selected cell in relation to basic admission control parameters. Uplink total interference is presented as $PrxTotal$ and the own-cell real time user and non-real time user loads are seen as L_RT and L_NRT. The admission control parameters are presented as $PrxNoise$ (noise floor), $PrxTarget$ (planned uplink interference) and $PrxOffset$.

The second screenshot shows the distribution of different handover types in a cell and the relation of successful and unsuccessful handovers per handover type. There can also

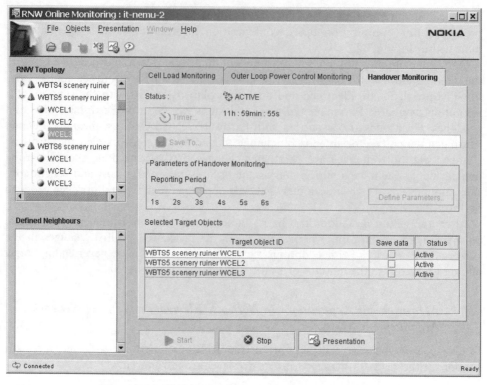

Figure 7.26   Cell-level online monitoring administration.

be other possibilities to be monitored real time like the distribution of different handover-triggering causes received from the UE (User Equipment) (handover reasons are shown in the screenshot).

7.3.4.2   Mobile Tracing Functionality

Measurements and online monitoring tell about service usage, traffic amounts, radio link characteristics and all kinds of errors detected in the radio network. But for locating problems for individual users/mobiles the operator must also have a means to pinpoint a specific user (IMSI) or a specific mobile (IMEI) from the network. This is done by the tracing functionality. Trace is a network-wide tool to collect all kinds of information for a user/mobile from all the network elements (BTS, RNC, MSC, etc.) and all the interfaces (RRC, Iub, Iur, etc.). Figure 7.29 depicts the trace activation mechanism.

Trace Activation
The activation of trace is not possible from any part of the network. Only the HLR and the Visitor Location Registers (VLRs) know if and when a certain UE is registered to

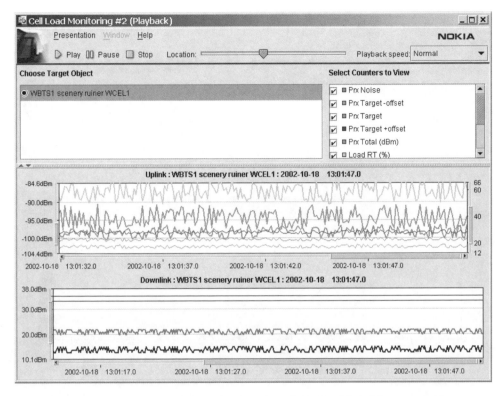

Figure 7.27   Cell-level online monitoring example for load behaviour monitoring.

the network. Therefore, the real activation of UE trace is done when a UE is registered to the network and there is a trace activation set 'on' in the HLR or a VLR.

Reference Numbers
Figure 7.29 shows that – in contradiction to non-real time measurements and online monitoring – trace data can come from any part of the UMTS (or GSM) network. In order to handle the data coming from different domains (NSS, PaCo, RAN) the system must use *reference numbers* to make it possible for the OS to gather data related to the same IMSI (International Mobile Subscriber Identity) or IMEI (International Mobile station Equipment Identity) that are sent from the network to the network management system.

Trace Records
The data are sent from the network to the network management system in trace records. A trace record contains the collected data for a UE. The contents of the trace data depend on trace activation and pre-defined trace record possibilities. Currently the trace record content is not specified by 3GPP but is planned for Release 6.

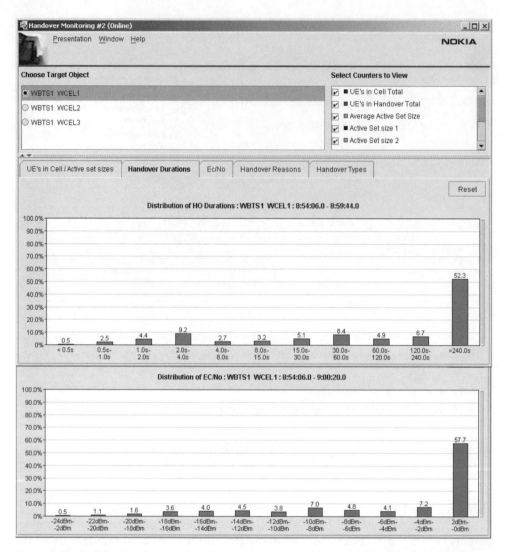

Figure 7.28 Cell-level online monitoring example for handover. Handover duration and E_c/N_0 trigger point views.

7.3.4.3 Reporting Tools in the Network Management System

Generally, the differentiating factor between the element management and network management function is related to data amounts. The reporting functions at the network management system level contain information from a multitude of network elements (like RNCs) network-wide. The requirement of being network-wide can limit the usability of certain data sources, because there are numerous data sources that do not provide naturally network-wide data – e.g., field measurement tools.

In addition to network-wide highly summarised data, the network management system supports drill-in functionality in order to zoom down to the needed details –

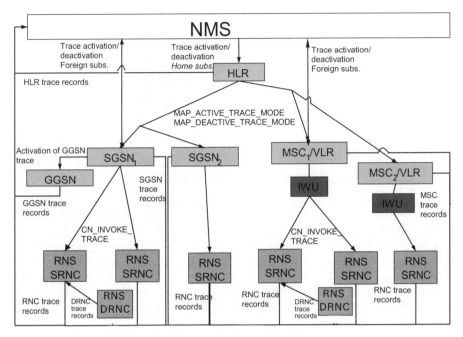

Figure 7.29   The activation of trace.

e.g., in troubleshooting cases. Summarisation can be done in terms of elements/objects, time or the measurement content. Element-level summarisation provides averaged reports over network elements (like RNC-level KPI reporting). Time axis summarisation provides, for example, daily average or busy hour KPIs instead of each value collected separately from the network elements. Measurement content summarisation is related to the grouping of measurements. In high-level reports, for example, the handover performance figure is presented as one value. In the case of poor handover performance, support is offered to zoom into the details of different handover types.

The power of the network management system lies in the fact that large amounts of data are available and that different views of these data can be provided. This is depicted in Figure 7.30. If the scope is the whole network (high number of objects), then it is obvious that the detail level of data in the report is small. Users should be supported so that they can easily find the relevant area and focus. In high-detail views the number of elements to be handled is typically lower but a different kind of data is available as well as time series or distributions.

If the level of detail is very high (high number of counters, KPIs, time series, etc.) then the number of objects should be really low. The reasons are (A) the user cannot abstract too much information at a time and (B) the data interfaces have a limited performance. For practical reasons the area (upper right in Figure 7.30) where we have all the objects within a region with best possible measurement detail level in reports is not feasible and thus not likely. Such reporting would cause high performance requirements of the interfaces from network elements to network management systems.

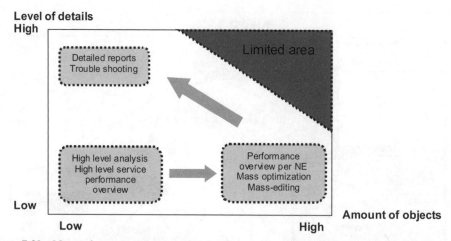

Figure 7.30 Network management system level reporting. In high-level analysis the data are heavily summarised and there are few objects visualised. If the number of objects is high the data are summarised and the number of different types of data to be visualised is low. If the number of elements is low then several key performance indicators, parameters, time series and distributions can be used. Arrows show the typical flow from one use case to another.

The support network management system level reporting tools provide is as follows:
● 'Drill-in' functionality to support the analysis of data by moving to a higher level of detail – e.g., through Web links in the result tables. This is used in data 'crunching' when moving from higher to lower object levels or from high to more detailed data summarisation levels, as depicted in Figure 7.30.
● Automatic creation of Top-N[1] lists into working sets from output-scheduled reports.
● Support for area/grouped selection of objects by the use of working sets. This is used in optimisation/troubleshooting to focus on a specific part of the network or a set of objects with a similar performance – like Top-N lists.
● Detailed reports at KPI level including trend graph and breakdown of KPI formulas showing the values of individual counters (used in optimisation/troubleshooting to understand sudden changes in the KPI trend).

Examples of network management system level reporting application output are in Figures 7.31 and 7.32.

Reporting solutions should be flexible towards scope selection. The size of the working set can be from a single element to a whole cluster. Another level of reporting is needed for SLA support and this is addressed in Chapter 8.

7.3.4.4   Active Service Monitoring

Traditional network management is based on monitoring of individual network elements. This gives a view of the network and service status, but does not always

---

[1] A 'Top-N' list is like a top 10 list.

DL DCH throughput per traffic class - RSRAN011 (2004.12.06 - 2004.12.19) Time Agg: Whole_period
Object Type: RNC, Object(s): '45306021' Object Aggregation:WCELL   (Row 1 ... 14 / 14)

Date	RNC name	WBTS Name	WCELL co_gid	WCELL Name	WCELL ID	Allocated DL DCH Capacity for CS Voice	Allocated Downlink Dedicated Channel Capacity for CS Voice	CS Conversational	CS Conversational, Erlangs	CS Conversational, Minutes	Allocated DL Dedicated Channel Capacity for CS Streaming	Allocated DL Dedicated Channel Capacity for Data Calls	Allocated DL Dedicated Channel Capacity for PS Streaming	Allocated DL Dedicated Channel Capacity for PS Interactive	Allocated DL Dedicated Channel Capacity for PS Background
Total	RNC4	Hippos	45321021	WHippos-1	1	0.00	0.00	0.00	0.00	0.00	0.00	0.00	0.00	0.00	0.00
Total	RNC4	Hippos	45325021	WHippos-5	5	6.47	6.39	7.54	0.12	105.97	0.00	3.81	0.00	0.00	0.10
Total	RNC4	Makkyla	45318021	WMakkyla-1	1	0.00	0.00	0.00	0.00	0.00	0.00	0.00	0.00	0.00	0.00
Total	RNC4	Makkyla	45319021	WMakkyla-3	3	0.00	0.00	0.00	0.00	0.00	0.00	0.00	0.00	0.00	0.09
Total	RNC4	Perkkaa	45329021	WPerkkaa-3	3	0.14	0.12	0.08	0.00	0.58	0.00	0.04	0.00	0.00	0.00
Total	RNC4	Sateri	45312021	WSateri-1	1	8603.56	8603.56	772.88	12.08	31156.78	0.00	56876.86	90366.18	35510.27	35526.31
Total	RNC4	Sateri	45313021	WSateri-2	2	2093.65	2093.65	4985.16	77.89	140207.74	0.00	58480.38	0.00	30450.46	62156.83
Total	RNC4	Sateri	45314021	WSateri-3	3	3476.66	3476.66	707.15	11.05	31158.69	0.00	126449.82	0.00	15040.34	136299.80
Total	RNC4	Sateri	45315021	WSateri-4	4	2120.55	2120.55	1146.02	17.91	124629.36	0.00	27236.25	44037.79	13186.06	22669.86
Total	RNC4	Sateri	45316021	WSateri-5	5	3019.22	3019.20	0.00	0.00	0.00	0.00	48969.89	8404.63	9304.65	51825.75
Total	RNC4	Sateri	45317021	WSateri-6	6	7832.44	7832.44	0.01	0.00	0.61	0.00	54506.60	114091.13	30334.46	41651.77
Total	RNC4	Sello	45330021	WSello-1	1	1463.63	1463.63	0.00	0.00	0.00	0.00	0.00	0.00	0.00	0.00
Total	RNC4	Sello	45331021	WSello-2	2	5428.94	5428.94	0.00	0.00	0.00	0.00	0.00	0.00	0.00	0.00
Total	RNC4	Sello	45332021	WSello-3	3	1.32	1.32	3.92	0.06	22.02	0.00	1.96	0.00	0.00	0.00

Figure 7.31 An example report generated by a network management system application. Key performance indicators are in columns, elements in rows.

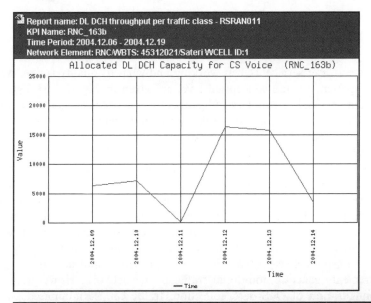

Figure 7.32 An example of drill-in into a specific key performance indicator and trend analysis. Starting point is the report in Figure 7.31. The first column of the table is opened to time series.

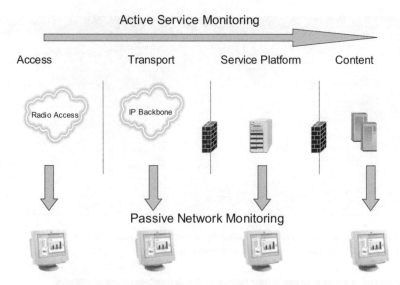

Figure 7.33   Active vs. passive network monitoring.

guarantee that the end-user service itself is working and what its quality is like. Active and passive measurements at the conceptual level are given in Figure 7.33 Typically these passive measurements, like counters and gauges from network elements, provide a statistical view of a component in the whole end-to-end chain. The network load situation can be analysed for business and resource (hardware/software) planning, and problems in the network may be found which currently have no impact on the service, but will have if not fixed in time. Active measurements are implemented with probes performing regular tests at scheduled intervals. Service usage is simulated as if an end-user uses the service. The probes are distributed around the network to gather results from all its parts. This method provides 24/7 statistics. With consistent test transactions performed at regular intervals, fault situations are immediately detected. Comprehensive statistical data are obtained as well, offering a reliable analysis of historical trends. Utilising active probing also allows proactive fault detection.

Active measurements generate additional traffic in the network. However, with a well-planned configuration of QoS data collection, traffic can be minimised, and in normal conditions the volume of the test traffic is likely to be very small compared with the actual end-user traffic. The amount of traffic generated using active measurements has an effect on the coverage of the solutions. Thus it needs to be planned carefully, taking into consideration the use case, all of which begs the question: Is it troubleshooting or general performance monitoring? The active service test can be seen as a simple, straightforward way to verify the service in the same way the customer will experience it. Due to the concept it is an easy way to monitor multi-vendor networks, as well as parts of the network that do not belong to the operator such as transmission lines, application server and IP backbone.

Passive network and active service monitoring should not be seen in opposition but as complementary methods to get the most out of your network. Monitoring services

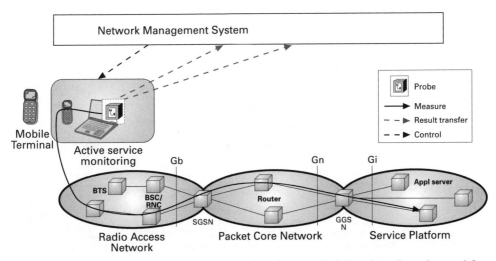

Figure 7.34 Probe(s) collect quality of service data from verified (monitored) services and forward the resulting data to the network management system, where monitoring, analysis and reporting products utilise the data.

using active monitoring is perceived as cost-effective because – with only a few or even a single probe – a service can be fully monitored end-to-end, probes can be centrally managed by the administration tool and can be easily distributed around the network. Measurements can be combined with other network management system level performance management applications. Active measurements provide a comprehensive solution for end-user service monitoring in mobile operator and service provider environments.

Probe(s) collect QoS data from verified (monitored) services and forward the resulting data to the network management system, where monitoring, analysis and reporting products utilise the data, as depicted in Figure 7.34. All probes are centrally configured and managed by a configurator running in the network management system. The probe verifies the service at regular intervals by simulating end-user behaviour. In Figure 7.34 the service is verified end to end, meaning from the mobile to the application server – e.g., a WWW or MMS server. Another possible scenario is verification of the service beginning from the Gb, Gn or Gi interfaces towards the application server. In case of Gb or Gn the BSS (Base Station System) – respectively, the SGSN – is emulated. This allows an isolated view of how the radio access network, packet core network or service platform behave for a certain service. Example use cases are given in Figure 7.35.

In some cases two probes are involved in a single service verification. This is the case for end-user services where an item is sent from one end-user to another, like MMS and SMS. Usually the probes will produce performance management and fault management data for each service verification. Fault management data are also known as 'alarms'. Alarms are generated based on configurable thresholds per probe. This information is sent to the network management system where several applications can use these data to conclude both network and service performance.

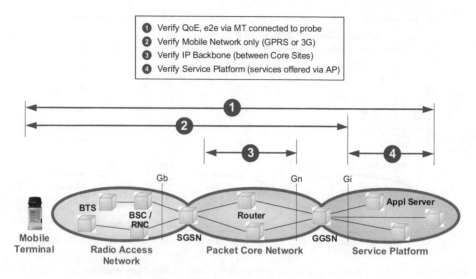

Figure 7.35   Example use cases.

7.3.5   *Optimisation Using Operations System Tools*

One of the themes of this chapter is automation and operational efficiency. Network management systems with workflow support and the availability of actual data from the network aim at OPEX-efficient optimisation solutions. Network management system level statistical optimisation aids the operator to control, visualise, analyse and optimise the mobile network using *both* performance data and configuration data on a network-wide basis.

As depicted in Figure 7.36 automation in connection with statistical optimisation includes automatic access to performance, fault and configuration data, functions to support network performance analysis and reporting, automated workflow support in order to move from one process phase to another. Further, algorithms for automated optimisation are provided (see Chapter 9) in addition to manual optimisation actions. Data availability enables effective diagnostics support for decision making in troubleshooting cases. The most visible differences between the extensive optimisation solution and service assurance tools is the utilisation of configuration data in the process and the possibility to react and change the configuration when needed. The optimisation solution utilises the performance management and configuration management issues discussed in earlier sections of this chapter.

KPI, alarm and configuration visualisation can happen in browser, GIS (map) or navigator view. The GIS functionality brings geographical aspects to the management system and enables *measured* KPI visualisation on cell dominance or cell icon, as well as visual verification of adjacency definitions for location and routing areas. This visualisation function can be used in the scope selection of the optimisation case. For example, the KPIs are sorted according to the performance level. Objects (or network elements) whose performance is below a set target are selected for further analysis and optimisation.

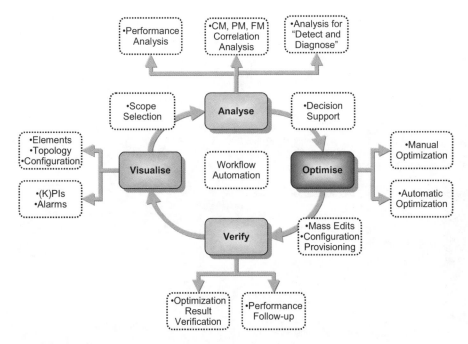

Figure 7.36 Workflow for optimisation supported by a network management system.

The analysis functionality contains further processing of the measurements and correlating the configuration with the performance, among other things. The possibility to visualise configuration management and related performance management simultaneously aids in troubleshooting activities. Often differences in performance can be explained simply by an anomaly in the configuration settings. More details of advanced analysis methods can be found in Chapter 9.

The optimisation tasks can proceed once the situation in the network is analysed. Based on the analysis results corrective actions can be taken. These actions can be either reactive or proactive, anticipating the changes required in order to guarantee stable quality in the network for the future. Optimisation can be done for individual objects and network elements or on a network-wide basis. Workflow supports the provisioning of the changes. In some cases the optimisation results are verified prior to provisioning. This is the case especially with automated optimisation. The proposed solution is further verified and analysed in order to gain confidence on the proposed changes. Another function used for verification is performance follow-up after the change has been made. An example realisation of an optimisation product in the network management system is given in Figure 7.37.

7.3.6 Field Measurement Tool

The optimisation process of a network is easily understood as something conducted purely in a centralised manner – e.g., by utilising network management systems that allow control of the network by, say, parameter adjustment. In many cases, however, a

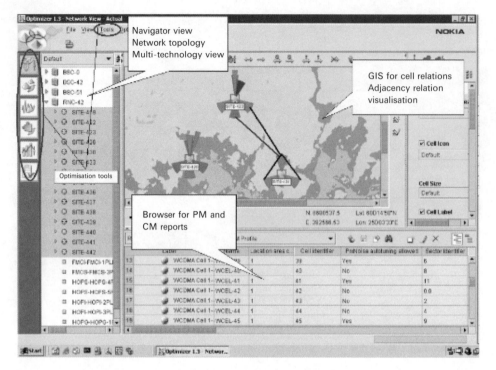

Figure 7.37 Realisation of an optimisation product utilising network management system level data. Navigator provides information on the element topology; browser can be used for configuration management and performance management data visualisation and analysis. The GIS functionality brings geographical aspects to network management. Automated optimisation functions can be found under the Tools menu.

network planner and optimiser must also study the network from a customer's point of view – i.e., conduct field measurements with a proper field measurement tool. In WCDMA, such a need may arise in, for example, the following cases:

• During a network launch phase, a set of field measurements to verify the basic performance of the network is required. In addition, the coverage areas of various services and bit rates need to be verified. Such tests should be conducted in both loaded and unloaded network conditions.

• In general, propagation maps used as input for planning tools in the network planning phase often contain inaccuracies, and need to be verified by means of field measurements to ensure proper network coverage in critical areas.

• Knowledge of soft handover areas is important, and is related to cell coverage measurements in general. Although soft handover is a desirable feature of WCDMA that adds reliability to the network, excessive soft handover (probability >35–40%) may decrease the downlink capacity due to additional downlink interference from soft handover connections. In addition, soft handover requires more system resources from the network (baseband in Node B, transmission and RNC).

- Measurements of Primary Common Pilot Channel (P-CPICH) coverage are also important, because the UE measures this channel for handover as well as cell selection and reselection purposes. Therefore, cell loads can be balanced by adjusting the P-CPICH power levels between different cells, as UEs from a cell with reduced P-CPICH power are more easily handed over to neighbouring cells. For network optimisation this is a challenge, due to the risk of improperly balanced P-CPICH powers. Field measurements are therefore needed to verify this balance in selected areas to avoid pilot pollution.
- Network expansion is also a challenge for network optimisation, since the original system performance must be maintained when adding new sites or more frequencies. Estimation of the new or changed coverage areas and QoS before the integration of the new site is also possible in this case only by conducting field measurements.

7.3.6.1   Basic Operating Characteristics

There are several commercial field measurement tools in the market for GSM/GPRS, and more tools are entering the market for 3G/WCDMA. The basic operating characteristics of these tools are usually almost identical, and the main differences are in the tool software and appearance, and data-logging features. The basic operating characteristics, which can also be regarded as requirements for a usable field measurement tool, can be listed as follows:

- *GPS*. A Global Positioning System (GPS) is required to properly align measurements with a digital map, which is usually provided with the measurement tool software. GPS accuracy nowadays is sufficient to enable tracking of measurements to street level accuracy (1–2 m). In special cases a differential GPS may also be used. This would enhance accuracy even further, but can be considered unnecessary for most applications.
- *PC hardware (laptop computer)*. Usually any field measurement tool is supposed to be portable in the sense that it could be operated without a vehicle, and carried around in, for example, city centres and office buildings. Any standard laptop computer should be sufficient to handle the tool software and allow for adequate data storage capacity.
- *Terminal (UE)*. Due to the interface between a terminal and the measurement software, it is not clear whether every 3G terminal can be used with every field measurement tool software. However, all commercial 3GPP-compliant terminals should support the measurements defined by the 3GPP standards [2].
- *Measurement tool software*. The terminal software operates the data-logging features, and in many cases also allows the user to plot the data and terminal location on a map with the help of the GPS signal. Logged data can be stored on the computer for further processing. Post-processing software can also be included in the measurement tool software package for offline processing of measurement data for later data analysis or demonstration purposes.
- *External antennas and cabling*. In many cases an external antenna may be useful in field measurements. In any case, in order to operate a GPS, a clear line of sight between the GPS antenna and the serving satellites is usually required. This may be difficult to achieve if the GPS receiver is located inside a measurement vehicle. On

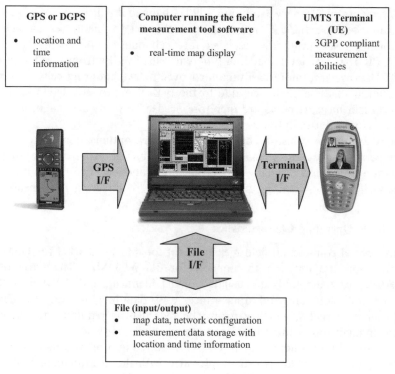

Figure 7.38 Typical field measurement tool composition with the main functional modules and interfaces.

the terminal side, the use of various antennas and antenna orientations can be investigated using external antenna setups, unless a fixed terminal antenna is specifically under study.

The typical modules of a field measurement tool for WCDMA networks are shown in Figure 7.38. The various interfaces present are also shown.

7.3.6.2   Parameters in Field Measurement Tool

To conduct optimisation tasks in WCDMA networks properly, the field measurement tool employed should support a number of parameters describing the network performance. In addition to direct quality-related parameters, such as number of call attempts, other parameters of a more supporting nature should be monitored by the tool.

The following list summarises some of the most relevant parameters to be measured from a WCDMA network using a field measurement tool. As mentioned, however, not all of these are directly linked to optimisation tasks. These supporting parameters can be used, for example, to investigate on a more detailed level a detected problem in the network, when other indicators are not providing enough information:

- *General information*. Service type (voice, data), Mobile Country Code (MCC), Mobile Network Code (MNC), downlink primary scrambling code number for active and neighbour set, carrier number, cell ID.
- *Coverage*. P-CPICH Rx E_c/I_0, UTRA cell signal strength (RSSI), P-CCPCH RSCP (Received Signal Code Power).
- *Signalling*. L2 messages (uplink/downlink), L3 messages (uplink/downlink).
- *Quality*. Downlink transport channel Block Error Rate (BLER), number of call attempts, call setup success ratio, call success ratio, uplink/downlink coding scheme.
- *Handovers (soft/hard)*. Number of handover attempts, handover success ratio, handover type, active set list, neighbour list, E_c/I_0 values of individual RAKE fingers.
- *PC*. UE transmit power per call, Dedicated Physical Channel (DPCH) SIR.

7.3.6.3 Field Measurement Sub-process of the Network Optimisation Process

In general, network optimisation is the part of the WCDMA network planning process that enables the availability of various network services and provides a defined service quality and performance. During the network launch period, however, troubleshooting is the main focus of pre-launch optimisation. This concentrates on locating problem areas and fixing them accordingly. Network expansion and/or traffic growth is another period in the network life cycle when field measurements are a vital part of the optimisation process.

Field measurements are naturally just a part of the whole network optimisation process, where network management system data and field measurements together are employed to determine the KPIs, and hence the QoS of the network.

The role of the field measurement tool is important during pre-operational mode optimisation, when there is only a limited amount of performance data available from the network elements, for the simple reason that the network carries little commercial traffic.

Espoo (Finland) Case Study

This section contains a single use case for the field measurement tool – i.e., evaluation of diversity gain in the WCDMA network. The basic layout for the experimental WCDMA system used in the measurements is shown in Figure 7.39.

In this case study the effect of uplink diversity on UE transmitted power level is demonstrated with field measurements. The propagation environment in the area is mainly suburban, rural areas can be found within the cell edges. Two routes were defined for the measurement campaigns, allowing various sub-types of these propagation environments to be investigated:

- *Route A*: This route goes through a semi-urban area (four-storey and five-storey buildings rather closely spaced) along a local main street. Towards the end of the route the buildings are clearly more widely spread. The maximum speed along this route is 50 km/h, although there are several traffic lights in the early part of the route, resulting in highly variable measurement–vehicle speeds along the route (from 0 to 50 km/h).

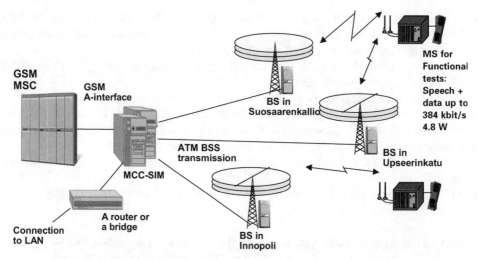

Figure 7.39   WCDMA experimental systems, located in Espoo, Finland.

- *Route B*: This route goes north along a regional highway as far as the limit of coverage. The speed limit is 70–80 km/h, with a few traffic lights. Route B is a rather open area and thus contains less multi-path than route A.

Two antenna configurations were investigated by measuring them over the same routes with otherwise similar settings in the network elements (UE, Node B and MCC-SIM). These two configurations included case VV with vertically polarised Rx and diversity receive branches, vertically polarised transmitter and case V with vertically polarised receive branch, but no receive diversity and vertically polarised transmitter. Both routes A and B are approximately aligned with Node B antenna directions. Example results for the two antenna configurations are shown in Figure 7.40(a)–(b) for route A (route B results are not separately presented).

The results of this limited comparison show that the antenna gain with and without uplink diversity is roughly 5 dB on route A and 2.7 dB on route B. The difference in the gains between routes A and B is due to different propagation channel characteristics. However, the results clearly emphasise the importance and benefit of uplink diversity schemes in adding network capacity in WCDMA systems.

7.4   Summary

In this chapter the optimisation process was discussed. Network management system standardisation focuses on a multi-vendor-capable management system by defining interfaces, a common resource model and by setting requirements for configuration management and performance management (not forgetting fault management issues). On the other hand, the TMF framework introduces the management layers from

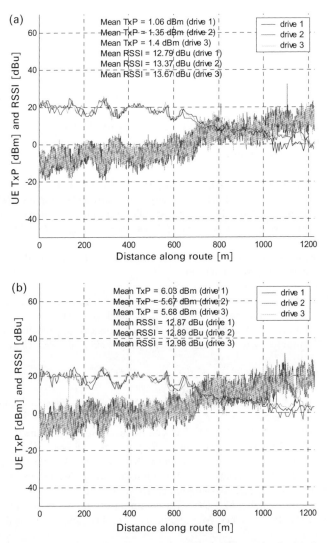

Figure 7.40 User equipment transmitter power [dBm] and received signal strength indicator [dBu] as a function of distance along route A, both with and without uplink diversity (shown in plots (a) and (b), respectively). Three identical drives for both antenna configurations are shown. The two Node B antenna configuration cases are VV and V – i.e., V-polarised receiver and receive diversity branches, and V-polarised receive branch and no uplink diversity, respectively. Note that 0 dBu corresponds to −115 dBm in the received signal strength indicator graphs.

element management through network management to service management and business management in operator organisation. Proper element management and network management are prerequisites for effective service management and service assurance and optimisation.

References

[1] C. Willard, T. Rochefolle, C.C.E. Baden, J.C.S Cheung, S.G. Chard, M.A. Beach, P. Constantinou and L. Cupido, Planning tools for mobile networks. *Electronics & Communication Engineering Journal*, **5**(5), October 1993, pp. 309–314.

[2] 3GPP, Technical Specification 25.215 v5.5.0 (2003-09), Physical Layer: Measurements (FDD) (Release 5).

[3] *Network Management Detailed Operations Map*, GB908, Evaluation Version 1.0, March 1999, TeleManagement Forum.

[4] *Telecom Operations Map*, GB910, Approved Version 2.1, March 2000, TeleManagement Forum.

[5] *Enhanced Telecom Operations Map*, Version 3.0, 2004, TeleManagement Forum.

[6] *SLA Management Handbook*, GB 917-2, Version 2.0, April 2004, TeleManagement Forum.

[7] W. Mende, E. Oppermann and L. Heitzer, Mobile radio network management supported by a planning tool. *Network Operations and Management Symp., NOMS 98*, Vol. 2, 1998, pp. 483–492.

[8] G. Riva, M. Frullone, C. Passerini and G. Falciasecca, Impact of the multiple access scheme on optimal site positioning. *Proc. of International Conf. on Universal Personal Communications*, Vol. 1, 1998, pp. 547–551.

[9] T. Binzer and F.M. Landstorfer, Radio network planning with neural networks. *IEEE VTS Proc. of Vehicular Technology Conf.*, Vol. 2, 2000, pp. 811–817.

[10] P. Calegari, F. Guidec, P. Kuonen and D. Wagner, Genetic approach to radio network optimisation for mobile systems. *IEEE VTS Proc. of Vehicular Technology Conf.*, Vol. 2, 1997, pp. 755–759.

[11] *Service Quality Management Business Agreement*, TMF506, Public Evaluation Version 1.5, February 2001, TeleManagement Forum.

[12] 3GPP, TS 32.101 v5.5.0 (2003-09), 3G Telecom Management: Principles and High Level Requirements (Release 5).

[13] 3GPP, TS 32.600 v5.0.1 (2003-06), Telecommunication Management; Configuration Management (CM); Concept and High-level Requirements (Release 5).

[14] 3GPP, TS 32.102 v5.6.0 (2004-03), 3G Telecom Management Architecture (Release 5).

[15] 3GPP, TS 32.401 v5.4.0 (2004-09), Telecommunication Management; Performance Management (PM); Concept and Requirements (Release 5).

[16] 3GPP, TS 32.403 v5.8.0 (2004-09), Telecommunication Management; Performance Management (PM); Performance Measurements – UMTS and Combined UMTS/GSM (Release 5).

[17] 3GPP, TS 32.622 v5.5.0 (2004-09), Telecommunication Management; Configuration Management (CM); Generic Network Resources Integration Reference Point (IRP); Network Resource Model (NRM) (Release 5).

[18] 3GPP, TS 32.642 v5.5.0 (2004-09), Telecommunication Management; Configuration Management (CM); UTRAN Network Resources Integration Reference Point (IRP): Network Resource Model (NRM) (Release 5).

[19] 3GPP, TS 32.644 v5.6.0 (2004-09), Telecommunication Management; Configuration Management (CM); UTRAN Network Resources Integration Reference Point (IRP): Common Management Information Protocol (CMIP) Solution Set (SS) (Release 5).

[20] 3GPP, TS 32.632 v5.5.0 (2003-12) Telecommunication Management; Configuration Management (CM); Core Network Resources Integration Reference Point (IRP): Network Resource Model (NRM) (Release 5).

[21] *Wireless Service Measurement, Key Quality Indicators*, GB923a, Version 1.6, 2002, TeleManagement Forum.

8

UMTS Quality of Service

Jaana Laiho, Vilho Räisänen and Nilmini Lokuge

In this chapter support for Quality of Service (QoS) in 3GPP systems is considered. The emphasis is on the Packet Switched (PS) part of the 3GPP framework. The focus is on UMTS and WCDMA QoS. A reader interested in General Packet Radio Service (GPRS) and GSM EDGE RAN (GERAN) QoS is referred to [19].

The definition of Quality of Service is first discussed, followed by a classification of end-user services. Then the characteristics and requirements of end-user services are addressed. Next, the 3GPP bearer concept and 3GPP QoS architecture are described. Service provisioning in 3GPP systems are discussed after that followed by a summary.

8.1 Definition of Quality of Service

The term 'Quality of Service' has been overloaded with multiple meanings. The International Telecommunications Union (ITU-T) enumerates the following viewpoints to QoS [4]:

- QoS requirements of customer;
- QoS planned by provider;
- QoS delivered by provider;
- QoS perceived by customer.

In this chapter, QoS is approached from the viewpoint of the customer (end-user) and expanded towards a framework encompassing both customer and provider. Sometimes QoS has also been used as an overarching term to cover mechanisms that are used for differentiated traffic treatment as well. To address the potential ambiguity here, one can differentiate between:

- service performance (end-to-end service quality);
- service quality support mechanisms.

'Service performance' may refer to both of the customer-oriented ITU-T viewpoints listed earlier. In what follows, the term 'service quality' is also used to refer to the former of the two categories. The term 'QoS' will be used where it is an established

Radio Network Planning and Optimisation for UMTS Second Edition
Edited by J. Laiho, A. Wacker and T. Novosad © 2006 John Wiley & Sons, Ltd

practice – e.g., '3GPP QoS profile'. As can be seen below, both end-user experience perspective and provider design perspective come into play. Estimation of end-user experience as such is not addressed – the interested reader is referred, for example, to [11] for further information.

There is some discussion in [19] about estimation of end-user experience of service performance based on technical aspects. However, psychological aspects also play a role in assessing end-user experience [20]. It is preferable that both viewpoints are taken into account [11].

From the viewpoint of an individual user, service quality can be either allocated on behalf of the end-user by the network provider, or requested by the terminal. An important point to observe here is that an end-user does not necessarily know – or should not need to know – which service performance to ask for. As is shown below, adequate service quality allocation is based on both the service in question and the planned service quality for an end-user.

8.2   End-user Service Classification

End-user services have a range of properties associated with them. In order to structure discussion, a conceptual framework ([7] and [11]) is used for discussing the relation of services to 3GPP QoS.

An individual end-user service may consist of multiple components, the characteristics of which may vary from component to component. An imaginary example is provided in [7] where a Voice over IP (VoIP) service is augmented with collaboration functionality. Service components are instantiated to individual end-users as service events with specific attributes. For example, the service could be instantiated for a business user. A service instance is composed of service events, which may be packet flows or transactions. Examples of composition of service instances are provided in [7] and [13].

Service events can be classified as service event types. Grouping of service events provides an aggregation, which can be used for service quality allocation purposes [13].

In what follows, end-user services are classified in five categories. The classification has been devised for the purpose of Service Management (SM):

- data transfer;
- interactive data transfer;
- messaging;
- streaming;
- conferencing media.

Data transfer refers to background transfer of a large amount of data.

Interactive data transfer entails end-user expectation of interactivity, and is typically based on request/reply pattern. Browsing is an archetypal example of this.

Messaging includes services that convey messages between end-users. A messaging service may or may not be session-based.

Streaming refers to transmission of streamed Real Time (RT) content such as video or audio from the provider to the consumer.

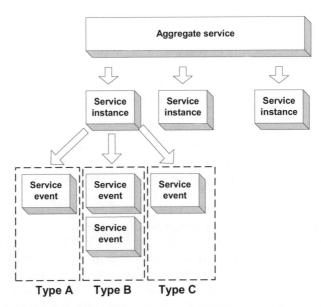

Figure 8.1 Relation between aggregate service, service instances and service events [11].

Conferencing media relate to audio and video sessions, which are used for conferencing purposes.

As depicted in Figure 8.1, an end-user service is typically composed of multiple service event types. The composition of end-user services in terms of service event types is discussed in Section 8.3. The classification will be put into the context of the 3GPP QoS model in Section 8.4.

8.3 Characteristics and Requirements of Services

Different aspects of services are relevant to allocating adequate resources for them. In what follows, they are called 'requirements' and 'characteristics'. Requirements cover the service quality support that is needed by the service in question, whereas the characteristics of a service describe its appearance.

Next, the requirements and characteristics for end-user service classes listed in the previous section are described. Thus, an end-user service including multiple service event types does not necessarily have a single set of requirements and characteristics associated with it, but may have sets associated with each service event type. To which extent this is true depends on the service-quality multiplexing strategy chosen by the mobile operator.

Below, a brief look at generic issues that are related to service quality is provided. Further sources for information about protocol interactions in IP (Internet Protocol) stacks are [8], [11], [15] and [19].

The classification is necessarily only a rough guide to IP-based services and may need to be applied in a modified form for special services.

8.3.1   Generic Issues

There are certain issues that span multiple service event types. These are discussed in the following. The layer numbers of the seven-layer ISO/OSI (International Organisation for Standardisation/Open Systems Interconnection) framework are used, so that L2 refers to the link layer, L3 refers to the network layer and L4 refers to the transport layer.

Layers above L4 are not discussed here – the reader interested in this is invited to study examples relating to browsing, MMS, streaming, gaming and Push To Talk (PTT) in [19]. The same source also includes discussion about header compression, which has peripheral relevance to service quality. Further examples about service performance analyses are provided in [17] and [18].

The Layer 4 protocol – e.g., TCP, UDP or SCTP – has an effect on issues such as reliability. These issues are not discussed here in detail, but some related remarks are made in passing below. Transmission Control Protocol (TCP) provides for reliable data transmission, User Datagram Protocol (UDP) does not provide reliability and Stream Control Transmission Protocol (SCTP) provides both reliable and non-reliable modes. All of the three L4 protocols listed above support flow multiplexing, allowing for sharing of an endpoint IP address through port numbers. If reliable data transmission is not provided on L4, it can be implemented on L2 or the application layer. Layer 3 – i.e., IP – does not provide reliable data transmission.

Typically TCP is used for data transfer, and provides both reliability and flow control functionality. TCP performs flow control based on the arrival pattern of packets at the receiving end using an acknowledgment mechanism. TCP typically cannot differentiate between Protocol Data Units (PDUs) lost because of congestion and PDUs lost for other reasons. In general, TCP performance is a function of packet loss and throughput [9]. TCP performs best when throughput is stable and packet loss is at a minimum. Further details about TCP behaviour can be found in [8].

Transmission of RT media such as telephony or streaming over IP networks requires supporting protocols. Real Time Protocol (RTP) and Real Time Control Protocol (RTCP) are usually used for this purpose in an open protocol environment, but there are also vendor-specific protocols available for this purpose.

A further issue is the policing scheme used for measuring the conformance of traffic to bit rates. In the Internet world, the Token Bucket algorithm is normally used [8].

Packet length has relevance to performance of scheduling ([11], [15] and [19]). The transmission time for a packet is dependent on its length. Delay variations incurred by longer packets can be alleviated with segmentation on the lower layer, but this brings into play reliability issues about the lower layer.

8.3.2   Data Transfer

Below, only background data transfer is considered. Interactive data transfer is discussed in Section 8.3.3.

Since background events are discussed here, the end-user viewpoint is less important than for the other service event types. There are certain cases when end-user viewpoints may be relevant, however. One of them is background loading of data, which will be used directly by the end-user, such as MP3 files. In such cases, it is desirable that the duration of downloading is predictable.

Involved service event types:
- Data transfer flow.
  Requirements:
  ○ reliable delivery;
  ○ stable throughput preferable;
  ○ end-to-end delay can be rather large.
  Characteristics:
  ○ typically uni-directional;
  ○ content size may be large.

8.3.3 *Interactive Data Transfer*

Interactive-data-transfer-type service events have an expectation of human participation related to them. Contrasting this type with the previous one, it is important that the service behaves responsively from the end-user viewpoint.

Individual service events may be carried on top of TCP or UDP. Running Wireless Application Protocol (WAP) over UDP is an example of the latter. When TCP is used, the considerations for data transfer from above apply as such. For UDP, reliability needs to be implemented on top of the UDP layer.

Interactive data transfer is typically based on a request/reply pattern. Two sub-types can be readily identified:
- reply is small;
- reply is large.

When reply is large, considerations of the kind discussed in the context of background data transfer are relevant, whereas for small replies, requirements are lighter. In particular, it is preferable that downloading time is predictable for large amounts of data.

Involved service event types:
- Request message.
- Data transfer flow.
  Requirements:
  ○ service instantiations relatively quick;
  ○ low packet loss preferable;
  ○ stable throughput preferable for reply messages containing large amounts of data.
  Characteristics:
  ○ typically bidirectional;
  ○ request typically small;
  ○ size of reply may vary a lot;
  ○ request and reply temporally close to each other.

8.3.4 Messaging

Messaging is a category of its own, and is related to a variety of end-user services – e.g., email, chat, group chat and PTT. The basic end-user expectation for messaging services is reliable delivery. When the messaging service is interactive in itself, which is the case of chat and PTT, timeliness considerations – discussed above within the context of interactive services – apply.

Involved service event types:

- Uplink message delivery.
- Downlink message delivery.
- Session-based messaging: session setup, session management.
  Requirements:
  ○ service initiation relatively quick;
  ○ reliable transfer of messages;
  ○ end-to-end delay can be large for non-session-based services but should be in 'interactive' category for session-based messaging.
  Characteristics:
  ○ bidirectional;
  ○ content size varies.

8.3.5 Streaming

Streaming refers to transmission of a continuous media stream across the network. Media thus transferred may be audio, video or both. Streaming relieves the terminal of having to store the entire media before being used. Streamed media is usually transported on top of UDP, but may also be transported on top of TCP – e.g., for firewall-traversal reasons.

The speech signal typically consists of 'talkspurts' – i.e., continuous periods of voice signal – and silence periods between them. During silence periods, a media stream is not transmitted but so-called comfort noise may be transmitted. The relative percentage of the two types depends on the coding used, audio hardware and background noise level, for example [11]. For music, an audio stream is typically continuous. A feature common to multiple-coding schemes, temporally correlated packet loss is typically more harmful than temporally distributed packet loss, even though the overall packet loss rate would be identical in the two cases.

Also a video stream is typically continuous in the sense that frames are transmitted at a pre-defined rate. A video signal created by a typical encoder such as H.263 is typically bursty for the reason that encoding is based on transmitting a 'full' picture at regular intervals and differences from the last full picture between them. A streamed video signal can be pre-shaped at the service provider's premises.

Involved service event types:

- Control messages.
- Streamed media (includes both media proper and flow control messages such as RTCP).
  Requirements:
  ○ service instantiation quick;

o packet loss can be tolerated, audio is more important than video;
o end-to-end latency can be large;
o delay variation of the order of seconds can be allowed for a media stream;
o stable throughput required for the media stream.
Characteristics:
o control signalling applied in irregular fashion;
o control signalling uplink, media stream downlink;
o media stream is continuous, but audio stream may have a temporal ON/OFF structure if it consists of speech;
o video stream may be bursty if it is not shaped.

8.3.6 Conferencing Media

Conferencing is an archetypal session-based service, and is here considered to contain at least a voice signal. In addition, it can contain a video signal, as well as collaborative components such as chat or data-sharing tools.

Regarding voice signals in conferencing that are specific to telephony, during silence periods, a media stream is not transmitted but so-called comfort noise may be transmitted. This leads to an ON/OFF-type media pattern, which is typical of telephony. There are also other telephony-specific issues such as delay jitter and its relation to packet loss. An example of the relation between delay jitter compensation and packet loss can be found in [21].

Involved service event types:

• Session management messaging.
• Media flows.
• Other session components, such as chat information sharing.
  Requirements:
  o availability is high;
  o reliable transport and low delay needed for control messages;
  o media stream requires stable throughput;
  o voice telephony signal requires low end-to-end delay;
  o packet loss allowable for media streams;
  o packet loss should be random rather than temporally correlated for media streams.
  Characteristics:
  o instantiation pattern is random;
  o service event invocation pattern random within sessions;
  o media stream is typically bi-directional.

8.4 3GPP Bearer Concept

The 3GPP system provides connectivity for services, where services can be of client–server type, or of connectivity type. Examples of the two types include content browsing and messaging, respectively. The services accessed may reside within a PLMN, or be external ones. Conceptually, service quality support provided for a service consists of a

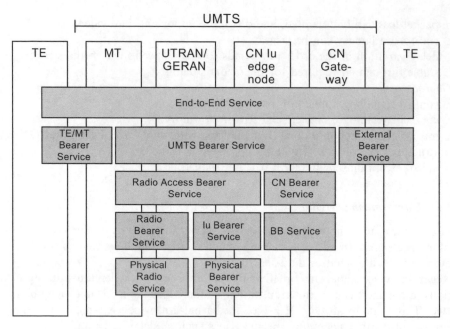

Figure 8.2 An illustration of 3GPP bearers for UMTS [1].

UMTS part as well as parts that may belong to the service provider's domain, external networks or relate to the particular equipment that the user is employing to access the service.

As discussed above, requirements and characteristics vary from service to service. Thus, the support provided by a UMTS network should favourably be adaptable to the type of service in question. In 3GPP architecture, the provision of differentiated treatment to services is represented using bearers. A bearer is associated with a particular service performance level. As shown in Figure 8.2, 3GPP models multiple levels of bearers. In what follows, these issues as well as related entities are discussed. As mentioned in the beginning, the focus is on packet switched entities in the explanations below.

8.4.1 Architectural Entities

3GPP differentiates between Mobile Terminal (MT) and Terminal Equipment (TE). The former of these is the endpoint of the bearer, and could be a UMTS-capable handset or a PC Modular Computer Interface Adapter (PCMCIA) card, for example. An MT may also be used together with a TE such as a laptop. The connection between TE and MT may affect end-to-end service, and is thus modelled as a bearer service in its own right.

The MT communicates with a Radio Access Network (RAN), which in the case of UMTS would be either UTRAN (UMTS Terrestrial Radio Access Network) or

GERAN. In the case of UTRAN, the RAN grouping includes a base station (Node B) and Radio Network Controller (RNC).

The Core Network (CN) consists logically of two types of entities – namely, the CN Iu edge node and CN gateway. The former interfaces towards the Iu interface connecting the CN to the RAN, whereas the latter acts as an interface for the entire UTRAN domain towards external networks. In practice, the CN Iu edge node is SGSN (Serving GPRS Support Node) and the CN gateway is a GGSN (Gateway GPRS Support Node).

8.4.2 Bearer Layers

The UMTS bearer service is defined between the MT and CN gateway and is supported by the set of elements it covers in Figure 8.2. Along with the TE/MT bearer, external service is not part of the UMTS bearer properties. Nevertheless, a 3GPP system typically supports related functionalities, as is shown later on.

The UMTS bearer service is supported by lower layer bearer services, including a radio bearer service, Iu bearer service and backbone service.

The 3GPP bearer is negotiated between the CN gateway and MT. The MT may specify a QoS profile in the bearer request, and also leave some parameters undefined. In the latter case the parameters will be filled in by the network. In the current 3GPP scheme, the mobile network may downgrade the service quality level requested by the MT, but may not upgrade it. Procedures have been specified for bearer establishment, bearer modification and bearer release. Bearer modification may be needed, for example, when new services or flows are mapped onto an existing bearer.

We now discuss how a 3GPP bearer is characterised by referring to realisation of the 3GPP bearer in packet switched domain – namely, PDP context. From the ISO/OSI architecture viewpoint, the properties of a bearer are defined between two Service Access Points (SAPs) in respective endpoints.

8.4.3 Packet Data Protocol Context Characterisation

A PDP context is used to transport packet-based data in a 3GPP system. Creation and modification of PDP contexts is requested by the terminal. A PDP context is associated with a PDP address. Starting with 3GPP Release '99, multiple PDP contexts may share a PDP address. The PDP context which was first created towards a PDP address is called primary PDP context, and subsequent ones sharing the same PDP address are called secondary PDP contexts. In general, multiple flows may share a PDP context. There are examples about PDP context multiplexing in [19].

A PDP context is associated with service quality parameters, which result from negotiation belonging to a PDP context activation procedure. The set of service quality parameters is called a QoS profile. The QoS profile consists of the parameters listed below. Most of the parameters are not important for the present description. The most relevant parameters for the current chapter are indicated in *italics* and will be described later:

- *traffic class*;
- *maximum bit rate*;
- *guaranteed bit rate*;
- *transfer delay*;
- *Traffic Handling Priority (THP)*;
- *Allocation/Retention Priority (ARP)*;
- delivery order (Boolean);
- maximum Service Data Unit (SDU) size;
- SDU format information;
- SDU error ratio;
- residual Bit Error Ratio (BER);
- delivery of erroneous SDUs (Boolean);
- source statistics descriptor.

There are inter-dependences between the parameters listed above, which will be discussed in the following.

Traffic classes represent a high-level categorisation of bearer services. The four categories of the 3GPP QoS framework are:

- *Conversational class*. This class preserves the relative order of SDUs and provides a low-latency UMTS service. Conversational class is suitable for multi-media conferencing.
- *Streaming class*. This class preserves the relative order of SDUs and is suitable for multimedia streaming.
- *Interactive class*. This class protects the payload from bit error and provides guaranteed delivery for request/response-type services. Suitable for interactive services such as browsing.
- *Background class*. This class protects the payload from bit error and provides guaranteed delivery. Suitable for email delivery.

Maximum bit rate is relevant for all traffic classes and defines the maximum momentary bit rate for the PDP context. Guaranteed bit rate is only relevant for RT traffic classes – namely, conversational and streaming classes. Enforcement of these constraints refers to particular policing algorithms (Token Bucket, see [8]), the details of which are omitted here.

Transfer delay specifies the 95th percentile of maximum delay of SDUs for delivery in the UMTS bearer during the lifetime of the bearer service.

THP can be used for specifying the relative urgency of different SDUs mapped to the same interactive class PDP context.

The ARP parameter defines the importance of a PDP context relative to other PDP contexts and can be used in admission control. The ARP parameter is a part of a subscription and thus cannot be negotiated for a PDP context.

8.4.4 Comments about 3GPP Bearers

From the viewpoint of end-to-end IP services, the 3GPP bearer and bearers beneath it represent L2. The 3GPP framework provides QoS attributes for specifying whether reliability is needed or not.

The mere properties of bearers are not sufficient for understanding service performance levels which can be achieved end to end. The provisioning of bearers, as well as the mapping of services to bearers, is important for end-to-end service performance. The service quality support architecture also needs to be understood. This brings us to our next topic, the 3GPP QoS architecture.

8.5 Overview of 3GPP Quality of Service Architecture

The 3GPP QoS architecture describes the way in which 3GPP network elements participate to support end-to-end service quality. Referring back to the division between service performance and service quality support mechanisms, the 3GPP QoS architecture provides a framework for negotiating service performance levels between the terminal and the network, as well as an architectural framework for service quality support mechanisms and mapping service performance to the support mechanisms.

In the following the core 3GPP QoS architecture is described first, followed by a description of Release 5 functionality related to IP Multimedia Sub-system (IMS) support. Finally, interworking towards external networks is covered.

8.5.1 3GPP Quality of Service Architecture

3GPP QoS functionalities can be classified into control layer and user layer ones. Below, the 3GPP R5 versions of figures depicting the respective functionalities are shown (Figures 8.3 and 8.4). We start with control layer functions.

The translation function interfaces the 3GPP domain towards non-3GPP signalling. The possibility of interfacing towards Resource Reservation Protocol (RSVP) signalling is described, but not used in practical networks.

UMTS bearer service managers in the mobile terminal, core network edge and gateway signal between each other during establishment or modification of a UMTS bearer service. Each of the UMTS bearer services interfaces towards the relevant lower

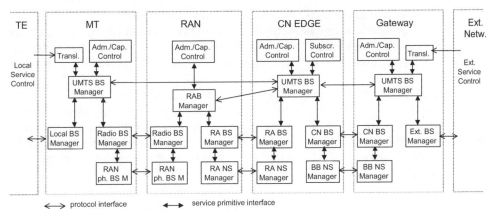

Figure 8.3 Quality of service management functions for UMTS bearer service (control layer) [1].

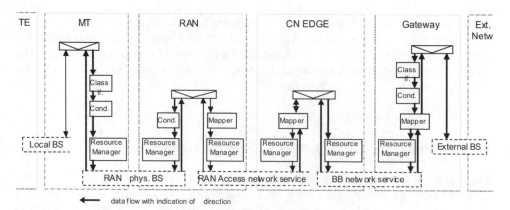

← data flow with indication of direction

Figure 8.4 Quality of service management functions for UMTS bearer service (user layer) [1].

layer functionalities within its domain. A service manager may perform attribute translation.

Admission/capability control maintains information about available resources within a network entity. It is used in bearer establishment modification.

Subscription control performs checking of bearer attributes against user profiles.

The 3GPP architecture is based on the concept of bearer. The 3GPP QoS architecture supports QoS provisioning per service and per end-user. An end-to-end service quality level is requested by the terminal as part of bearer activation or modification. The allowable range for service quality is provided to Access Point Names (APNs) in the CN edge – i.e., GGSN. The QoS profile is stored in the Home Location Register (HLR) for each subscriber, and defines the allowable service quality level according to the APN. During bearer activation, the APN is detected based on the destination IP address.

The 3GPP QoS architecture allows for end-to-end QoS signalling. Support for RSVP has been described in [2]. In practice, end-to-end signalling is only used within 3GPP systems. Towards external IP domains (Gi interface), service quality support can be provided by mapping bearer QoS attributes to Differentiated Services (DiffServ) Per Hop Behaviours (PHBs). This is discussed in Section 8.5.3 below.

The 3GPP QoS architecture does not describe implementation details, and they are left vendor-specific. What is standardised is the high-level framework, including bearer-level service quality management. Next, some implementation aspects of the 3GPP QoS framework are addressed.

In Release '99 of UMTS, transport between Node B, the RNC and SGSN is based on Asynchronous Transfer Mode (ATM). From Release 4 onwards, IP-based transport between the RNC and SGSN for the packet-based interface (Iu-PS) is also supported. Packet-based data are allowed in the Gn interface – i.e., between SGSN and GGSN. Service quality support can be provided by mapping PDP context QoS attributes to DiffServ PHBs. An example of this is provided in Section 8.5.3.

The mobility gateways of the 3GPP architecture, SGSN and GGSN, are central elements from the viewpoint of service quality as well. Implementation of the 3GPP QoS framework from the viewpoint of these two elements is next discussed. SGSN and

GGSN implement the core network edge and gateway functionalities of the 3GPP QoS framework, respectively.

Both SGSN and GGSN have functions relating to control layer and user layer functions, as explained in the context of the framework. Control functions relate to resource monitoring and reservation, as well as to admission control. User layer functions include traffic policing, shaping, packet classification, marking and scheduling. Control layer functions are not discussed further here, but rather focus is put on user layer functions.

Packet classification maps downlink packets to correct PDP contexts. PDP contexts sharing a single IP address are differentiated with Traffic Flow Templates (TFTs). The secondary PDP context mechanism uses TFTs.

Traffic policing monitors the data rate of flows against the QoS attributes of the PDP context. Measurement is performed using the Token Bucket algorithm [8]. Packets exceeding the maximum bit rate are discarded.

Traffic shaping can be used to smooth out momentary bandwidth peaks by means of buffering. Traffic shaping is a better option for UDP-based traffic than discarding of packets, since it avoids end-to-end retransmissions of packets. It should be noted that buffering incurs delay to packets, however.

Packet marking can be used in the Gn interface for marking the DiffServ Code Point (DSCP) corresponding to the DiffServ PHB in the Gn transport domain.

Packet buffering is used for storing PDUs awaiting transmission. Multiple queues can be differentiated between different service quality support classes. Buffer management algorithms are needed for cases when the buffer becomes full.

Packet scheduling determines when packets are transmitted and from which buffer they are taken.

8.5.2 Support for the IP Multimedia Sub-system

3GPP Release 5 brings with it the capability of dynamically linking the bearer QoS to characteristics of IP Multimedia Sub-system (IMS) sessions. The functionality of IMS – in general – is covered, for example, in [10] and will not be repeated here. The IMS architecture is described in [3], and [2] contains a description of the interworking of IMS with GGSN and the UE. An outline of the basic operation of Release 5 architecture is provided. 3GPP Release 6 extends the functionality further, but is beyond the scope of the present work.

In Release 5, session-specific service quality is needed for Session Initiation Protocol (SIP) services. The characteristics of the session – and hence potentially also the QoS profile of the bearer associated with the session – may vary between different sessions. The reason for this is that the coding scheme used for media streams, as well as some other parameters such as bit rate, are negotiated between the originating and receiving parties of the call. The SIP framework employs Session Description Protocol (SDP) for conveying the parameters. It is in the interest of the mobile network operator – and also of the user of the service – to request the correct QoS profile for the 3GPP bearer which supports the IMS session in the PLMN.

In what follows, the steps undertaken during bearer establishment are described. The following phases may be identified during a session setup between the originating terminal (UE1) and terminating terminal (UE2):

1. UE1 requests a SIP session with UE2 using standard IMS procedures.
2. UE1 and UE2 negotiate SDP parameters for the SIP session. The Proxy Call State Control Function (P-CSCF) of the originating IMS domain learns the SDP parameters.
3. The P-CSCF of the originating IMS domain informs the Policy Decision Function (PDF) of the originating domain of the relevant SDP parameters.
4. The PDF returns to the P-CSCF an authorisation token relating to the session in question.
5. The P-CSCF conveys the authorisation token to UE1.
6. UE1 requests bearer activation for the media stream of the session from the GGSN of the originating domain and sends the authorisation domain as part of the bearer activation request.
7. By referring to the authorisation token, the GGSN checks from the PDF that the requested QoS profile corresponds to the SDP parameters of the session.
8. The PDF acknowledges parameters with the GGSN.
9. A normal 3GPP bearer activation procedure follows.

The phases relating to bearer negotiation (6–9) are illustrated in Figure 8.5. Please note that the numbering in the figure does not correspond to the above numbers, as the figure is borrowed from a 3GPP specification.

It should be noted that the linking between the IMS and the mobile network via the PDF is also used for other purposes which do not relate to service quality – namely,

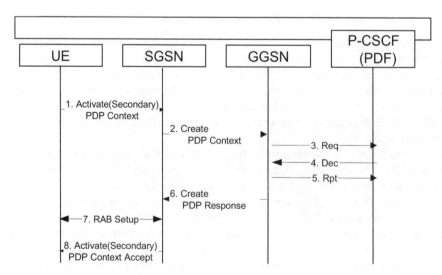

Figure 8.5   An illustration of secondary PDP context activation procedures [2].

gating and charging. An overview of these in 3GPP can be found in [10]. Service control examples in [7] contain some extensions to standard 3GPP procedures:

- Generally speaking, bearer authorisation is part of the Service Based Local Policy (SBLP) concept of 3GPP. A PDF is used for SBLP in Release 5. Only P-CSCF control for the PDF has been specified in Release 5, but the SBLP is extended in Release 6.
- The 3GPP Release 5 standard also allows an operator not to enable SBLP. In this case, an authorisation token and PDF are not used.

8.5.3   External Bearer Service

For external bearers, mapping to the DSCP based on bearer QoS attributes is typically performed. A description of such mapping can be found in the GSM Association's recommendation for GPRS Roaming Exchange (GRX) service quality. GRX is actually used in the Gn interface between the SGSN of the visited domain and the GGSN of the home domain, but the principle is the same as in the Gi interface (between the GGSN and external IP resources).

An example of the mapping between QoS attributes and PHBs is shown in Table 8.1.

Table 8.1 An example mapping between bearer quality of service attributes and DiffServ per-hop behaviours.

Traffic class	Per-hop behaviour
Conversational	Expedited forwarding
Streaming	Assured forwarding 1, low drop precedence
Interactive (THP1)	Assured forwarding 2, low drop precedence
Interactive (THP2)	Assured forwarding 3, low drop precedence
Interactive (THP3)	Assured forwarding 4, low drop precedence
Background	Best effort forwarding

DiffServ PHBs, identified by the DSCP, can be used for mapping traffic to lower-layer transport QoS parameters. For example, packets marked with a particular DSCP can be mapped onto a particular IEEE 802.1Q priority class. For Multi-Protocol Label Switching (MPLS) networks, Forwarding Equivalence Class (FEC) mapping can be done based on PHB. Some examples can be found in [11].

8.6   Quality of Service Management in UMTS

QoS management in UMTS networks is discussed in this section. Regarding the 3GPP QoS architecture, UMTS networks provide a particular implementation of service quality support mechanisms, which is more advanced than in Release '99 GPRS, for example.

8.6.1   Introduction to Quality of Service Management Challenges

In this section QoS is looked at from the management point of view. QoS is an enabler for different strategies and services and the scope of this section is to demonstrate the use cases QoS can support.

The UMTS QoS class concept supporting conversational class, streaming class, interactive class and background class provide the mechanisms to treat traffic differently. One of the main distinguishing factors between these QoS classes is how delay-sensitive the traffic is: conversational class is mainly for traffic which is delay-sensitive while background class is the most delay-insensitive traffic class. More details can be found in [5]. Conversational and streaming classes are mainly used to carry RT traffic flows. Conversational RT services, like video telephony, are the most delay-sensitive applications and these data streams should be carried in conversational class. Interactive and background classes are mainly used by traditional Internet applications like the WWW, email, Telnet, FTP and news. Due to lower delay requirements, as compared with conversational and streaming classes, both provide better error rate by means of channel coding and retransmission. The main difference between interactive and background class is that the former is mainly used by interactive applications – e.g., interactive email or interactive web browsing – while background class is meant for background traffic – e.g., background download of emails or background file downloading. Separating interactive and background traffic applications ensures responsiveness for interactive applications. For example, traffic in the interactive class can be given higher priority in scheduling than background class traffic.

Within 3GPP-specified radio access networks and core networks, UMTS bearer QoS provision and resource reservation are based on PDP context activation. Every PDP context is associated with a set of QoS attributes, which are negotiated in conjunction with the context activation procedure. This set of QoS attributes is referred to as a QoS profile. It includes those essential parameters for the user application to define the QoS required for its media streams/bearer services. These parameters are, for example, traffic class, target transfer delay, reliability, guaranteed bit rate and priority. 3GPP provides support for multiple PDP contexts per IP address with each context being capable of acquiring a separate QoS profile in the negotiation phase. This enables a multimedia session to define the QoS requirements for each individual media stream. One UMTS bearer is equivalent to one PDP context. QoS management (transport) layers are depicted in Figure 8.6. To translate end-user services to an actual physical channel, many mappings have to be performed.

One of the most significant advances for UMTS is its support for a richer variety of services and a higher level of service personalisation, enabled by the UMTS QoS framework as a result of 3GPP standardisation. Network services are end to end, this means from one terminal equipment to another terminal equipment or from a content server to a terminal equipment. An end-to-end service typically requires a certain QoS, which is then provided for the user by the equipment in the end-to-end chain. Quality of end-user Experience (QoE) is partly affected by subjective factors. Ultimately, it is the end-user who decides whether he is satisfied with the provided service or not.

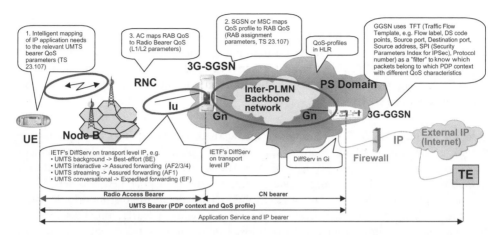

Figure 8.6   3GPP UMTS layered quality of service architecture.

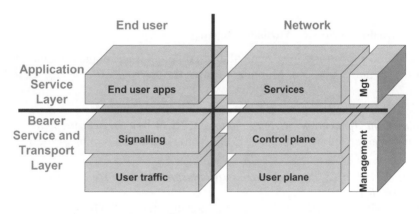

Figure 8.7   A conceptual model of quality of service functionality.

To understand the complete picture of end-to-end QoS, one needs to examine the issue on a more abstract level. Figure 8.7 shows an overview of end-to-end QoS architecture, depicting the essential components involved: application service layer, bearer service (network transport) layer domain and management layer. An important aspect of this model, and also the principle of UMTS bearer services, is the separation of application service and network transport service domains. The role of the former is to provide QoS control to the latter. The network transport layer's role, on the other hand, is to provide DiffServ quality with transport technologies. On the abstract level, the role of each domain for QoS can be described from the following important aspects:

- *Network service QoS*: RAN, mobility core and IP transport network including backbone network, service capabilities and QoS control.
- *Endpoint QoS*: terminal, end-user and application service QoS awareness and QoS control.

- *Network QoS control*: a control functionality that administers network resources in order to achieve consistent service behaviour when a specific service QoS is needed.
- *QoS management*: functionalities on the management plane (network and NMS) to provide a statistical feedback loop for network QoS control – e.g., through service quality management and various functionalities in the network, service and customer care layers in eTOM.

Within a PLMN, QoS functionality can be further divided into User plane (user plane) and Control plane (control plane). User plane includes telecom/datacom processing and forwarding services of user traffic, while C-plane provides telecom/datacom control and signalling services. Finally, the transport network service and protocols also span the UE. This gives mobile terminals an interesting double role in offering both application services and network services. Clearly, intelligence in the terminals plays an important role in optimised mapping of service requirements onto transport QoS.

8.6.1.1 Quality of Service Attribute Mapping

This section demonstrates how an attribute in different levels of QoS (within the UMTS network) should be mapped. Mapping towards the external world was given in Section 8.5.3.

Mapping from Application Attributes onto UMTS Bearer Service Attributes
Mapping from an application onto UMTS bearer attributes is implementation-specific.

Mapping from UMTS Bearer Service Attributes onto Radio Access Bearer
Service Attributes
When establishing a UMTS bearer and the underlying RAB to support a service request, some attributes on the UMTS level typically do not have the same values as the corresponding attributes on the RAB level. For example, the requested transfer delay for the UMTS bearer will typically be larger than the requested transfer delay for the RAB, as transport through the core network will use a part of the acceptable delay. For the following attributes/settings, the attribute value for the UMTS bearer will normally be the same as the corresponding attribute values for the RAB:

- traffic class;
- traffic handling priority;
- allocation/retention priority;
- maximum bit rate;
- guaranteed bit rate;
- delivery order;
- delivery of erroneous SDUs (if set to 'yes', handling of error indications on the UMTS bearer level and the RAB level will differ);
- maximum SDU size;
- SDU format information (note that the list of exact sizes of SDUs will be the same; an exact format for SDU payload does not exist on the UMTS bearer level).

These are the important QoS attributes derived from the quality requirements of the end-user service. For example, in order to provide 'PTT over cellular' or video-conferencing with adequate quality, the traffic class, as well as other QoS attributes, is determined by the application. This can be called 'inherent service quality requirement'. The application provided to the end-user determines the settings in this case.

For the following attributes, the attribute value for the UMTS bearer will normally be different from the corresponding attribute value for the RAB. The relationship between the attribute values for the UMTS bearer service and the RAB service depends on the implementation and, for example, on network dimensioning.

- The residual BER for the RAB service will be reduced as a consequence of the bit errors introduced in the core network by the core network bearer service.
- The SDU error ratio for the RAB service will be reduced as a consequence of the errors introduced in the core network by the core network bearer service.
- The transfer delay for the RAB service will be reduced as a consequence of the delay introduced in the core network – e.g., on transmission links or in a codec resident in the core network.

The following attributes/settings only exist on the RAB level:

- SDU format information: the exact format of the SDU payload is retrieved from the codec integrated in the core network.
- The source statistics descriptor is set to speech if the RAB transports compressed speech generated by the codec integrated in the core network.

Mapping from UMTS Bearer Service Attributes onto Core Network Bearer Service Attributes

Mapping from UMTS bearer service attributes to core network bearer service attributes is based on operator choice.

8.6.2 Radio Bearer Mapping of UMTS Traffic Classes

In Figure 8.8 radio bearer mapping onto UMTS traffic classes is depicted. Each radio bearer is associated with a Transport Channel (TrCH), whose transport format set defines the allocated bandwidth, Transmission Time Interval (TTI – delay over the air) and error correction method (channel coding and rate matching). Details of this mapping are operator-specific.

Flexible shared media access can be implemented in the form of a scheduled Dedicated Transport Channel (DCH), common transport channel (RACH/FACH) or in the form of a High-speed Downlink Shared Transport Channel (HS-DSCH), which is the TrCH type used in the High-speed Downlink Packet Access (HSDPA) concept. Shared media access is intended for traffic classes that do not impose strict delay and jitter requirements – e.g., interactive and background traffic. The RAB (or PDP context) is not allocated a fixed amount of bandwidth, but rather incoming packets are put into a queue and serviced as soon as there is resource available. Shared media access potentially allows instant access to a very high bandwidth

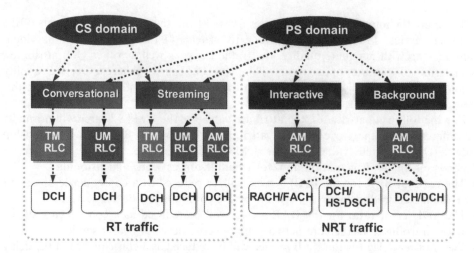

Figure 8.8 Radio bearer mapping of UMTS traffic classes.

transfer capability, and thus low packet delay, without causing high blocking to other traffic in the cell.

Such a mapping has a positive effect on provided and perceived QoS. For example, for conversational traffic class packet switched RABs the RNC always uses Unacknowledged Mode (UM) Radio Link Control (RLC), but for streaming traffic class packet switched RABs the RNC again uses UM RLC for very tight transfer delays, while if the transfer delay requirement is not that tight, then Acknowledged Mode (AM) RLC can be used. The same principle applies for interactive and background traffic class RABs.

8.6.3 Utilisation of Quality of Service in the UMTS Domain

The task of a multi-service communication system is to provide an appropriate transport service for delivering relevant types of traffic streams for different users. It can sometimes be difficult to define the exact technical parameters required to ensure such delivery, as different end-users' perceptions of 'service quality' may vary. Furthermore, for different user groups different service packages are offered and thus different QoS and services are promised in the first place. The 3GPP QoS concept of traffic classes and other QoS parameters offers a variety of implementation options. It can even be difficult to understand how the QoS model and related parameters can actually be used in networks and terminals. Figure 8.9 captures the different items related to practical utilisation of QoS capabilities. In this section operator QoS strategies, control plane functions and PDP context negotiation issues are discussed. Description of the PDF is found in Section 8.5.2. The roles of network elements (user plane) are introduced in Section 8.6.3.5.

As for the DiffServ marking for IP QoS and lower-level QoS mappings, the reader is referred to Section 8.5.3.

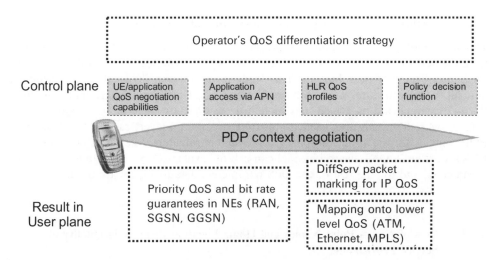

Figure 8.9   A quality of service model for packet switched differentiation.

8.6.3.1   Operator Quality of Service Strategy

The 3GPP QoS framework allows for different operator strategies, in terms of possibilities for service or user differentiation. An example of this is demonstrated in Figure 8.10. The 3GPP concept allows for user or service differentiation, or a mixture of both. It is important to note that Configuration Management (CM) and Performance Management (PM) content and functions should also support the monitoring of such segmentation.

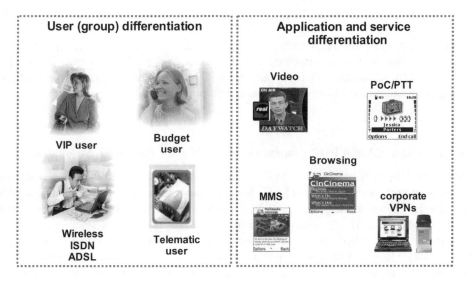

Figure 8.10   Conceptual segmentation based on different quality of service profile examples.

8.6.3.2   Quality of Service Negotiation Capabilities of Terminals

The 3GPP QoS framework allows a terminal to leave bearer QoS attributes blank and have the network fill them in. In the latter case, QoS attributes are generated based on the subscriber/APN combination. This leads to a situation in which some terminals may fill in QoS attributes where others will not do so. Furthermore, terminals may host third-party applications, which may have little or no knowledge of 3GPP QoS.

QoS-aware terminals and applications (like streaming player) can request their own PDP contexts for the application for QoS-related parameters (like traffic class, bit rate, BER). Further, a QoS-aware terminal can ask for explicit QoS values only for certain parameters, and for the rest of the QoS parameters default values are used.

8.6.3.3   Role of Access Point Names and Home Location Register in Quality of Service Provisioning

In addition to the QoS awareness of the terminal's applications, the APN settings related to the terminal client also have an impact on PDP context usage and the QoS differentiation scenarios for both terminal types.

Based on the standards, QoS differentiation for delay-critical applications (e.g., terminal embedded streaming players or PoC/PTT clients) in GPRS, EGPRS (Enhanced GPRS) and WCDMA networks can be provided in one of two ways. First, applications with delay-critical QoS requirements are associated with different APNs and the QoS parameters for each combination of subscriber and APN are forced from the HLR. The second option is to make terminal clients for QoS-aware applications activate parallel primary or secondary PDP contexts (with different QoS parameters) to the same APN. In this case, the same QoS parameters stored in the HLR are used as the upper limit on the QoS given by the network for the parameters that are explicitly requested. Other parameters are enforced from the HLR for QoS-aware applications and terminals as well.

When configuring the HLR QoS profile parameters it is good to keep in mind that HLR is in a way technology-independent. The same QoS profiles are used in EGPRS and WCDMA networks and thus each HLR QoS parameter should be considered taking the requirements from both systems into consideration.

8.6.3.4   Standard-based PDP Context Quality of Service Parameter Negotiation

Within the UMTS network the UMTS bearer service consists of the RAB service and the core network bearer service. The UTRAN is responsible for the radio bearer service and the Iu bearer service. The SGSN, GGSN and the possible IP-based network between those network elements provide the core network bearer service.

In the packet switched part of the network the UMTS bearer service is realised by a PDP context. Every PDP context has a set of QoS attributes associated with it. This QoS profile is negotiated during the PDP context activation procedure. It describes the quality of the UMTS bearer service offered to the user.

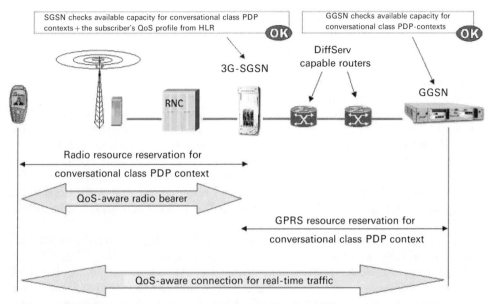

Figure 8.11 Quality of service components in the third-generation network.

Within the 3GPP-specified radio access network and core network, QoS provision and resource reservation are based on PDP context activation. Every PDP context is associated with a set of QoS attributes to be negotiated in conjunction with the context activation procedure. The set of QoS attributes is referred to as a 'QoS profile'. It includes those essential parameters for the application to describe the QoS need of its media streams – e.g., traffic classes, target transfer delay, reliability, guaranteed bit rate, priority, etc.

3GPP supports multiple PDP contexts of the same IP address with different QoS requirements. This enables a multimedia session to describe the QoS needs of individual media streams. One UMTS bearer corresponds to one PDP context. The complete QoS-activation chain between the UE and the GGSN is illustrated (Figure 8.11) and described below:

- The terminal starts the setup of the PDP context. In the case of a non-QoS-aware terminal application the terminal will request a primary PDP context with subscribed or default values for all QoS parameters. Should the terminal be QoS-aware the traffic class and guaranteed bit rate would be explicitly requested from the network in the primary or secondary PDP context activation. The terminal requests a PDP context activation from the 3G SGSN.

- 3G SGSN receives the request from the terminal application. It compares the request against its own resource reservation level and the individual subscriber's QoS profile from the HLR associated with the requested APN. The HLR converts the internal database information to the 3GPP-defined format and sends the information over the standard interface to the 3G SGSN. For all QoS parameters related to the PDP context and requested by the terminal as subscribed or default, the values received from the HLR are used as such. In case of QoS-aware terminals, the values of the parameters explicitly requested by the terminal application are compared against the ones stored in the HLR and the smaller of the two values is used for each individual parameter. Both of the above-mentioned variables might serve as criteria for the 3G SGSN to suggest to the terminal a lower QoS than requested as a part of PDP context activation.

- Provided that the QoS request conforms to the subscriber's QoS profile and that the 3G SGSN has capacity available, the 3G SGSN signals using the Tunnelling Protocol Control Plane (GTP-C) Protocol the request for PDP context activation to the GGSN with the requested QoS parameters. The GGSN returns the activation request to the SGSN. The returned information is either an acknowledgement on the successfully set up radio bearer or a proposal for a lower QoS than requested. The RAN performs internal admission control and resource reservation. If successful, a RAB will be set up. The SGSN now makes a PDP context activation request to the GGSN.
- The 3G SGSN continues the PDP context setup by sending the RNC a request for RAB setup with the requested PDP context QoS parameters. The 3G SGSN uses Radio Access Network Application Protocol (RANAP) for this signalling. The RNC either acknowledges a successful setup of a radio bearer or rejects the requested bearer.
- In the case when the 3G SGSN or GGSN has requested a lower QoS than requested by the terminal, the 3G SGSN sends the QoS degradation request to the terminal. In the case that the 3G SGSN, the RNC and the GGSN have all accepted the PDP context with requested QoS parameters, the 3G SGSN completes the PDP context setup negotiation with the terminal.

8.6.3.5   Practical Realisation for End-to-end Service Management

Until recent years, the development of the telecommunication industry has been technology-driven. Main focus was on systems, technical solutions and their implementation rather than on the service-handling capabilities. Today, the situation has changed and services and content are becoming the driving force, as often the same service types can be delivered using several different access technologies (GPRS, EDGE and WCDMA, for example). The transition from voice service to a variety of packet-based mobile services provides possibilities and opportunities, but at the same time sets new challenges for vendors and operators.

To aid and speed up this transition, new tools and concepts are needed. Therefore, the evolution of the Operations Support System (OSS) in the service management area ought to be technology-transparent and provide service management support independent of the access technology. Furthermore, the abstraction level of service management needs to allow the operator to focus on service management rather than on network element configuration.

The new packet-based services pose multiple challenges for network management and are based on:

- balancing the allocation of capacity between circuit switched and packet switched traffic;
- differing kinds of requirements that the new packet-based services pose on the network bearer – e.g., in terms of throughput.

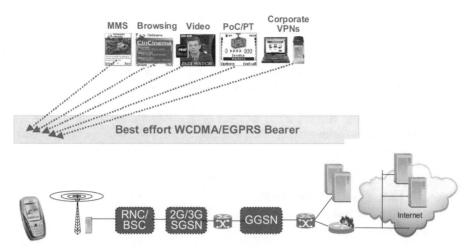

Figure 8.12 Services run over best effort bearers.

Catering for the requirements of the most needy services – such as the streaming type of services, requiring 20 kbps on average – may require considerable investment in network capacity – especially if the services are run over a best effort network.

Today the same GPRS/WCDMA packet bearers are used regardless of the application or end-user service. There is no service differentiation in network planning, optimisation or in the radio network Key Performance Indicator (KPI) data. This is illustrated in Figure 8.12.

Service Provisioning Utilising 3GPP Quality of Service

The implementation of 3GPP QoS provides means to operators to change from best effort radio network planning, provisioning, optimisation and monitoring to a more service-oriented approach. Ideally, the same QoS model applies to both (E)GPRS and WCDMA, but, in practice, differences will remain for some time because of network element time lines.

This section describes the concept of *priority treatment*, which mobile network providers can use to manage QoS differentiation when providing a wide range of value-added IP connectivity services to end-users, while wishing to generate the best revenue streams from the available capacity.

The need for different treatment for service arises from the simple fact that each service type sets its own requirements and priority requirements when it comes to 3GPP QoS attributes. Those services that are delay-sensitive need to be served first by the network elements in order to keep the QoE at an acceptable and predictable level. Further, in reality, the mobile network is not homogeneous from the QoS point of view. Instead, it is divided into sub-domains – each sub-domain having different capabilities for QoS control. As sub-domain QoS characteristics must be in line with end-to-end characteristics and requirements, a management concept is needed in order to provide consistent treatment and adequate quality for the mobile services.

The proposed *priority treatment* framework offers a simplified synthesis of different QoS management capabilities in the network and portrays them in a format that is relatively easy to comprehend. This framework allows operators to logically divide their network into 'pipes', offering differing characteristics – e.g., in terms of throughput. Individual end-user services can then be mapped into these pipes according to their needs.

The *priority treatment* framework is a subset of 3GPP QoS attributers. The selection is 3GPP-compliant and allows adequate differentiation for the pipes. Current attribute selection consists of:

- UMTS traffic class;
- Traffic Handling Priority;
- Allocation/Retention Priority (ARP);
- maximum bit rate; and
- guaranteed bit rate.

As mentioned in Section 8.6.1.1 these attributes are such that the attribute value for the UMTS bearer will normally be the same as the corresponding attribute value for the RAB. So no additional mapping from UMTS bearer to RAB is required. It is also possible to enhance the concept to support the whole QoS attribute set.

All end-to-end services over a mobile network may be classified into priority pipes. All traffic needing similar treatment in terms of QoS should be treated in each element. This *priority treatment* framework is also an inter-connecting concept between service- and network-layer policies. An example of implementation of the *priority treatment* framework is given in Figure 8.13. In the example case the attributes to differentiate the pipes are traffic class: background, interactive and streaming. In addition the interactive class's THP setting is in use. The mapping of an application onto the *priority treatment*

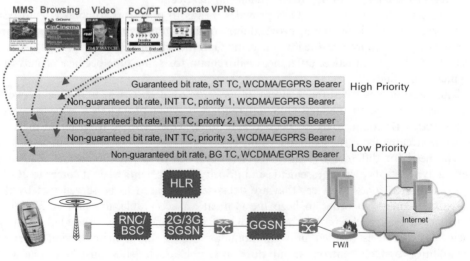

Figure 8.13 An example mapping of services onto priority pipes. Differentiation with services.

TC = Traffic Class; ST = Streaming; INT = Interactive; BG = Background.

pipe will then determine the value for the guaranteed bit rate and maximum bit rate. For example, in the case of PTT over Cellular (PoC) the guaranteed bit rate would be 8 kbps and for Multimedia Message Service (MMS) 20 kbps. The full solution would consist of ten different priority classes, utilising a full set of traffic class and ARP definitions. For simplicity it is proposed that THP = ARP.

An operator undertakes *priority treatment* planning based on service offering, QoS differentiation and APN strategy. Traffic forecast and network dimensioning are taken into account when planning the mapping of services onto the priority pipes. The configuration should be done based on the expected traffic volumes, the traffic mixture and the QoS requirements for the services and thus for the priority pipes.

Also performance monitoring should be done supporting the same *priority treatment* framework. Priority treatment-based service monitoring is the enabler for service-level optimisation of network parameters.

The 3GPP QoS compliance of the *priority treatment* concept sets requirements for both the control plane and user plane. The control plane handles the mapping of the services and subscribers onto relevant priority pipes. This mapping is based on the terminal, terminal application client and core network capabilities. The HLR, 2G SGSN, 3G SGSN and GGSN are expected to have a role in this process.

As a result of user plane traffic having different QoS requirements, thus needing different *priority treatment*, this traffic is handled differently based on the QoS pipe it belongs to. In the GSM Base Station Controller (BSC) or RNC the Radio Resource Management (RRM) algorithms implement QoS differentiation. The resources in access transport and air interface are limited, and in order to efficiently utilise the resources priorities are needed.

In core network elements QoS differentiation is done by queuing systems and in the IP backbone QoS differentiation is based on the usage of DiffServ. As the core elements are connected by high-capacity fibre links, the queuing delays on the core network are not expected to be significant parts of the end-to-end delay budgets – especially when compared with the radio parts.

Quality of Service Control in the Radio Network Controller

QoS control is a part of RRM functions. In this section we present QoS management functions for UTRAN: admission control, load control and packet scheduler. Handovers are based on the same discrimination principle. More details on RRM functionality can be found in Chapter 4. 3GPP does not specify the QoS support implementation details in network elements, thus only an example of possible scenarios is introduced here. The examples are based on the original work published in [23] and [24]. An enhanced version of the simulator used can be found in [25].

Admission Control

The admission control algorithm estimates the load increase, which is caused by the establishment or modification of the bearer in the RAN. In the decision phase the admission control will use threshold values, initially assigned during radio network planning. In order to decide whether the admission control accepts the request, the current load situation of the surrounding cells in the network has to be known and the additional load due to the requested service has to be estimated. In a power-based admission control case the overload situation for the downlink is defined as in Equation (8.1) where P_{NGB} and P_{GB} are power levels used for Non-GB (NGB) and GB traffic, respectively. Further, $P_{TxTarget}$ is a threshold and $P_{TxOffset}$ is an offset, see also Section 4.4.4 and Figure 4.14. Another admission criterion is defined in Equation (8.2). If either one of these inequalities is true a GB user is not admitted. In Equation (8.2) ΔP_{GB} is the estimated change in transmission power if the user is admitted and $P_{TxNGBCapacity}$ is the amount of resources reserved for NGB users from the total capacity pool. Similar logic can be applied for the uplink direction:

$$P_{TxTotal} = P_{NGB} + P_{GB} > P_{TxTarget} + P_{TxOffset} \qquad (8.1)$$

$$P_{GB} + \Delta P_{GB} > P_{TxTarget} - P_{TxNGBCapacity} \qquad (8.2)$$

Example of differentiation functionality in admission control: Each Radio Resource (RR) request is assigned a *Resource Request Priority* (RRP) based on RAB attributes provided by the CN when the RAB is set up. The attributes are ARP, traffic class and THP for bearer services of the interactive class.

In case of overload, RR requests are arranged in a queue and served following a priority principle and, at the given priority, having taken into account the corresponding arrival times (FIFO). Four reasons and four queue types are considered to be present: RAB admission, handovers, NGB traffic bit rate rescheduling and overload actions. Each queue is serve-event-based with a maximum resolution defined by the Radio Resource Indication (RRI) period. RRI sets the frequency of measurement report messages by the Base Station (BS). At the BS, loading information is collected. RR requests cannot stay in the queue longer than the *Maximum allowed Queuing Time* (MQT), and are immediately rejected if the queue length exceeds a pre-defined threshold (*Maximum Queue Length* – MQL). Except for the overload situation, NGB traffic is always admitted (but not scheduled).

Load Control
The main functionality of load control can be divided into two tasks: preventive actions and overload actions. During preventive actions load control works together with admission control and the packet scheduler in order to keep the system is a stable state. In an overload situation, load control is responsible for reducing the load and thereby bringing the network back into the desired operating area defined by radio network planning.

Example of differentiation functionality in load control: The only congestion control action supported during the simulator is the reduction of the bit rates of NGB bearer services when Equation (8.1) is satisfied. The bit rates are downgraded starting from the bearer services with lowest priority and, at the given priority, based on their arrival times (FIFO), but none of the sessions is released. From the QoS point of view the impact of the *Minimum Allowed Bit Rate* (MAB) setting has an effect on QoS differentiation.

Packet Scheduler
WCDMA packet access is controlled by the packet scheduler. The functions of the packet scheduler determine the available radio interface resources for Non-Real Time

(NRT) radio bearers, share the available radio interface resources between NRT radio bearers and monitor the allocations for NRT radio bearers. Admission control takes care of admission and release of the RAB. RRs are not reserved for the whole time of a connection but only when there are actual data to transmit. The packet scheduler allocates appropriate RRs for the duration of a packet call – i.e., active data transmission. More about packet scheduler functionality can be found in Section 4.4.3.

Example of differentiation functionality in the packet scheduler: The bit rate of admitted NGB services is allocated based on RRPs and the capacity request arrival time (FIFO), starting from the *minimum allowed bit rate* (MAB). The bit rates are then increased in rotation until all requested maximum bit rates are satisfied or all the capacity (wideband power) left by GB traffic is exploited:

$$P_{Allowed}^{NGB} = P_{TxTarget} - (P_{NGB} + P_{GB}) \qquad (8.3)$$

The maximum transmission rate (R_{Max}), at a given location of the terminal, is computed as a function of the geometry factor (G) and the downlink required transmission power (P_{RL}) for that particular radio link:

$$R_{Max} = \frac{W}{E_b/N_0} \cdot \frac{P_{RL}}{P_{TxTotal}} \left[\frac{1}{1 - \alpha + 1/G} \right] \qquad (8.4)$$

where W is the modulation bandwidth; and α is the orthogonality value.

Example Results on Radio Resource Management Quality of Service Differentiation Capabilities
These results are based on the simulations presented in [23] and [24]. An example of both service and user differentiation is presented.

Case 1   Simulation results and discussion [23]

Case 1 investigates the impact of MAB values on the differentiation of NGB *users* within the interactive and within the background class. In this case, the results of two simulations are compared. The first simulation was run with the MAB and priority values reported in Table 8.2. Interactive users are served before background users. Then, within the same traffic class, Gold users have top priority, followed by Silver and Bronze users, respectively. In the comparison case, the same RRP values are used as in the above scenario, but here all NGB users have equal MABs – i.e., 32 kbps.

Table 8.2   Case 1: Radio resource priority and minimum allowed bit rate (kbps) values.

Call/Session type	Gold (ARP = 1) RRP/MAB		Silver (ARP = 2) RRP/MAB		Bronze (ARP = 3) RRP/MAB	
Interactive (HTTP, UDP)	5	128	6	64	7	32
Background (emails,						
FTP/Napster)	8	32	9	32	10	32

Reproduced by permission of IEEE.

Simulation results are depicted in Figures 8.14 through 8.16. Figure 8.14 shows that only the deployment of different MAB values for each user group provides noticeable differentiation in terms of Active Session Throughput (AST). The RRP setting alone does not allow diverse treatment between the users. When the quality is monitored in terms of transfer delay (Figure 8.15), the measured values in all prioritised cases are better than the corresponding performance with the non-prioritised scenario. This improvement comes at the expense of the bit rates allocated to background users. According to the study the deterioration of such application services remains however tolerable in all analysed traffic conditions. Further, when using different MAB values, web page delays for Gold and Silver users, with relatively high traffic conditions (around eight sessions per minute), are significantly improved with respect to the corresponding values in the non-prioritised scenario. Results for Bronze users are practically independent of MAB usage. Furthermore, in both cases, at low traffic volumes, differentiation between users of the same traffic class turns out to be moderate.

Figure 8.16 shows the drawback of MAB differentiation in terms of cell throughput. In fact, with different MAB settings, throughput is up to 18% lower, depending on the traffic conditions. This is due to the fact that in this case the available power budget is distributed to users with larger granularity than in the 32 kbps case. As a result, the free capacity during bit rate allocation is not exploited as effectively as with smaller step size.

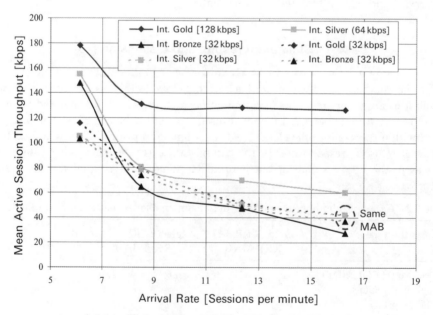

Figure 8.14   Case 1: Mean active session throughput.

Reproduced by permission of IEEE.

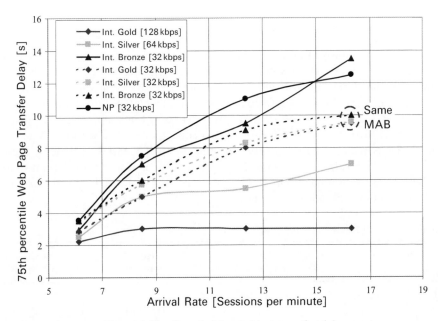

Figure 8.15 Case 1: Mean object transfer delay.

Reproduced by permission of IEEE.

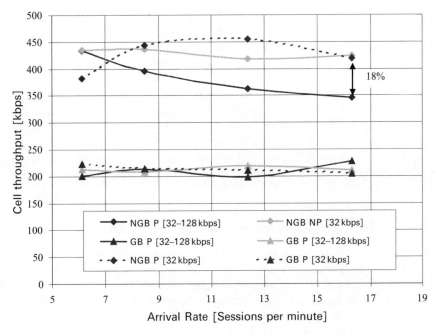

Figure 8.16 Case 1: Cell throughput.

(N)GB = (non) guaranteed bit rate; (N)P = (non) prioritised case. Reproduced by permission of IEEE.

Case 2   Simulation results and discussion [23]

In this second case study, a policy that improves NGB throughput at the expense of GB service accessibility is introduced. A target performance for GB traffic is set as follows: 2% blocking for speech and 8% blocking for collision detection and streaming. These high-level performance targets translated into new parameter settings lower the MAB for NGB users from 64 to 32 kbps and reserves 2 dB for NGB-dedicated capacity. For NGB-dedicated capacity see also Section 9.3.6.

The MAB is reduced based on the fact that a smaller MAB provides a higher resource utilisation rate yet maintains an adequate degree of user satisfaction. In this case, best effort streaming (UDP traffic) is served first, followed by web browsing (http), emails and FTP/Napster, respectively (see Table 8.3).

Table 8.3   Case 2: Radio resource priority values.

Call/Session type		Gold (ARP = 1)	Silver (ARP = 2)	Bronze (ARP = 3)
Interactive	THP1	5 (UDP)	—	—
	THP2	—	6 (HTTP)	—
	THP3	—	—	7 (emails)
Background		—	—	8 (FTP/Napster)

Reproduced by permission of IEEE.

The impact of NGB capacity reservation on GB performance in terms of a Call Block Ratio (CBR) is illustrated in Figure 8.17. Should the NGB reservation be too high, GB blocking could increase to an unacceptable level.

Application of the above rule, which slightly degrades the CBR of GB services, improves NGB accessibility. As a matter of fact, a 44% gain can be derived from Figure 8.18, where the average number of simultaneous active sessions during the simulation period is displayed as a function of $P_{TxNGBCapacity}$. The effect comes partly from the reduced MAB and partly from the capacity reserved for NGB services. Although not shown in this section, when performing a similar comparison in terms of NGB Capacity Request Rejection Ratio (CRRR), it is seen that the average CRRR reduces from 23% to 12% after policy deployment.

Case 3   Simulation results and discussion [24]

This third case provides an example of preferential treatment between services of interactive and background traffic classes. In this case, the QoS profile of Gold, Silver and Bronze users is generated with equal probability. The MAB for NGB bearer services has been set to 64 kbps in all prioritised and non-prioritised scenarios.

The mixture of NGB traffic consists of web browsing (65%), FTP/Napster (20%), emails (10%) and UDP data (5%). During the simulation period, each user may make more calls or open more sessions of the same type depending on

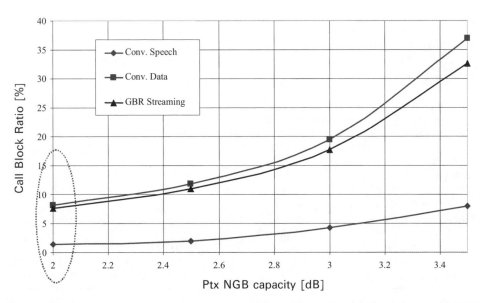

Figure 8.17   Case 2: Impact of dedicated non-guaranteed-bit-rate capacity on guaranteed-bit-rate performance.

Reproduced by permission of IEEE.

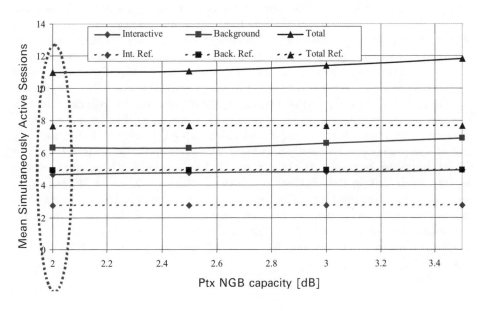

Figure 8.18   Case 2: Impact of dedicated non-guaranteed-bit-rate capacity on the amount of active non-guaranteed-bit-rate sessions.

Reproduced by permission of IEEE.

its traffic profile. All bearer services are served based on the resulting RRP values
(the lower the RRP the higher the priority of the RAB in question). These values are
reported in Table 8.4. For the latter case, the mapping recommended in [1] for
UMTS–GPRS interworking is used.

Table 8.4 Case 3: Radio resource priority values.

Call/Session type		ARP = 1	ARP = 2	ARP = 3
Signalling		9	9	9
Emergency call		1	1	1
Conversational speech		2	2	2
Conversational data		3	3	3
Streaming		4	4	4
Interactive	THP1	5 (UDP)	—	—
	THP2	—	6 (HTTP)	—
	THP3	—	—	7 (emails)
Background		—	—	8 (FTP/Napster)

Reproduced by permission of IEEE.

The service differentiation results are presented in Figures 8.20 through 8.21. The
performance for each application service is presented in terms of average active
session throughput, mean and maximum object delay for the 90th percentile of
the distribution of delays for all delivered objects during the lifetime of the
different bearer services. The bit rate provided to UDP applications, having the
highest priority among interactive and background services, remains above
74 kbps (see Figure 8.20). 74 kbps is high enough to carry, for example, best effort
audio or video streaming; even the quality of experience of HTTP users benefits from
the lower bit rates allocated to email and FTP/Napster users. This is demonstrated in
Figures 8.20 and 8.21, where prioritised and non-prioritised (dashed lines) results are
compared. In fact, in the prioritised scenario, the maximum web page delay for the
75th and 90th percentile stays below 10 s and 18 s, respectively. Furthermore, even in
this case, from Figure 8.21, setting a maximum web page delay of 10 s, a capacity
gain of about 80% can be estimated. In fact, at a given QoE, 80% more UDP and
HTTP users can be served at the expense of lower priority applications, whose
degradation is still up to standard, as shown in Figures 8.20 and 8.21. The mean
object delay for email and FTP/Napster use is less than 23 s and 95 s, respectively, in
all analysed traffic mix scenarios. The measured metrics are slightly worse in Figure
8.21, where the 90th percentile of the delay distributions is presented. In this case the
maximum object delay is less than 40 s for emails and below 140 s for FTP/Napster
users.

The results presented in this section are based on simplistic simulations, but the
positive indication on QoS differentiation capabilities is encouraging. It is shown that
QoS differentiation enables both user and service differentiation strategies and usage of

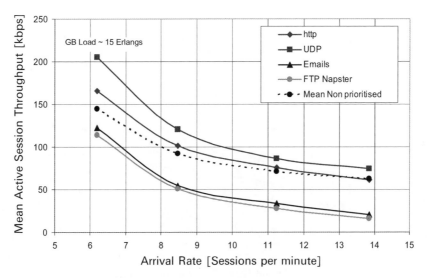

Figure 8.19   Case 3: Active session throughput.

Reproduced by permission of IEEE.

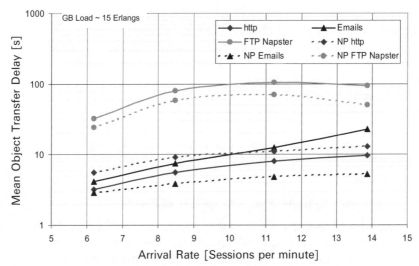

Figure 8.20   Case 3: Mean object transfer delay.

Reproduced by permission of IEEE.

a QoS mechanism aids the operator to utilise network resources more effectively when the traffic volume or mix changes. QoS mechanisms guarantee the service availability to remain tolerable for all traffic types. Furthermore, these cases prove that QoS policies can be used to fine-tune the system to serve more users, and thus to increase the utilisation of network resources.

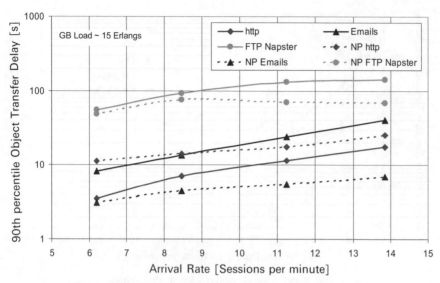

Figure 8.21   Case 3: Object transfer delay (90th percentile).

Reproduced by permission of IEEE.

Packet Core Elements' Roles in Quality of Service Control

The 3GPP QoS functions residing in mobility gateways (SGSN and GGSN) accounted for in Section 8.5.1 can be used for QoS control. Examples include:

- admission control policy in gateways;
- packet-marking policy in gateways for Gn transport;
- buffer management parameters for Gn transport.

In addition, further parameters – such as packet scheduling weights in gateways – can be used for QoS control.

8.6.3.6   Management Systems' Role with Quality of Service

In earlier sections of this chapter the UMTS QoS mechanism was described and in Chapter 7 the optimisation process was introduced. The main scope of this section is to introduce the QoS management layers and demonstrate the role of statistical information in the QoS management loop. Furthermore, the relationship between network performance and QoS is discussed. It is important to note that there is a distinct difference between the user/service QoS requirements defined in Service Level Agreements (SLAs), network level QoS/network performance and QoS enabling mechanisms.

3G networks and services will bring new possibilities but also challenges to operators, service providers and vendors. Service Quality Management (SQM) will enable providers to manage QoS against objectives set out in customer SLAs, and will also enable customers to compare the service offerings of different service providers.

The issues in moving from traditional network provisioning and optimisation to QoS management are:

- Operators need support for service differentiation: fast introduction of new services, service fulfilment and assurance with relation to their business targets. It is possible to use SLAs as a differentiating factor.
- Management solutions should provide flexible fulfilment and assurance tools in 3G network/bearer services for end-user services and applications with different QoS requirements.
- Service quality must be defined in a standardised way, consisting of KPIs from different domains and levels of the service network and applications.
- Service quality between providers and between systems that cooperatively implement the services has to be managed.
- Information must be available via open interfaces for utilisation in other tasks and processes in operator organisation.
- Management systems must provide means to prioritise and schedule corrective maintenance activities according to the level of impact on a service.
- Management systems should support service and service-based network planning with information and analysis tools, and these management processes should be automated where feasible.
- Management systems should be able to combine the capacity and QoS views. Service provisioning is often a tradeoff between quality and capacity. A conceptual view of such a service optimisation is given in Figure 8.22.

In Figure 8.23 a framework for service management is depicted. It consists of familiar conceptual parts, which were touched upon in the context of optimisation in Chapter 7. Now the additional challenge is the increased granularity in network element

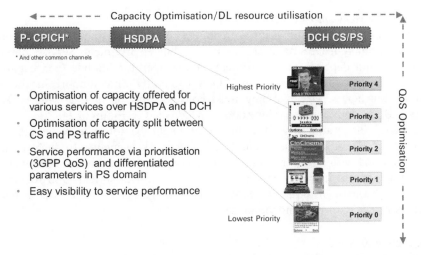

Figure 8.22 Service optimisation concepts in management systems. Quality of service and capacity views are needed. The capacity optimisation issues can be found in Section 9.3.6. High-speed downlink packet access is in Section 4.6.

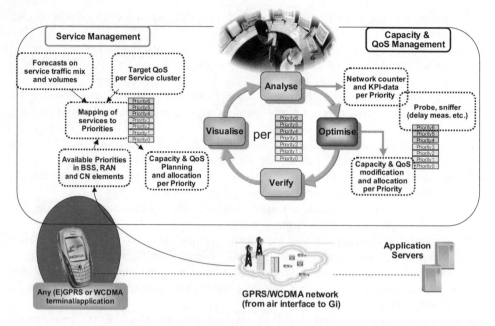

Figure 8.23   Service management overview.

configuration and performance monitoring. In addition, a linkage from service management and service priorities to the network management layer is introduced. The need for this linkage was discussed in Section 7.2, where the TMF eTOM model is introduced. Subscription and device management issues are not discussed; the reader is referred to [7].

Mapping of Services onto Priority Classes
As depicted in Figure 8.13 the priority treatment concept offers a possibility to provide services with different QoS requirements in a mobile environment. Service mapping is configurable and depends on the strategy of the operator, mobile QoS capabilities, available core network solution, and QoS capability and support in the access networks. Further, service mapping must consider and be in line with HLR QoS profiles and the APN strategy. An example is depicted in Figure 8.24.

In practice the granularity of Figure 8.24 might be too fine when taking into consideration the support and capabilities in network elements. In such a case the services depicted within one transmission convergence will get similar treatment.

With this concept it is encouraged to map services with similar QoS requirements onto the same treatment pipe (like MMS, FTP, email). The concept deals with service categories rather than with individual services.

Configuration of Services
Service configuration relates to both network and service layer functionalities. In the service layer the user-facing services are configured using Service Configurator, as discussed in Chapter 7. Service QoS requirements translate into QoS-relevant

	Conversational	Streaming	Interactive	Background
ARP/THP = 1	PS Video Call	SWIS	Corporate VNS 	
ARP/THP = 2	Action games VoIP call 	Poc/PTT 	RT gaming Chat	
ARP/THP = 3		Streaming 	Web browsing 	File transfer MMS Email Delay tolerant games

Figure 8.24   Example of service mapping using a subset of 3GPP quality of service attributes.

parameter settings in network elements like queue weights in SGSN and MAB or RRP in RNCs.

The granularity of network element configuration capabilities dictates service mapping onto the network layer. For example, in a WCDMA network where only transmission convergence level decisions are made the service mapping of Figure 8.24 would contain only one row, since the RRM decisions in admission control and the packet scheduler do not necessarily utilise ARP nor THP values in decision making.

Current autotuning and automatic optimisation algorithms (network management system level) focus on finding a new configuration for network elements in terms of reduced interference, improved capacity, etc. This functionality provides a good base on which service level optimisation can be built. In addition to moving from the network layer to the service layer according to the TeleManagement Forum (TMF) model the transition to service level optimisation requires an end-to-end aspect, combining the configuration management and performance management of all the domains within one TMF layer.

Service configuration is done partly using policies and partly using more traditional parameter/parameter group handling. Thus optimisation also consists of more that one aspect. Clearly, there is a need for optimisation mechanisms like the present day aim of good performance for all services and users without any differentiation. In addition to that, there is a need to optimise within a configuration controlled by a set of policies. In this case the service assumptions, rules resulting in configuration settings and traffic volume expectations are used as the framework for optimisation. An optimisation tool can propose an improved configuration set, so that it does not conflict with the assumptions and rules used for the initial configuration.

It is not possible to optimise within this framework in all cases. The reasons for this can be wrong assumptions (conflict between assumed traffic volume per service and the measured volumes, for example) or a non-optimal rule set. In this case the rules themselves are the target for optimisation. It is important to notice that service management and traffic management cannot be separated:

- Traffic management functions are mainly concerned with the management of network resources with the purpose of accommodating offered traffic in an optimal fashion.
- Service management functions deal with the handling of customer service requests, trying to maximise incoming traffic, in terms of number of contracts and throughput, while respecting the service provider's commitment on agreed service level guarantees.

Quality of Service management ties the two functions together in being very much based on the expected or measured traffic load for each service in the network.

Introduction of a new service or just repricing an old one will cause a change in end-user behaviour and thus a change in traffic volumes. In order to be able to optimise services, one needs to understand and take the traffic volume impact into account. It is important to estimate how service level changes influence the overall traffic situation and service performance.

Automated QoS management holds an enormous potential for the future. Automation increases network availability by reducing operator and configuration errors. Furthermore, tighter linkage between traffic management and service management requires more automated tools.

Service Performance Management

Service performance is used to capture technical performance and customer satisfaction. Service performance definition provides a possibility to balance between customer expectations, price and provisioned service quality. In [16] the differentiation of service and network performance was discussed.

The six main factors contributing collectively to the overall service performance perceived by the user of a telecommunication service are depicted in Figure 8.25.

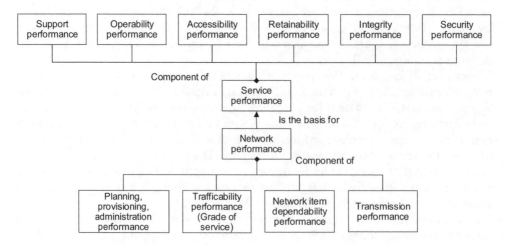

Figure 8.25 Service performance vs. network performance [16].

Each factor should be seen as a concept characterised by many measurements or parameters.

Service support performance is the ability of an organisation to provide a service and assist in its utilisation. An example of service support performance is the ability to provide assistance in commissioning a basic service or a supplementary service such as the call-waiting service or directory enquiries service. Typical measures include mean service provisioning time, billing error probability, incorrect charging or accounting probability, etc.

Service operability performance is the ability of a service to be successfully and easily operated by a user. Typical measures are related to service-user mistake probability, dialling mistake probability, call abandonment probability, etc.

Service accessibility performance is the ability of a service to be obtained, within given tolerances when requested by a user. Measures include items such as access probability, mean service access delay, network accessibility, connection accessibility, mean access delay, etc.

Service retainability performance is the ability of a service, once obtained, to continue to be provided under given conditions for a requested duration. Typically items like service retainability, connection retainability, premature release probability, release failure probability, etc. are monitored.

Service integrity performance is the degree to which a service is provided without excessive impairments (once obtained). Items like interruption of a service, time between interruptions, interruption duration, mean time between interruptions, mean interruption duration are followed. Service security performance is the protection provided against unauthorised monitoring, misuse, fraudulent use, natural disaster, etc.

Network performance is composed of planning, provisioning and administrative performance. Further, trafficability performance, transmission performance and network item dependability performance are part of network performance. Various combinations of these factors provide the needed service performance support.

Planning, provisioning and administrative performance is the degree to which these activities enable the network to respond to current and emerging requirements. All actions related to RAN optimisation belong to this category.

Trafficability performance is the degree to which the capacity of the network components meets the offered network traffic under specified conditions.

Transmission performance is related to the reliability of reproduction of a signal offered to a telecommunication system, under given conditions, when this system is in an in-service state.

Network item dependability performance is the collective term used to describe availability performance and its influencing factors – reliability performance, maintainability performance and maintenance support performance.

Network performance is a conceptual framework that enables network characteristics to be defined, measured and controlled so that network operators can achieve the targeted service performance. A service provider creates a network with network performance levels that are sufficient to enable the service provider to meet its business objectives while satisfying customer requirements. Usually this involves compromise between cost, the capabilities of the network and the levels of performance that the network can support.

An essential difference between service and network performance parameters is that service performance parameters are user-oriented while network performance parameters are network provider- and technology-oriented. Thus, service parameters focus on user-perceivable effects and network performance parameters focus on the efficiency of the network providing the service to the customers.

Service Availability

Service availability as such is not present in the 'service performance' definition, but has turned out to be one of the key parameters related to customer perception and customer satisfaction [16]. Although definitions for network and element availability exist, service availability as such does not have an agreed technical definition. This leads easily to misunderstandings, false expectations and customer dissatisfaction.

In Figure 8.26 the combined items from accessibility, retainability and integrity performance are identified as components of service availability performance.

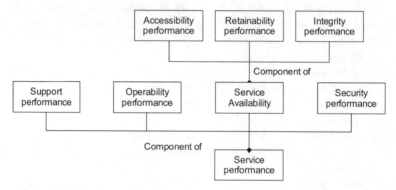

Figure 8.26   Relationship of service availability to service performance.

In [16] examples how to compute service availability measures are shown. Further, in [16] the 3GPP contribution for the definitions of Key Quality Indicators (KQIs) can be found. [16] does not provide direct support for measurement grouping in order to conclude service availability, though. The practical realisation of service availability monitoring is discussed in the next section.

Service Quality Monitoring

The variety of mobile services brings new challenges for operators in monitoring, optimising and managing their networks through services. Service quality management should support service level processes by providing operators with up-to-date views of service quality based on QoS KPIs collected from the network. The performance information should be provided service by service and prioritised for each service package for effective and correctly targeted optimisation.

Involvement of OSI Layer 1, 2 and 3 methods of controlling service performance is required, so that end-user QoS requirements can be translated into technology-specific delivered service performance/network performance measurements and parameters, including QoS distribution and transactions between carriers and systems forming part of any connection. Thus service monitoring as well as good network planning are needed, and the close coupling of traffic engineering and service and network performance cannot be overemphasised.

Performance-related information from the mobile network and services should be collected and classified for further utilisation in reporting and optimisation tools. Network-dependent factors for a mobile service may cover:

- radio access performance;
- core network performance;
- transmission system performance data;
- call detail records;
- network probes;
- services and service systems data.

Different views and reports about PM information should support an operator's network and service planning:

- 3G UMTS service classes (UMTS bearer);
- individual services (aggregate);
- customer's service class/profile;
- geographical location;
- time of day, day of week, etc.;
- IP QoS measures, L1, L2 measures;
- terminal equipment type.

It should also be possible to trace the calls and connections of individual users (see Chapter 7). One source for QoS KPIs is service-specific agents that can be used for monitoring Performance Indicators (PIs) for different services. Active measurement of service quality verification implies testing of the actual communication service, in contrast to passive collection of data from network elements. Network measurements can be collected from different network elements to perform regular testing of the services. Special probes can be used to perform simulated transaction requests at scheduled intervals. By installing the probes at the edge of the IP network, the compound effects of network, server and application delays on the service can be measured, providing an end-user perception of the QoS.

In order to conclude service performance an end-to-end view is important. A network management system entity that is able to combine measurements from different data sources is required. The Service Quality Manager (SQM) concept effectively supports the service monitoring and service assurance process. All service-relevant information that is available in the operator environment can be collected. The information forwarded to the SQM is used to determine the current status of defined services. The current service level is calculated by service-specific correlation rules. Different correlation rules for different types of services (e.g., MMS, WAP, streaming services) are provided.

Figure 8.27 illustrates the general concept of SQM and its interfaces to collect relevant data from other measurement entities and products.

Passive data provide information about the alarm situation (fault management) and performance (performance management) within individual network elements.

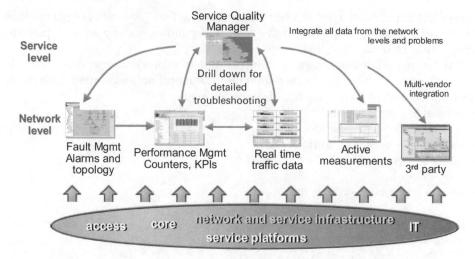

Figure 8.27 Service quality manager and data sources.

Performance management data in terms of network element measurements and KPIs are discussed in Chapter 7. Real time traffic data from charging and billing records provide additional information, which can be utilised to have a very detailed view towards specific services. Active measurements (probing) complement the previous data sources well, providing a snapshot on service usage from the customer perspective.

All these different data sources can be integrated in SQM. SQM correlates the data from different origins to provide a global view towards the network from the customer perspective. SQM's drill-down functionality to all underlying systems at the network level allows efficient troubleshooting and root cause analysis. SQM can be configured to provide information of the service availability. Measurements from different sources are collected and correlated with service availability-related rules. An example of SQM output related to service availability is given in Figure 8.28.

The ability to calculate profiled values using Service Quality Manager provides a powerful mechanism to discover abnormal service behaviour or malfunctions in the network. Further, a rule set to indicate the severeness of a service-related fault or performance degradation can be defined and the distribution of the different levels of faults can be monitored. This severity-based sorting helps the operator to put right the priority of corrective actions. An example of service degradation output is given in Figure 8.29.

The SQM concept bridges the gap between network performance and service performance. With operator-definable correlation rules and the capability to utilise measurements of a different nature, service performance can be monitored and concluded. Thus technical network and technology-facing measurements can be translated to measures that provide an indication of end-to-end performance and end-user satisfaction.

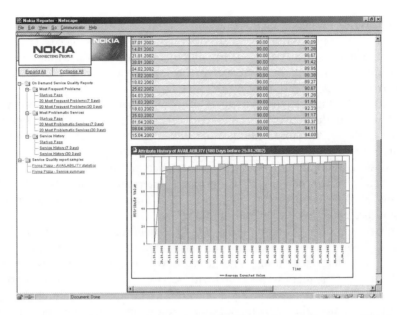

Figure 8.28 Service quality manager report: availability of a service over 180 days. The definition of service availability is operator-specific and contains items from the service performance framework (see Figure 8.26).

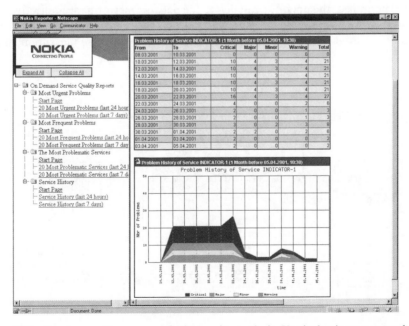

Figure 8.29 Service quality manager fault severity analysis. Vertical axis represents the number of problems, horizontal axis is time and colour coding indicates whether the service problem is critical, major, minor or warning. The classification is operator-specific.

Quality of Service Feedback Loops

In Chapter 7 the optimisation feedback loop concept is introduced. The interfaces, configuration management and performance management data availability and the management system role are discussed. In this section the same concept is applied, but now from the QoS point of view.

The most important requirement for QoS management is the ability to verify the provided quality in the network. Second requirement is then the ability to guarantee the provided quality. Therefore, monitoring and post-processing tools play a very important role in QoS management. A post-processing system needs to be able to present massive and complex network performance and quality data both in textual and in highly advanced graphical formats. Interrelationships between different viewpoints of QoS are presented in Figure 8.30. The picture comes from [26], and the version presented here is slightly modified.

The figure captures the complexity of QoS management very well. From the optimisation point of view, there are at least three main loops, which constitute a challenge for management tools. The network-level optimisation loop (on the right side of Figure 8.30) is mainly concerned with service assurance. Network performance objectives have been set based on QoS-related criteria, and the main challenge of the operator is to monitor the performance objectives by deriving the network status from network performance measurements.

The optimisation loop from service level to network level covers the process from determination of QoS/application performance-related criteria to QoS/application performance offered to the subscriber. Once application performance-related criteria have been determined, the operator can derive the network performance objectives. The network is then monitored and measured based on these objectives, and application performance achieved in the network can be interpreted from these measurements. At

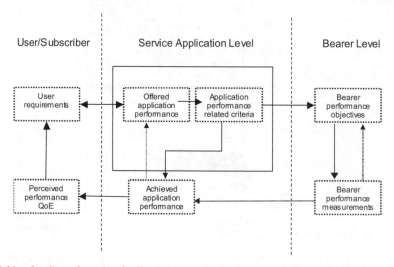

Figure 8.30 Quality of service feedback loops. Dashed arrows indicate feedback; solid arrows indicate activity and flow [26].

this point, there may be a gap between the offered and the achieved application performance. Depending on the types of difference, application performance-related criteria or the network configuration might need fine-tuning.

Further, there can be application-related performance and usability deficiencies that cannot be fixed by retuning the network.

The third optimisation loop involves optimisation of QoS perceived by the subscriber, who has certain requirements for the quality of the application used. The QoS offered to the subscriber depends on application needs and actual network capacity and capability. The perceived quality depends on the quality available from the network and from the applications, including usability aspects. Subscriber satisfaction then depends on the difference between his/her expectations and the perceived quality. The ultimate optimisation goal is to optimise the QoS that the subscriber perceives – i.e., the QoE.

8.7   Concluding Remarks

The classification of end-user services was discussed within a framework allowing for detailed analysis. Requirements and characteristics for services were discussed within this framework.

The 3GPP QoS architecture is a versatile and comprehensive basis for providing future services as well. From the viewpoint of the service management process, there are certain issues which the standardised architecture does not solve. The support provided by the 3GPP standard architecture to service configuration was discussed above. The reverse direction requires that it is possible to map specific counters in network elements onto service-specific KQIs. The Wireless Service Measurement Team of TeleManagement Forum has defined a set of KQIs and KPIs [6] and submitted it to the SA5 working group in 3GPP. Many of the counters have been standardised by 3GPP, but they are not – and indeed should not be – associated with particular services. Thus, conceptually one needs a service assurance 'middleware' layer for mapping element-specific counters to end-to-end service performance levels, as depicted in Figure 8.31.

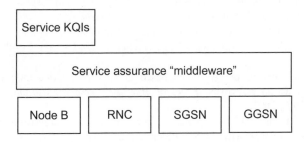

Figure 8.31   An illustration of service-oriented assurance.

References

[1] 3GPP, TS 23.107, v5.12.0 (2004-03), QoS Concept and Architecture, March 2004.

[2] 3GPP, TS 23.207, v5.9.0 (2004-03), End-to-end QoS Concept and Architecture, March 2004.

[3] 3GPP, TS 23.228, v5.13.0, IP Multimedia Subsystem (IMS), December 2004.

[4] Communications Quality of Service: A Framework and Definitions, ITU-T Recommendation G.1000, November 2001.

[5] End-user Multimedia QoS Categories, ITU-T Recommendation G.1010, November 2001.

[6] *Wireless Service Measurement, Key Quality Indicators*, GB 923A, v1.5, April 2004, TeleManagement Forum.

[7] Koivukoski, U. and Räisänen, V. (eds), *Managing Mobile Services: Technologies and Business Practices*, John Wiley & Sons, 2005.

[8] McDysan, D., *QoS and Traffic Management in IP and ATM Networks*, McGraw-Hill, 2000.

[9] Padhye, J., Firoiu, V., Towsley, D. and Kurose, J., Modelling TCP reno performance. *IEEE/ACM Transactions on Networking*, **8**, 2000.

[10] Poikselkä, M., Mayer, G., Khartabil, H. and Niemi, A., *IMS: IP Multimedia Concepts and Services in the Mobile Domain*, John Wiley & Sons, 2004.

[11] Räisänen, V., *Implementing Service Quality in IP Networks*, John Wiley & Sons, 2003.

[12] Räisänen, V., Service quality support: An overview. *Computer Communications*, **27**, pp. 1539ff., 2004.

[13] Räisänen, V., A framework for service quality, submitted to IEEE.

[14] Schulzrinne, H., Casner, S., Frederick, R. and Jacobson, V., *RTP: A Transport Protocol for Real-time Applications*, RFC 1889, January 1996, Internet Engineering Task Force.

[15] Armitage, G., *Quality of Service in IP Networks*, MacMillan Technical Publishing, 2000.

[16] *SLA Management Handbook, Volume 2, Concepts and Principles*, GB917-2, TeleManagement Forum, April 2004.

[17] Cuny, R., End-to-end performance analysis of push to talk over cellular (PoC) over WCDMA. *Communication Systems and Networks, September 2004, Marbella, Spain*. International Association of Science and Technology for Development.

[18] Antila, J. and Lakkakorpi, J., On the effect of reduced Quality of Service in multi-player online games. *International Journal of Intelligent Games and Simulations*, **2**, pp. 89ff., 2003.

[19] Halonen, T., Romero, J. and Melero, J., *GSM, GPRS, and EDGE Performance: Evolution towards 3G/UMTS*, John Wiley & Sons, 2003.

[20] Bouch, A., Sasse, M.A. and DeMeer, H., Of packets and people: A user-centred approach to Quality of Service. *Proc. IWQoS '00, Pittsburgh, June 2000*, IEEE.

[21] Lakaniemi, A., Rosti, J. and Räisänen, V., Subjective VoIP speech quality evaluation based on network measurements. *Proc. ICC '01, Helsinki, June 2001*, IEEE.

[22] 3GPP, TS 32.403, v5.8.0 (2004-09), Telecommunication Management; Performance Management (PM); Performance Measurements – UMTS and Combined UMTS/GSM (Release 5).

[23] Laiho, J. and Soldani, D., A policy based Quality of Service management system for UMTS radio access networks. *Proc. of Wireless Personal Multimedia Communications (WPMC) Conf.*, 2003.

[24] Soldani, D. and Laiho, J., User perceived performance of interactive and background data in WCDMA networks with QoS differentiation. *Proc. of Wireless Personal Multimedia Communications (WPMC) Conf.*, 2003.

[25] Soldani, D., Wacker, A. and Sipilä, K., An enhanced virtual time simulator for studying QoS provisioning of multimedia services in UTRAN. *Proc. of MMNS 2004 Conf., San Diego, California, October 2004*, pp. 241–254.
[26] *Wireless Service Measurement Handbook*, GB923, v3.0, March 2004, TeleManagement Forum.

9

Advanced Analysis Methods and Radio Access Network Autotuning

Jaana Laiho, Pekko Vehviläinen, Albert Höglund, Mikko Kylväjä, Kimmo Valkealahti and Ted Buot

9.1 Introduction

Introduction of third generation (3G) cellular systems will offer numerous possibilities for operators. The introduction of General Packet Radio Service (GPRS) into Global System for Mobile communications (GSM) networks is already changing the operation environment from circuit switched to the combination of Real Time (RT) and Non-Real Time (NRT) services. The 3G traffic classes (conversational, interactive, streaming, background), Quality of Service (QoS) provisioning mechanisms and QoS differentiation possibilities, together with the joint management and traffic sharing between second generation (2G) and 3G networks provide a challenging playground on one hand for vendors, and on the other hand for service providers and network operators. To be able to fully utilise the resources and to focus on the service provisioning rather than troubleshooting tasks, advanced analysis and visualisation methods for the optimisation process are required. Further, automation in terms of data retrieval, workflow support and algorithms is of essence.

In Chapter 7 Network Management System (NMS) level statistical optimisation and its components were introduced. These components are depicted in Figure 9.1. In this chapter the focus is on analysis, data visualisation means and automated optimisation.

Once a WCDMA network is built and launched, an important part of its operation and maintenance is to monitor and analyse performance or quality characteristics and to change configuration parameter settings in order to improve performance. The automated parameter control mechanism can be simple but it requires objectively defined Performance Indicators (PIs) and Key Performance Indicators (KPIs) that unambiguously tell whether performance is improving or deteriorating.

Radio Network Planning and Optimisation for UMTS Second Edition
Edited by J. Laiho, A. Wacker and T. Novosad © 2006 John Wiley & Sons, Ltd

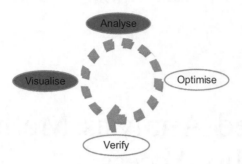

Figure 9.1 Different tasks in optimisation workflow. This section focuses on analysis and data visualisation. Optimisation is in Section 9.3.

To ease optimisation, or provide robust autotuning, a way of identifying similarly behaving cell groups or clusters, which can have their own parameter settings, is introduced in this chapter. Advanced monitoring – i.e., data mining and visualisation methods such as anomaly detection, classification trees and self-organising maps – are also presented.

Further, this chapter introduces possible autotuning features such as coverage–capacity tradeoff management in congestion control. With this feature the operator only has to set quality and capacity targets and costs that regulate the quality–capacity tradeoff.

The target of autotuning is not necessarily the best quality as traditionally defined. In some cases it might be that slightly degraded quality with the possibility of offering more traffic is more beneficial for an operator's business case than quality-driven optimisation. A high-level objective is also to integrate WCDMA automation with other systems such as EDGE and WLAN. Autotuning of neighbour cell lists is presented in this chapter as an example of inter-system automation.

9.2 Advanced Analysis Methods for Cellular Networks

The scope of the following sections is to introduce examples of how advanced analysis methods – such as anomaly detection, data mining methods and data exploration – benefit operators in monitoring and visualisation tasks. Example cases are provided using data from GSM networks and WCDMA simulations.

9.2.1 Introduction to Data Mining

Subscribers, connected to the network via their UEs (User Equipment), expect network availability, connection throughput and affordability. Moreover, the connection should not degrade or be lost abruptly as the user moves within the network area. User expectations constitute QoS, specified as 'the collective effect of service performances, which determine the degree of satisfaction of a user of a service' [1]. The operating personnel have to measure the network in terms of QoS. By analysing the information they get from their measurements, they can manage and improve the quality of their services.

However, because operating staff are easily overwhelmed by hundreds of measurements, the measurements are aggregated as KPIs.

Personnel expertise with the KPIs and the problems occurring in the cells of the network vary widely, but at least the personnel know the desirable KPI value range. Their knowledge may be based on simple rules such as 'if any of the KPIs is unacceptable, then the state of a cell is unacceptable.' The acceptance limits of the KPIs and the labelling rules are part of the *a priori* knowledge for analysis.

Information needed to analyse QoS issues exists in KPI data, but sometimes it is not easy to recognise. The techniques of Knowledge Discovery in Databases (KDD) and data mining help to find useful information in the data.

The most important criterion for selecting data mining methods for use in this chapter was their suitability as tools for the operating staff of a digital mobile telecommunications network to alleviate their task of interpreting QoS-related information from measured data. Two methods were chosen that fulfilled the criterion: classification trees and Self-Organising Map (SOM) type neural networks.

In particular, the automatic inclusion of prior knowledge in preparing the data is a novelty because *a priori* knowledge has so far been overlooked [2].

9.2.2　Knowledge Discovery in Databases and Data Mining

KDD, a multi-step, interactive and iterative process requiring human involvement [3], aims to find new knowledge about an application domain.

9.2.2.1　Knowledge Discovery in Databases

The KDD process [2] consists of consecutive tasks, out of which data mining produces the patterns of information for interpretation (see Figure 9.2). The results of data mining then have to be evaluated and interpreted in the resulting interpretation phase before we can decide whether the mined information qualifies as knowledge [3].

The discovery process is repeated until new knowledge is extracted from the data. Iteration distinguishes KDD from the straightforward knowledge acquisition by measurement.

9.2.2.2　Data Mining

Data mining is a partially automated KDD sub-process, whose purpose is to non-trivially extract implicit and potentially useful patterns of information from large datasets [2]. Specifically, data mining for QoS analysis of mobile telecommunications networks involves five consecutive steps (Figure 9.3), four of them closely related to the use of data mining methods: attribute construction, method selection, pre-processing and preparation.

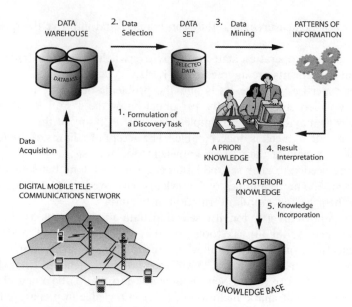

Figure 9.2   Knowledge discovery in databases for quality of service analysis of a network is an interactive and iterative process in five consecutive steps [2].

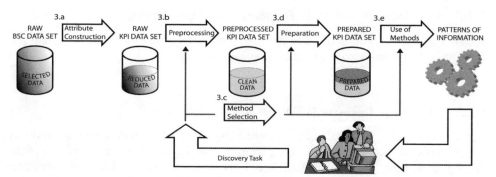

Figure 9.3   Data mining for quality of service analysis of mobile telecommunication network steps [2].

9.2.2.3   Attribute Construction: Quality Key Performance Indicators

A KPI is considered an important performance measurement, constructed from several raw measurements. In network management, KPIs may be used for several purposes; thus selecting KPIs for analysis is a subjective matter. QoS-related KPIs in this subsection are based on the measurements of Standalone Dedicated Control Channel (SDCCH), Traffic Channels (TCHs), logical channels and handovers. The performance management process and KPIs were further discussed in Chapter 7.

Intrinsic QoS analysis depends on quality-related KPI measurements available from Network Elements (NEs). The intrinsic QoS of a bearer service means that the network's radio coverage is available for the subscriber outdoors and indoors.

However, availability of the network is necessary for mobile applications; therefore, KPI data contain information about those cells where the bearer service or end-to-end service is degraded.

9.2.2.4   KPI Limits Based on *A Priori* Knowledge

An optimisation expert knows roughly the good, normal, bad and unacceptable range of KPI values. For instance, his a priori knowledge of *SDCCH Success* is that it is normal for KPI values to be close to 100. He also knows that if the value drops below 100, a problem ensues because the signalling channels should be available all the time. To ensure that his a priori knowledge is justified, the analyst can plot the KPIs' Probability Density Function (PDF) estimates, assuming that the data are acquired from a network that has been under normal operational control. PDF estimates are plotted so that variable data are divided into slots along the horizontal axis, which represents a KPI's value. Each slot has an equal number of data points, which means that the height of the slot is proportional to the density of data points over the range of one slot. The PDF plot for *SDCCH Success* is shown in Figure 9.4.

Based on the limits and his *a priori* knowledge, the operator can then write out his rules to interpret the data as a labelling function.

When the analyst scrutinises the plotted KPI PDFs, he can justify and possibly refine the limits of a good, normal, bad and unacceptable KPI.

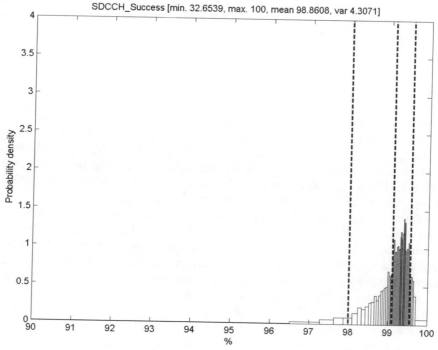

Figure 9.4   *A priori* limits of value ranges of key performance indicator *SDCCH Success*.

9.2.2.5 Pre-processing

The main objective of the pre-processing phase is to ensure that analysis methods are able to extract the correct and needed information from the data [4].

Neural network methods are multi-variate methods that study the combination of variables – i.e., their joint distribution. Before they can be applied to the data, the data have to be prepared for analysis in the pre-processing phase. Pre-processing has to filter out noise, handle the problem of missing values and balance different variables and their value ranges. What needs to be done originates from the current information need. For example, in network analysis one can either be interested in bad cells with abnormal indicator values in order to be able to fix them or in the behaviour of the best cell in order to copy its configuration to other corresponding cells.

It is not feasible to severely alter the dataset straightaway, since useful information could be lost. However, noise, missing values, and inconsistencies are features that are not accepted in any dataset, and one should, if possible, correct these unwanted features before one selects the data mining methods [2].

In order to extract the correct information from network data the used variables must be balanced by scaling. The most common method to do the balancing is to normalise the variance of each variable to 1. Normalisation might be skewed if there are outliers in variable value series. If the average normal behaviour is studied, the usual solution is to remove outliers or to replace them with an estimated normal or correct value. If outliers carry interesting information, for example – as is the case in our study in Section 9.2.9, where they can be signs of network problems that are searched for – it is possible to keep outliers but not let their large values dominate the analysis results. This can be done by using some sort of conversion function like tanh (or log) before normalisation of the variance.

9.2.2.6 Preparation of Data: Labelling Function

A labelling function is necessary for labelling observations with a decision indicator value, which in turn is necessary for a supervised learning algorithm. The function can be thought of as a formulated inference rule of the operator judging the behaviour of the network. The inference and its limits (see Table 9.1) are the operator's *a priori* knowledge. The values of the rest of the limits resulted from subjective inference from the PDF estimate distributions in the previous section.

The function makes use of logical inference based on the predetermined limits of the PIs. As a result, it labels each observation as good, normal, bad or unacceptable. It does not include information about the causes of changes in the observations but indicates simply whether a cell is in a more or less acceptable state (good, normal, bad) or whether a state requires immediate attention (unacceptable). The labelling function is a set of four rules on the seven quality-related KPIs – i.e., *SDCCH Access, SDCCH Success, TCH Access, TCH Success, HandOver (HO) Failure, HO Failure Due to Blocking* and *TCH Drops*. The labelling function labels the observations in the KPI dataset according to the following four rules, which are applied in descending order so that the label is the one that first applies. Thus the state of the network is:

Table 9.1 Discretised key performance indicator values [%] with corresponding discretisation limits. The *a priori* limits given by a domain expert are greyed out.

KPI	Unacceptable	Bad	Normal	Good
SDCCH Access	≤99.00	—	>99.00	—
SDCCH Success	≤98.00	≤99.10	≤99.56	>99.56
TCH Access	≤99.00	—	>99.00	—
TCH Success	≤98.00	≤98.75	≤99.35	>99.35
HO Failure	≥5.00	≥2.08	≥0.91	<0.91
HO Failure Due to Blocking	≥5.00	≥0.23	≥0.08	<0.08
TCH Drops	≥2.00	≥0.57	≥0.19	<0.19

- *unacceptable* if any quality-related KPI is rated as unacceptable;
- *bad* if any quality-related KPI is rated bad;
- *good* if KPIs *SDCCH Access* and *TCH Access* are classified as normal and KPIs *SDCCH Success, TCH Success, HO Failure, HO Failure Due to Blocking* and *TCH Drops* are rated good
- *normal* if KPIs *SDCCH Access* and *TCH Access* are classified as normal and KPIs *SDCCH Success, TCH Success, HO Failure, HO Failure Due to Blocking* and *TCH Drops* are rated either normal or good.

The labels can be coded numerically as in Table 9.2.

Table 9.2 Labels of the decision class indicator.

State of a cell	Decision class indicator
Good	1
Normal	2
Bad	3
Unacceptable	4

9.2.3 Classification Trees

In data mining, a common classification method is the identification of a classification tree [5] that suits both classification and prediction. In this section the application of the Classification and Regression Trees (CART) algorithm is applied to the QoS KPIs.

The benefits of binary splitting, a simple splitting condition, and CART's ability to process both numerical (KPI data) and nominal values, were the main criteria why CART was chosen for the classification tree algorithm.

9.2.3.1 Application

Before analysis with CART, the KPI dataset was pre-processed by removing observations with missing values and prepared by subjecting the data to the labelling function.

With the aid of the tree-growing theory [8], the whole KPI dataset of 3069 observations was analysed with the CART algorithm. The Gini index of diversity – see Equation (9.1) – was chosen as the score function, and tree growing was set to terminate if any further growth reduced the observations in a node to less than 20 observations:

$$Gini(t) = 1 - \sum_v p^2(v \mid t) \tag{9.1}$$

where $p(v \mid t)$ is the estimated probability that a KPI observation is of class v (good, normal, bad, unacceptable), given that it falls into node t.

The CART algorithm resulted in the tree structure shown in Figure 9.5. The tree has 9 levels and 27 nodes, 14 of which are terminal nodes and 13 splitting nodes. The nodes are numbered from 1 to 27 with their identification number increasing from left to right and moving up to the next level after passing the rightmost node on a level.

The higher the split node number of the KPI, the less important the KPI is in separating large pure groups of observations within the dataset.

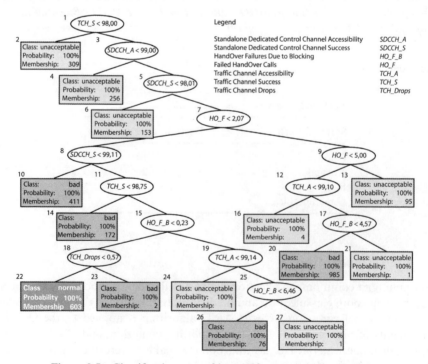

Figure 9.5 Classification tree of key performance indicator data.

Table 9.3   Splits of a pruned tree vs. key performance indicator discretisation limits [%].

KPI	Unacceptable	Bad	Normal	Good
SDCCH Access	≤99.00 (node 3)	—	>99.00	—
SDCCH Success	≤98.00 (node 5)	≤99.10 (node 8)	≤99.56	>99.56
TCH Access	≤99.00	—	>99.00	—
TCH Success	≤98.00 (node 1)	≤98.75 (node 11)	≤99.35	>99.35
HO Failure	≥5.00 (node 9)	≥2.08 (node 7)	≥0.91	<0.91
HO Failure Due to Blocking	≥5.00	≥0.23 (node 15)	≥0.08	<0.08
TCH Drops	≥2.00	≥0.57	≥0.19	<0.19

Examining the 14 terminal nodes, one can notice that they are all pure nodes (with 100% class probability in each terminal node). Seven of the nodes are classified as unacceptable (nodes 2, 4, 6, 13, 16, 21, 24 and 27), five bad (nodes 10, 14, 20, 23 and 26), and one normal (node 22). The tree had no good terminal nodes.

Examination of the oval-shaped split nodes in Figure 9.5 reveals that most splits (8 out of 13) are based on KPI PDF estimates' label range boundaries (see Table 9.1). This is not surprising because the tree is structured according to the decision indicator, which in turn is based on the labelling function (Section 9.2.2.6), which again pre-classifies observations according to label range boundaries.

Is this circular reasoning? Yes, if one is interested only in boundary values, but no, if one seeks to identify those KPIs and their corresponding boundaries that separate the observation groups in the dataset. Splits along the label range boundaries have been added in Table 9.3 (derived from Table 9.1), and the alignment is indicated with a node number in parentheses.

9.2.4   Anomaly (Outlier) Detection with Classification Tree

An outlier is defined by [7] as a single, or very low frequency, occurrence of the value of a variable that is far away from the bulk of the values of the variable. The 5 splits that are not along the discretisation boundaries mark off the data points and are reflected by the number of observations in nodes 16, 21, 23, 24 and 27 (Figure 9.5). They all seem to contain a few (one to four) outliers, which are clearly separable from the rest of the data and should thus be analysed separately.

9.2.5   Self-Organising Map

If there is only limited *a priori* knowledge, or one needs to check one's prior knowledge on the data, one has to apply an unsupervised or self-organised learning method to look for features that are not known before the analysis but that describe the data. One such method is the Self-Organising Map (SOM), an unsupervised neural network, introduced by Professor Teuvo Kohonen in 1982. SOM-based methods have been applied in the analysis of process data – e.g., in the steel and forest industries ([12]–[16]).

9.2.5.1 Concepts

The SOM provides a powerful visualisation method for data. The SOM algorithm creates a set of prototype vectors, which represent a training dataset, and projects the prototype vectors from the n-dimensional input space – n being the number of variables in the dataset – onto a low-dimensional grid. The resulting grid structure is then used as a visualisation surface to show features in the data [9].

The created prototype vectors are called neurons, connected via neighbourhood relations. The training phase of a SOM exploits the neighbourhood relation in that parameters are updated for a neuron and its neighbouring units.

The neurons of a SOM are organised in a low-dimensional grid with a local lattice topology. The most common combination of local and global structures is the two-dimensional hexagonal lattice sheet, which is preferred in this example case as well.

9.2.5.2 Theory

Let $x \in \Re''$ be a randomly chosen observation from dataset X. Now, the SOM can be thought of as a non-linear mapping of the probability density function $p(x)$ onto the observation vector space on a lower (two in our case) dimensional support space. Observation x is compared with all the weight vectors w_i of the map's neurons, using the Euclidean distance measure $\|x - w_i\|$.

Among all the weight vectors, the closest match w_c is chosen based on Euclidean distance, to observation x and call neuron c (c is the neuron's identification number on the map grid) related to w_c the Best Matching Unit (BMU):

$$\|x - w_c\| = \min_i\|x - w_i\| \tag{9.2}$$

After the BMU is found, denoted by c, its weight vector w_c is updated so that it moves closer to observation x in the input space. The update rule for all the weights of the SOM is:

$$w_i(t+1) = w_i(t) + \alpha(t)h_{ci}(t)[x - w_i(t)] \tag{9.3}$$

where t is an integer-discrete time index; $\alpha(t)$ the learning rate function; $h_{ci}(t)$ the neighbourhood function; and x a randomly drawn observation from the input dataset. Note that $h_{ci}(t)$ is calculated separately in the map dimension (two), whereas x and weight vectors w_i have the dimension of the input space (seven in our case).

The learning rate is chosen so that the update effect decreases during the SOM's training phase. One such rate is:

$$\alpha(t) = \frac{\alpha_0}{1 + (kt)/T} \tag{9.4}$$

where α_0 is the initial value of the learning rate function; k some arbitrarily chosen coefficient; and T the training length.

The neighbourhood kernel around the BMU can be defined in several ways, one possibility is the Gaussian function denoted by:

$$h_{ci}(t) = \exp(-\|r_c - r_i\|/2\sigma_i^2) \tag{9.5}$$

where σ_t is the kernel radius at time t; r_c the map coordinates of the BMU; and r_i the map coordinates of the nodes in the neighbourhood.

9.2.6 *Performance Monitoring Using the Self-Organising Map: GSM Network*

Like with the CART, the dataset was pre-processed by removing the missing values since they are problematic in the SOM algorithm [12]. The variables in the training dataset must be rescaled. Should the data have very different scales, the variables with high values are likely to dominate the training when the SOM algorithm minimises the Euclidean distance measure between weight vectors and observations [19].

The variables are commonly scaled so that the variance of each variable is 1. But since the ranges of the variables were known *a priori*, that information was used for scaling [2].

To present SOM information in an easily interpretable form, the value of each variable is shown on the map in a variable-specific figure instead of showing all variables in one figure. Such separate figures are called 'component planes'.

Each component plane has a relative distribution of one KPI. The values in component planes are visualised in shades of grey. These values were scaled so that white or light shading represents preferable KPI values and black or dark shading unwanted KPI values. On the side of each component plane is placed a grey scale to link the shading and actual KPI values. Note that the shading is specific to each component plane. The component planes of the trained SOM are shown in Figure 9.6. In addition, the component planes show the *a priori* information of the labelling function – i.e., the value of the decision variable of observations with the most occurrences in the node.

One can immediately see that the unwanted values of *SDCCH Success*, *TCH Success* and *TCH Drops* of the right side of the component planes are almost black.

SDCCH Success may also take unwanted values separately from *TCH Success* and *TCH Drops*, since the nodes in the top left corner are dark, whereas the component planes of *TCH Success* and *TCH Drops* are light in those nodes.

Furthermore, one can see that *TCH Access* correlates with *HO Failure Due to Blocking*, since the nodes in the low left corner are dark in both planes.

HO Failure has its worst values in the nodes in the bottom right corner, which are dark. *HO Failure* is somewhat connected to *SDCCH Access*, because its component plane is grey in the same nodes. *SDCCH Access* has its worst values quite independently of the rest of the KPIs.

Hit hexagons (see Figure 9.6) show that most observations were distributed among the top and bottom rows of the map and in the middle. The *a priori* knowledge seems to match the component planes well, for the nodes that match normal states are located in the top middle section of the map. The worst observations fall on the left and right sides and in the bottom corners of the map.

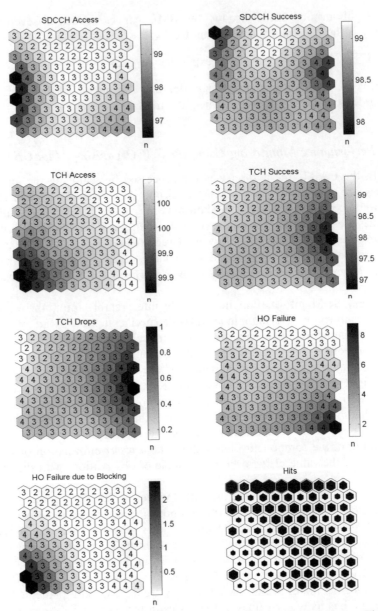

Figure 9.6 Self-organising map component planes and relative hit counts of nodes. The numbers are labels from the labelling function.

9.2.6.1 Anomaly Detection with Clustering Methods

Clustering methods, such as the SOM, introduced in the previous section, can also form a part of a method to detect anomalous or abnormal performance of NEs – e.g., BSs and RNCs. The principle of the method is as follows:

- Select an NE type to be monitored.
- Select variables or PIs to monitor. One observation of these variables forms a data vector.
- For each element to be monitored:
  1. Store n data vectors that describe the functioning (normal behaviour) of the element during a certain time period.
  2. Use the vectors as input data to a clustering method (such as SOM or k-means, both introduced later in this chapter) to train a profile for each element, consisting of nodes.
  3. For each data vector used in training the profile, calculate the distances to the closest node in the profile using a distance measure (usually Euclidean distance) to obtain a distance distribution.
  4. To test whether a new data vector is abnormal, calculate the distance to the closest node in the profile.
  5. The new observation can be considered abnormal if its distance exceeds a certain percentage (e.g., 0.5%) of distances in the distance distribution of the profile.
  6. The most abnormal variables or PIs can be calculated by examining their contribution to the deviating distance. The biggest contribution means the most abnormal variable, etc.

For details of the anomaly detection method, see [6]. Figure 9.7 shows an example of anomaly detection using hourly data for a GSM base station. The training period in this case was quite short: the 14 day period prior to the observation was tested. The profile was retrained daily for the previous 14 days. Eight PIs were monitored in addition to two time components (two needed for a daily repetitive pattern). The indicators describe dropping, blocking, traffic, success and requests on the TCH and SDCCH. The indicators were normalised before analysis and plotting in Figure 9.7. The first set of anomalies on day 14 seems to be due to high SDCCH dropping. The anomaly on day 16 seems to have been caused by relatively high SDCCH blocking and dropping while the SDCCH traffic was relatively low. The anomaly on day 23 seems to have been caused by high SDCCH blocking under heavy SDDCH traffic.

The main advantage of the anomaly detection method is that it detects abnormal variable or indicator combinations in addition to abnormal values of individual variables or indicators. The method is therefore very useful in network monitoring and much easier to use than manual setting and updating of thresholds.

9.2.7 Performance Monitoring Using the Self-Organising Map: UMTS Network

Mobile network data consist of parameters of NEs and quality information from circuit switched or packet switched calls. For analysis, one can build either:
- a model of the network using state vectors with parameters from all mobile cells; or
- a general one-cell model trained using one-cell state vectors from all cells.

In both methods further analysis is needed. In the first method the distributions of parameters of one cell can be compared with those of the others, while the second

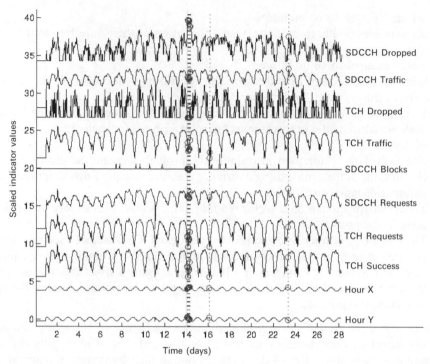

Figure 9.7 Scaled GSM performance indicators analysed using moving 14 day profile training. The anomalies are marked with dotted vertical lines. The first 14 days were used only for profile training and are therefore not analysed.

method can compare how well the general model represents each cell. Cell grouping can be used for optimisation and automation purposes. The grouping is based on state space – i.e., KPI values, parameter values, physical coordinates, etc. The role of this feature would be to support usage of cell grouping in autotuning and parameter optimisation. In this example case SOM is used to analyse and conclude cell types from measured data alone.

9.2.7.1   Network Scenario and Data Used in SOM Analysis

The SOM method used in this chapter uses both uplink and downlink data from the micro-cellular network scenario depicted in Figure 9.8. Results are provided for the micro-cellular scenario since it represents a more challenging environment from the propagation point of view. Further it is foreseen that the high capacity requirements of data services will require a small-cell environment.

The WCDMA radio networks used in this study have been planned to provide 64 kbps service with 95% coverage probability and with reasonable (2%) blocking. The ray-tracing model was used for propagation loss estimation and an additional indoor loss of 12 dB was applied in areas inside buildings. The network layout comprises 46 omni-directional base station sites. The selected antenna installation

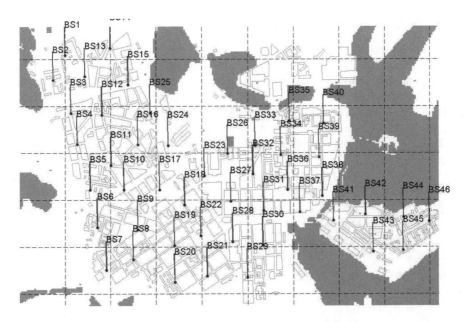

Figure 9.8 Micro-cellular scenario used in simulations.

height was on average 10 m. Due to the lack of measured data from live networks at the time of this study simulated data were used in the advanced analysis cases. The data used in this work have been generated using a WCDMA radio network simulator [17]. During simulations the multi-path channel profile of the ITU Outdoor-to-Indoor A channel was assumed. The system features used in the simulations are according to 3GPP. A detailed description of the network parameters can be found in [4]. The users in the network were using 64 kbps service and admission control was parameterised so that the uplink loading/interference level did not limit the admission decision.

During the simulations several KPIs were monitored and the KPIs in Table 9.4 were selected for SOM analysis. KPIs were collected for each cell. In the case of real network measurements more KPIs can be added to the clustering analysis. This analysis also serves as a KPI correlation indicator, since it is possible to identify those KPIs that have the largest impact on the cluster formation. A KPI that does not change the clustering correlates with another KPI used in the analysis.

Table 9.4 Key performance indicators collected during simulations and used for the purpose of analysis in Section 9.2.8.2.

Parameter name	Description
dlFer	Frame error rate value for downlink
dlTxp	Transmit power per link, downlink direction
nUsr	Number of users in a cell
ulFer	Frame error rate value for uplink
ulANR	Average noise rise for uplink

9.2.7.2   Data Monitoring Using Self-Organising Maps

In this section the usage of advanced neural methods in WCDMA cellular network analysis is presented. The motivation for the introduction of neural analysis on network performance data is to provide an effective means of handling multiple KPIs simultaneously. Furthermore, effective analysis methods reduce operators' troubleshooting efforts, speed up the optimisation cycle and thus increase the network utilisation rate. As mentioned SOM has several beneficial features, especially the possibility to cluster cells based on performance and visualise the mapping in two-dimensional views (as shown in this section).

The method described in [4] and [10] has been used to analyse both the uplink and downlink direction in micro-cellular and macro-cellular network scenarios. The example presented here focuses on the micro-cellular case only. The presented method consists of the following steps:

- target selection;
- data pre-processing;
- cluster analysis;
- result interpretation.

The first step in the process is target selection. This includes selection of the geographical area, network objects (BSs, RNCs, routers, etc.) and visualisation task specification. The selection of network objects and the visualisation task have a strong impact on the selection of the measurements and KPIs to be analysed. Naturally each object in the network has its own specific measurements. The visualisation task can be more generic or problem-oriented. General performance analysis requires a different set of measurements than a specific troubleshooting case.

Data pre-processing was introduced in this chapter in Section 9.2.2.5. The data vectors of all the cells are clustered using the two-phase clustering algorithm. First, the SOM is trained using data vectors. Next, the clustering algorithm is run for SOM codebook vectors so that exact clusters can be defined. When the data clusters of the cells are formed the dynamic simulator provides the input data for SOM. In this work the data clusters are further analysed by automatically generated rules in order to find the most qualitative description for the cells within a cluster. An example of this type of data presentation is given in Figure 9.9. In this case k-means clustering is used. More about clustering techniques can be found in [25].

In order to analyse a sequence of data samples instead of a single data point a histogram map is computed. Histograms consist of proportions of data samples falling in each of the data clusters. These histograms describe the long-term behaviour of data sequences; they are used in cell classification. A new SOM is generated using the histogram information as the training set. By using a clustering algorithm exact behavioural clusters can be generated. An example of this is given in Figure 9.10.

For the analysis of the combined uplink and downlink directions in the micro-cellular scenario, five variables (KPIs in this case) have been selected: number of users ($nUsr$), uplink average noise rise relative to basic noise floor ($ulANR$), uplink frame error rate

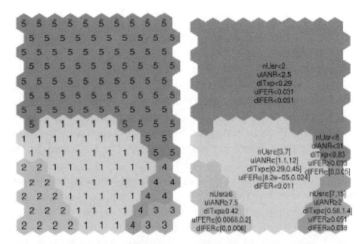

Figure 9.9 Clustered self-organising map for combined uplink and downlink case and rules for clusters in the micro-cellular scenario [4] *Data Clusters*. Grey shades or numbers in the figure have no other meaning than to show the areas of the clusters.

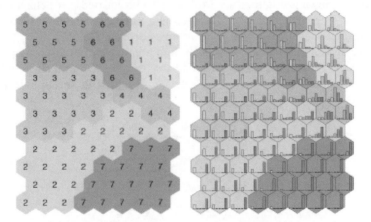

Figure 9.10 Histogram map for both uplink and downlink data of the micro-cellular scenario ([4] and [10]). *Behavioural clusters*.

(*ulFER*), downlink average transmission power (*dlTxp*) and downlink frame error rate (*dlFER*). The frame error rate values are pre-processed using a tanh function to be able to see the changes at a lower error rate level as well. All the parameters are also normalised to zero mean and a variance of 1.

Figure 9.9 shows the clustered SOM. The data samples are divided into five data clusters, of which cluster 3 in the lower right corner represents data samples with high *dlFER* (downlink quality problems) and cluster 4 data samples with acceptable *dlFER* but high *ulFER* (uplink quality problems).

In Figure 9.10 the corresponding histogram map and behavioural clusters for combined uplink and downlink data for the micro-cellular scenario are shown. The bars in the histograms indicate the number of samples in the data clusters of Figure 9.9. The first bar in the histogram is characterised with the rules of data cluster 1 in Figure 9.9.

The highest proportion of samples that fall in data clusters 3 and 4 in Figure 9.9 is in behavioural cluster 4 on the histogram map – i.e., in Figure 9.10. This can be found by looking for the map nodes (i.e., hexagons) in which the third and fourth bar are highest. Also, two map nodes in behavioural cluster 1 indicate the high number of samples in data cluster 3 (i.e., third bar in histogram – samples with the highest *dlFER* values in Figure 9.9). Other characteristics for data cluster 3 can be found in the lower right corner of Figure 9.9.

In the combined uplink/downlink case the dominant behavioural clusters are 2, 3 and 7. Typical for these clusters is the number of users ranging from low to medium, high correlation of the number of users and the used resources (i.e., good control of external interference) and good FER performance. Each of the cells in this area is capable of serving users with high probability and good quality. As can be seen from Figure 9.9 these cells fit the rules for data clusters 1 and 5 in Figure 9.10. The geographical locations of the clustered cells are depicted in Figure 9.12.

Figure 9.11 shows how the data samples from each mobile cell have been distributed in the clusters shown in Figure 9.12. Mobile cell 44 is located in a behavioural cluster 1 near cluster 4. There is a high proportion of data samples in data cluster 3, indicating a lot of high values for *dlFER* – i.e., performance problems.

When the downlink information is taken into account in the clustering process, it can be seen that the geographical area covered by cells in behavioural clusters 2, 3 and 7 is very similar to the area covered by clusters 1, 2 and 6 in the uplink analysis case

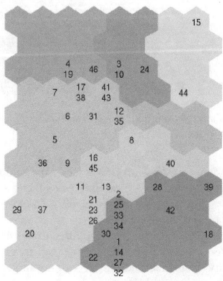

Figure 9.11   Mobile cell clustering, Each cell is mapped to the corresponding self-organising map cluster. Using this position information on the performance of a cell can be concluded and compared with another cell.

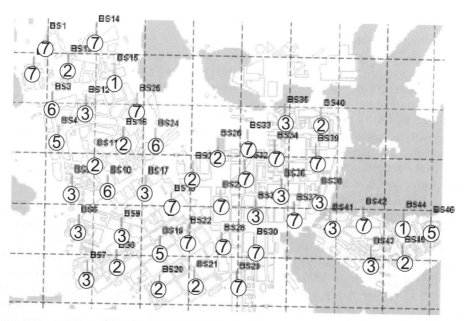

Figure 9.12 Locations of classified cells ([4] and [10]). Numbers refer to the cluster the cell belongs to. Same number indicates similar performance and behaviour in cells.

presented in [4]. This indicates that adding the downlink information to the analysis did not bring significant new findings. This is due to the fact that the service used for generation of the input data was symmetric in uplink and downlink directions. Furthermore, the performance in the micro-cellular network is well balanced between the links. Should the services be asymmetric, the clustering results for the uplink case as well as the combined uplink and downlink case were different.

In order to further analyse the behaviour of some mobile cells in the micro-cellular scenario in both uplink and downlink directions, the behaviour as a function of time – i.e., trajectories of the cells – can be obtained. Figure 9.13 shows the trajectories for cells 8, 14 and 44; both uplink and downlink performance is included in the analysis. Cell 8

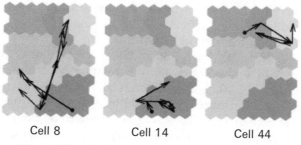

Cell 8   Cell 14   Cell 44

Figure 9.13 Trajectories of the cells ([4] and [10]).

operates initially in behavioural cluster 7 on the histogram map with almost all of the samples in data cluster 5. As can be seen from Figure 9.9 the data cluster 5 represents data samples having a very small number of users. Then, cell 8 visits the area in which data samples are distributed almost equally to data clusters 1 and 5. This is explained by a small increase in the number of users. Cell 8 also visits behavioural cluster 1 briefly in the upper part of the histogram map, indicating a peak in the number of samples with high *ulANR*. Cell 14 operates in behavioural cluster 7 with very low load through the whole analysis session, since almost all of the data samples are located in data cluster 5. The low number of users is one strong characteristic of this cell. Cell 44 operates very close to the problem area – i.e., behavioural cluster 4 and lower part of cluster 1 on the histogram map. In these clusters, a high proportion of samples is distributed in data cluster 3 with the highest *dlFER*.

The strength of the SOM is seen once the user has learned the meaning and content of the behavioural clusters. It is easy to distinguish the good and bad performance clusters on SOM and focus on the cells in the bad performance area. For example, in Figure 9.10 the area of cluster 4 and lower edge of cluster 1 is the area of unacceptable performance. All the cells in this performance area are optimisation targets. In Figure 9.13 cell 44 makes a visit to the bad performance area. Whether this is severe is for the operator to decide. Furthermore, it is possible to define an own set of performance measures and to use them during the training of the SOM. Thus behavioural clustering is more customised and fits better the wanted performance targets than with the case that is presented here.

More about usage of SOM and trend analysis during optimisation can be found in Section 9.2.8.3.

9.2.7.3　Cell Grouping in Optimisation

The scope of this section is to discuss further how to utilise the clustering results based on SOM. As demonstrated in the earlier section SOM is an efficient tool for visualisation, monitoring and clustering of multi-dimensional data. Since the number of parameters that control the RAN is very large, it is easy to understand that finding an optimum set of parameters for each cell manually is a tedious task when the number of cells can be thousands. The additional complication to the optimisation process arises from the fact that the network is optimised based on measurements collected from NEs. The number of these 'raw' measurements is thousands. For an operator to provide the maximum capacity (with required quality) supporting multiple traffic mixes, more advanced analysis methods are required to support configuration parameter settings. In addition, effective means to monitor and classify cells and to identify problem areas in the network are needed.

In this section, use of the SOM in the optimisation process is described (for details see also [18]). Figure 9.14 demonstrates the optimisation process utilising the SOM-generated performance spectrum. This feature makes it much easier for the operator to optimise the cell-specific parameters. With the help of SOM (or some other clustering method, example in Section 9.2.9) the cells can be clustered or grouped based on traffic profile and density, propagation conditions, cell types, etc. Grouping based on multiple criteria instead of just one (like cell type) is more accurate and the operation of the

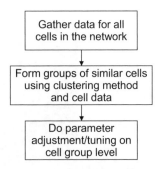

Figure 9.14   Flowchart for the methodology.

network will benefit from this. First the network is started with default parameter settings. After the network has been operational in this sub-optimal mode, measurements from cells are collected. With the help of a clustering method each cell is automatically assigned to a cluster, the number of clusters being well under the number of cells in the network.

Selection of the input data is done on a functional area basis. An example of a functional area is availability. For clustering purposes availability-related measurements are used as the input space for the SOM. Clusters (performance spectrum) that highlight the *availability performance space* are generated. Each cell's behaviour is now compared with the performance spectrum and grouped accordingly. Each cell in a cell group behaves similarly, has similar symptoms and thus should use the same configuration parameter values. This simplifies and eases the optimisation process greatly. The optimisation phase will concentrate on the optimisation/automation of a cell group owning a parameter set, rather than optimising each individual cell with its own selection of configuration parameter settings. This method also reduces the possibility of human error in the parameter settings and parameter provisioning, owing to the fact that part of this process – e.g., the selection of target cells – can be automated. Cell clusters can also be utilised to optimise only a sub-set of the configuration parameters. In the troubleshooting case, problematic cells can be found rapidly using some clustering method and visualisation of the clusters. Additionally, using these visualisation properties, the operator can easily analyse what kind of cell types he has in his network with respect to certain PIs and variables and combine the results with geographical relationships.

In addition to the cell grouping for parameter provisioning purposes the performance spectrum can be used as an indicator for further optimisation or autotuning activities, see also [11]. Figure 9.15 illustrates the case. The cells in the lower left corner are in the problem area of the performance spectrum. These cells are automatically chosen for an optimisation task. This type of approach requires that the performance spectrum is connected to a set of configuration parameters. In other words, the performance spectrum demonstrating cells' admission control performance ought to be linked to parameters controlling the admission process.

The performance spectrum also offers powerful means for optimisation verification or network trend analysis. Trend analysis can be performed using data averaged over

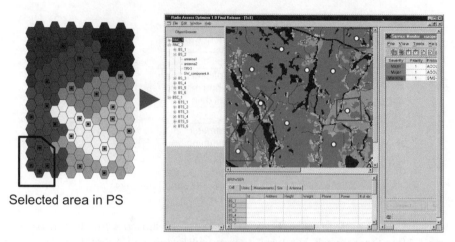

Figure 9.15   Selection of cells to be optimised/autotuned.

PS = Performance Spectrum.

various time periods, ranging from tens of seconds to days. One could, for example, follow one cell movement in SOM during peak traffic hours, assuming that networks are able to report cell performance frequently enough. Another possibility is to analyse network behaviour using data collected during a whole year.

Figure 9.16 shows the trend analysis for 32 cells, all in separate displays. There are three main groups in Figure 9.16 highlighted with different grey shades. For some cells the group membership varies during the monitored period. The advantage of this method is a highly visual representation of changes. Furthermore, the cells' behaviour can be visualised as a function of time – e.g., over 24 hours. Depending on the traffic mix and traffic density in the network, the performance will be different. On the performance spectrum the areas of bad performance are known, and it can be easily seen whether the monitored performance stays away from unwanted areas. Compared with traditional analysis methods, it is easier and faster to understand the characteristics of cell behaviour if this kind of function is used.

Another application for trend analysis is related to the network optimisation phase. When the NE configuration is changed, the operator normally wishes to see the effect of the change on performance. The procedure to improve network performance with SOM basically involves:

1. Collect performance data.
2. Train SOM with the data.
3. Analyse (this step can be done several times over different time periods).
4. Adjust parameters if needed to correct the possible problem.
5. Verify adjustment effect on performance using SOM.

If once more the lower left corner is assumed to indicate malfunctioning cells, the change in the position of the cells on the performance spectrum can be detected after the optimisation, provided that the optimisation has been successful. Figure 9.17 illustrates the example.

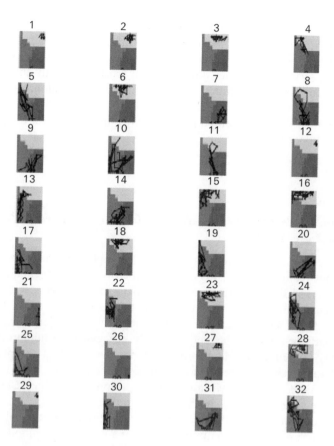

Figure 9.16 Trend analysis for 32 cells, all in separate displays [11].

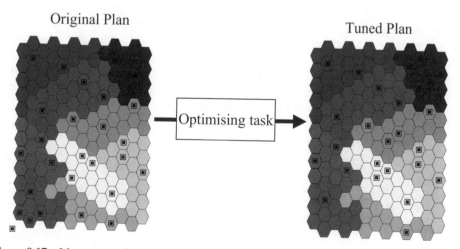

Figure 9.17 Movement of the cells in the performance spectrum as a result of optimisation or autotuning.

SOM-based analysis and optimisation aids support archiving requirements for effective WCDMA capacity utilisation. The optimisation challenge will be much more multi-dimensional than with current networks. In the case of WCDMA there will be multiple services, customer differentiation (customers with different priorities) and multiple radio access technologies to be managed simultaneously, using one resource pool.

It is worth noting that the essential part of SOM utilisation lies in the pre-processing of the measured data used as input.

9.2.8 *High-level Performance Analysis by Clustering Network Performance Data: Case Study*

Operating a mobile network involves a series of complicated tasks [20]. The analysis method presented aims at supporting experts in their routine performance management of GSM networks. The performance management expert must decide on operative actions on element or network parameters so that performance will be satisfactory for end-users. At present, the expert must laboriously collect large amounts of data. Once armed with the data the challenge is that it is excessive for human understanding. An expert has to manually find the relevant part of the data for present decision making, and that task is repeated several times during a day. This is an inefficient use of experts' time, especially without tools to automate routine tasks and to focus their attention on the most relevant part of the data.

Mobile-network-monitoring data can be stored in three-dimensional data matrices. Several indicators are stored for a number of NE at a pre-defined sampling interval. For performance management purposes data are usually summarised in time so that hourly or daily averages are used. Each KPI is summarised from several original measurement variables with a pre-defined formula [21]. The dimensions for summarisation are depicted in Figure 9.18.

This section presents a method for high-level performance analysis and a case study with results. The method provides summarisation – in the element dimension – by presenting the behaviour of problem classes rather than individual elements. The indicators are further summarised by pointing out the ones that most significantly distinguish the element's class from other problem classes.

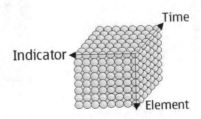

Figure 9.18   Summarisation of information can be done in three dimensions.

9.2.8.1 Data Cube Aggregation Strategies

The challenge of trying to reduce the massive amount of data to a level that is manageable for even human experts is that the reduction method should not lose any important information.

Data reduction strategies and the concept were presented in [5]. The different strategies presented for data reduction and examples in network management are:

- *Data aggregation*: it is common in network management that performance data are aggregated for parent objects in the network hierarchy. Another common aggregation in performance data is summarisation to longer time periods.
- *Dimension reduction*: because the number of available network performance measurements is big (several hundreds) it is important to select only the ones that are relevant to the aims of the expert. In NMSs there are typically several ready-made reports that contain indicators for only one network functionality (e.g., GRPS or capacity).
- *Data compression*: data compression is used especially in the radio interface, where the number of bits used to express a message is minimised.
- *Numerosity reduction*: clustering similarly behaving elements to groups is a typical example to reduce numerosity.
- *Discretisation and hierarchy generation*: an example of hierarchical data reduction is the definition of high-level PIs from a combination of lower level indicators.

9.2.8.2 Decision Support System

This section describes the phases of a method to automate the classification of NEs according to their state and to summarise the relevant information. The method provides the user with easy access to essential information and supports the expert in his task of identifying the root cause of the problem.

The method is targeted, but not limited, to a daily performance analysis. Data from the previous day are fed into the system and the NEs requiring actions are shown in groups that have the same type of problem.

The method consists of three phases (Figure 9.19), which classify NEs by their type, performance and behaviour.

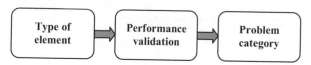

Figure 9.19 Three-phase method that results in problematic elements clustered as categories.

Data Cleaning and Scaling

Pre-processing is an essential part of data mining or the analysis process. Its methods affect the results of the actual analysis ([5], [22] and [23]). Data cleaning and scaling are the pre-processing steps performed on the data before starting the actual analysis. Data cleaning includes removing clearly erroneous values and handling missing values. Erroneous values are those outside the measurement range. Many PIs are on the percentage scale, thus the normal range is from 0 to 100 and any values outside this range are errors. These can originate from errors or malfunctions in data retrieval or storage. Erroneous values are removed from the data and treated as missing values.

The decision support system should be able to run the analysis even if the data contain missing values. Samples including missing values can be dropped or the missing values can be filled in with estimates.

Element quality indicators are scaled continuously and piecewise linearly to interval [0, 1] extremes – 0 meaning the worst and 1 the best performance, respectively. The mapping was constructed on the basis of *a priori* information of network experts. Four values of PIs are defined: *worst possible, very poor, satisfactory, best possible*. These values are scaled to 0, 0.2, 0.9 and 1, respectively, and the scaling function is created using linear interpolation. Scaling function parameters can be adjusted to different PIs, different networks and target performance levels. After the scaling all PIs are within the same limits, and the same value refers to the same level of performance in each indicator. More about pre-processing can also be found in Section 9.2.2.5.

Typification of Network Elements

The first phase classifies elements by their physical and behavioural type within the network. Examples of type categories are indoor/outdoor cells or cells along a busy highway, elements from different network domains (radio/core network), and different network types (GSM/WCDMA).

This phase enables individual treatment of all element types. For example, the PIs used are different and possibly gathered from different sources. Performance target and scaling function parameters can also be different. The network operator may set different performance criteria for a cell serving a hotspot location in a city centre than for a cell located in a rural area.

The method divides the cells by their type into three groups: indoor, micro and macro. Macro-cells are further divided into four categories by dividing both total traffic and number of handovers into two groups. The means of traffic and handovers in available history is calculated for each cell. 2/3 quantile of these means is used to split the cells into groups of 'high traffic', 'low traffic', 'high numbers of handovers and low numbers of handovers'.

The method allows typification based on different parameters according to the operator's approach to performance management.

Performance Validation

The second phase filters out the elements either performing well enough, according to the operator's requirements, or having no major problems. This group of elements – usually constituting the majority of all elements – require no further attention and thus the amount of data relevant in the third phase is drastically reduced.

The decision whether an element is performing well or does require further attention can be based on different approaches. For example, the operator may want to select a given number or percentile of the worst performing elements to be analysed in detail. The selection could also be based on comparison with certain fixed performance targets or comparison with elements' performance histories.

Performance validation is performed with scaled data; the Euclidean distance from the best theoretical performance in the N-dimensional space can be used when ordering elements by performance.

Problem Clustering

In the third phase, the problematic elements are clustered based on their behaviour expressed with the scaled indicators. Clustering is done by the agglomerative hierarchical method with Ward linkage [24]. The optimal number of problem clusters is tested with the Davies–Boulding index [26] and the mean silhouette method [27]. Thus, the number of problem clusters is dynamic, based on the input data and the problem data in it.

Clustering reveals the typical problem categories. Problem categorisation helps the expert to continue the analysis and to identify the root causes of each problem class. The elements in one problem category have similar behaviour according to the input data. If the indicators in input data are selected properly – i.e., they are such that they have high relevance – the further actions performed by the expert can be the same for all elements in the problem cluster.

The actions done by the expert are, for example, querying more detailed information about the problem, generating a report and forwarding the problem or changing the configuration parameters of the elements.

9.2.8.3 Case Study Results

Data Used in Case Study

The data used in the case study were from a network performance database from a commercial European GSM operator. The database contained radio network performance counters, the most important of which were used. The performance data were collected from 2385 GSM radio cells. The measurement period was 6 weeks. The data were aggregated so that for each counter one sample per day was available.

The data contained numerous invalid values and missing samples. Also possible changes in network configuration invalidates the data and causes anomalies in measurements. This is a typical situation in network management, but the analysis system should recover from errors and anomalies in the dataset. In the dataset used, there were a total of 14 cells that contained invalid data or data were partially missing.

Performance Indicators

To create the high-level cluster analysis of radio network performance and of possible performance degradations, the PI set – i.e., feature vector – should cover the key functionality of GSM networks. The analysis method presented in this section was designed to support any feature vector, but only one feature vector was used in the case study.

The data used for typification of the elements were:

- *Cell type*: this parameter defines whether the cell is a macro-, micro- or indoor cell.
- *Handover amount per day*: this indicator describes how many cell handovers were made into or from the cell during a day.
- *Traffic amount per day*: the amount of calls in Erlang capacity units.

The feature vector describing each element for performance validation and problem clustering contained the following PIs:

- *Dropped Call Ratio (DCR)*: describes what percentage of calls was dropped during a day.
- *Handover Success (HO_succ)*: describes what percentage of handovers into or from the cells was successful during a day.
- *Congestion*: describes how many seconds of the day the element was in a state such that no new calls could be accepted due to lack of resources.
- *RXQUAL in classes 1 to 4 (RX_DLqual)*: this indicator describes what percentage of measured radio quality samples was in good-quality classes.
- *Average Downlink Signal Strength (DL_lev)*: this is the average signal strength received by the mobiles served by the cell during a day. The unit is dBm.
- *Call Setup Success Ratio (CSSR)*: describes what percentage of the call setup process was successful during a day.

Typification of Case Study Data
The numbers of macro-, micro- and indoor cells were 2168, 191 and 26, respectively. Because the numbers of micro- and indoor cells were rather low, a more detailed typification would not bring any benefit.

Macro-cells were further divided into four types based on traffic and handover amount. The amount of low traffic and low number of handover cells were 1298, low traffic and high handover amount 147, high traffic and low handover amount 146, high traffic and high handover amount 577. The result of typification can be seen in Figure 9.20.

Performance Validation of the Elements
In the case study, performance verification was executed so that 15% of the worst performing cells were considered as problematic and thus taken to the forwarding step.

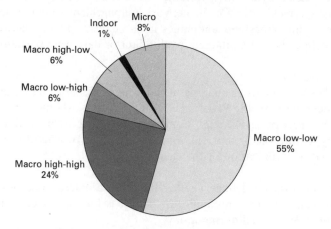

Figure 9.20   The distribution of elements after typification.

The number of micro-cells defined as problematic was 28, which is 14.7% of the total. The number of indoor cells selected was 3 (11.5%). The number of macro-cells that were defined as problematic was 327 (15.1%).

Problem Clustering the Case Study Data

Table 9.5 shows the statistics after the third and final phase of the method. The number of problem clusters was 5.

Table 9.5 The number of cells in problem clusters (1–5) after the analysis process. Macro (low, low) refers to macro-cells with low traffic and low number of handovers.

	Sum	Micro	Indoor	Macro (low, low)	Macro (low, high)	Macro (high, low)	Macro (high, high)
Sum		28	3	189	18	3	117
1	37	9	1	11	1	0	15
2	9	1	0	3	2	0	3
3	21	3	0	12	1	0	5
4	172	6	0	152	9	1	4
5	119	9	2	11	5	2	90

Figure 9.21 shows the centroids of the five problem clusters. The graph shows that the characteristics of each problem cluster are different.

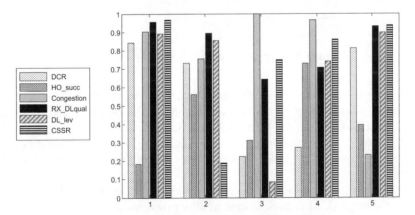

Figure 9.21 The centroids of the problem clusters.

Problem Clustering Results Analysed by an Expert

The box plots [28] of the problem cluster were shown to a radio network expert. This section describes the expert's analysis of the results.

Problem cluster 1 in Figure 9.22 most probably contains cells with a poor adjacency plan (an adjacency plan defines the neighbouring cells where handover is possible). As a corrective action the definition of a new adjacency plan should fix the majority of problems.

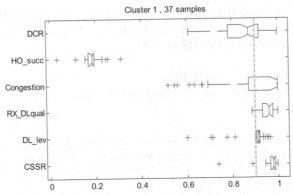

Figure 9.22   Problem cluster 1 from the case study.

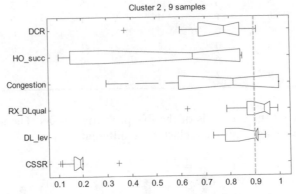

Figure 9.23   Problem cluster 2 from the case study.

Because the number of cells in problem cluster 2 in Figure 9.23 is low and the call setup success ratio is poor a possible root cause for the problems may be hardware failure in the element or in the transmission links. Another possible cause is a lack of signalling capacity in the elements. Corrective action for these cells could be checking hardware and transmission links. In some cases resetting the element may solve the problems.

Most probably, the problems in cluster 3 in Figure 9.24 are caused by poor coverage (low signal strengths). This causes some quality problems and dropped calls. To solve these problems adding new cells or changing the antenna bearings or tilts might be a solution.

In cluster 4 in Figure 9.25 there are slightly low signal levels, but the main root cause for the poor performance is radio quality problems. Typically, changing the frequencies of the problematic cells solves these problems. In order to find the problematic frequencies, the situation should be studied on a geographical map.

Problem cluster 5 in Figure 9.26 is very similar to cluster 2. The difference is that here the call setup success ratio is acceptable. The assumption is that the problems in this cluster are caused by congestion, which leads to failures in handovers. Corrective action would be adding more radio capacity to elements or possibly changing the capacity configuration parameters.

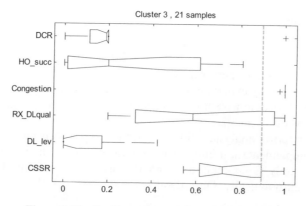

Figure 9.24 Problem cluster 3 from the case study.

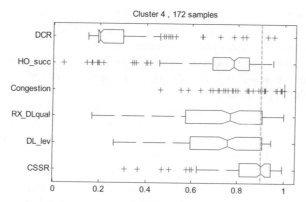

Figure 9.25 Problem cluster 4 from the case study.

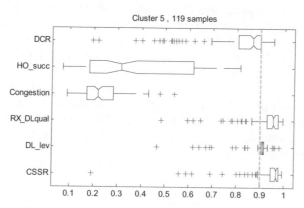

Figure 9.26 Problem cluster 5 from the case study.

9.2.8.4   Conclusions

The three-phase method developed in this research can be used to summarise information about a cellular mobile network. It finds well-performing elements and it clusters elements with unsatisfactory performance.

The decision making by experts is supported by showing the main characteristics of problem clusters. A decision on how to improve network performance can be made collectively for several elements.

This method supports analysing different element types with any set of indicators. In the current implementation the corner points of the scaling function for each indicator are the only input required from the user. Those supplied the procedure can be run fully automatically. Once defined for the network the same corner points can be reused.

9.3   Automatic Optimisation

The complex operating environment and service level capacity–quality tradeoff management set certain requirements for a system providing autotuning features. Chapter 7 introduced a hierarchical solution with a statistical feedback loop to optimise the performance of the two fast Radio Resource Management (RRM) loops. The main scope of this section is to present a concept and tools to aid optimisation at the uppermost layer.

Currently there are hundreds of configuration parameters (independent of the multiple access technique) which control the quality and capacity of the RAN. Clearly, finding an optimum set of parameters for each cell manually would be a tedious task when there could be thousands of cells.

An additional complication to the optimisation process arises from the fact that the network is optimised based on numerous network measurements. Currently, the number of raw network measurements can be close to 1000. For an operator to provide the maximum capacity with required quality supporting multiple traffic mixes, more advanced methods are required to support the setting of optimal configuration parameters.

The elements and requirements for optimisation were discussed in Chapter 7, including the availability of configuration data, performance data and online access to management databases in order to ensure the correctness of the input data for the optimisation task. In addition to the monitoring and analysis of KPIs and performance visualisation support there is a possibility to automate optimisation cases. In this section some of these cases are introduced and discussed. In Figure 9.27, the operator's workflow is presented and the optimisation task highlighted.

The need for operational efficiency in the workflow and the fact that the cellular multi-radio environment increases the complexity of management tasks are the main drivers for the automation of optimisation. The role of optimisation is to provide automated or manual means to improve the performance of the network. Furthermore the task of optimisation is to understand and translate the relationship between measured network performance and set QoS targets. The definition of performance

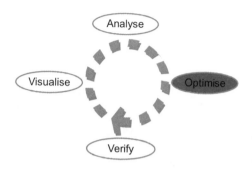

Figure 9.27 Focus of this section: Optimise. In some cases the verification phase is embedded in the optimisation algorithm.

in the case of 3G is changing: it must be capacity–quality tradeoff management, rather than traditional performance improvement.

With the statistical loop and cost function approach it is possible to offer automatic coverage–capacity tradeoff-based network management. With this concept the operator has to set quality and capacity targets and related costs that regulate the quality–capacity tradeoff. A new aspect in this area is the fact that the target of autotuning is not the best quality as traditionally defined. In some cases it might be that slightly degraded quality and the possibility to offer more traffic is more beneficial for the operator's business case than quality-driven optimisation.

For an operator it is essential to utilise all available resources to improve the capacity and QoS of the radio network and for this an overall control function is required. This control function provides centralised quality monitoring, which monitors optimisation and automation sub-systems. In addition, a mechanism for cost function minimisation for optimal capacity, performance and operator revenue is required. Once the cost function is minimised, the task for the NMS is to provision optimal configuration parameters at the network level. To guarantee the optimum performance of a cellular network the operator ought to have flexible means to set the QoS target based on the system KPIs and/or a cost function derived from them. In multi-radio environments (GSM-WCDMA, WLAN) it is important to have the possibility to pool the resources of the networks for optimised *capacity* and *quality*. This also requires an overall control functionality at the highest hierarchy level.

Currently manufacturers propose default values for most parameters. These are not optimal for all conditions. The operator's task is to optimise the network cell cluster by cell cluster. This proposed concept will make the initial parameter settings less crucial: for example, at the beginning of network operation admission control and handover control could work with looser limits admitting high numbers of users to the network. Based on the current QoS situation as measured using KPIs and using set QoS targets the relevant parameters can then be autotuned. After the parameter change the new situation is compared with KPI history data and the tuned parameters are accepted if the change in QoS performance is improved.

The mathematical formulation of the task can be seen as finding such a combination of air interface configuration parameters based on which the KPIs are as close to the

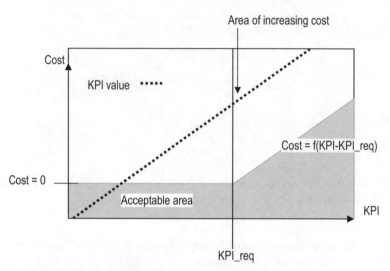

Figure 9.28 Example of a key performance indicator cost function.

desired area as possible. First, the operator sets capacity requirements for certain capacity KPIs denoted *KPI_C*. The requirements have 'req' in the sub-index. Correspondingly, the operator sets quality requirements for certain *KPI_Q*s. Quality and capacity costs can then be calculated as in Equation (8.1):

$$\left.\begin{array}{l} QualityCost = \displaystyle\sum_{cells \in CLUSTER} \sum_i \alpha_i * f(KPI\_Q_i - KPI\_Q_{i,req}) \\[3mm] CapacityCost = \displaystyle\sum_{cells \in CLUSTER} \sum_i \beta_i * g(KPI\_C_i - KPI\_C_{i,req}) \end{array}\right\} \quad (9.6)$$

Figure 9.28 shows an example of a KPI cost functions f and g. In this example the cost of KPI values higher than *KPI_req* is increasing linearly. The cost functions can also take other shapes.

The total cost function to be minimised is presented in Equation (9.7). Capacity and quality tradeoff can be made using the parameter W. Minimisation is done by adjusting the configuration parameters, see Equation (9.8). KPI values also depend on service distribution – e.g., different costs and parameter settings will be achieved depending on service distribution:

$$\left.\begin{array}{l} KPI\_C_i = f(Configuration\ parameters, Service\ distribution) \\[2mm] KPI\_Q_i = g(Configuration\ parameters, Service\ distribution) \end{array}\right\} \quad (9.7)$$

$$TotalCost = W * QualityCost + (1 - W) * CapacityCost \quad (9.8)$$

Factors affecting automated optimisation outcome are traffic profile (service mix), traffic density, pricing of each service, etc. Ultimate goals when minimising the total cost include:

- optimise revenue;
- minimise Capital and Operational Expenditures (CAPEX and OPEX);
- maintaining the good reputation of the operator.

Clearly, many parameters will not be autotuned, and autotuning cannot always fix or optimise the network – e.g., troubleshooting tools should be used to spot and solve hardware and other problems.

9.3.1 Examples of Automated Optimisation

Currently, in both GSM and WCDMA areas, many different autotuning algorithms are under development (e.g., [34]–[40]). Current autotuning algorithms have been developed primarily to control single parameters or sub-sets of parameter values only. PIs or KPIs are used for manual optimisation work and similar methods can be automated. The mathematical formulation of the task can be seen as finding a combination of air interface configuration parameters based on which the KPIs are as close to the desired area as possible.

In order to structure the optimisation tasks the RAN is split into sub-systems that are as independent and orthogonal as possible. In this section a proposal for the sub-systems and optimisation cases for these sub-systems is presented. Examples are provided for mobility management and power resource sub-systems. Capacity allocation-related issues are also addressed. Sub-systems could consist of items like:

- mobility management sub-system;
- automatic definition of neighbouring cells;
- automatic adjacency synchronisation between vendors;
- automated optimisation of soft handover parameters using cost function;
- Location Area (LA) considerations;
- call admission control sub-system;
- power-based admission control optimisation using cost function;
- uplink and downlink issues;
- capacity optimisation and traffic balancing;
- pilot power autotuning using cost function;
- capacity split between circuit switched and packet switched traffic;
- capacity reservation for High-speed Downlink Packet Access (HSDPA);
- Congestion relief using handovers.

Several algorithms utilise the cost function approach and the same network scenario. The scenario is introduced in Section 9.3.2. In Section 9.3.3 the relevant KPIs and costs used in the algorithms are introduced. For the other optimisation cases the decision logic is presented in own sections, together with the problem definition.

9.3.2 Network Scenario and Data Used in Algorithms with Cost Function

The automated control methods were verified with an advanced WCDMA radio network simulator [17]. A set of mobile terminals moved in the area with constant speed and, with random intervals, made calls that utilised different services: voice, circuit switched data and packet switched data. The main differences between voice and circuit switched data calls were in the bit rates and that the former had talkspurt silence periods. The data rates of voice and circuit switched data calls were fixed but packet data rates could vary. The simulation step was one frame or 10 ms, at which the

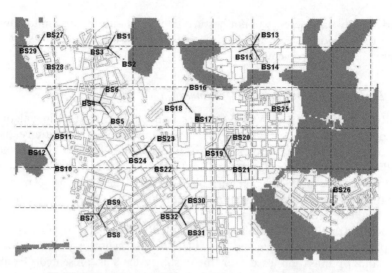

Figure 9.29   Macro-cellular scenario used in the studies.

transmission powers, received interferences and signal-to-interference ratios were re-calculated for each connection in the uplink and downlink. The method of [29] was used to obtain the correctness of received frames from signal-to-interference ratios.

Three main simulation scenarios were used in this study. They were a micro-cell scenario with omni-directional sites (see Figure 9.8) and a macro-cell scenario with 32 cells (Figure 9.29). Also a mixed scenario was included in the study consisting of both micro- and macro-cells.

The main parameters used in the simulations, the simulator details and information of the services used in the study are listed in [30]. During the simulations the traffic was uniformly distributed along the streets of the simulated area and they made new calls according to a Poisson inter-arrival distribution. The packet size of packet calls was generated according to a Pareto distribution. The call parameters were selected to produce a high system load and load the system up to the target level of downlink transmission power and uplink total received power. The simulation time was 300 s. In the simulations, the planned pilot power was 1 W in macro-cells and 200 mW in micro-cells, which was 5% of the maximum BS power.

9.3.3   Measurements and Costs Used in Algorithms

Several algorithm proposals utilise the same cost function structure. In this section the measurements and costs used by those autotuning algorithms are introduced.

9.3.3.1   Uplink and Downlink Poor Quality

The general call quality in a cell was described as the ratio of active connections suffering from an increased block error rate among all active connections in the cell. The strongest cell in the active set determined the cell to which the measurement was associated. The connections have specific Block Error Rate (BLER) targets that depend on the service used. If the ratio of connections significantly exceeding their BLER targets was significantly higher than a pre-defined allowed level, the cell showed increased poor call quality. Only RT services with planned coverage in the entire cell were monitored for call quality. The significance of increased poor call quality in a cell was analysed with the formula:

$$S_{PQ} = \frac{R_{PQ} - T_{PQ}}{\sqrt{T_{PQ}(1 - T_{PQ})/N_{PQ}}} \tag{9.9}$$

which describes the sample size (N_{PQ}) corrected deviation of ratio R_{PQ} from the allowed ratio T_{PQ} of connections with increased BLER. The formula basically divides the ratio difference estimate by its expected standard deviation giving, thus, the number of standard deviations that the measured ratio deviates from the allowed ratio. The criteria used for determining whether a cell suffered from significant poor quality was that S_{PQ} should be higher than 2 and N_{PQ} higher than $5/R_{PQ}$. The former condition was based on the 98th percentile of the normal distribution, giving a 2% probability of incorrectly showing poor quality, and the latter condition indicated that there was a sufficient sample size for estimating the ratio R_{PQ} accurately. The computation of poor quality was similar in the uplink and downlink. However, no terminal reporting of the correctness of received blocks in the downlink was assumed. Instead, the transmitted frames during link power outage were considered erroneous. The allowed level of the poor-quality ratio was 2% in the uplink and downlink.

9.3.3.2 Uplink and Downlink Congestion

Congestion was described using ratios between blocked RT calls and queued packet calls. The blocking ratio was the ratio of blocked RT calls to the total number of RT call admission requests in the cell. The blocking ratio was measured separately for the uplink and downlink. The uplink blocking ratio was based on calls blocked due to admission control. The downlink blocking ratio comprised calls blocked due to an unsatisfied admission control criterion, but for certain optimisation methods lack of hardware channels or scrambling codes were also included into the statistics. The packet-queuing ratio was the ratio of queued packet calls to the total number of packet users in the cell measured at specific intervals. Packet-queuing ratios were measured separately in the uplink and downlink.

In small amounts call blocking and packet queuing were regarded as normal. Congestion occurred if blocking or queuing ratios significantly exceeded certain allowed levels. The significance of increased blocking or queuing ratios was measured with formulas similar to Equation (9.9). If Equation (9.9) exceeded 2 and the sample size was sufficient then the cell showed increased congestion with the particular blocking or queuing ratio. The allowed level of blocking ratio was 2% and that of queuing ratio 10%. Blocking statistics were denoted $S_{BL,UL}$ and $S_{BL,DL}$ for the uplink and downlink, respectively. The queuing statistics were denoted $S_{QE,UL}$ and $S_{QE,DL}$.

9.3.3.3 Common Pilot Coverage

In the UMTS, the terminal measures and reports the received level of the primary common pilot energy-per-chip-to-total-wideband-interference-density ratio, E_c/I_0, for the selection of the cell to which a call is set up or to which cell handovers are performed. The primary common pilot power determines the cell coverage area and thus the average number of terminals connected to the cell. If the pilots of all cells are too weak for a terminal to decode any of their signals, call setup is not possible. The minimum pilot strength is specific to the receiver electronics. Therefore, the specifications of Third Generation Partnership Project (3GPP) require that the terminal must be able to decode the pilot from a signal with an E_c/I_0 of $-20\,$dB. Quality receivers can cope with ratios several decibels lower than that. Too good coverage is not desirable either as it indicates unnecessarily high common pilot powers, consuming the limited power capacity of cells. Consumption is even compounded if the powers of other common channels are scaled with respect to the common pilot power. Pilot coverage was measured as the ratio of reported E_c/I_0 values exceeding $-18\,$dB. Only the strongest E_c/I_0 was included in the coverage ratio from a single report of a terminal. A 98% coverage ratio was considered as an optimal target. The test statistic of the difference between the cell coverage ratio and target coverage ratio was derived using Equation (9.9). The common pilot coverage was regarded as significantly smaller or larger than the target if the test statistic was lower than -2 or higher than 2, respectively, and the sample size was sufficient.

9.3.3.4 Uplink Load, Downlink Load and Load Balance Statistics

Cell-specific uplink and downlink loads were measured using the geometric averages of cell received power (*PrxTotal*) and transmitted power (*PtxTotal*), respectively. A load balance statistic was calculated for each cell, in order to test whether a load in a certain cell differed from the load in neighbouring cells. The statistic was based on the *PtxTotal* measurement divided by the target transmission power *PtxTarget*. The target power depends on the maximum transmission power of the cell and differs among macro-cells and micro-cells. Cell load was thus commensurate among different cell layers. As capacity is in general more limited in the downlink than in the uplink, it was decided that only the measuring of downlink load balance suffices. The load was sampled separately for each cell. Three counters were kept for each cell. The first collected the number of samples, the second collected the sum of the sample values and the third collected the sum of the squared sample values. Let us denote the counter values of cell i by N_i, S_i and T_i, respectively. The sample mean and variance of the load in cell k was obtained using:

$$m_1 = \frac{S_k}{N_k} \tag{9.10}$$

and

$$v_1 = \frac{T_k}{N_k} - m_1^2 \tag{9.11}$$

The statistics for the load in the neighbouring cells of cell i were obtained using:

$$m_2 = \frac{\sum_i S_i}{\sum_i N_i}, \qquad i \neq k \tag{9.12}$$

and

$$v_2 = \frac{\sum_i T_i}{\sum_i N_i} - m_2^2, \qquad i \neq k \tag{9.13}$$

Neighbour cells are normally defined during network planning and optimisation. The test statistic of the difference between the own-cell and neighbour cell loads was attained using:

$$t' = \frac{m_1 - m_2}{\sqrt{\dfrac{v_1}{N_k} + \dfrac{v_2}{\sum\limits_{i \neq k} N_i}}} \tag{9.14}$$

9.3.3.5 Costs Used in the Algorithms

Costs were defined to enable tradeoffs between capacity and quality. The cost of congestion was based on the blocking and queuing statistics:

$$CostCongestion(x) = f(S_{BL,x}) + 0.25 * f(S_{QE,x}) \tag{9.15}$$

in which $x = UL, DL$ depending on the link direction. The threshold function $f(S)$ was defined as:

$$f(S) = \max(S - 2, 0) \tag{9.16}$$

in which the maximum operator and subtraction with two zeroed insignificant values of the statistic. Equation (9.14) shows that the cost of packet queuing was only one-quarter of the cost of blocking an RT user. The cost of poor call quality was defined to be five times higher than the cost of blocking:

$$CostQuality(x) = 5 * f(S_{PQ,x}) \tag{9.17}$$

If *CostCongestion* is greater than *CostQuality* then congestion is more costly than poor quality in a cell. If the reverse is true then poor quality is more costly than congestion. These costs can be operator-definable, depending on the strategy and weighting that the operator wants to emphasise.

9.3.4 Autotuning in the Mobility Management Sub-system

This section describes the automatic optimisation issues related to mobility management. Autotuning of neighbouring cell lists, considerations related to the LAs and routing areas and the synchronisation of adjacency definitions between different vendors are discussed.

9.3.4.1   Automatic Definition of Neighbouring Cells

Correct adjacency definitions are the basic requirement for mobility. Optimisation of neighbour cell lists saves BS and MS transmission powers, since MSs are connected to optimal cells. Also, the number of dropped calls is reduced (see [31] for simulation results). The adjacencies need to be defined correctly within the WCDMA system and between WCDMA and GSM networks. The maximum number of neighbouring cells that can be configured in a WCDMA Radio Network Controller (RNC) are:
- 32 intra-frequency neighbours;
- 32 inter-frequency neighbours (max. 32 on the same frequency);
- 32 inter-system (GSM) neighbours.

In addition to these the total number of adjacency definitions is limited and is vendor-specific. Handover adjacency definitions are initially created during the planning phase and they are based on geographical locations and estimated relationships between the cells. In practice, prediction-based neighbour cell lists are not necessarily optimal. This means that the terminals of the considered cell might receive too low Primary Common Pilot Channel (P-CPICH) E_c/I_0 values from the planned neighbour cell, which therefore would not need to belong to the neighbour cell list. On the other hand, there might be a cell whose P-CPICH E_c/I_0 is good enough for it to be considered as a potential candidate for handover and thus to the neighbour cell list even though this cell is not a planned neighbour for the given cell. Additions of new sites and new buildings also motivate the autotuning of neighbour cell lists. Owing to these facts it is practical that adjacency lists can be updated automatically and that they are always optimised based on measurements rather than predictions with limited accuracy. In addition to the accuracy requirements operational efficiency plays an important role. With this feature the overall planning effort can be reduced when building a WCDMA network on top of GSM, for example. Incorrectness of planned data can be fixed quickly with an automated optimisation process and network performance can be improved to meet the target level.

RNCs and Base Station Controllers (BSCs) should be able to collect the reported statistics of handovers per adjacency definition. Measurements should be collected from other frequencies and even between radio access technologies. At the end of the measurement period RNCs and BSCs report the statistics to the management layer for further analysis and optimisation. It is possible to have automated functionality to evaluate the situation and to generate optimised adjacency lists for intra-frequency, inter-frequency and inter-system (GSM) adjacencies in the NMS. After verification the neighbour cell lists can be automatically downloaded to the network.

The deletion of cells from the neighbour list is based on the usage of the respective adjacency definitions. Equation (9.18) is used in the evaluation of the adjacencies. Definitions that are not utilised are removed from the neighbour cell lists. Such cleaning of the lists is important since the lengths of the neighbour cell lists are limited.

Finding missing adjacencies is more challenging; the process is illustrated in Figure 9.30. The idea behind the proposed method is to use predictions for the initial neighbour and candidate cell lists of each cell in the network. The candidate cells are formed with looser criteria than the actual neighbour cell list and include cells within a

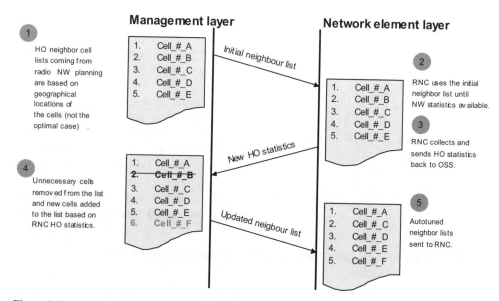

Figure 9.30   Example of the operation of the automatic definition of neighbouring cells feature.

larger area around the cell in question. The candidate cell list consists of good potential cells to be added to the actual neighbour cell list. The actual neighbour cell list has a space reservation for possible new candidate cells, and the cells are added to the actual neighbour list in iterations. An iterative process is required because of the limited lengths of the neighbour cell lists.

After each iteration all the cells in the current actual neighbour cell list are evaluated based on handover statistics. The updated neighbour cell list consists of those cells to which handovers most frequently happen. For this decision a definition of 'good neighbour' is needed: see Equation (9.18). Based on the 'good neighbour' score, the cells are sorted and those cells that are better than the set threshold are accepted in the new neighbour cell list:

$$\frac{\text{HOs made to this cell}}{\text{HOs made to all cells (in this layer)}} \geq \text{User-given parameter} \qquad (9.18)$$

Rotation to add new candidates to the actual neighbour cell list is performed until statistics from all neighbour and candidate list cells are gathered and the adjacency importance evaluated. In addition to handover statistics, knowledge about the received E_c/I_0-level of the cell's pilot can be used in the evaluation.

Normally a terminal is searching only for those scrambling codes associated with its neighbour cells, but the RNC can also order a terminal to do a full intra-frequency code search – i.e., to search also for those scrambling codes that do not belong to the cells in the neighbour cell list in the same frequency. This option would offer an alternative optimisation algorithm for finding intra-frequency adjacencies, and rotation of the neighbour cell lists could be omitted.

Autotuning can be performed under the supervision of the operator – i.e., the operator can accept, reject or modify the neighbour cell relations suggested by the method. Alternatively, autotuning can be completely automatic. A good location for the autotuning method is the management layer, since then RNC–RNC border areas can also be handled.

There are attempts to perform automated optimisation of neighbouring cells using field measurement data. The challenge with this method is the limited amount of performance and traffic data. Additionally it does not contain information on the relevance of the neighbour (no traffic-weighted information on the amount and success of handovers). Finally, in order to have accurate information for reliable statistics on P-CPICH E_c/I_0, extensive measurement campaigns need to be organised. This method can be used in troubleshooting and finding important missing adjacencies, but it is not an efficient method for overall adjacency optimisation.

9.3.4.2 Automatic Adjacency Synchronisation between Networks from Different Vendors

Today the situation with many operators is such that their GSM and WCDMA networks are from multiple vendors. In such a case inter-system handover-related adjacencies and relevant control parameters are defined between different vendor networks. In most cases both networks are managed independently by their respective NMSs (or OMCs), but inter-system handover would require parameter-level information about the target cells in the other system. Parameters can be listed in the rollout phase consistently, but network evolution, site additions, subsequent frequency changes in the Base Station System (BSS), etc. will require adaptation in the other vendor's network and in the adjacency definitions as well. Each change on the GSM side related to the attributes in adjacency definition have to be reflected on the WCDMA side. Additional challenges arise from the fact that, despite the efforts in 3GPP, actual implementation of the handover control mechanisms in NEs is vendor-specific.

Inconsistent inter-system handover data will have a strong impact on handover success and service continuation, therefore these data are subject to synchronisation in both networks on a daily basis. In order to perform synchronisation a master database is needed. This database consists of the correct definitions for the adjacencies. Typically WCDMA–GSM handovers are seen more critical, owing to the fact that WCDMA coverage is limited, thus handovers to GSM are mandatory requirements for mobility. In this case the master definition for the adjacencies comes from the WCDMA side. Alternatively, instead of choosing one of the NMSs to be master a reference database can be used. This database is for global management and is updated to reflect recent changes in both networks in order to synchronise.

For synchronisation it is important to agree on how to handle discrepancies in adjacency definitions. In other words: If GSM cell 1 is defined as an adjacency to WCDMA cell 2, but not vice versa, should the definition be added to the GSM side also, deleted from the WCDMA side or should the discrepancy be accepted and have a unidirectional adjacency? Currently the most common option is to use WCDMA definitions as master: if GSM cell 1 is defined as adjacency for WCDMA cell 2 this definition is also added to the GSM side if missing. If WCDMA cell 3 is defined as an adjacency for GSM cell 4, but this cell is not used as an adjacency definition on the WCDMA side, the definition is deleted. The rationale behind this is that the coverage of the WCDMA network is expanding and thus coverage reason handover boundaries are changing. It is easiest for the operator to make the adjacency definition and optimisation on the WCDMA side and copy the definitions during synchronisation to the GSM side as well.

Changes on the GSM side related to adjacency identifications should automatically be reflected in WCDMA–GSM adjacency definitions. WCDMA cells are updated on the basis of knowledge from the GSM target cell related to, for example, cell global identity, BCCH carrier number or frequency, cell type (GSM900, GSM1800, etc.), the threshold that triggers the handover (and possibly hysteresis), minimum level that the handover can be performed (minimum access level), maximum allowed transmit power for the terminal at access and so on.

For an update on the GSM side, target cell items like cell global identity, frequency (channel number), the threshold for the 'quality' measure, E_c/I_0, for handovers to WCDMA (and possibly hysteresis) and naturally the scrambling code are defined.

The exact content of target cell information is vendor-specific and only an incomplete example was provided here. Further, a similar method can be applied between WCDMA-only or GSM-only cases.

9.3.4.3   Automated Optimisation of Soft Handover Parameters Using a Cost Function

In this section a method controlling two of the parameters related to Active Set Updates (ASUs) is presented – namely, *AdditionWindow* and *DropWindow*. As discussed in Section 4.3 *AdditionWindow* determines cell addition to the active set of a terminal. If the active set is not full and the received pilot signal is higher than that of the strongest cell in the active set minus *AdditionWindow*, addition is performed. *DropWindow* determines when cell dropping is done from the active set. An active set cell is dropped if its received pilot signal is lower than that of the strongest cell minus *DropWindow*. The parameters have an effect on the average size of a terminal's active set and on the average level of soft handover overhead. If *AdditionWindow* is set to too high a value, then the active set sizes of the terminals are too large on average, which can cause increased downlink-based congestion due to:

- insufficient physical (channel elements) and logical (codes) resources; and
- increased BS total transmission powers due to many links.

If *AdditionWindow* is set to too low a value, the active set sizes of the terminals are too small on average, which can cause increased uplink interference, poor quality or congestion.[1]

Soft handover parameter control was a compromise between downlink congestion on one hand and uplink blocking and bad quality on the other hand. *AdditionWindow* and

---

[1] For more on the influence of *AdditionWindow* and *DropWindow* see Section 4.7.2.1.

DropWindow were increased by 0.5 dB if:

$$CostQuality(UL) + CostCongestion(UL) > CostCongestion(DL) \qquad (9.19)$$

If the cost balance was reversed, the parameters were decreased by 0.5 dB. The *ReplacementWindow* and handover timers were not tuned in this study, since controlling these parameters had a low impact on the studied WCDMA.

According to the study presented in [30] and [41], optimisation of *AdditionWindow* and *DropWindow* decreased the number of blocked speech service calls by 10% and circuit switched data calls by 11% while the quality of the calls did not deteriorate at all. Optimisation was also repeated in a hardware limited scenario, in which the number of simultaneous channels was limited to 32 per cell. This was to verify the supposition that *AdditionWindow* control reduces blocking due to insufficient channels, which was the anticipated limiting resource in the first WCDMA networks. The results were similar to those in the power limited scenario. The number of blocked speech calls decreased by 11% and the number of blocked circuit switched data calls decreased by 13%. The total increase in utilised capacity due to optimisation was 15%. Quality was not a problem, as the power resources were sufficient owing to the limited number of users in the cells.

9.3.4.4 Location Area Considerations

Experience in GSM networks has shown that the handover success between BSC borders (inter-BSC handover) is lower than that for intra-BSC HO. The same applies to inter-RNC handovers. Further, the additional paging-related signalling load that results from scattered location or RA definitions is another motivation for optimisation.

Part of the mobility management functionality is the mechanism to identify and address users and their terminals. In addition to this, mobility management requires structuring of the network in order to locate the terminals. These concepts are introduced next.

Area Concepts

For mobility functionality, four different area concepts are used. Location Areas (LAs) and Routing Areas (RAs) are used in the Core Network (CN). UMTS Terrestrial Radio Access Network (UTRAN) Registration Areas (URAs) and Cell Areas (UCAs) are used in the RAN. LAs are related to circuit switched services. RAs are related to packet switched services.

One LA is handled by one CN node. For a terminal that is registered in an LA, this implies that the terminal is registered in the specific CN node handling this specific LA. An LA is used, for example, at CN-initiated paging related to circuit switched services. A circuit switched service-related temporary identity, circuit switched Temporary Mobile Subscriber Identity (TMSI), may be allocated to the terminal. This temporary identity is then unique within an LA. An LA is used by the 3G Mobile Switching Centre/Visitor Location Register (MSC/VLR) for paging the terminal. An RNC may include many LAs or an LA may span over many RNC areas; in this last case, the RNCs have to be connected to the same MSC/VLR.

One RA is handled by one CN node. For a terminal that is registered in an RA, this implies that the terminal is registered in the specific CN node handling this specific RA. An RA is used by the 3G Serving GPRS Support Node (SGSN) for paging the terminal. An RA is used, for example, at CN-initiated paging related to packet switched services. A packet switched service-related temporary identity, PS-TMSI, may be allocated to the terminal. This temporary identity is then unique within the RA. An RNC may include many RAs or an RA may span over many RNC areas but not over many LAs; in the latter case, the RNCs have to be connected to the same SGSN.

Mapping between one LA and RNCs is handled within the MSC/VLR owning this LA. That between one RA and RNCs is handled within the SGSN owning this RA. Mapping between one LA and cells (respectively between one RA and cells) is handled within the RNC.

URAs and UCAs are only visible in the RAN, not at the CN level; they are used in RRC-Connected Mode. Each cell in the network is assigned at least one URA identifier (URAid) and thus overlapping URAs are possible. The possibility of having overlapping URAs reduces the number of URA updates for a given terminal. A URA consists of a number of cells belonging to one RNC and is used to avoid a high number of cell updates for high-mobility terminals. The terminal will change its state from Cell_PCH to URA_PCH based on certain parameters defined in the RNC. Thus terminals only perform URA updates, not cell updates.

A cell is the smallest entity in the UTRAN and is unknown in the CN. Cell update is a Radio Resource Control (RRC) procedure. A cell update takes place if the user equipment (UE) crosses the cell border while it is in Cell_FACH or Cell_PCH state. A cell update can also happen in Cell_DCH state if the criteria for radio link failure are met. Cell update is also done to notify UTRAN after re-entering the service area in the Cell_FACH or Cell_PCH state, or during a paging response.

All area updates can also be done periodically. The following area relations exist (see Figure 9.31):

- there does not need to be any relation between a URA and LA (respectively between a URA and RA);
- one LA consists of a number of cells belonging to RNCs that are connected to the same CN node;

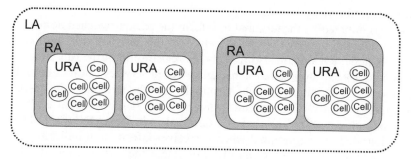

Figure 9.31 Area relations.

LA = Location Area; RA = Routing Area; URA = UTRAN Registration Area.

- one RA consists of a number of cells belonging to RNCs that are connected to the same CN node;
- one LA is handled by only one CN serving node – i.e., one combined MSC and SGSN or one 3G MSC/VLR;
- one RA is handled by only one CN serving node – i.e., one combined MSC and SGSN or one 3G SGSN.

In addition to the mobility and paging control areas a Service Area (SA) is defined and used for Location Services (LCS) and cell broadcast services. SAs are not as such related to LAs or RAs. There is no SA update message sent between the UE and RAN. The SA broadcast feature enables the information provider to submit short messages for broadcasting to a specified SA within the Public Land Mobile Network (PLMN). These messages could be used for informing of, for example, PLMN news, emergencies, and traffic reports, road accidents, delayed trains, weather reports, theatre programmes, telephone numbers or tariffs. The Service Area Identifier (SAI) identifies an area consisting of one or more cells that belong to the same LA.

Service Registration and Location Update
Service registration (attach) in the respective CN domain is done initially (after a terminal is detached due to, say, power-off). When a URA is changed a location update is performed. In addition, periodic registration can be performed. Descriptions of when the respective CN registration area is changed now follow.

LOCATION AREA UPDATE
LA update is initiated by the terminal to inform the circuit switched domain of the CN that the terminal has entered a new LA. In case the new LA is in an area served by another CN, the LA update also triggers the registration of the subscriber in the new CN and a location update for circuit switched services towards the Home Location Register (HLR).
   LA update is only initiated by the terminal when the terminal is in CS-IDLE state, and this independently of the packet switched state. If the terminal is CS-IDLE but RRC-connected, which means that the terminal is in PS-CONNECTED state, LA update is initiated by the terminal when it receives information indicating a new LA.

ROUTING AREA UPDATE
RA update is initiated by the terminal to inform the packet switched domain of the CN that the terminal has entered a new RA. In case the new RA is in an area served by another CN node, RA update also triggers the registration of the subscriber in the new CN node and a location update for packet switched services towards the HLR.
   RA update is only initiated by the terminal when the terminal is in PS-IDLE state, and this independently of the circuit switched state. If the terminal is PS-IDLE but RRC-connected, which means that the terminal is in CS-CONNECTED state, RA update is initiated by the terminal when it receives information indicating a new RA.
   When the terminal is in PS-CONNECTED state the terminal initiates RA update when the Routing Area Identifier (RAI) in mobility management system information changes.

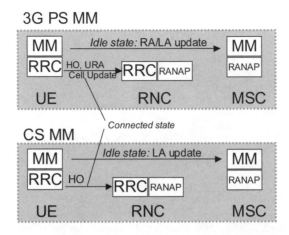

Figure 9.32 Different update procedures in idle and connected state.

In CS-IDLE mode, the CN knows the location of a terminal to the accuracy of LA, and correspondingly in PS-IDLE mode, the CN knows the locations of a terminal to the accuracy of LA/RA. In connected mode (CS-CONNECTED or PS-CONNECTED) the location of the terminal is given to the CN by the Serving RNC (SRNC). In 3G there is no LA or RA update when a terminal is in connected mode, see Figure 9.32.

Optimisation of Area Concepts

At the moment operators are mainly using geographical information from maps to decide in which RNC a new site/cell should be placed. Further, during network evolution the cells are attached to an RNC (and LA, RA) where there is physical space, thus from the mobility perspective not necessarily optimally. As a consequence of sub-optimal choices a high number of inter-RNC handovers can be created (see Figure 9.33). A high number of inter-RNC handovers is detrimental to network performance as this handover type has a higher probability of resulting in dropped calls than an RNC-controlled handover. Further, inter-RNC handover causes excess signalling and it can happen that the signalling load between LAs (RNCs) is not equally distributed. For these reasons mechanisms and decision logic to reduce the number of inter-RNC handovers are needed.

In addition to optimisation of the cell's LA/RA in order to minimise inter-RNC handovers and paging load another case can be identified – namely, RNC split.

In splitting case sites are rehosted when there is a need for more capacity, and therefore a new RNC is integrated in the network. In order to keep inter-RNC handovers to a minimum, BTS sites in the same area should be under the control of the same RNC. To achieve this target, a number of already existing sites are moved from being under control of the current RNC (source RNC) to being under control of the new RNC (target RNC). During the WCDMA BTS site reparenting operation, one must modify not only the radio network configuration, but also the Asynchronous Transfer Mode (ATM) layer configuration as well as IP network configuration data.

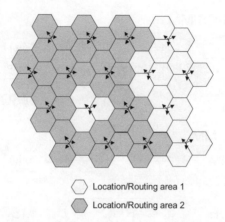

Figure 9.33 Scattered location area planning. Increased number of inter-RNC handovers.

The algorithm advises which cells should be reparented to which RNC (LA or RA) from the radio network point of view. Both incoming and outgoing handovers are counted and used in the decision making. Intra-RNC handover and inter-RNC handover shares are calculated for each cell including all neighbouring RNCs:

$$HO\_Share = Inter\text{-}RNC\ HO\ to\ the\ neighbour\ RNC\ in\ question$$

$$-\ Intra\text{-}RNC\ HO \tag{9.20}$$

If $HO\_Share > 0$ this means that reparenting the cell to the RNC in question reduces the number of inter-RNC handovers. In the reparenting case the statistics of intra-RNC HOs will become inter-RNC statistics, and vice versa. Once the HO_Share for each RNC taking part in the inter-RNC handover related to that cell is analysed the results are sorted and a new proposal for the parent RNC is identified.

In addition to the HO_share the number of actual location updates as well as the cell traffic can be used to weight the HO_share analysis. If the traffic in a cell is very low the urgency of reparenting is less than for a cell that is highly loaded and causes large numbers of MSC-controlled handovers.

Planning of LAs and RAs parallel with major roads and railways should be avoided. Pingponging will create a heavy update-messaging load. Similarly RAs that run perpendicular to major roads will create excessive update messaging on the cells serving that area. In early network operation phase LAs and RAs could be defined such that the RA and LA update will happen at the same place. One LA/RA could then correspond to one RNC area.

9.3.5 Autotuning in the Call Admission Control Sub-system

In this section methods for improving the power-based admission control performance is introduced. Optimisation is based on tradeoff management between capacity and quality. In addition to admission control optimisation a section is dedicated to autotuning of P-CPICH power.

9.3.5.1   Autotuning of Total Received Power Using a Cost Function

In WCDMA, power-based admission control can be used in both the uplink and downlink. There is a cell-based target for total received power level (*PrxTarget*) in the uplink and for total transmitted power level (*PtxTarget*) in the downlink. These targets determine whether or not to admit a new user and also how to schedule packets (see Chapter 4). Accordingly, they determine how much traffic is allowed in the cell. To have the correct targets is important. If the targets are too low, not all the capacity of the network is utilised. On the other hand, if the targets are too high, too many connections are admitted in the cell and the increased interference causes bad quality or, at worst, dropped calls. This is caused by power outage of the mobiles in uplink and downlink connections that hit the maximum connection-specific transmission powers, reducing cell coverage and degrading call quality. The function of *PrxTarget* is illustrated in Figures 9.34 and 9.35 for low and normally loaded cells.

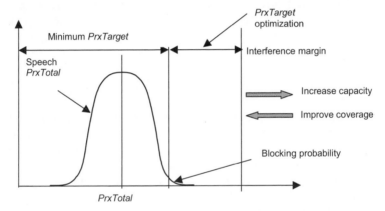

Figure 9.34   Setting of *PrxTarget* for speech traffic at low loads.

It is important to note that applying this type of autotuning to the admission control or packet scheduling the process will provide similar performance improvement for systems utilising throughput-based admission control.

Background
Let us look more closely at the effects of the loading to the main KPIs, which are the blocking probability and the bad call probability or the dropped call ratio. The main uplink parameter that controls the admission of calls or scheduling of packets is *PrxTarget*. By virtue of load control and admission control, a call is accepted if the estimated *PrxTotal*, the total uplink interference level, is less than *PrxTarget*, otherwise the call is not admitted. Thus, the higher *PrxTarget*, the higher will be the allowed capacity for the cell. However, one cannot allow a very high *PrxTarget*, since every cell has a limited interference margin. To show the range of the *PrxTarget* setting, Figure 9.34 will help. This shows that *PrxTarget* can be set freely within the minimum and maximum limits in order to balance coverage and capacity.

For example, one can set *PrxTarget* below the interference margin in order to improve the coverage of other bearers with a higher bit rate in exchange for lower capacity. On the other hand, one can increase *PrxTarget* to allow higher capacity but with some degradation in coverage performance. This depends on the decision of the network planner. The maximum limit for *PrxTarget* is equal or close to the interference margin for the planned bit rate (e.g., speech bearer). Its minimum value should be large enough not to cause unnecessary blocking while the cell is at normal load. For example, it should allow at least 99% of the calls to be admitted if there is no congestion.

The next consideration in the setting of *PrxTarget* is when there is a presence of packet traffic. In this case, one needs to take into account the fact that the capacity of NRT or packet bearers is what is left over from the circuit switched or RT bearers, if there is no dedicated NRT traffic capacity. NRT bearers come in a set of bit rates. Therefore, when considering outage probability one should make sure that both the speech and the minimum NRT bearer can be supported with acceptable coverage probability. For example, if the minimum packet data bit rate is 16 kbps at a required E_b/N_0 of 2 dB and the speech bit rate is 8 kbps with a required E_b/N_0 of 7 dB, the limiting bearer for setting *PrxTarget* will be:

$$\max\left\{10 \cdot \log_{10}\left(\frac{16}{8}\right) + 2, 10 \cdot \log_{10}(1) + 7\right\} \tag{9.21}$$

which is 7 dB. This means that in this case the speech bearer is the limiting link in setting the maximum allowed *PrxTarget*. Note also that the network planner may decide to set *PrxTarget* beyond the interference margin just to achieve the capacity requirement in exchange for a degradation in coverage performance.

The next step in choosing *PrxTarget* is to decide the allowed capacity for the NRT traffic with the given KPIs. Normally, NRT services require some maximum queuing probability. This depends on how different *PrxTarget* is from the average *PrxTotal* which is the combination of the RT and NRT traffic. In Figure 9.35, for example, the average *PrxTotal* is below *PrxTarget*, which means that the cell in not overloaded. Only in some instances is *PrxTarget* reached and exceeded. When *PrxTotal* exceeds *PrxTarget*, newly arriving packets are expected to be queued. Therefore the task is to balance the NRT traffic capacity and the corresponding queuing probability.

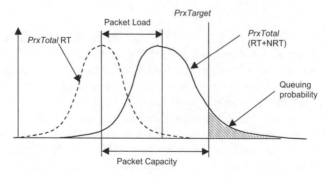

Figure 9.35 *PrxTarget* setting for a normally loaded cell.

Impact of Loading to the Network in Uplink

Another important factor to consider when setting *PrxTarget* is the variance of *PrxTotal* within the allowed range. In both the uplink and downlink, the link budget is normally expressed in decibels [dB]. In the uplink, noise rise is an exponential function with respect to the load factor; the variance of *PrxTotal* increases if the network is allowed to operate at higher loads. This makes it more challenging to control *PrxTotal*. This is true especially for capacity limited cells (e.g., small cells) that can allow high values of *PrxTotal*. However, a small gain comes from multiplexing at higher loads. At lower load factors, the change in noise rise with respect to load factor is relatively lower, but the load factor has a higher deviation owing to lower multiplexing gain. On the other hand, higher load factors result in higher multiplexing gains, but the change in noise rise with respect to load factor (i.e., $\Delta NR/\Delta LF$) is relatively higher, which is why it is important to approximate the load factor distribution before setting *PrxTarget*.

Real Time Traffic Sensitivity with PrxTarget Setting

When RT and NRT traffic are mixed as in a typical cell, *PrxTotal* is the sum of both traffic types. As *PrxTotal* increases, terminal power outage probability increases, causing either the dropped call ratio or the bad-quality call ratio to increase. The outage probability is higher for higher bit rates. In the case of a loaded cell, *PrxTotal* is always close to *PrxTarget*, making *PrxTarget* a very sensitive parameter. However, if *PrxTarget* is set to a small value, then the call blocking probability can be high. Therefore, there is a tradeoff between call blocking and call dropping/bad quality in the setting of *PrxTarget*. Its value depends on the interference margin or cell radius. If the QoS for the RT data traffic is optimised, the results are as shown in Figure 9.36.

Figure 9.36 shows that small values of *PrxTarget* can dramatically increase the blocking probability. This is true for all RT bearers. When *PrxTarget* is increased, the bad quality and dropped call ratio will slowly increase. Therefore, there is a region where the KPIs for the RT bearers are optimal. When *PrxTarget* is higher than optimal, the cell becomes coverage limited for some RT data bearers. When considering NRT traffic in the optimisation, one should remember that the higher

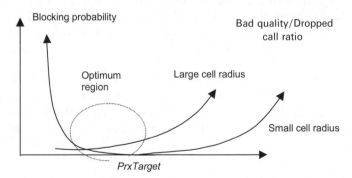

Figure 9.36 Sensitivity of real time data to *PrxTarget*.

the value of *PrxTarget*, the higher the capacity for NRT traffic, so tradeoffs between all the KPIs for both RT and NRT need to be considered.

Downlink Parameter Sensitivities
The proposed downlink admission control is similar to that of the uplink except that the reference point for the maximum downlink link power is the pilot (P-CPICH) power, and the maximum load is identified by the maximum downlink transmission power. The behaviour of the downlink transmission power is among others controlled by RRM parameters: connection power target and total power target (*PtxTarget* and *PrxTarget*). The goal is to have stable downlink power while achieving low enough outage probability per link.

Advantages of Autotuning
Autotuning makes it much easier for the operator to optimise the cell-specific uplink and downlink total power targets (*PrxTarget* and *PtxTarget*). The optimal target values vary from cell to cell depending on the radio propagation environment and size of the cell. In some cells where the propagation environment is good, there is more room for increasing the targets to allow more users in the cell, because in these kinds of cells the quality of calls does not degrade much when the targets are increased. In some other cells with a poor propagation environment, power targets should be kept somewhat lower in order to meet the quality requirements of the cell. The different environments of the cells are difficult to take into consideration with a planning tool, but autotuning algorithms use cell-specific measurement data. This is very useful, since optimal targets can also change over time.

Method for Autotuning
The *PrxTarget* of a cell is autotuned using quality measurements from that specific cell gathered during high uplink load. The *PtxTarget* of a cell is correspondingly autotuned using quality measurements gathered when the cell has been under high downlink load. The targets are autotuned so that they are as high as possible, when taking into account the quality of calls, packet queuing and blocking of calls. If the bad-quality situation is worse than the blocking and queuing situation, the target is lowered. On the other hand, if bad quality has a lower cost than the cost of queuing plus blocking, the target is increased – i.e., autotuned in a direction that increases capacity. This provides a good capacity vs. quality or coverage tradeoff. The tradeoff can be adjusted by adjusting the costs and allowed levels of bad quality, call blocking and packet queuing. By giving a lower cost to bad call quality or allowing poorer quality, the capacity is increased, thus in particular a higher throughput and/or lower blocking of calls is enabled, while it is correspondingly decreased by giving a high cost to bad quality or allowing a very small amount of bad quality. Table 9.6 shows the 27 different states possible in power target autotuning and the corresponding adjustments. The different quality indicators can be significantly below the allowed level, significantly above the allowed level and within the confidence margins of the allowed level. The adjustment is then either no adjustment, upward adjustment, downward adjustment or up (+) or downward (−) adjustment after comparison of costs. Compare Cost+ means checking whether *CostCongestion(UL)* is greater than *CostQuality(UL)* and, if true,

Table 9.6 Rules for autotuning. The 27 different states of the system. Each quality indicator can be significantly below the allowed level, significantly above the allowed level or within the confidence margins of the allowed level. The adjustment is then no adjustment, upward adjustment, downward adjustment, or upward or downward adjustment ($+/-$), optionally after comparison of costs.

Bad quality	Blocking	Queuing	Adjustment
Below	Below	Below	No adjustment
Below	Below	Within	No adjustment
Below	Below	Above	Increase
Below	Within	Below	No adjustment
Below	Within	Within	No adjustment
Below	Within	Above	Increase
Below	Above	Below	Increase
Below	Above	Within	Increase
Below	Above	Above	Increase
Within	Below	Below	No adjustment
Within	Below	Within	No adjustment
Within	Below	Above	Compare Cost+
Within	Within	Below	No adjustment
Within	Within	Within	No adjustment
Within	Within	Above	Compare Cost+
Within	Above	Below	Compare Cost+
Within	Above	Within	Compare Cost+
Within	Above	Above	Compare Cost+
Above	Below	Below	Decrease
Above	Below	Within	Compare Cost−
Above	Below	Above	Compare Cost+/−
Above	Within	Below	Compare Cost−
Above	Within	Within	Compare Cost−
Above	Within	Above	Compare Cost+/−
Above	Above	Below	Compare Cost+/−
Above	Above	Within	Compare Cost+/−
Above	Above	Above	Compare Cost+/−

increasing the *PrxTarget*. Compare Cost− correspondingly means checking whether *CostCongestion*(*UL*) is greater than *CostQuality*(*UL*) and, if false, decreasing the *PrxTarget*. Compare Cost+/− means that the *PrxTarget* is increased or decreased if the condition mentioned above is true or false, respectively. The definitions for bad quality, queuing and congestion can be found in Section 9.3.3.

It is very important to cope with mobility. The quality measures to be associated with a cell should be derived only from the locally generated parts of those calls that are connected to the cell in question. For example, poor quality of parts of calls that start far away but end in the cell to be autotuned should not affect the autotuning of the cell's power targets. Also, diversity handover issues must be taken into account when evaluating the quality of calls. A possible addition to this proposed method is that, before raising the power target in a cell, a check as to whether adjacent cells are suffering from poor quality should be made; instead of just optimising on a cell-by-cell basis cell clusters can also be considered.

Also, for throughput-based algorithms it is possible to use quality criteria to maximise load-level targets. This is due to the fact that with this type of autotuning the admission control principle (power/throughput/number of connections) is not critical; all methods can provide the same capacity–quality tradeoff.

This feature allows inaccurate or even incorrect target values to be avoided. Moreover, it provides correct cell-specific *PrxTarget* values. Thus, the capacity of the network is increased, since usually a network is initially set with too low target values in order to ensure that the required quality criteria are achieved. However, sometimes in a network the target values might be set too high, which results in excessive interference levels in a cell, causing poor call quality. In this case, the targets are reduced in order to achieve the required level of quality.

Results from Simulations

In [42] the detailed results of the study are presented. One simulation was done using the macro-cell scenario and two simulations with different service mixes using the micro-cell scenario as introduced in Section 9.3.2. In general the results improved the performance significantly compared with fixed parameter settings and in particular for cautiously made parameter settings. The results are shown in Table 9.7 and indicate that the setting of the uplink power targets has a significant impact on capacity and quality uplink throughput increased by 48% to 57% compared with the conservative *PrxTarget* of 4 dB (60% loading), by 16% to 25% compared with the 6 dB *PrxTarget* (75% loading) and by 2% to 10% compared with the 8 dB *PrxTarget* (84% loading). The degradation in RT call quality was minor, while RT call blocking decreased significantly.

Table 9.7 Simulation results for *PrxTarget* optimisation.

Scenario	Improvement with respect to fixed *PrxTarget*		
	4 dB	6 dB	8 dB
Macrocell	57%	25%	10%
Microcell 1 (NRT PS and RT speech)	52%	21%	6%
Microcell 2 (NRT PS and RT CS data)	48%	16%	2%

9.3.5.2 Automatic Optimisation of Downlink Power Resource Parameters Using a Cost Function

The admission control parameters selected for power resource optimisation were *PrxTarget*, *PtxTarget*, *CPICHToRefRABOffset* and *PtxDLAbsMax*. Optimisation of the *PrxTarget* is in Section 9.3.5.1. Complete studies on the optimisation of these parameters showing significant capacity gains were reported in [42] and [43]. Definitions for the measurements and costs used in the study are in Section 9.3.3.

Method for Autotuning

PtxTarget determines the amount of downlink traffic allowed in the cell. If the targets are too low, the capacity of the network is not fully utilised. On the other hand, if the targets are too high, too many connections are admitted in the cell with increased interference as a consequence. Increased interference causes poor call quality or even dropping of calls. In the downlink, poor quality occurs in connections whose required link transmission power exceeds the maximum connection-specific link power. Moreover, outage of total cell power causes poor call quality. The parameter *CPICHToRefRABOffset* defines the maximum link transmission power for a selected reference Radio Access Bearer (RAB) – e.g., 12.2 kbps speech service – as a corresponding fraction of the cell's common pilot power, *PtxPrimaryCPICH*. The maximum link powers of other services are obtained by scaling the reference maximum power with the ratio of bit rates, R/R_{ref}, and the ratio of downlink E_b/N_0 requirements, ρ/ρ_{ref}, between the reference and a particular service:

$$PtxMax = \frac{PtxPrimaryCPICH}{CPICHToRefRABOffset} \frac{\rho \cdot R}{\rho_{ref} \cdot R_{ref}} \qquad (9.22)$$

Formula (9.22) produces similar cell coverage for all services it is applied to. The coverage of high bit rate services can be limited with the parameter *PtxDLAbsMax*, which defines the absolute maximum link power for any service.

The parameter *CPICHToRefRABOffset* for downlink link power maximum determination was adjusted if there was either significant downlink poor quality or congestion. Tuning was done using cell-specific quality measurements gathered during high downlink load, taking into account the quality of calls, packet queuing and power blocking of calls. The simplified rule was that *CPICHToRefRABOffset* was increased or decreased by 0.5 dB if *CostCongestion(DL)* was higher or lower than *CostQuality(DL)*, respectively. The rules actually implemented and described in [43] and presented in Table 9.6 are somewhat more complex but in practice not significantly different from the simple rule above.

PtxDLAbsMax was set equal to the maximum power, *PtxMax* of Equation (9.22), of the service with the highest power requirement planned to have coverage in the entire cell, which was the 64 kbps circuit switched data service. *PtxDLAbsMax* was calculated online using the current optimised *CPICHToRefRABOffset* in Equation (9.22).

The selection of *PtxTarget* was based on the assumption of *PtxTotal* being normally distributed and an allowed 2% probability of total power outage. *PtxTarget* was set equal to the maximum BS power minus two times the measured standard deviation of *PtxTotal*.

Table 9.6 shows the different states possible in autotuning and the corresponding adjustment. The different quality indicators can be significantly below the allowed level, significantly above the allowed level and within the confidence margins of the allowed level. The adjustment is then either no adjustment, upward adjustment, downward adjustment or up (+) or downward (−) adjustment after comparison of costs. Compare Cost+ means checking whether *CostCongestion(DL)* is greater than

CostQuality(DL) and, if true, increasing the *CPICHToRefRABOffset*. Compare Cost– correspondingly means checking whether *CostCongestion(DL)* is greater than *CostQuality(DL)* and, if false, decreasing the *CPICHToRefRABOffset*. Compare Cost+/– means that the *CPICHToRefRABOffset* is increased or decreased if the condition mentioned above is true or false, respectively. The confidence margins were calculated using binomial confidence intervals. If good quality was achieved, *PtxTarget* was increased. On the other hand, if the quality was poor, *PtxTarget* was decreased.

Even in this case it is necessary to associate to a cell only the quality measures of those parts of the call that the call is connected to that cell in question. It would not be good, for example, if the poor-quality periods of calls that started far away but ended in the autotuned cell affected the autotuning of the cell's power downlink link maxima and power targets. Also diversity handover issues must be taken into account when evaluating the quality of calls, so that poor quality is associated with the strongest or if required with all cells in the active set of the UE. A possible addition before raising the power target in a cell is to check whether adjacent cells are suffering from poor quality.

Results from Simulations for Automated Optimisation of CPICHToRefRABOffset, PtxDLAbsMax and PtxTarget

In [43] a full set of results for the optimisation of *CPICHToRefRABOffset* and *PtxTarget* is presented. Two fixed settings of *CPICHToRefRABOffset* were validated using the micro-cellular scenario introduced in Section 9.3.2. The results show that even if the default parameter setting was somewhat incorrect (case 2 in [43]), the proposed method improved network performance. In comparison with fixed *CPICHToRefRABOffset* values, autotuning decreased bad quality significantly, which made it possible to increase the *PtxTarget*. The adjustment of *PtxTarget* together with the autotuning of *CPICHToRefRABOffset*, made a significant increase in throughput possible (up to 39% compared with default parameter settings) in addition to a quality improvement.

In Table 9.8 results from a micro-cellular case are presented. This is case 1 in [43], which has a more incorrect initial offset setting than in case 2. Both cases showed similar results.

Only the downlink was simulated due to the fact that downlink autotuning methods were validated and the downlink was assumed to be the limiting link. The soft handover overhead was about 50% for both the circuit switched traffic and packet traffic in all simulation cases.

Results (case 1 [48]) show that system performance with autotuning turned on improved when the *CPICHToRefRABOffset* was set to a somewhat conservative value of 5.5 dB. In comparison with the fixed *CPICHToRefRABOffset*, autotuning decreased bad quality significantly, which made it possible to increase the *PtxTarget*. Increasing the *PtxTarget* produced very poor quality with a **fixed** *CPICHToRefRABOffset* of 5.5 dB, which corresponds to a 250 mW downlink link maximum for the circuit switched RT 64 kbps service. The blocking probability was very high due to a very high rate of arriving calls, which was selected in order to load the system up to the target level.

Table 9.8   Micro 46-cell scenario: results for circuit switched speech and packet switched traffic.

Measure	Parameter setting			
	PtxTarget 33 dBm, fixed offset 5.5 dB	PtxTarget 33 dBm, DL link maxima tuned	PtxTarget 35.5 dBm, fixed offset 5.5 dB	PtxTarget 35.5 dBm, DL link maxima tuned
Number of ended RT CS calls	14472	13696	15057	14674
Probability of degraded RT CS BLER	7.0%	2.0%	18.9%	3.8%
RT CS blocking probability	8.2%	13%	4.5%	6.9%
RT CS throughput [kbps/cell]	500	473	520	507
NRT PS throughput [kbps/cell]	264	255	542	494
DL total throughput [kbps/cell]	764	728	1063	1001

The conclusion drawn from the results is that the autotuning of cell-based downlink link maxima and load targets improves significantly the system performance as measured with throughput particularly in comparison with cautious or incorrect parameter settings. Therefore, the feature is a promising candidate for implementation into the NMS.

9.3.6   Capacity Optimisation and Traffic Balancing

In the following sections some mechanisms to share resources between circuit switched and packet switched traffic or between cells is discussed. A similar logic can be applied to GPRS as demonstrated in Chapter 10. The traffic control mechanism between systems, inter-system handover, is introduced in Chapter 4.

9.3.6.1   Autotuning of P-CPICH Power

The primary objective of the methods presented in this section is to minimise the usage of power resources for the P-CPICH, while ensuring good enough P-CPICH coverage. This is even more important if the power levels of all other common channels are set with respect to P-CPICH power – i.e., higher amounts of power resources can be saved and more traffic served. The original work is presented in [36].

Method for Autotuning
P-CPICH defines the power of the P-CPICH in the cell. Increasing or decreasing the pilot power makes the cell larger or smaller. Thus, the tuning of pilot powers can be applied to balance cell load among neighbouring cells and, additionally, to provide sufficient signal reception for the terminals. The common pilot coverage issues are discussed in Section 9.3.3.3.

   In the rule-based method of [36] the pilot power of a cell was increased or decreased by 0.5 dB if the cell load was significantly lower or higher than the neighbour cell load as indicated by statistics in Section 9.3.3, Equation (9.14). If the load was not

Table 9.9 Pilot power control actions.

Load balance, t	Coverage balance, c	Pilot power change and counter reset
−1	−1	Increase pilot and reset all counters
−1	0	Increase pilot and reset load counters
−1	1	Increase pilot and reset load counters
0	−1	Increase pilot and reset coverage counters
0	0	No change and no reset
0	1	Decrease pilot and reset coverage counters
1	−1	Decrease pilot and reset load counters
1	0	Decrease pilot and reset load counters
1	1	Decrease pilot and reset all counters

significantly unbalanced among the cells, but the pilot signal reception was significantly lower or higher than the target, the pilot power was increased or decreased by 0.5 dB, respectively. Pilot power was limited between 3% and 15% of the maximum BS power. Pilot power control actions are presented in Table 9.9.

In Table 9.9 t (load balance) and c (coverage balance) are calculated as follows. The test statistic of the difference between own-cell and neighbour cell loads was obtained using:

$$t' = \frac{m_1 - m_2}{\sqrt{\dfrac{v_1}{N_k} + \dfrac{v_2}{\displaystyle\sum_{i \neq k} N_i}}} \qquad (9.23)$$

Statistic t' was quantified to three levels of load balance:

$$t = \begin{cases} -1, & t' < -2 \\ 0, & -2 \leq t' \leq 2 \\ 1, & t' > 2 \end{cases} \qquad (9.24)$$

The terminals in the sector reported the received E_c/I_0 of the pilot. For each reported E_c/I_0 cell-specific counter N_{ecio} was incremented. If E_c/I_0 exceeded -18 dB, counter N_{over} was also incremented. The counters were reset at the point of pilot power adjustment as shown in Table 9.9.

The test statistic of the difference between cell coverage and target coverage, C, was calculated according to:

$$c' = \frac{N_{over} - N_{ecio} \cdot C}{\sqrt{N_{ecio} \cdot C \cdot (1 - C)}} \qquad (9.25)$$

The target coverage was set to $C = 0.98$. As above, statistic c' was quantified to three levels of coverage balance:

$$c = \begin{cases} -1, & c' < -2 \\ 0, & -2 \leq c' \leq 2 \\ 1, & c' > 2 \end{cases} \qquad (9.26)$$

Results from Simulations

A full set of results for the optimisation of pilot power using a rule-based method is presented in [36]. The mixed macro-cell and micro-cell scenario depicted in Section 9.3.2 was used in the simulations. Table 9.10 shows the improvement of downlink packet data performance measures obtained with the autotuning method. Table 9.11 shows that the average downlink total transmission powers (*PtxTotal*) increased slightly when rule-based optimisation was applied. The initial P-CPICH power values were 5% of the total maximum transmission power in each cell – i.e., 200 mW for micro-cells and 1 W for macro-cells. Increased total BS powers explain the improved performance of rule-based optimisation. This can be taken as an indication that the load was more evenly distributed. As the target pilot coverage was 98%, the results in Table 9.11 show that the coverage deteriorated with autotuning. Coverage could be improved by increasing its weight in the cost function and by adjusting the rule priorities.

The results corroborated that the fixed setting of the pilot power –i.e., by default to 5% of the maximum BS power – is a warranted choice. Coverage was sufficient and the packet data performance was close to that obtained with the pilot control. Thus, pilot power control may only benefit performance in congested cells. However, load balancing, which was clearly attained, can benefit single cells whose performance is not reflected in the total network performance but which subjectively can be highly significant. The total BS powers show the effect of load balancing obtained with the pilot power control. Mean macro-cell total power moved closer to the target of 10 W and the standard deviation of the power decreased. The decrease of standard deviation was shown in micro-cells as well.

Table 9.10   Improvement of packet data performance with pilot power optimisation compared with initial pilot power setting.

	Rule-based method [%]
Total throughput	4
Active session throughput	21
Allowed bit rate	31
95th percentile of packet delay	5

Table 9.11   Performance results of pilot power optimisation.

	No optimisation	Rule-based method
Macro *PtxTotal* (std) [W]	9.0 (2.0)	9.4 (1.3)
Micro *PtxTotal* (std) [W]	1.9 (0.3)	1.9 (0.2)
Macro pilot power (std) [W]	1 (0)	1.6 (0.9)
Micro pilot power (std) [W]	0.2 (0)	0.24 (0.15)
1 − Macro coverage [%]	1.6	2.3
1 − Micro coverage [%]	1.3	3.5

These results suggest that, first, the balancing of load among cells and aiming to achieve a specific coverage level is feasible using the simple heuristic rules that control pilot power. Second, the pilot power control method improves the air interface performance. Finally, the method is a valid means for improving network operability by its automation.

In [19] another optimisation technique for adjusting pilot powers in CDMA systems is presented. The idea in this technique is to reduce the unused pilot signals seen by mobiles, thus reducing the number of pilot powers within a certain margin relative to the strongest pilot. The results in [19] showed that the lowering of pilot power pollution gives some improvement in downlink link coverage and capacity in addition to reduction in deployment efforts spent in optimising pilot powers.

9.3.6.2 Autotuning of Dedicated Capacity for Non-Real Time Services or for High-speed Downlink Packet Access

The basic RRM without the dedicated capacity for NRT services allows only one threshold for any traffic when performing admission control for the new entering RAB, when modifying an existing RAB or when performing packet scheduling. This means that RT and NRT traffic will use the same entry criteria and in case the cell is fully loaded with RT traffic there will be no room for NRT traffic at all. With the dedicated NRT traffic capacity feature the operator can guarantee at least some capacity for NRT traffic as well.

The dedicated NRT traffic capacity feature provides uplink and downlink target power thresholds for RT and NRT traffic separately. This feature improves the QoS, because it provides a possibility to guarantee some capacity for NRT traffic on a cell-by-cell basis. The capacity reservation for NRT traffic requires support in NEs in terms of algorithms and configuration parameters to do the resource reservations in practice.

In Figure 9.37 the idea underlying dedicated NRT capacity reservation is presented. In phases A, B and C both RT and NRT traffic are getting the needed capacity, and there are no traffic restrictions. In phase D NRT traffic experiences capacity shortage

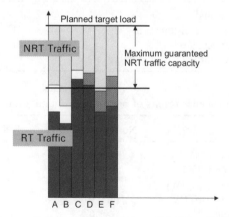

Figure 9.37 Conceptual presentation of the operation of dedicated non-real time traffic capacity reservation.

and new RT RAB setups are rejected until NRT traffic gets the capacity it requires. At point E RT traffic experiences blocking. NRT traffic is allocated the capacity that is left over from the RT traffic. At F both traffic types experience blocking, new RT RAB setups are rejected and NRT traffic is given the maximum guaranteed capacity.

Optimisation of dedicated NRT traffic capacity would take care that the threshold controlling the size of the dedicated territory would be adaptive. Some of the resources available in the uplink and downlink can be dedicated to NRT traffic. During heavy load a tradeoff between RT traffic blocking and NRT traffic queuing can be performed. One possible method is to attach costs to blocked and queued bearers. Autotuning can be done so that the dedicated NRT traffic capacity is increased if the cost of queued bearers is significantly higher than the cost of blocked bearers, and correspondingly decreased if the cost of queued bearers is significantly lower than the cost of blocked bearers.

Similarly, HSDPA functionality shares the physical and logical resources in terms of power and codes with Dedicated Channels (DCHs). Should the HSDPA performance be degraded due to lack of power resources or codes, a similar method to the one above should reallocate physical and logical resources optimising the performance of HSDPA channels and of DCHs.

9.3.6.3 Intra-frequency Traffic Balancing Using Cell Individual Offsets

As discussed in Chapter 4 handovers within the UTRA-FDD system can be classified as intra-frequency handovers and inter-frequency handovers. In intra-frequency soft handover, an MS is allowed to connect simultaneously to several BSs, which are added or removed from the terminal's active set by applying relative handover thresholds. The most important ones are the addition threshold, the addition timer, the dropping threshold and the dropping timer. In principle, if a received P-CPICH E_c/I_0 from a new BS is within a window defined by the addition threshold relative to the best serving BS's E_c/I_0 for a time period longer than the addition timer, it is added into the user's active set. When the P-CPICH E_c/I_0 from a BS in the active set is lower than the P-CPICH E_c/I_0 of the best serving BS by a margin defined by the dropping threshold that BS is removed from the active set. Typically, the measurement quantity is P-CPICH E_c/I_0 but it can also be path loss.

Further, a Cell Individual Offset (CIO) value can be used to make one neighbouring cell more attractive than another. This is demonstrated with Figure 9.38. In order to make the handover to a cell with P-CPICH 3 happen earlier, an offset is applied to manipulate the terminal's decision. The offset raises the P-CPICH 3 curve.

The terminal measures the E_c/I_0 levels of the pilot signals of neighbouring cells. The terminal initiates changing of the active set by sending a measurement report and an ASU request to the RNC. The reporting conditions have the following general form:

$$CPICH(monitored) + AdjsEcNoOffset(best, monitored)$$
$$> ReportingCriterion(CPICH(best)) \quad (9.27)$$

where $CPICH(monitored)$ and $CPICH(best)$ are the measurement results (E_c/I_0) of the monitored cell and the best active set cell, respectively; and

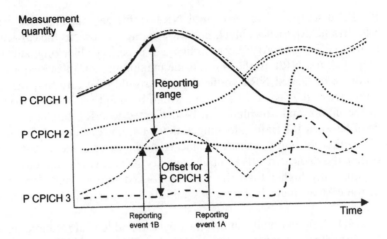

Figure 9.38 Cell individual offset. A positive offset is applied to P-CPICH 3 before event evaluation in the terminal.

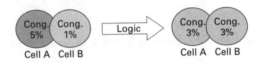

Figure 9.39 Conceptual presentation of the congestion relief logic. After optimisation the average behaviour is the same, but the blocking performance is improved for the highly congested cell.

AdjsEcIoOffset(*best, monitored*) is the cell individual offset added to the measured E_c/I_0 of the monitored cell. It is specific to the primary cell in the active set – i.e., there is an offset for each neighbour of a cell.

The neighbour set for a specific combination of cells in the active set is a union or intersection of the neighbour sets of the individual active set cells formed using a particular method. The maximum size of the combined neighbour set is 32 cells. When an ASU is made, the terminal gets signalled the new neighbour set and CIO-related information.

The idea of congestion relief is to utilise these CIOs to force traffic from a highly congested cell to neighbouring cells that are less loaded (see Figure 9.39). Such a situation applies, for example, in the business areas of city centres. A few highly loaded cells might serve certain office buildings, and the surrounded cells are low loaded during business hours. Thus these surrounding cells can be taken to serve the business complex with the proposed method. The offsets between two cells *A* and *B* – i.e., *AdjsEcIoOffset*(*A, B*) and *AdjsEcIoOffset*(*B, A*) are adjusted if the ratios of blocked calls differ significantly between the cells. Blocking can be measured and evaluated in several ways – for instance, as:

- soft blocking due to insufficient power resources (downlink total transmission power exceeding its target level);

- hard blocking due to insufficient hardware or logical (codes) resources; or
- hard and soft blocking combined, soft handover overhead-related information – if abnormal, increased blocking may be due to a poorly set *AdditionWindow* which is indicated to the user. The user then may wish to perform *AdditionWindow* optimisation first.

The proposed control method gives the best gain if insufficient hardware resources caused the blocking but the method is able to balance the load from soft-blocked cells to other cells as well. The combination of hard and soft blocking as the blocking measure is the best solution.

The algorithm collects blocking statistics during specified hours and/or load conditions and, as soon as significant differences between the blocking of a cell pair is detected, handover event-triggering parameters between the cells are adjusted in order to balance the load with handover actions. With conservative setting of control method parameters, the method reacts slowly to differing blocking ratios. If the average blocking ratio in a cell pair is 2%, the number of samples required for detecting a blocking ratio difference with sufficient statistical accuracy is some hundreds in both cells.

When a blocking ratio difference bigger than a certain threshold is detected, the parameters for event-triggered measurement reporting are changed slightly; the change in the CIO values is a function of the difference between the blocking ratio of the cells in question.

The control method inherently assumes that the downlink is the limiting direction with respect to power resources. If the load is high in the uplink, the control actions of the method can cause terminals to run out of power. Thus, control actions should be made cautiously with the operator monitoring the cells with high uplink load.

Load-balancing Process

CIOs are a tool to move the cell border. Thus, adjusting the offsets can reduce traffic in congested cells and increase traffic in low loaded cells. Traffic in congested cell *A* is moved to a neighbouring less loaded cell *B* by decreasing *AdjsEcIoOffset(B, A)* and increasing *AdjsEcIoOffset(A, B)*. Decreasing *AdjsEcIoOffset(B, A)* inhibits soft handovers from cell *B* to cell *A* – i.e., cell *A* is more difficultly added to and more easily dropped from the active set when the user is close to cell *B*. Increasing *AdjsEcIoOffset(A, B)* makes users close to cell *A* favour cell *B* in the soft handover, which moves traffic from other neighbours of cell *A* to cell *B* as well. The algorithm may have the following internal parameters that the user can adjust:

- *Normal* – the level of blocking ratio that can still be considered normal. Its value must not be zero.
- *Step* – the adjustment step of the CIOs in decibels. Its range begins with 0.1 dB; it can be a function of the differences in cell-blocking ratios.
- *Max* – the maximum absolute value of the CIOs.
- *Threshold* – the threshold for indicating a significant difference in the blocking ratios. The parameter determines the sensitivity required to make an offset adjustment. Its conservative values lie between 2% to 3% in absolute terms. However, setting it closer to zero can increase adaptation speed without significant adverse effects.

Control is performed for the selected group of cells periodically. The actual value of the period is not crucial and it can be as often as is practically possible; a change is made when a significant difference in the blocking rates is detected. The control algorithm is described using the steps in the following list, but, before optimisation, the *planned* CIOs of a selected group of cells, C, are stored in the reference configuration management database:

1. Iterate Steps 2 to 10 for all cell pairs (c_1, c_2) in the cell group selection. *Note:* Do not repeat steps for a cell pair (c_2, c_1) if (c_1, c_2) is already processed.
2. Obtain the measured blocking ratios for cells (c_1, c_2) from the performance management database and KPI calculation engine.
3. Obtain the current CIO values from the configuration management database.
4. Check for situations when blocking ratios are not greater than *Normal*. If the blocking statistics show that cell blocking is at an allowed level and the difference in blocking ratios between cells is less than a pre-defined threshold then move the offsets towards their initial planned values in the reference database (provided that values differ) and continue with Step 10. Otherwise, continue with the next step.
5. Retrieve the KPI for the average blocking ratio.
6. Compute the deviation, D, using the blocking ratios (B_1, B_2 for cells 1 and 2, respectively; N_1, N_2 being the number of samples, N_{limit} being, e.g., 5). This is obtained using:

$$D = \frac{B_1 - B_2}{\sqrt{\bar{B}(1 - \bar{B})\left(\dfrac{1}{N_1} + \dfrac{1}{N_2}\right)}} \tag{9.28}$$

where $\bar{B}$ is the average blocking. However, D is zero if:

$$\min[\bar{B}, (1 - \bar{B})] \cdot N_1 < N_{limit} \quad \text{or} \quad \min[\bar{B}, (1 - \bar{B})] \cdot N_2 < N_{limit} \tag{9.29}$$

7. Compute the change in the CIOs:

$$\Delta Offset = \begin{cases} Step & \text{if } D > Threshold \\ -Step & \text{if } D < -Threshold \\ 0 & \text{otherwise} \end{cases} \tag{9.30}$$

in which *Step* is the CIO adaptation step.
8. Compute the new CIOs with range checking.
9. If both computed new CIOs of cells c_1, c_2 differ from their current setting, change the CIOs and provision the change to network.
10. If there are unprocessed cell pairs, take the next one and continue with Step 2.
11. Reset the congestion measurement counters and KPIs of the cells whose CIOs were changed.

It is the operators' choice what level of congestion to allow and tolerate even after the congestion relief algorithm. The proposed method is not an answer to all blocking-related problems, but it can solve certain traffic hotspot-type situations.

Another identified application area for CIO optimisation is for areas along highways. The soft handovers of a mobile user can be controlled by prioritising the adjacency definitions using offset values. Cells intended to cover the highway are higher prioritised in handover evaluation and thus unnecessary ASUs (cell addition and immediate deletion again), involving cells aside the highway not intended for highway usage but which can be locally received, can be avoided. This would reduce signalling load and would be especially beneficial for fast-moving mobiles.

9.4 Summary

In this chapter advanced analysis methods for cellular networks were introduced. NMS level intelligence is needed in order to cope with the challenges arising from the increased amount of traffic and new mobile services. Further, some WCDMA-specific automation examples were presented. The presented methods bring first of all operational efficiency, owing to the high level of process, analysis and decision logic automation. Second, with automation network performance is improved and network resources are used more efficiently.

References

[1] ETSI, TS 100.908, v8.10.0, GSM Technical Specification 05.02: Digital Cellular Telecommunications System (Phase 2+); Multiplexing and Multiple Access on the Radio Path, 2001.

[2] Vehviläinen, P. (2004). Data mining for managing intrinsic quality of service in digital mobile telecommunications networks. Thesis (Doc.Tech.), Tampere University of Technology.

[3] Fayyad, U., Piatetsky-Shapiro, G. and Smyth, P., From data mining to knowledge discovery: An overview. In: U. Fayyad, G. Piatetsky-Shapiro, P. Smyth and R. Uthurusamy (eds), *Advances in Knowledge Discovery and Data Mining*, pp. 1–34, MIT Press, 1996.

[4] Laiho, J., Raivio, K., Lehtimäki, P., Hätönen, K. and Simula, O., *Advanced Analysis Methods for 3G Cellular Networks*, Report A65, Publications in Computer and Information Science, Helsinki University of Technology, 2002. Modified version resubmitted to *IEEE Transactions on Wireless Communications* end 4/2002.

[5] Han, J. and Kamber, M., *Data Mining: Concepts and Techniques*, Morgan Kaufmann, 2001.

[6] Höglund, A.J., Hatonen, K. and Sorvari, A.S., A computer host-based user anomaly detection system using the self-organising map. *Proc. IEEE-INNS-ENNS International Joint Conf. on Neural Networks (IJCNN 2000)*, Vol. 5, pp. 411–416, 2000.

[7] Pyle, D., *Data Preparation for Data Mining*, Morgan Kaufmann, 1999.

[8] Breiman, L., Friedman, J., Olshen, R. and Stone, C., *Classification and Regression Trees*, Chapman & Hall/CRC Press, 1984.

[9] Vesanto, J. and Alhoniemi, E., Clustering of the self-organising map, *IEEE Transactions on Neural Networks*, **11**(3), pp. 586–600, 2000.

[10] Laiho, J., Radio network planning and optimisation for WCDMA, Thesis (Doc. Tech.), Radio Laboratory, Helsinki University of Technology, July 2002.

[11] Laiho, J., Kylväjä, M. and Höglund, A., Utilisation of advanced analysis methods in 3G networks, *IEEE VTS Proc. of Vehicular Technology Conf., Spring 2002, Birmingham, Alabama*, pp. 726–730.

[12] Kohonen, T., *Self-organising Maps*, Springer-Verlag, 1995.

[13] Kohonen, T., Analysis of processes and large data sets by a self-organising method, *Proc. of 2nd International Conf. on Intelligent Processing and Manufacturing of Materials, 1999*, vol. 1, pp. 27–36.

[14] Kohonen, T., Oja, E., Simula, O., Visa, A. and Kangas, J., Engineering applications of the self-organising map, *Proceedings of the IEEE*, **84**(10), October 1996, pp. 1358–1384.

[15] Kohonen, T., New developments and applications of self-organising maps, *Proc. of International Workshop on Neural Networks for Identification, Control, Robotics, and Signal/ Image Processing, 1996*, pp. 164–172.

[16] Ahola, J. Alhoniemi, E. and Simula, O., Monitoring industrial processes using the self-organising map, *Proc. of IEEE Midnight Sun Workshop on Soft Computing Methods in Industrial Applications, 1999*, pp. 22–27.

[17] Hämäläinen, S., Holma, H. and Sipilä, K., Advanced WCDMA radio network simulator, *Proc. of PIMRC 1999, Aalborg, Denmark, October 1997*, pp. 509–604.

[18] Raivio, K., Simula, O. and Laiho J., Analysis of mobile radio access network using the self-organising map, *Proc. of IEEE International Conf. on Data Mining, San Jose, California, November/December 2001*, pp. 457–464.

[19] Vesanto, J., Himberg, J., Alhoniemi, E. and Parhankangas, J., *SOM Toolbox for Matlab 5*, Report A57, Helsinki University of Technology, 2000.

[20] Vehviläinen, P, Hätönen, K. and Kumpulainen, P., Data mining in quality analysis of digital mobile telecommunications network, *Proc. of XVII IMEKO World Congress, Dubrovnik, Croatia, June 22–27, 2003*, pp. 684–688.

[21] Suutarinen, J., Performance measurements of GSM base station system. Thesis (Lic.Tech.), Tampere University of Technology, 1994.

[22] Hätönen, K., Kumpulainen, P. and Vehviläinen, P., Pre- and post-processing for mobile network performance data, In: R. Tuokko (ed.), *Automaatio03, Seminaaripäivät [Automation Makes it Work]: Automaation sovellukset ja käyttökokemukset, September 9–11, 2003*, pp. 311–316, Finnish Society of Automation.

[23] Hätönen, K., Laine, S. and Similä, T., Using the LogSig-function to integrate expert knowledge to Self-Organising Map (SOM) based analysis, *IEEE International Workshop on Soft Computing in Industrial Applications, Birmingham University, New York, June 23–25, 2003*, pp. 145–150.

[24] Johnson, R.A. and Wichern, D.W., *Applied Multivariate Statistical Analysis* (4th edn), Prentice Hall, 1998.

[25] Everitt, B.S., *Cluster Analysis*, Edward Arnold, 1993.

[26] Davies, D.L. and Bouldin, D.W., A cluster separation measure, *IEEE Transactions on Pattern Analysis and Machine Intelligence*, **1**(2), pp. 224–227, April 1979.

[27] Rousseeuw, P.J., Silhouettes: A graphical aid to the interpretation and validation of cluster analysis, *Journal of Computational and Applied Mathematics*, **20**, November 1987, 53–65.

[28] McGill, R., Tukey, J.W. and Larsen, W.A., Variations of boxplots, *The American Statistician*, **32**, pp. 12–16, 1978.

[29] Hämäläinen, S., Slavina, P., Hartmann, M., Lappetelainen, A., Holma, H. and Salonaho, O., A novel interface between link and system level simulations, *Proc. ACTS Summit 1997, Aalborg, Denmark, October 1997*, pp. 599–604.

[30] Höglund, A. and Valkealahti, K. Automated optimisation of key WCDMA parameters, *Journal of Wireless Communications and Mobile Computing*, in press.

[31] Olofsson, H., Magnusson, S. and Almgren, M., A concept for dynamic neighbor cell list planning in a cellular system, *Proc. 7th IEEE International Symposium on Personal, Indoor and Mobile Radio Communications (PIMRC'96)*, pp. 138–142.

[32] Love, R.T., Beshir, K.A,., Schaeffer, D. and Nikides, R.S, A pilot optimisation technique for CDMA cellular systems. *IEEE VTS 50th Vehicular Technology Conf., VTC 1999, Fall 1999*, Vol. 4, pp. 2238–2242.

[33] 3GPP, TS 25.133, v3.50, Requirements for Support of Radio Resource Management, 2001.

[34] Valkealahti, K., Höglund A. and Novosad, T., UMTS radio network multi-parameter control, *Proc. IEEE PIMRC 2003*, pp. 616–621.

[35] Valkealahti, K., Höglund, A., Parkkinen, J. and Flanagan, A., WCDMA common pilot power control with cost function minimisation, *Proc. IEEE 56th VTC Fall 2002*, Vol. 4, pp. 2244–2247.

[36] Valkealahti, K., Höglund, A., Parkkinen, J. and Hämäläinen, A., WCDMA common pilot power control for load and coverage balancing, *Proc. IEEE PIMRC 2002*, Vol. 3, pp. 1412–1416.

[37] Wacker, A., Sipilä, K. and Kuurne, A., Automated and remote optimisation of antenna subsystem based on radio network performance, *Proc. IEEE 5th WPMC 2002, October 2002, Honolulu, Hawaii*, pp. 752–756.

[38] Yang, J. and Lin, J., Optimisation of power management in a CDMA radio network, *Proc. IEEE 52nd VTC Fall 2000*, Vol. 6, pp. 2642–2647.

[39] Zhu, H., Buot, T., Nagaike, R. and Schreuder, H., Load balancing in WCDMA systems by adjusting pilot power, *Proc. 5th International Symposium on Wireless Personal Multimedia Communications 2002*, Vol. 3, pp. 936–940.

[40] Lee, W.C.Y. and Lee, D.J.Y, Optimise CDMA system capacity with location, *Proc. 54th IEEE VTC, Atlantic City, NJ, October 2001*, Vol. 2, pp. 1015–1019.

[41] Flanagan, A. and Novosad, T., Automatic selection of *AdditionWindow* in a WCDMA radio network based on cost function minimisation, *Proc. IEEE ISSSTA, September 2002, Prague*, Vol. 3, pp. 672–676.

[42] Höglund, A., Pöllönen, J., Valkealahti, K. and Laiho, J., Quality-based auto-tuning of cell uplink load level targets in WCDMA, *Proc. IEEE VTC, Spring 2003*, Vol. 4, pp. 2847–2851.

[43] Höglund, A. and Valkealahti, K., Quality-based tuning of cell downlink load target and link power maxima in WCDMA, *Proc. IEEE 56th VTC, Fall 2002*, Vol. 4, pp. 2248–2252.

10

Other 3G Radio Access Technologies

Jussi Reunanen, Simon Browne, Pauliina Erätuuli,
Ann-Louise Johansson, Martin Kristensson, Jaana Laiho,
Mats Larsson, Tomáš Novosad and Jussi Sipola

This chapter deals with two technologies that are different from UMTS Terrestrial Radio Access Frequency Division Duplex (UTRA FDD). Section 10.1 discusses the General Packet Radio Service (GPRS) used in Global System for Mobile communications (GSM) technology. GPRS brings variable-rate packet data traffic into the air interface of what was originally a Circuit Switched (CS) and single data rate service-oriented technology. Thus, this technique is now paving the way in packet data communications towards Third Generation (3G) and Universal Mobile Telecommunications System (UMTS). From the planning point of view, GPRS and its modifications are affected by the variability in user data rate in a somewhat similar way to Wideband Code Division Multiple Access (WCDMA).

The second new technology, covered in Section 10.2, is the Time Division Duplex (TDD) mode of WCDMA (UTRA TDD), a potentially interesting technology for high data rate indoor users. Unlike FDD, TDD does not need a paired spectrum, but it presents several potential coexistence problems with FDD due to mutual interference, as discussed later in this chapter.

10.1   GSM Packet Data Services

This part of the chapter deals with issues relating to the planning of GPRS and Enhanced GPRS (EGPRS) services on the GSM network. Data rate variability has introduced another variable into network planning. This also affects air interface and transmission issues.

Radio Network Planning and Optimisation for UMTS Second Edition
Edited by J. Laiho, A. Wacker and T. Novosad © 2006 John Wiley & Sons, Ltd

Table 10.1 Standard convergence.

Specification status (standardisation group)	GSM (ETSI SMG2)	TIA/EIA 136 (TIA TR 45.3)
End 1997	200 kHz, GMSK circuit-switched infrastructure max. 14.4 kbps	30 kHz, DQPSK circuit-switched infrastructure max. 9.6 kbps
End 1998 + GPRS	200 kHz, GMSK GPRS infrastructure max. 171.2 kbps	30 kHz, 8-PSK GPRS infrastructure max. 52.2 kbps
End 1999 + EDGE	200-kHz, 8-PSK GPRS infrastructure max. 473.6 kbps	

10.1.1 Introduction

Packet data services were introduced in GSM with the GPRS. By May 2001 the majority of the existing GSM networks supported or were about to support the GPRS service. GPRS will enable packet data rates of up to 20 kbps per timeslot over the existing GSM network; four possible coding schemes are implemented.

Due to the ever-increasing pressure to boost throughput and data service speed, an enhancement to GPRS – namely, EGPRS, which is part of EDGE (Enhanced Data for Global Evolution) – was developed and standardised during 1999. EGPRS will enhance data throughput up to around 60 kbps per timeslot.

EDGE is a common convergence of two standards, from the US Telecommunications Industry Association (TIA) and the European Telecommunications Standards Institute (ETSI) (lately ETSI specification work was transferred to 3GPP) Standards (Table 10.1).

EDGE services will be carried over GSM/GPRS networks, utilising their existing control channels and traffic channels. Interim Standard 136 (IS-136 or TDMA in the US) is based on a different approach, providing only data services in a relatively small freed dedicated spectrum.

In this chapter some general ideas about how to plan and optimise GPRS and EDGE (over the GSM/GPRS) networks are discussed. It should be noted that GPRS refers to features (or issues) applicable only to GPRS, and EDGE refers to features (or issues) applicable only to EDGE. The EDGE issues discussed in this chapter are only considered for the EDGE system over GSM system.

10.1.2 Modulation and Coding Schemes

GPRS uses Gaussian Minimum Shift Keying (GMSK) modulation. Four coding schemes are defined: CS-1 to CS-4. CS-1 offers the highest level of error protection, while CS-4 offers no error protection of user content, as shown in Table 10.2 [6]. In

Table 10.2 GPRS coding schemes.

Scheme	Code rate	Radio block[a]	Coded bits	Punctured bits	Data rate [kbps]
CS-1	1/2	181	456	0	9.05
CS-2	≈2/3	268	588	132	13.4
CS-3	≈3/4	312	676	220	15.6
CS-4	1	428	456	—	21.4

[a] Excludes uplink state flag and binary coded signal bits.

Table 10.3 EDGE modulation and coding schemes.

Scheme	Code rate	Header code rate	Modulation	RLC blocks per radio block [20 ms]	Raw data within one radio block	Raw data rate [kbps]
MCS-9	1.0	0.36	8-PSK	2	2 × 592	59.2
MCS-8	0.92	0.36	8-PSK	2	2 × 544	54.4
MCS-7	0.76	0.36	8-PSK	2	2 × 448	44.8
MCS-6	0.49	1/3	8-PSK	1	592	29.6
MCS-5	0.37	1/3	8-PSK	1	448	22.4
MCS-4	1.0	0.53	GMSK	1	352	17.6
MCS-3	0.80	0.53	GMSK	1	296	14.8
MCS-2	0.66	0.53	GMSK	1	224	11.2
MCS-1	0.53	0.53	GMSK	1	176	8.8

consequence, the user data rate increases with higher coding schemes, at the expense of an increasing signal-to-interference level requirement.

For EDGE, both GMSK and 8-Phase Shift Keying (8-PSK) are defined as modulation schemes, and for both of these there are several different code rates: see Table 10.3 [6]. EDGE offers user bit rates between 8.8 kbps and 59.2 kbps per radio timeslot. The use of 8-PSK allows for a trebling of the air interface bit rate when compared with GMSK, albeit with increased signal-to-interference ratio requirements [7].

10.1.2.1 Protocol Stack

The GPRS or EDGE air interface consists of a layered protocol structure that provides control procedures, such as error correction and retransmission, to user data. The protocols that should be considered when analysing the air interface performance are shown in Figure 10.1, and are used between the Mobile Station (MS) and the Base Station Controller/Serving GPRS Support Node (BSC/SGSN).

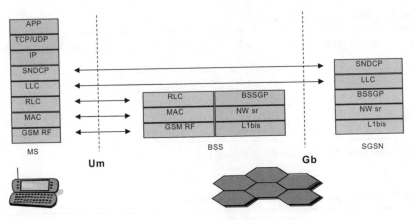

Figure 10.1   SGSN–MS protocol stack.

A brief explanation of the function of the protocols is given below:

- *Sub-Network Dependent Convergence Protocol (SNDCP)*. Maps the network-level Packet Data Units (N-PDUs) to the underlying Logical Link Control (LLC) layer. Also provides optional compression functionality, of both the Transmission Control Protocol/Internet Protocol (TCP/IP) header and the data content.
- *Logical Link Control (LLC) layer*. Provides a reliable ciphered link between the SGSN and the MS. This protocol is independent of underlying radio interface protocols. The layer can be operated in both acknowledged and unacknowledged modes, and this is one of the parameters defined by the *reliability class* field present in a Packet Data Protocol (PDP) context Quality of Service (QoS) profile [8].
- *Radio Link Control (RLC) layer*. A key layer in the air interface, this provides reliable transmission of data using optional Automatic Repeat reQuest (ARQ) \bar functionality. In addition, segmentation/desegmentation of data from/to the LLC layer is performed. The RLC layer can be operated in both acknowledged and \bar unacknowledged modes, and, as with LLC, this is defined by the reliability class in the QoS.
- *Medium Access Control (MAC) layer*. This layer controls MS access to the common air interface and provides scheduling of the associated signalling.
- *GSM Radio Frequency (RF)*. The GSM Time Division Multiple Access (TDMA) physical interface. Bit interleaving, modulation/demodulation and power control are examples of functionality within this layer.

Figure 10.2 shows the typical data block format within the MS–GPRS protocol stack. Application level data may split into multiple TCP and IP blocks, depending on data volume and TCP packet size. Typically, each TCP/IP packet then maps one to one through the SNDCP and LLC layers before being split into a number of RLC blocks. These have a header added in the MAC layer before being sent on the air interface in four bursts, which are sent over four consecutive TDMA frames, with an average duration of 20 ms.

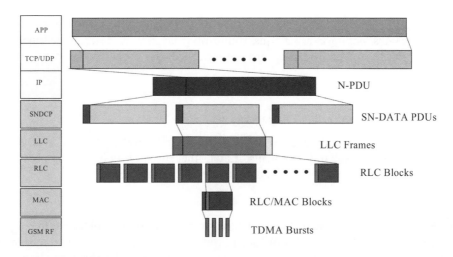

Figure 10.2   Data blocks in the MS–GPRS protocol stack.

Performance over the GPRS network is very dependent on the interaction between the different layers in the protocol stack, and these interactions must be understood if the network is to be optimised.

In particular, the TCP layer is found to interact strongly with RLC [3]: reliability at the RLC level can have a serious impact on TCP throughput and this implies the use of RLC acknowledged mode where TCP is used. In addition, high RLC block error rates can lead to significant delay at LLC, which, in turn, can cause TCP congestion avoidance algorithms to trigger. The use of User Datagram Protocol (UDP) rather than TCP as a transport protocol allows for a more transparent mode of service, and, if limited packet loss can be tolerated, permits the RLC and LLC layers to be operated in unacknowledged mode.

10.1.3   EDGE Radio Link Performance

In this section, EDGE radio link performance is analysed using link-level simulations [1]. The input to the simulation is a characterisation of the radio channel, including signal-to-noise or Carrier-to-Interference (C/I) ratio, delay profile and fading characteristics. The most important quantity to be analysed is throughput, but also delay is discussed.

10.1.3.1   Simulation Assumptions

The simulations in this section are made assuming one MS and one timeslot. The channel is a typical urban channel with the MS moving at 3 km/h, in the 900 MHz band. Ideal frequency hopping is used. These simulations study the performance in an interference limited network, and C/I is the ratio of the corresponding mean powers. One continuous interferer is modelled using the same modulation and channel statistics as the wanted signal. Simulation length is 40000 bursts.

The RLC protocol is modelled with the following parameters: transmitter window 128 RLC blocks, polling every 320 ms, roundtrip delay 220 ms. Temporary Block Flow (TBF) establishments and releases are not modelled, but a continuous data stream is assumed.

10.1.3.2 Performance without Enhancements

The purpose of this analysis is to show the basic link level performance of EDGE, without any link level improvements. Later, the effect of two such improvements (incremental redundancy and link adaptation) will be analysed.

Block Error Rate (BLER)

The RLC layer produces RLC blocks, which are mapped to radio blocks at the physical layer. Each radio block consists of four normal bursts, and the average duration of a radio block is 20 ms. A radio block contains one or two RLC blocks depending on the coding scheme (see Table 10.3).

BLER is the percentage of erroneously received RLC blocks. The ARQ protocol causes a retransmission for each erroneously received block. Therefore, BLER is also the ratio of the number of retransmissions to the number of all transmissions.

Figure 10.3(a) shows the BLER of the 8-PSK coding schemes. As can be seen, a higher Modulation and Coding Scheme (MCS) always has a higher BLER.

Throughput

Throughput is the amount of information transferred per time unit. In this link performance analysis, throughput is measured per timeslot. In practice, each timeslot

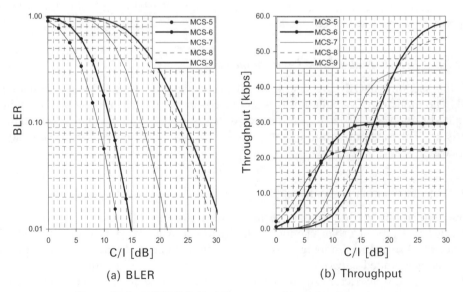

(a) BLER   (b) Throughput

Figure 10.3 EGPRS link performance without improvements.

can have multiple users and/or a user may use multiple timeslots. Therefore, the throughput experienced by the user is not necessarily the same as the throughput per timeslot.

If the BLER is known, throughput can be calculated as $tp = tp_{peak} \cdot (1 - \text{BLER})$, where tp_{peak} is the maximum throughput of the MCS. The throughput curves in Figure 10.3(b) show how the optimal MCS (that giving the best throughput) depends on the C/I value.

10.1.3.3 Incremental Redundancy (IR)

IR combines channel coding and ARQ protocol. It is based on soft combining different transmissions of the same block at the receiver, thereby increasing the probability of correct reception of retransmissions. Original transmissions are not affected and therefore IR has no effect unless ARQ is used.

The EDGE standard supports IR operation by specifying separately coded headers to identify blocks before channel decoding, different puncturing schemes for each coding scheme and a requirement for IR combining capability at the MS receiver.

Operation

Figure 10.4 shows the operation of the transmitter and the receiver when IR is used. At the transmitter, the payload is encoded using convolution coding at the rate 1/3. The codeword is punctured by removing two of the three bits. The remaining bits are transmitted with an effective code rate of 1. Retransmissions are performed in the same way, except that different sets of bits are punctured. This means changing the puncturing scheme. For each transmission, the effective code rate is 1, since the number of bits transmitted equals the number of bits in the payload. In EGPRS, coding schemes MCS-4 and MCS-9 have this property. For other MCSs, fewer bits are punctured but the same principle still applies.

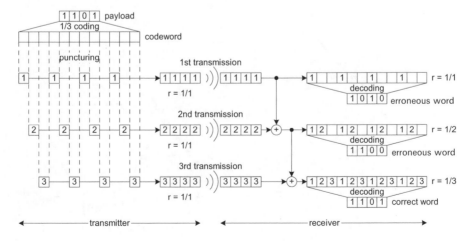

Figure 10.4 Operation of incremental redundancy.

At the receiver, the first transmission is received normally by performing de-puncturing and decoding. In this example, the decoding fails, so a new transmission is performed. However, the received bits are now combined with those of the first transmission, yielding an effective code rate of 1/2. This combination is then fed to the channel decoder. The increased redundancy increases the probability of correct reception. If, however, the second transmission still fails, there will be a third transmission, which will be combined with the first two transmissions.

Benefits of Incremental Redundancy
The gain of IR is significant when observing the throughput of an individual coding scheme. Figure 10.5(a) shows the throughput with and without IR. The gain is very high when throughput is low relative to the peak throughput. For example, at 20 kbps the gain is 9 dB, but at 50 kbps it is less than 1 dB. The reason is that IR only improves retransmissions, and at a lower throughput there are more of them.

The gain at low throughputs means that a high MCS can achieve the same throughput as a low coding scheme even at a low C/I. This is seen in Figure 10.5(b), which illustrates all the 8-PSK coding schemes of EDGE. This also makes MCS selection much easier, which is also a big benefit of IR.

Effect of Finite Receiver Memory
The receiver needs to store the soft values of the incorrectly received blocks until the blocks are correctly received. Because of the nature of the selective ARQ and depending on the polling frequency and roundtrip delay, tens of blocks can be incomplete at a given time instant. Therefore, the memory consumption of the receiver increases along with the number of required retransmissions.

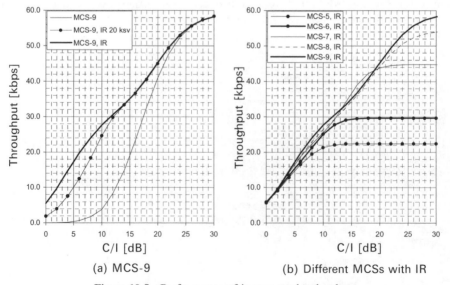

(a) MCS-9   (b) Different MCSs with IR

Figure 10.5 Performance of incremental redundancy.

Figure 10.5(a) also shows the performance with finite receiver memory. In the curve labelled '20 ksv', the receiver is assumed to be able to store 20480 soft values (soft decisions at the output of the equaliser). For low C/I, the high number of retransmissions means that not all blocks can be stored in memory, therefore the throughput is decreased. For high C/I, ideal IR performance is achieved. Finite receiver memory is one of the reasons for using a lower MCS when C/I is low.

10.1.3.4 Link Adaptation (LA)

The task of LA is to select the best performing coding scheme for each channel condition. The need for link adaptation is evident if incremental redundancy is not used (see Figure 10.3). If IR is used, a high-coding scheme such as MCS-9 could always be used, except that the high number of retransmissions causes a long delay and high-memory consumption.

Ideal Link Adaptation

Ideal LA is a conceptual algorithm that selects the coding scheme with best throughput for each C/I value. Link simulations do not model shadowing or path loss, but their effect is assumed to be included in the C/I value. Therefore, ideal LA will adapt to shadowing and path loss. Fast fading is not included in the C/I value but it is really simulated, so ideal LA does not adapt to fast fading. In terms of throughput plots, the throughput of ideal LA is the envelope of all the coding schemes.

Sometimes it is even possible to achieve a better performance than with ideal LA, if fast fading can be taken into account, or if IR is utilised efficiently by using a higher coding scheme for retransmissions than for original transmissions. The performance of a real LA algorithm can be evaluated by comparing it with ideal LA.

Bit Error Probability (BEP) Measurements

The EDGE radio standard defines BEP measurement for the MS. BEP is defined as the probability of a bit error during a burst in a given channel instance. BEP measurement includes both the mean and variance of burstwise BEP. Variance is calculated during four bursts of a radio block, and therefore reflects any fast changes in the channel – e.g., due to frequency hopping or high mobile speeds.

Simulation Results with Incremental Redundancy and Link Adaptation

Figure 10.6(a) shows the throughput of EDGE with both link-level enhancements – IR and LA. A realistic (20480 soft values) receiver memory is assumed. It can be seen that LA is able to select the optimal MCS is most cases.

Delay Analysis

Delay of an LLC frame is measured from the radio transmission of the first bits of the frame to the time when the block is completely and correctly received. In this analysis, LLC frames are 8192 bits. The delay experienced by the user will additionally include any queuing in various buffers as well as delays caused by the core network or higher layers, but these are not taken into account here.

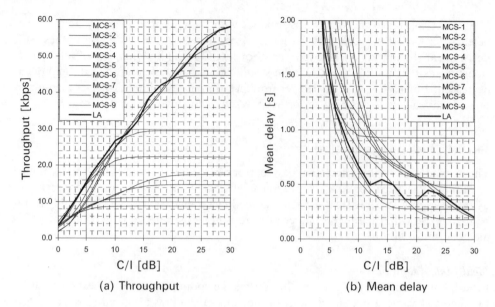

(a) Throughput   (b) Mean delay

Figure 10.6   Performance of bit error probability-based link adaptation in typical urban channel 3 with intelligent frequency hopping, incremental redundancy memory size 20 ksv.

Figure 10.6(b) shows the delays of individual coding schemes and LA. The main criterion for LA is throughput and not delay, but the delay achieved with LA is still rather low, usually less than 0.5 s.

10.1.4   GPRS Radio Link Performance

In this section, the performance of the GPRS radio link is analysed using link-level simulations. Input to the simulation is a characterisation of the radio channel, including signal-to-noise or C/I ratio, delay profile and fading characteristics. The most important quantity to be analysed is throughput.

The actual performance of each of the four GPRS coding schemes is dependent upon the channel C/I. The results shown in Figure 10.7 are based on a frequency-hopping environment and the typical urban 3 km/h channel type (TU3). In a non-hopping case the crossing points of the coding scheme are different.

Ideal LA ensures that the coding scheme changes from one coding scheme to another as the C/I increases or decreases to maximise the user data rate. In practice the BLER of a TBF is obtained by recording the number of RLC blocks that require retransmission. This is used as the basis to make the LA decision to change from one coding scheme to another.

The maximum throughput per timeslot depends on the number of coding schemes implemented. If only CS-1 and CS-2 have been implemented, it is clear that CS-2 provides a better throughput in most cases. The C/I has to be lower than 6.5 dB before CS-1 will provide the highest data rate. For a non-hopping cell this crossing

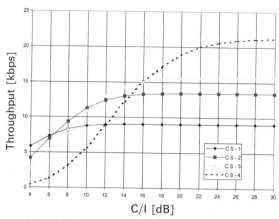

Figure 10.7 GPRS data throughput vs. carrier-to-interference ratio for CS-1 to CS-4 (TU3, FH).

point is even lower, hence CS-2 will provide a better throughput in all cases when a connection can be maintained.

10.1.5 Coverage

In GPRS and especially in EDGE the coverage area shrinks when the throughput per timeslot increases as the required E_s/N_0 (E_b/N_0 for GPRS) increases when higher coding schemes are required. Coverage can be increased by using link enhancement features. Cell ranges without and with enhancements for different coding schemes (CS-1 to CS-4 for GPRS; MCS-1 to MCS-9 for EDGE) are discussed in this section.

The C/I – i.e., the quality of coverage – gives another dimension to the achievable throughput (this was discussed in Section 10.1.3).

10.1.5.1 Input Parameters

The coverage discussion presented in this section is based on link simulations (Section 10.1.3.1) and the simple coverage assumptions listed below:
- outdoor urban environment considered;
- frequency band used: 900 MHz;
- slow fading margin: 7.36 dB;
- Base Transceiver Station (BTS) output power: 43 dBm (41 dBm with 8-PSK);
- MS noise figure: 10 dB;
- Antenna gain (MS, BTS): 2 dBi, 18 dBi (+diversity gain: 3 dB for BTS receiving end);
- propagation model: COST231 Okumura–Hata with 4 dB area type correction factor;
- BTS/MS antenna height: 25 m/1.5 m;
- Required E_b/N_0 (GMSK) and E_s/N_0 (8-PSK) are simulated by using a simulator with a TU3 non-frequency hopping multi-path profile.

Using the above parameters, the normal (GMSK) speech cell range is calculated as 3.9 km.

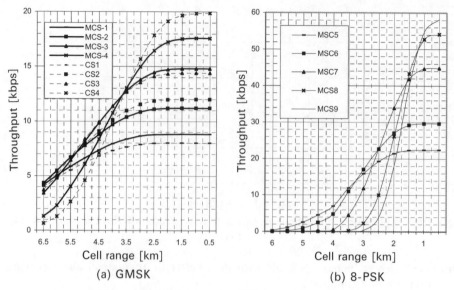

Figure 10.8 Cell range for different coding schemes.

10.1.5.2 Cell Range without Enhancements

In Figure 10.8 cell range vs. throughput per timeslot is presented. Graph (a) is for GMSK modulation and graph (b) for 8-PSK modulation. Both graphs are without any improvement effects.

As can be seen from Figure 10.8, cell range decreases rapidly with increasing data rate, especially when 8-PSK modulation is used. If the network was designed for an outdoor speech service with a 7.36 dB slow fading margin, the achieved maximum throughput per timeslot from the network (only the coverage dimension included) is about 11 kbps for GMSK and about 8 kbps for 8-PSK. When comparing the GMSK and 8-PSK cell ranges for certain throughputs per timeslot, it should be remembered that there is 2 dB backoff in transmitted power when 8-PSK is used in order to secure transmitter linearity. When comparing CS-1 to CS-4 cell ranges with MCS-1 to MCS-4 ranges, refer to Tables 10.2 and 10.3 to see the difference in coding rates.

10.1.5.3 Incremental Redundancy

IR increases the achieved throughput as explained in Section 10.1.3.3. In Figure 10.9 the effect of IR on the throughput vs. cell range curves is shown. Input parameters are the same as mentioned in Section 10.1.5.1. It should be noted that IR is only applicable for EDGE.

Figure 10.9 shows that IR enhances throughput per timeslot for ranges above about 1 km. Comparing the range with the speech cell range of 3.9 km, note that the achievable throughput per timeslot is about 12 kbps with GMSK and about 11 kbps with 8-PSK. The increase after introducing IR was only 1 kbps per timeslot for GMSK modulation but 3 kbps per timeslot for 8-PSK.

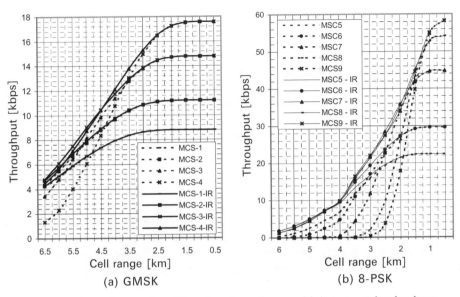

Figure 10.9 Cell range for different coding schemes with incremental redundancy.

10.1.5.4 Downlink Diversity

One way to enhance coverage is to introduce downlink diversity methods – e.g., by using two transmitters with a certain symbol delay between the two transmissions. Downlink diversity will add 3 dB to received power, and in addition the symbol delay will add or subtract an extra −1 dB to +1 dB (depending on the environment) to the link budget. Figure 10.10 represents throughput per cell range with downlink

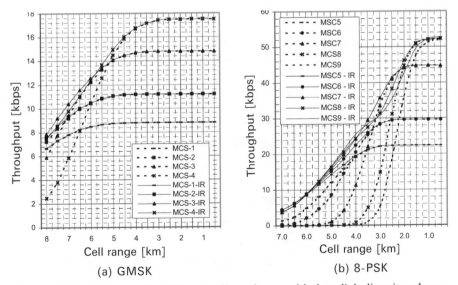

Figure 10.10 Cell range for different coding schemes with downlink diversity scheme.

diversity. The E_b/N_0 and E_s/N_0 values are taken from simulations provided by the simulator presented in Section 10.1.3.1.

As can be seen from Figure 10.10, downlink diversity increases throughput per timeslot for large ranges. For the 3.9 km cell range the achievable throughput per timeslot for 8-PSK is about 26 kbps with downlink diversity compared with 11 kbps with only IR. For GMSK the corresponding throughputs per timeslot are 16.5 kbps and 13.5 kbps. By introducing IR and downlink diversity, throughput can be trebled for the speech cell range from the original without any enhancements.

10.1.6 Capacity Planning

Once link-level results are available, it is possible to estimate the air interface performance and perform network dimensioning based on a given network topology. This would typically involve the use of a planning tool, and this section considers some of the issues that would need to be considered when developing such a tool, or the associated processes. Both traffic and signalling capacity are considered.

10.1.6.1 Traffic

GPRS and EDGE traffic and GSM CS traffic use a common air interface resource. The challenge in dimensioning a network for capacity is therefore in dividing this capacity while offering a satisfactory grade of service to both user types.

Figure 10.11 illustrates one way that traffic resource within a cell (here TRX 2) can be divided into CS, GPRS or EDGE territories. TRX 1 carries all the signalling channels.

In [6] it is stated that that the capacity available to GPRS and EDGE may be either fixed or follow the *capacity on demand* principle. The two cases are now considered.

Fixed GPRS Capacity
It is possible to assign *fixed* (or *dedicated*) *GPRS capacity*, where one or more timeslots are allocated on a permanent basis to GPRS. These timeslots are always configured for GPRS and cannot be used by CS traffic. This ensures that GPRS capacity is always available in a cell. The drawback with this approach is that, for a given cell

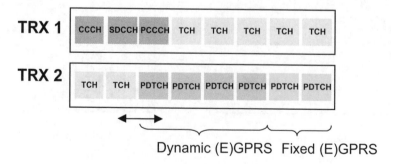

Figure 10.11 Illustration of cell resource.

configuration, blocking levels for CS traffic will increase since the number of available channels is reduced.

The decision on whether to assign fixed GPRS or EDGE territory is a tradeoff between providing a minimum level of GPRS and EDGE service and increasing the blocking for CS services. This decision needs to take into account network operator priorities, network performance and predicted GPRS and EDGE usage levels.

Dynamic GPRS Capacity

Dynamic capacity, or capacity on demand, can be used for offered GPRS traffic when the CS load level permits.

The actual implementation of dynamic capacity varies between infrastructure vendors. It is possible to have dynamic capacity that is, by default, allocated to GPRS and EDGE. The capacity for packet traffic is only removed when required for CS services. The converse is also possible: that is, capacity that is by default allocated for CS use but allowed for EDGE and GPRS use when the packet traffic load is such that extra capacity would be desirable and the CS load is low. Although at first thought there may appear to be little difference between these two cases, there can be implications for EDGE and GPRS, in terms of the air interface performance due to the algorithms used to trigger the capacity switch and to EDGE and GPRS capacity at the Packet Control Unit (PCU) in the BSC. Some implementations use both types of dynamic capacity simultaneously.

Depending on the CS load level and the traffic profiles for CS and packet traffic, dynamic capacity will typically form a substantial part of the EDGE or GPRS capable resource, particularly while EDGE traffic levels are at low/medium levels. Dedicating significant resource to EDGE or GPRS is typically costly to network operators, and therefore the use of dynamic capacity in the main is generally preferred initially. As traffic levels grow, however, the balance is likely to change with increasing use of dedicated resources. The trend of increasingly higher dedicated resources is clear in case of frequent multi-slot devices like EDGE capable laptop cards, etc.

Available EDGE or GPRS Resources within the Circuit Switched Design

Since EDGE and GPRS typically make use of the resource not carrying CS traffic, the first step in packet capacity planning is determination of the CS load. The design for CS capacity typically involves application of the Erlang B formula:

$$P = \frac{\dfrac{A^N}{N!}}{\sum_{i=0}^{N} \dfrac{A^i}{i!}} \tag{10.1}$$

where P is blocking probability [%]; A is traffic load [Erl]; and N is number of channels [timeslots]. The actual traffic carried, L, is then given by:

$$L = A \cdot (1 - P) \tag{10.2}$$

A system designed for CS traffic will usually allow for a basic (E)GPRS throughput: since the system has been designed with a sufficient margin to permit a low blocking

Table 10.4 Mean number of GPRS timeslots available.

TRX (TCH)/cell	CS TCH load @ 1% blocking	CS TCH load @ 2% blocking	Mean (E)GPRS available TCH (CS load @ 1% blocking)	Mean (E)GPRS available TCH (CS load @ 2% blocking)
1 (6)	1.9	2.2	4.1	3.8
2 (14)	7.3	8.0	6.7	6.0
3 (21)	12.7	13.8	8.3	7.2
4 (29)	19.3	20.6	9.7	8.4
5 (36)	25.3	26.8	10.7	9.2
6 (44)	32.2	34.0	11.8	10.0

level, there is typically spare instantaneous capacity that can be utilised for packet data transmission. As long as the packet traffic can be temporarily interrupted to accommodate peaks in CS traffic, then no degradation in CS services will result.

Let us illustrate this with an example: a cell offering a CS load of 14 Erlangs with 21 traffic channels will, on average, have 7.2 spare circuits. These could carry packet data on an on-demand basis, relinquishing the channels for CS traffic when required. In this way, the blocking probability of the CS facility is not degraded, even though the traffic channels are subject to higher overall utilisation.

Table 10.4 shows the mean number of timeslots (TCHs) available for EDGE and GPRS for different numbers of TRXs per cell and for CS blocking probabilities of 1% and 2%. Due to the trunking efficiency for CS traffic, the capacity available to EDGE and GPRS does not increase linearly with increasing cell configurations, as can be seen.

Having determined the capacity available to EDGE and GPRS, it is necessary to be able to dimension that capacity in order to provide a satisfactory user grade of service. Since it is generally accepted that packet data traffic is very different from CS traffic, being rather bursty in nature and typically represented by a succession of short transactions, the use of the Erlang B equation is not suitable. The actual design must take into account the traffic profile, in particular the applications being used.

Estimation of the actual average maximum supportable cell throughput for EDGE or GPRS is given by simply multiplying the average capacity available by the data rate per timeslot.

Taking the previous example of 7.1 timeslots available, and assuming the use of EDGE MCS-7 at a 10% BLER, the average available cell throughput is given by:

$$Throughput = EDGE\_timeslots \cdot Mean\_data\_rate / TSL$$

$$= 7.1 \cdot 44.8 \cdot (1 - 0.1) = 286.2 \text{ kbps} \tag{10.1}$$

The dimensioning process is a way of producing, from the available cell throughput, the throughput that can be achieved while preserving a satisfactory user grade of service. As such, network level simulations are required for the anticipated traffic profile and these should form the basis of the planning and optimisation processes and tools.

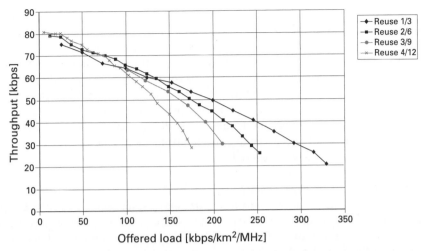

Figure 10.12   Net EGPRS throughput vs. offered load (IR + LA).

Figures 10.12 and 10.13 show example output from such simulations [2], giving net throughput and LLC frame delay vs. offered EGPRS load. Simulations are based on the link-level simulation results presented in Section 10.1.3. A network of 75 cells is the basis for the work, and the performance includes the effects of LA and IR (see Sections 10.1.3.3 and 10.1.3.4). Four reuse schemes were simulated: 1/3, 2/6, 3/9 and 4/12. Packet Switched (PS) downlink traffic was simulated, with MS multi-slot capability of three timeslots.

The key to the capacity-dimensioning task is in determination of the upper load threshold that will support a satisfactory user grade of service. This threshold maybe determined from network-level simulation results, such as the above, after design criteria are defined. This typically involves the required net throughput or maximum allowable mean LLC delay. For example, using the graphs above, and a re-use of 4/12,

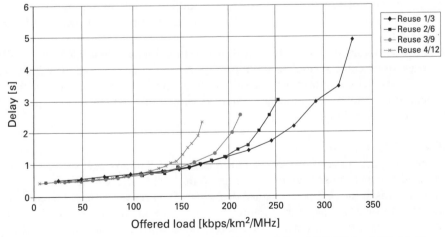

Figure 10.13   Logical link control frame delay vs. offered EDGE load (IR + LA).

the maximum offered load for the given network should not exceed $200\,\text{kbps}/\text{km}^2/\text{MHz}$ if a mean net throughput of 50 kbps is to be achieved. Conversion between the units used in the offered load results may be required – e.g., to cell offered load in kbps.

The actual design target will depend heavily on the traffic mix and the sensitivity of user applications to delay and/or compromised throughputs. In addition to the exponentially increasing packet delays at high load, some interaction with TCP (where used) may be anticipated and this will compound the issue [3]. The mechanism to monitor network performance is clearly very important when traffic levels are increasing, and the necessary mechanisms should be available to upgrade network capacity as and when required.

Increasing EDGE and GPRS Capacity

As shown, a typical network will, at most times, have some limited available capacity for EDGE and GPRS traffic. To increase the available capacity, further resources will be required in the form of traffic channels. These can be obtained by dedicating resources to EDGE and GPRS (at the expense of CS blocking), by upgrading the cell with additional TRX capacity, or by the use of network traffic management features. Another way is to apply the Adaptive Multi Rate (AMR) codec with substantial half-rate usage for speech traffic. The feature gains additional capacity for speech due to the fact that half-rate usage accommodates two speech users in one timeslot. AMR requires AMR capable mobiles and, thus, the effect depends upon AMR mobile penetration. Half-rate usage can only be applied in good radio conditions. If one considers a spectrum limited scenario AMR mode brings substantial spectral capacity gain [22].

Dedicating EDGE or GPRS Resources

If a cell is found to be performing within the CS-blocking criterion, however, packet performance is compromised, then dedication of resources to packet traffic could be considered. The impact of such changes on the CS service should be evaluated, as blocking levels can change substantially. Figure 10.14 shows the effect of dedicating different numbers of timeslots on blocking, for different cell configurations and a CS offered load resulting in a 2% blocking level (in the absence of a dedicated EDGE or GPRS resource).

It can be seen, for example, that dedicating two timeslots in a four TRX cell results in an increase in blocking from 2% to almost 4% for the same CS load. The increase in blocking in smaller configurations is clearly even greater. Therefore, it may be the case that dedicating resources should be combined with traffic management features that reduce end-user blocking.

Additional TRX Capacity

By monitoring the grade of service offered to both the CS and packet services, it can be determined whether greater capacity is required in a cell. If dedicating resources would result in unsatisfactory CS performance then extra TCH resources may be required, by means of cell capacity expansion. Assuming the CS load remains constant when a TRX is added, the increase in capacity for packet traffic will be significant. For example, based on Table 10.4, in a five TRX cell EDGE or GPRS available resources will

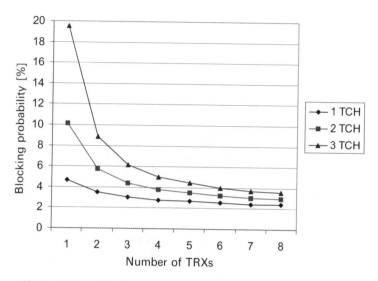

Figure 10.14   Effect of dedicating GPRS timeslots on circuit switched blocking.

increase from 9.2 timeslots to 17.2 timeslots with the expansion to a six TRX cell, the additional TRX offers an additional eight timeslots. EDGE capacity expansion is crucial in case of wide usage of multi-slot EDGE laptop cards. The end user cannot use these devices efficiently in case of capacity deficiencies.

Traffic Management

As an alternative, in some cases, to capacity expansion, traffic management features may be used in a network to give capacity gain. Typically, these offer a means to distribute CS traffic between cells where coverage overlaps exist. The net effect of this is that peaks in the CS load are smoothed in each cell and therefore an increase in average load can result.

An example of such functionality is the BSC-controlled Traffic Reason Handover (TRHO) feature. This is a load based algorithm that changes the power budget threshold for one or more outgoing adjacencies for a given source cell, depending on its load. If the load exceeds a given level, and the load in a neighbouring cell is below another given level, then the power budget threshold for this adjacency is reduced, with, in effect, the source cell shrinking in coverage area. Load level discrepancies are therefore balanced between the two cells; overall blocking levels can therefore be reduced and/or greater utilisation of the cell resource achieved (hence the term 'capacity gain').

The same principles can be used to attempt to improve packet resource availability, by the use of appropriate thresholds.

Figure 10.15 shows the TRHO concept with associated parameters. In this example, if the load level on the source cell exceeds 75% while that on the neighbour is less than 50%, the normal power budget handover margin is changed from +6 dB to −10 dB. Note that the received level from the target cell must exceed −80 dBm for the handover to be allowed.

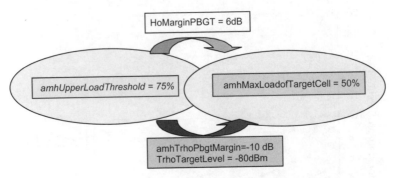

Figure 10.15 Illustration of BSC controlled traffic reason handover.

In this way, each cell attempts to retain the CS load within 75% of the cell Traffic Channel (TCH) resource, thus allowing 25% of resource to be available to packet traffic.

10.1.6.2 Signalling Capacity

The signalling associated with EDGE or GPRS data transfer can utilise either existing Common Control Channel (CCCH) resources or be carried on the dedicated packet signalling channel (PCCCH) if this channel is deployed.

The deployment of the PCCCH, typically in a separate timeslot, allows the signalling for the packet traffic to be removed from the CCCH. This is clearly of benefit to a network operator whose CCCH (usually the downlink PCH/AGCH) capacity is limited. A typical configuration for the logical signalling channels is shown in Figure 10.16.

The physical channel supporting the Packet Broadcast Control Channel (PBCCH)/ PCCCH also supports the use of the Packet Data Traffic Channel (PDTCH) – i.e., traffic carrying capability. The actual capacity offered by the PCCCH depends on the configuration of this channel: it is possible to set a usage split by means of a number of parameters and, in addition, where resource logical channels are allocated dynamically based on demand, the actual capacity will depend on the respective priorities. The mapping of downlink logical channels is done according to the following rules [20]:

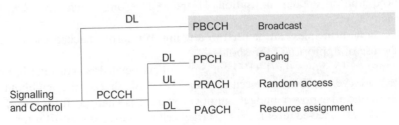

Figure 10.16 Packet signalling channel configuration.

- The PBCCH is mapped onto the 52-multiframe. The parameter *BS_PBCCH_BLKS* specifies the number of radio blocks allocated to the PBCCH.
- Radio blocks which are not available for paging, defined by the operator parameter *BS_PAG_BLKS_RES*, are allocated. These blocks can be used for the Packet Access Grant Channel (PAGCH), PDTCH and Packet Associate Control Channel (PACCH) and can therefore carry assignment messages, packet data and TBF-associated signalling.
- The remainder of the radio blocks in the 52-multiframe can be used for the Packet Paging Channel (PPCH), PAGCH, PDTCH and PACCH. These blocks can carry paging messages in addition to the messages specified in the previous point.

The 52-multiframe divides into 12 blocks (each of four frames, and labelled B0–B11) assigned to the above logical channels and four idle frames.

Figure 10.17 shows an example of downlink PBCCH/PCCCH mapping, where *BS_PBCCH_BLKS* = 3 and *BS_PAG_BLKS_RES* = 4.

PBCCH	PAGCH, PDTCH or PACCH	PPCH, PAGCH, PDTCH or PACCH	PBCCH	PAGCH, PDTCH or PACCH	PPCH, PAGCH, PDTCH or PACCH	PBCCH	PAGCH, PDTCH or PACCH	PPCH, PAGCH, PDTCH or PACCH	PAGCH, PDTCH or PACCH	PPCH, PAGCH, PDTCH or PACCH	PPCH, PAGCH, PDTCH or PACCH
B0	B1	B2	B3	B4	B5	B6	B7	B8	B9	B10	B11

Figure 10.17 An example of downlink PBCCH/PCCCH mapping onto the 52-multiframe.

For the uplink case, the only signalling resource is the Physical Random Access Channel (PRACH). Radio blocks can carry PDTCH traffic in addition to this. By use of the parameter *BS_PRACH_BLKS*, it is possible to define those blocks that can only carry PRACH. The remaining blocks can carry PRACH or PDTCH, subject to demand and priority. Figure 10.18 shows an example, where *BS_PRACH_BLKS* = 5.

PRACH (fixed)	PRACH (fixed)		PRACH (fixed)			PRACH (fixed)			PRACH (fixed)		
B0	B1	B2	B3	B4	B5	B6	B7	B8	B9	B10	B11

Figure 10.18 An example of fixed PRACH mapping onto the 52-multiframe.

The priority for each type of message on a given block determines the instantaneous configuration, and this is typically scheduled in the PCU.

The downside to PCCCH deployment is the removal of one timeslot that can carry CS traffic, so this must be taken into account. However, the higher signalling capacity and packet specific idle mode cell reselection criteria have to be taken into account.

If the PBCCH/PCCCH physical timeslot is not deployed then the existing CCCH signalling channels are used. These are Random Access Channel (RACH), Access Grant Channel (AGCH) and Paging Channel (PCH). Typically the AGCH/PCH common resource is that which has the greatest capacity constraint, and therefore the additional load generated by EDGE and GPRS should be taken into account.

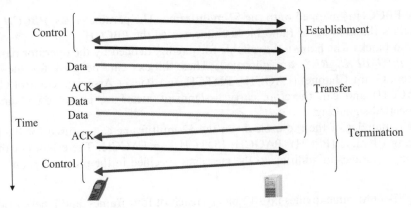

Figure 10.19 TCP message flow.

Signalling Traffic Estimation

In the initial deployments of GPRS it was found that packet data flow required rather heavy usage of signalling channels. This can be understood by consideration of Figure 10.19, which illustrates a typical message flow associated with a TCP session.

The figure represents the transfer of a small data volume (1.5 kB), and illustrates the level of handshaking performed. A total of ten blocks of data are transferred, with at least nine associated TBFs. The use of the CCCH for the establishment of each TBF therefore required extensive use of RACH and AGCH capacity. Consequently a number of change requests were submitted to ETSI with the aim of reducing the use of the CCCH and, instead, using the PACCH for TBF establishment where possible, based on the fact that the time between subsequent data flows was typically small (up to several hundred milliseconds). In addition to the reduction in required signalling capacity, the changes also led to significantly reduced data transfer times.

Initial observations based on specific TBF establishment and release procedures have shown a significant reduction in usage of the CCCH, by a factor of at least 75%, for a transfer such as the above. Because the actual capacity requirement is so critically dependent on radio resource algorithms, which are somewhat vendor-specific (subject to the constraints of the standards) and may change in the short term, it is not considered appropriate to give absolute values that may be anticipated. Rather, it is assumed that monitoring of signalling resource usage will be performed on a network and on capacity dimensioning, based on the results of this together with the data traffic growth trend.

The actual capacity in terms of the absolute number of signalling messages conveyed is unlikely to differ significantly between the use of the CCCH and PCCCH, and therefore the load seen on the CCCH can be used as the basis for the load to be expected on the PCCCH, if this is to be deployed at a later stage.

10.1.7 Mobility Management

Mobility management in GSM is controlled by C1 and C2 parameters, which are broadcast on the Broadcast Control Channel (BCCH). These parameters define cell reselection in GSM idle mode. There is no cell reselection in GSM dedicated mode, as serving cell selection is controlled by handover procedures.

ETSI defines [4] three Network Control (NC) order parameters, which determine measurement reporting and network control on the MS:

- NC0: MS controlled cell reselection, no measurement reporting;
- NC1: MS controlled cell reselection, MS sends measurement reports;
- NC2: network-controlled cell reselection, MS sends measurement reports.

In EDGE or GPRS the cell reselection method depends on the network control order. In NC0 and NC1 cell reselection is controlled by the MS, based on C1/C2 or C31/C32 [4]. Hence, cell reselection is done by the MS in both standby and ready states, even during data transfer. The serving cell of a class A MS in dedicated mode is an exception, and is determined by the network according to handover procedures.

Only NC2 allows for network-controlled cell reselection. In this case it is possible for the network to order MSs to send measurement reports to the network and to suspend their normal cell reselection. The MSs then apply the network made decisions for cell reselection in the ready state [5]. The relevant parameters are sent on the PBCCH but can be overridden for an individual MS by sending new ones on the PACCH.

10.1.7.1   C1/C2

The idea behind using C1 is that the MS compares the field strength levels of different cells defined in the idle mode BA (BCCH allocation) list and selects the strongest one using the C1 criteria according to Equation (10.3). The C2 parameter – see Equation (10.4) – can be utilised together with the C1 parameter to provide greater traffic management capability. The C2 parameter was designed for use in a layered architecture – namely, micro-cell and macro-cell or dual-band networks.

The MS calculates the value of C1 and C2 for the serving cell and will recalculate C1 and C2 values for the neighbouring cells every 5 s. The MS will then check whether the path loss criterion (C1) for the current serving cell falls below zero for a period of 5 s. This indicates that the path loss to the cell has become too high. The MS will also check whether the calculated value of C2 for a non-serving suitable cell exceeds the value of C2 for the serving cell for a period of 5 s.

If the new cell belongs to a different location area or, for an EDGE or GPRS capable MS, to a different routing area or always on for an EDGE or GPRS MS in ready state, the C2 value for the new cell must exceed the C2 value of the serving cell by at least the cell reselect hysteresis value.

C1 and C2 are defined by ETSI [4] using the following equations:

$$C1 = A - Max(B, 0) \tag{10.3}$$

where:

- $A =$ The received signal level (suitably averaged) $-p1$;
- $B = p2 -$ Maximum power of MS;
- $p1 = RXLEV\_ACCESS\_MIN$; and
- $p2 = MS\_TxPWR\_MAX\_CCH$;

$$C2 = C1 + CELL\_RESELECT\_OFFSET - TEMPORARY\_OFFSET$$
$$* H(PENALTY\_TIME - T) \quad \text{for } PENALTY\_TIME \neq 11111 \qquad (10.4)$$

or

$$C2 = C1 - CELL\_RESELECT\_OFFSET \quad \text{for } PENALTY\_TIME = 11111$$

where:
- $PENALTY\_TIME$ describes the time delay before the final comparison is made between two cells;
- $TEMPORARY\_OFFSET$ describes how much field strength could have been dropped during this penalty time; and
- $CELL\_RESELECT\_OFFSET$ describes an offset to cell reselection.

10.1.7.2 C31 and C32

With the introduction of a separate PBCCH, it is possible to use separate cell selection criteria for EDGE and GPRS – namely, C31 and C32 – which are broadcast on the PBCCH. C31 is the signal strength criterion that determines whether prioritised cell reselection will be used. C31 is defined by ETSI [4] with the following equations, where s refers to the serving cell and n to neighbour cells:

$$\left.\begin{array}{l} C31(s) = RLA\_P(s) - HCS\_THR(s) \qquad\qquad \text{for serving cell} \\ C31(n) = RLA\_P(n) - HCS\_THR(n) - TO(n) \cdot L(n) \quad \text{for neighbour cell} \end{array}\right\} \quad (10.5)$$

where:

- HCS_THR is the signal threshold for applying hierarchical cell structure (HCS) GPRS and Localised Service Area (LSA) reselection. It is broadcast on the PBCCH of the serving cell.
- RLA_P is received signal level average for each of the carriers in BA list for GPRS.
- $TO(n) = GPRS\_TEMPORARY\_OFFSET(n) * H(GPRS\_PENALTY\_TIME(n) - T(n))$;
- $L(n) = 0$ if $PRIORITY\_CLASS(n) = PRIORITY\_CLASS(s)$;
- $L(n) = 1$ if $PRIORITY\_CLASS(n) \neq PRIORITY\_CLASS(s)$;
- $H(x) = 0$ for $x < 0$; and
- $H(x) = 1$ for $x \geq 0$.

C31 is used to determine whether prioritised hierarchical (E)GPRS and LSA cell selection will apply. The MS will make a cell reselection if C1 for the serving cell falls below zero and there is a suitable neighbour cell that is better than the serving cell. In this case C1 is used in the same way as for GSM idle mode, except that A and B in Equation (10.3) use EDGE and GPRS-specific values for the received signal level and maximum transmit power.

C32 is a development from C2. It applies an individual offset and hysteresis value to each pair of cells, as well as the same temporary offsets as for C2. Additional hysteresis values apply for cell reselection that requires cell or routing area update. The best cell is selected using C32 from the cells that have the highest priority class and for which

$C31 \geq 0$. If no cell fulfils this criterion, then C32 is applied to all cells to select the best one. C32 is defined by ETSI [4] using the following formula:

$$C32(s) = C1(s) \text{ for serving cell} \qquad (10.6)$$

or

$$C32(n) = C1(n)$$
$$+ GPRS\_RESELECT\_OFFSET(n) - TO(n) * (1 - L(n)) \quad \text{for neighbour cell}$$

where the symbol meaning is the same as in the previous equation.

In the above, *GPRS_RESELECT_OFFSET* applies an offset and hysteresis value to each cell. *GPRS_TEMPORARY_OFFSET* applies a negative offset to C31/C32 for the duration of *GPRS_PENALTY_TIME* after the timer *T* has started for that cell.

GPRS_RESELECT_OFFSET, *PRIORITY_CLASS*, *GPRS_TEMPORARY_OFFSET* and *GPRS_PENALTY_TIME* are broadcast on the PBCCH of the serving cell. If the parameter *C32_QUAL* is set, positive *GPRS_RESELECT_OFFSET* values will only be applied to the neighbour cell with the highest *RLA_P* value of those cells with which C32 is compared.

10.1.7.3 Optimising C1/C2 and C31/C32

If EDGE and GPRS are implemented using C1 and C2, mobility management for GSM idle mode and both the EDGE or GPRS CS part of the network have to be optimised using the same parameters. This has an effect on existing GSM as well as EDGE and GPRS parameter settings and performance. In order to ensure a packet service without any gaps, EDGE and GPRS must be enabled on all cells, otherwise there will be areas where network cells have coverage but there is no packet service. This is due to the fact that the serving cell is selected according to C1/C2 and the decision is based on field strength and not on whether EDGE or GPRS has been enabled in the cell.

Features such as cell load based handovers (such as TRHO) and directed retry allow the operator to move Circuit Switched (CSW) traffic between different frequency bands and hierarchical layers as necessary to either improve network quality or reduce blocking. These features are not supported in EDGE or GPRS, because there are no handovers for packet services. Implementing EDGE or GPRS with C1/C2 may require changes – e.g., in the existing GSM handover strategy, in order to ensure that there will be capacity available for packet traffic. The C2 parameter can be configured to make a layer look artificially better and therefore have higher traffic absorption. It is possible to prefer, for example, the 1800 MHz layer, in which case most call setups will take place on that layer. Handovers or directed retry can then be used to direct CSW calls down, in case the 1800 MHz layer does not have any capacity available. As these mechanisms do not work for EDGE and GPRS, a full 1800 MHz band inevitably means blocking for packet traffic. Hence it is important to ensure that there is some capacity available for EDGE or GPRS on the layer pointed to by C1/C2 . This can be done by dedicating timeslots for EDGE or GPRS use or using handovers for offloading the 1800 MHz layer.

If EDGE and GPRS are implemented using C31 and C32, mobility management for GSM idle mode and EDGE and GPRS can be optimised independently. This allows a more flexible approach. C1/C2 can make different network layers look more attractive than C31/C32, making it easier to ensure that channels are available for EDGE and GPRS. This also reduces the need to use handovers for clearing capacity for packet traffic. This makes it possible to enable EDGE or GPRS only on some cells or a certain layer in case of dense networks. The 1800 MHz layer can be a good choice for EDGE or GPRS if there are more frequencies available and if a better C/I can be provided.

Both C1/C2 and C31/C32 offer a more limited capability for coping with high cell traffic loads than handovers. These criteria do not take into account the cell load. Hence it is important to adjust the existing handover strategy so that cell loads allow for EDGE and GPRS capacity. An EDGE or GPRS capable MS will make a cell reselection even if the new serving cell is full and cannot provide any service. During data transfer this causes lower user throughput. Hence good end-user perception of the network may require dedicating timeslots for EDGE and GPRS.

10.1.8   Frequency Hopping Techniques

Frequency hopping is a well-known technique to increase the spectrum efficiency of GSM networks. The techniques are also applied for packet data services.

10.1.8.1   Baseband Frequency Hopping

EDGE and GPRS have similar restrictions concerning frequency hopping as High Speed Circuit Switched Data (HSCSD). When baseband (BB) hopping is used the timeslots for multi-slot mobiles have to be allocated from the same hopping group. Hence the BCCH timeslot cannot be allocated to multi-slot usage. Due to this limitation the EDGE territory may exclude the initial timeslot (BL0) from cells that use BB hopping, depending on the vendor's implementation. This results in a maximum EDGE territory size of seven timeslots per TRX. However, the EDGE or GPRS territory can still be extended across several TRXs.

Deactivation of intra-cell handovers is commonly performed with baseband hopping. If the implementation of EDGE or GPRS territory is based on dynamic capacity, intra-cell handovers might be used to free the territory from CS traffic or allow for an expansion of the territory. The intra-cell handovers caused by EDGE or GPRS should be enabled despite deactivation of normal intra-cell handovers.

10.1.8.2   RF (Synthesised) Frequency Hopping

RF hopping differs from BB hopping in that the BCCH carrier is non-hopping. Therefore, the BCCH and non-BCCH TRXs will behave differently for EDGE and GPRS. Figure 10.20(a) shows the RF hopping performance for GPRS CS-1 and CS-2, with cyclic hopping over eight frequencies. Taking CS-2, it can be seen that there is a crossing point at around 10 dB C/I where throughput for hopping and

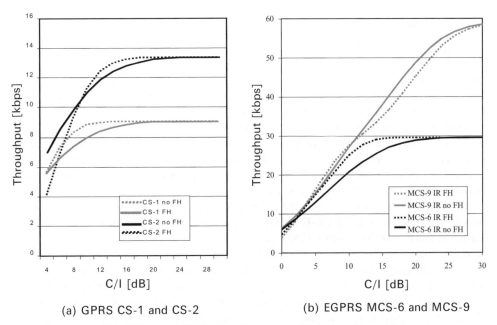

(a) GPRS CS-1 and CS-2   (b) EGPRS MCS-6 and MCS-9

Figure 10.20   GPRS and EGPRS performance with and without RF hopping.

non-hopping is identical. Below this level there is degradation in throughput, so, in effect, a hopping loss. For CS-1, hopping gain extends all the way down to 4 dB C/I. Hence, the hopping loss or gain is clearly network and cell dependent, as it is dictated by the C/I.

EDGE demonstrates a similar behaviour in simulations with frequency hopping over uncorrelated frequencies and with one interferer having the same hopping sequence as the carrier signal. Figure 10.20(b) shows the performance with IR in cases of ideal hopping and no hopping for MCS-6 and MCS-9. The possible hopping gain depends on the C/I range and is hence cell and network dependent, as for GPRS.

CS-1 and CS-2 perform differently with frequency hopping than without hopping. This results in different BLER thresholds for link adaptation for hopping and non-hopping cells. It is possible that implementations that provide link adaptation take account of the hopping state of the cell and use the non-hopping thresholds on the BCCH layer of an RF hopping cell and the hopping thresholds on non-BCCH TRXs.

If RF hopping is used, it is possible to prefer the BCCH carrier for EDGE and GPRS traffic. BCCH timeslots are broadcast on full power, even if there is no traffic on the BCCH carrier. If EDGE or GPRS have no downlink power control, preferring the BCCH carrier for EDGE and GPRS can result in better overall network performance. Also throughput might be better if BCCH band frequencies are cleaner than the hopping frequencies. On the other hand there will be no frequency hopping gain for EDGE or GPRS on the BCCH carrier, but as the frequency hopping gain for EDGE or GPRS depends on the C/I of the link this is not necessarily a disadvantage.

10.1.9   Conclusion

Both EDGE and GPRS coverage and performance (in terms of user throughput) depend heavily on the E_s/N_0 and C/I experienced by the UE. Packet data service coverage was shown to decrease very rapidly when throughput was increased, though with certain enhancements (as demonstrated) EGPRS throughput performance against both C/I and E_s/N_0 becomes very attractive.

As shown in the capacity calculation section, increasing traffic (number of EDGE and GPRS users) will play a very important role when planning the throughput for individual subscribers. Also how timeslots are allocated between CS and PS traffic inside the cell must be very well considered and optimised in order to get the best performance from the EDGE and GPRS network.

All the above-mentioned issues should be taken into account by the planner when he or she starts to plan the EDGE or GPRS network or to introduce the EDGE and GPRS feature into the existing GSM network.

While planning the network for services demanding high bit rates and and high spectral efficiency, one must carefully select the sites and antenna locations and direction in order to avoid cell overlap and therefore bad C/I performance. Even with the existing network, before launching any data services the network quality should be investigated and the average and maximum throughput for different services predicted. Also, all possible low throughput areas (i.e., bad quality areas) and high blocking areas (due to high traffic demand) should be investigated and fixed before launching data services, by utilising planning tools as well as information from existing networks. Possible ways of fixing those low quality areas could include but are not limited to:

- antenna downtilting;
- antenna direction change;
- antenna height lowering;
- site relocation.

After launching the data services, traffic and throughput demand changes should be followed constantly by monitoring some of the Key Performance Indicators (KPIs), like cell throughput, so that it is possible to react to the changing capacity demand quickly enough without any noticeable degradation of quality. As for the performance of EDGE or GPRS networks, the whole group of Performance Indicators (PIs) should be considered.

10.1.10   EDGE Performance Assessment

There is not a single simple indicator that is able on its own to assess EDGE or GPRS performance. To answer how the packet domain of the network is performing one needs to consider several performance items or areas. One should consider EGPRS territory behaviour, cell reselection and routing area update issues, radio performance, and end-to-end performance. The performance areas to consider are briefly discussed below. A more detailed description of EDGE/GPRS network performance can be found in [22].

10.1.10.1 EDGE Territory Behaviour

The EDGE or GPRS area of consideration includes a cluster of parameters – i.e., Packet Data Channel (PDCH) availability, PDCH blocking, voice pre-emption ratio and channel occupancy. These parameters address various aspects of the availability of radio network resources relative to the volume of PS and CS traffic.

PDCH availability shows the number of timeslots that can be upgraded for PS traffic. This needs to be seen however in relation to other parameters like territory size and the total number of GPRS or EDGE enabled timeslots. One should monitor this performance indicators on a recurring basis in order to secure a picture of the stability and evolution of PDCH availability and correlate these findings with PS territory downgrades, upgrades and their success ratios in order to understand the end-user impact of this on PDCH availability.

PDCH blocking is another view of availability. This is also interpreted as the territory upgrade reject ratio. PDCH blocking also contains the PCU data carrying capacity along with the interfaces.

Voice pre-emption is another metric to quantify the degree to which voice traffic may be degrading data services. As the volume of PS traffic grows, operators should monitor this PI at the cell level in order to identify 'hotspots' of CS/PS contention that may not be immediately visible from aggregate statistics at the BSC level.

Channel occupancy provides a picture of the available radio resources occupied by data users. The PI provides information about data usage in the cells. However, the result should not be taken in isolation.

Another metric to be examined in relation to channel occupancy is the maximum number of EDGE or GPRS timeslots used, which can provide an additional view of the real loading. Also, with the increase in packet traffic, hourly cell level statistics should be studied to secure real data load and channel occupancy.

Temporary Block Flow Usage

Temporary Block Flow (TBF) usage PIs are the maximum number of TBFs per used timeslot, successful transfer ratio and TBF establishment time. Maximum number of TBF per timeslot gives a view of TBF multiplexing on a slot basis and TBF establishment time gives an idea of system performance.

The successful transfer ratio KPI shows the percentage of TBFs which end correctly from all established TBFs and measures the quality of the radio interface for TBF sessions. Nevertheless this should not be confused as a measure of successful file transfer ratio or session success ratio by higher layers. A session can be completed by a number of TBFs. The value of the PI is in finding whether there are any RF-related issues with the site or not.

EDGE Abis Pool Analysis

With enhanced data rates per radio timeslot varying between 8.8 and 59.2 kbps, the traditional static Abis allocation does not use transmission resources efficiently. Thus, different kinds of dynamic Abis capacity allocation schemes have been introduced to optimise the loading by dividing Abis capacity into permanent one for CS and signalling traffic and by providing a shared pool for the data traffic.

However, a shared capacity need is very sensitive to the increase of EDGE traffic and introduction of multi-slot mobiles (like laptop cards) in the market. Therefore, it is essential to anticipate data traffic growth and do proper planning in order to prevent a congestion situation to ensure satisfactory end-user performance. It is crucial to dimension the Abis interface in such a way that the payload offered by the data radio timeslot can be accommodated.

Other important factors that need to be considered when defining EDGE dynamic shared capacity on the Abis interface are:

- EDGE terminals capabilities;
- data service spectrum;
- average MCS in use, determined by average radio conditions in the network;
- expected grade of service for each of the particular services.

10.1.10.2   Cell Reselection and Routing Area Update Performance

The radio, TCP/IP and UDP/IP outage times for cell reselection with and without RAU are worthy of consideration. The outages are impacting namely delay sensitive services with mobility. Local area and routing area update procedures are mostly done in a purely sequential order. It is possible to reduce outage time by implementing a combined location area + routing area update process in a network. However, this needs to implement the Gs interface (between SGSN and MSC) and make use of the PBCCH.

10.1.10.3   Radio Performance

This means the overall relationship between throughput per timeslot, BLER and MCS distribution in both the uplink and downlink direction.

Uplink BLER is normally higher than downlink BLER because of the power control in uplink, while full power is used in downlink. Uplink power control in PS is not as effective as it is for CS because uplink PS data traffic are typically characterised by bursty transmissions. Thus, uplink TBFs are predominantly bursty with a very short duration, driving the BLER to very high values in bad radio conditions.

MCS distribution
The MCS PI indicates the radio link condition. One should monitor this PI in conjunction with EDGE Dynamic Abis Pool occupancy results and the profile of transmissions.

RLC Throughput
The PI addressing RLC throughput per timeslot is expressed in kbps and indicates the timeslot capacity at the radio interface. It is important to understand that this PI does not necessarily provide end-user perception. Timeslot capacity might be very high (45 kbps) but the throughput experienced by a particular user could be very low due to the multiplexing effect. Throughput, however, is a good indicator of air interface limitations. End-user throughput per timeslot cannot be higher than this value.

10.1.10.4   End-to-end Analysis

End-to-end perception has a direct correlation with the end-user perception of that specific network. The picture is however influenced by the whole chain including client and server settings. A reliable figure can be obtained from several metrics. First of all FTP and UDP application throughput comparison for stationary and drive tests is useful. Another measure is real network latency from Ping tests, which gives a figure for the roundtrip time in the network.

Another important indicator is the UDP packet loss from a stationary and a drive test. UDP packet loss ratios for a drive test could be higher than the packet loss ratios for a stationary test. The behaviour is due to outage times during the cell reselections that take place in a drive test.

End-to-end delay analysis is of importance if delay-sensitive services are used in the network – like Push over Cellular – to achieve good end-user perception of such services.

An introduction to management level tools for end-to-end monitoring can be found in Section 10.1.11.

10.1.11   Quality of Service and Service Optimisation for EGPRS

In Chapter 8, priority treatment using 3GPP QoS definitions, the service optimisation concept of capacity and the QoS view were introduced. In this section we take a brief look at how a similar concept can be utilised in GSM/EGPRS networks. Since capacity and QoS management is based on tradeoffs, service optimisation consists of views for both. First we focus on the capacity view.

10.1.11.1   Capacity View in Service Optimizer for EGPRS

The capacity view provides information on the allocation of radio resources. The capacity-related configuration view at the cell/BTS level in Service Optimizer displays the TRXs (rectangles in the screenshot in Figure 10.21) in the given BTSs, visualising how capacity has been allocated currently on timeslot level. Capacity allocation is indicated by coloured timeslots (see *www.wiley.com/go/laiho* for colour images) so that the user can see the type of use allocated to any given timeslot. Service Optimizer visualises timeslot utilisation and indicates the timeslot usage for the BCCH, SDCCH, GPRS fixed territory and GPRS dynamic territory. For more details on these territories see Section 10.1.6.

PS territories are defined for each cell but settings are done on TRX level. GPRS dedicated territories consist of consecutive timeslots that cannot be used for the CS service. Depending on the traffic, the default GPRS territory can be used for both the PS and CS service, but CS traffic has priority. Associated configuration parameters that express the number of timeslots allocated for the GPRS, those for dedicated and default territories, as well as those for control channels are needed. An example of the capacity view is given in Figure 10.21.

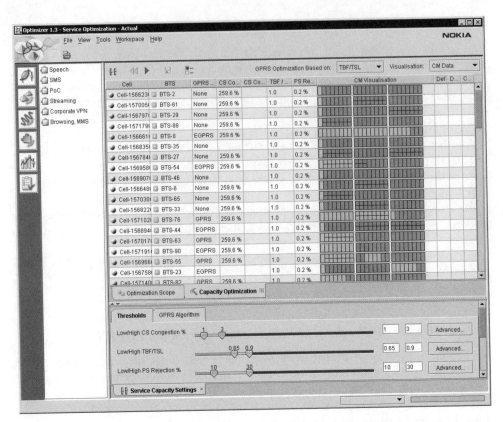

Figure 10.21 Capacity configuration data visualisation. Timeslots allocated for GPRS or circuit switched traffic can be visualised. Also fixed vs. dynamic territories can be seen.

Part of the service optimisation concept is the correct allocation of PS capacity, and GPRS dedicated territory optimisation based on performance data is introduced to the service optimisation concept. The target is to maximise the capacity of the network taking into consideration CS and PS data performance and capacity requirements. The main principle of the algorithm is that GPRS (PS) uses the excess capacity that remains from CS traffic. Optimisation logic is defined by parameters and rules. GPRS territory upgrade rejection and CS TCH congestion are the main drivers. If the PS traffic queuing and blocking rate and the territory update rejection rate are high, but the CS congestion indicators show low values, more timeslots most likely can be allocated for PS traffic. Additional measurements can be, for example, the amount of CS and PS traffic. PIs for optimisation are recommended to be measured during busy hour periods. More details on performance data can be found in Section 10.1.10.

Before making the decision to update the dedicated territory setting, it is important that neither CS nor PS performance should conflict with the set thresholds. Further, before reducing any timeslots from CS territory, estimation of the congestion of the new CS capacity and measured traffic is performed. If such an estimation predicts congestion that is less than the user defined threshold the number of timeslots for

GPRS traffic is increased. If congestion is estimated to be higher than allowed no upgrade is implemented and TRX addition is proposed.

10.1.11.2 QoS View in Service Optimizer for EGPRS

The QoS framework introduced in Chapter 8 and the implementation and support of traffic classes and other QoS attributes in radio access and packet core networks enable service differentiation in EDGE networks as well. Mechanisms to admit and schedule bearers which take priority treatment into account ensure adequate QoS for each application. How the QoS is implemented in EDGE radio access and the related packet core is however vendor dependent.

In the QoS view of service optimisation (Figure 10.22), the priority treatment concept as explained in Chapter 8 is utilised. Services are allocated to priority classes and the throughput of those classes can be measured and compared with the set target. Target throughputs are given separately for uplink and downlink traffic, and the definition needs to equal the requirement that the particular traffic type demands from the network at the end-user level. For example, if the Service Optimizer user needs to dedicate one priority class for Push over Cellular, then the requirements/targets he or she needs to define both for uplink and downlink traffic at the user level are ~8 kbps. Further, were there such a priority class in the network that the measured throughput is less than the target value, then the configuration parameter controlling the allocated

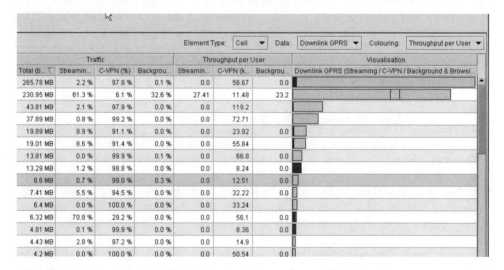

Figure 10.22 Conceptual view of quality of service and service performance monitoring. The length of the bar on the right represents cell total throughput. Colouring (see *www.wiley.com/go/laiho* for colour images) indicates whether the provided throughout per service is at target, too good relative to the target or unacceptable – i.e., below target. Measured throughput values can also be found in the table for each service. The network configuration for service differentiation has been achieved by using the differentiation capabilities provided by 3GPP QoS attributes.

bandwidth for that priority class can be tuned, provided that the other services achieve or exceed the target for QoS performance.

10.2   Time Division Duplex Mode of WCDMA (UTRA TDD)

The TDD radio access mode of UTRA and its potential usage is introduced in this part. TDD mode is partly a result of the original UMTS spectrum allocation, consisting of one paired and two unpaired bands. This led to an ETSI decision in 1998 that not just one but two of the proposed access technologies should be adopted for the UMTS standard. Hence, FDD should be used in the paired band and TDD in the unpaired band.

The two modes differ in a number of ways in the physical layer, but for compatibility and implementation reasons they are harmonised as far as possible, especially in higher layers. More details on the differences and distinctions can be found in [19]. Harmonisation enables the same services to be offered over both modes, while the differences lead to one mode being best utilised in certain system scenarios while the other mode may perform better in other scenarios. Furthermore, due to the current spectrum allocation, inter-mode interference needs to be considered when planning networks.

The net effect of TDD mode properties and spectrum allocation is that TDD is considered mainly as a 'capacity booster' for areas with high traffic density. The basic idea is therefore not to build standalone wide area TDD networks, but rather to let FDD and possibly GSM provide continuous wide area coverage while TDD would serve as a separate capacity-enhancing layer in the network. In particular, because of the possibility of varying uplink/downlink asymmetry, TDD is also better suited than FDD to offering services with very high bit rates – i.e., up to 2 Mbps – and in later phases possibly even higher.

10.2.1   Some Time Division Duplex Specific Properties

The following short list highlights three key characteristics of UTRA TDD:
- The multiple access scheme is a combination of time division and code division multiple access.
- Uplink and downlink traffic connections are separated via time division duplex, meaning that they coexist on the same carrier but are located in different timeslots.
- The physical layer procedure applies short spreading and scrambling codes.

From a consideration of the above three properties, several of the other differences between TDD and FDD follow. Next, some of these additional differences are pointed out, followed by a brief list of parameter values in FDD and TDD in Table 10.5. This section does not describe any of the physical layer details. These may be found by visiting the 3GPP Standard website or by consulting [15].

Because of the short channelisation and scrambling codes in TDD, it may be possible to construct multi-user detectors that are of reasonable implementation complexity. With an efficient multi-user detector, a large part of intra-cell interference is

Table 10.5 Some of the key parameters for UTRA frequency division duplex and time division duplex modes.

	UTRA FDD	UTRA TDD
Frame structure	15 slots/frame	15 slots/frame
Frame length	10 ms	10 ms
Chip rate	3.84 Mcps	3.84 Mcps
Uplink spreading factors	4 to 512	1 up to 16
Number of parallel uplink codes per user		1 or 2
Downlink spreading factors	4 to 512	1 or 16
Number of parallel downlink codes		1 up to 16
Modulation	QPSK	QPSK
Power control update rate	1500 Hz	Theoretically up to 800 Hz; in practice, only 100 Hz in DL, and 100 Hz or possibly 200 Hz in UL
Handover	Soft and hard	Hard only
Dynamic channel allocation	N/A	Slow and fast

cancelled. This indicates that the limiting factor in a TDD network may be inter-cell interference and not so much intra-cell interference. A drawback with the rather low spreading factors is that for high data rates it may not be possible to reuse all the timeslots in all the cells. This leads to the conclusion that the network must somehow control which slots and directions are used in which cells.

The function that handles the long-term allocation of slots and their directions in each cell is commonly referred to as 'slow dynamic channel allocation'. Within the limits set by this function, the so-called 'fast dynamic channel allocation' algorithm tries to find the best combination of slots for each user in each cell. The slow dynamic channel allocation algorithm determines which timeslots may be used for uplink, downlink or an optional direction in each cell. The fast dynamic channel allocation algorithm distributes users in an efficient way within the limits set by the slow-dynamic channel allocation algorithm.

Since the smallest resource unit that can be allocated to an MS is one channel code per timeslot, the power control procedure must work independently for each timeslot. In addition, since a terminal often transmits in only one timeslot in each timeframe, the updated power control parameters may be applied only once per frame (every 10 ms). With this in mind, it is easy to conclude that the TDD power control mechanism will not be able to follow the fast fading pattern as well as the FDD power control mechanism does.

The timeslot structure is also one of the reasons why TDD does not support soft handover.

10.2.2 System Scenarios

In TDD systems, network planning and Base Station (BS) deployment procedures are more complicated than in FDD because of TDD-specific interference scenarios.

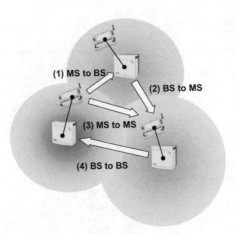

Figure 10.23   Possible TDD–TDD interference scenarios between cells.

Furthermore, introducing a TDD network in an area where FDD is already deployed (or vice versa) brings additional challenges and new network planning considerations. In general, interference scenarios, methods and recommendations are described in [21], but the most important cases are tackled in the following sub-sections.

10.2.2.1   TDD–TDD Interference Scenarios for Synchronised Cells with the Same Asymmetry

The main challenge in UTRA TDD is in controlling interference, since the same frequency band is used in both uplink and downlink, creating possible BS–BS, BS–MS and MS–MS interference scenarios. If all cells in a system are slot synchronised and the same asymmetry is applied – i.e., the same slots are used for uplink and downlink transmission, respectively, in the cells – then the possible interference scenario can be depicted in cases (1) and (2) of Figure 10.23. In this scenario all MSs in the system transmit and receive at the same time, hence they will not interfere with each other. The same is true for the BSs. However, the BS in one cell will interfere with MSs in surrounding cells and MSs in one cell will interfere with BSs in other cells. This is also valid for slot synchronised systems with the same asymmetry operating in the same area and using adjacent frequency bands – e.g., systems belonging to different operators.

10.2.2.2   TDD–TDD Interference Scenarios for Unsynchronised Cells or Cells with Different Asymmetry

When uplink and downlink are applied in the same timeslots in different cells, the interference scenario shown in cases (3) and (4) of Figure 10.23 is experienced. In this situation an MS in one cell is transmitting at the same time as an MS in another cell is receiving, and the same situation occurs for the second BS.

A combination of the two described scenarios occurs when slots are not synchronised. The severity of the 'asynchronism' depends heavily on the specific uplink/downlink slot allocation and the related switching points. The acceptable amount of out-of-band emission is specified by an emission mask and Adjacent Channel Leakage power Ratio (ACLR). On the receiver side, Adjacent Channel Selectivity (ACS) defines the impact of the adjacent channel power that results from imperfections in the receiver filters.

10.2.2.3 TDD–FDD Interference Scenarios

Besides the interference scenarios that occur within a single TDD system or between different TDD systems, there is also the risk of potential interference from a UTRA FDD system that is operating in the same area. The spectrum that is allocated for the uplink in FDD is adjacent to the spectrum allocated for TDD, as shown in Figure 10.24. Hence, especially those operators who operate systems using the TDD and FDD (uplink) frequency bands next to each other will experience interference. Nevertheless, interference between systems allocated to frequency bands two or even more channels apart may experience relatively high interference depending on the location of the BSs in relation to each other.

Figure 10.25 shows how the interference is directed between FDD and TDD systems. Since FDD is a paired system and the duplex spacing between uplink and downlink is large, as seen in Figure 10.24, there is no interference from an FDD BS (downlink) to the TDD system, and FDD MSs are not affected by TDD. However, FDD MSs (uplink) interfere with both TDD BSs and TDD MSs, and, similarly, FDD BSs will experience bursty interference from both TDD BSs and MSs.

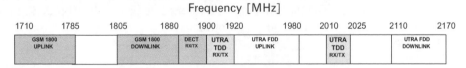

Figure 10.24 Current UMTS spectrum allocation.

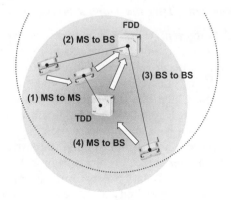

Figure 10.25 TDD–FDD interference scenarios.

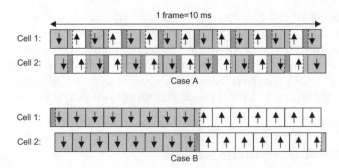

Figure 10.26 Downlink/uplink slot allocation with multiple (case A) and single (case B) switching points, respectively. The same uplink/downlink asymmetry factor (7/8) and the same level of asynchronism are used in both cases. The interfered fractions of each slot are marked.

10.2.3 Synchronisation of Cells

As has been noted already, synchronisation of cells – i.e., ensuring that frames and timeslots of different cells do not overlap but start and end at the same point in time – is important, especially for TDD uplink capacity. If the load in interfering cells is high and the cells within the system are not synchronised – causing the considered uplink and downlink slots to overlap – the uplink capacity in a slot may be significantly reduced, potentially to almost zero. Consider an uplink/downlink split of the slots, as shown in case A of Figure 10.26. Here, a synchronisation error causes overlap in downlink and uplink slots at every switching point, resulting in large capacity losses. However, if the number of switching points is reduced, as in case B of Figure 10.26, the result may not be as bad for the whole system, even though the slots bordering the switching points would still have a great problem.

10.2.4 Single-operator Time Division Duplex Networks

Synchronisation between adjacent cells in a network is very important to reduce possible interference scenarios, as noted above. Capacity tends to decrease in cases where synchronisation errors occur. However, an operator's whole system does not have to be synchronised, just the cells that are adjacent and cause interference between each other. Therefore, the BSs that interfere with each other can also listen to each other's signals, and synchronisation can be obtained over the air.

Synchronisation errors do not have a major impact on the performance of downlink, which changes the interference source from BSs only, via partly BSs and partly MSs, to MSs only. Instead, the main interference problems from lack of synchronisation occur in uplink. Obviously, however, if isolation between adjacent cells is high, such as in certain indoor environments, the possibility of being able to use opposing timeslots in adjacent, or unsynchronised, cells is much higher.

10.2.5 Synchronisation in Multi-operator Time Division Duplex Networks

The same interference scenarios that occur within a single operator's system may also occur between two or more operators' systems that are operating in adjacent channels within the same area. The possible interference between MSs and BSs operating on adjacent carriers is, however, suppressed by the transmitter–receiver requirements that are set in UTRA TDD radio specifications ([9] and [10]). The ACLR is defined as the ratio of transmitted power to power leaking to the adjacent channel. Hence, it is a measure of out-of-band emission. On the receiver side ACS defines the ability of the receiver to suppress adjacent channel power. This is the ratio of filter attenuation of the desired frequency band compared with the adjacent frequency band. The specified ACS and ACLR for the BS and the MS are summarised in Table 10.6.

The combined effect of the ACLR and ACS can be given as the Adjacent Channel Interference power Ratio (ACIR), which is defined as:

$$ACIR = -10 \cdot \log_{10}[10^{-ACLR/10} + 10^{-ACS/10}] \qquad (10.7)$$

The ACIR then determines the net effect of the interference power from an MS or BS operating in an adjacent channel, taking both transmitter and receiver imperfections into account.

Co-siting of different operators' BSs is feasible only if the BSs are synchronised and the same asymmetry is used in both systems. This is possibly an undesirable requirement for operators, though some level of coordination in network planning is needed if they do not synchronise their systems while covering the same area. Studies have shown that a coupling loss of about 70–80 dB is needed for satisfactory performance when BSs use the first adjacent channel. In an indoor environment this corresponds to about 5–15 m. For urban deployments, however, it is very difficult to state an absolute distance, since there are so many different ways of placing antennas and the huge impact from surrounding buildings. Detailed site surveys and measurements will therefore play an important role here. Other results [16] show that the relative distance between the BSs of two different operators should be at least half a cell radius to obtain a BS outage probability of less than 5%. Similar results are presented in [17], where only minor capacity reduction is experienced for a BS

Table 10.6 Adjacent channel leakage power ratio and adjacent channel selectivity requirements for the first and second adjacent carriers.

	ACLR		ACS
	±5 MHz	±10 MHz	
Mobile station	33 dB	43 dB	33 dB
Base station	45 dB	55 dB	45 dB[a]
BS close to another BS	70 dB	70 dB	

[a] This number is derived from the specifications [10, §7.4.1.1] as: $ACS =$ Interference signal − Wanted signal $- [PG_{user} - (E_b/N_0)_{user} -$ Margin], where PG is processing gain. Applying the suggested values to the formula results in $ACS = -52 - (-109 + 6) - [12 - 3.6 - 2] = 44.6$ dB.

separation of more than half a cell radius if the load of the interfering system is not too high. It is interesting to note that if BS separation is increased to one cell radius the highest relative capacity figures in uplink are obtained for the case where the timeslots in the systems are in opposing directions. The reason for this is that if operators have the same asymmetry, there is now a risk that an MS located at the cell edge will cause very high interference to the other operator's BS if the MS transmits when the BS receives.

If two operators do not want to synchronise their systems or use the same asymmetry, the BSs of the two systems should preferably be more than half a cell radius apart.

10.2.6 Erlang Capacity for Time Division Duplex Networks: A Simple Way of Estimating Capacity per Cell

In a TDD system, traffic channels are allocated to different slots, and each slot may be either code limited or interference limited. Since, in a general case, load is not uniform over all slots, theoretical capacity assessments are not very straightforward. Therefore, a simplified way of making such assessments will be presented here. It applies in principle only to CS services, but at the same time can at least give an indication of obtainable throughput in terms of bits per second.

A straightforward way of finding a good estimate of the number of available channels per slot in various environments is to use static system level simulations. It can then be established when it is realistic to assume code limitation and when interference limitation, including reasonable loading per slot, is the better choice.

After defining the number of available channels per service in a BS, the average capacity for each service may accordingly be estimated from a pre-defined allowed blocking probability and the number of slots allocated to the respective service. This is then accomplished simply by determining the offered traffic in a system using the Erlang-B formula in the same way as for traditional CS networks.

10.2.7 Co-existing Time Division Duplex and Frequency Division Duplex Networks

Both TDD and FDD can be used as coverage and capacity increasing systems in various environments. However, due to the adjacent channel interference instances shown in Figure 10.25, when both system modes are deployed in the same geographical area or, for indoor systems, in the same building, care needs to be taken in network planning. This statement is valid regardless of whether one or several operators are running the two modes.

It is the uplink band of the FDD system that is interfered with by the TDD system, and therefore it is FDD MSs that will interfere with both the BSs and UEs of the TDD system. Conversely, both the BSs and UEs of the TDD system will interfere only with BSs of the FDD system.[1] The preferred TDD–FDD deployment is that TDD and FDD

---

[1] The described interference instances relate to the present spectrum allocation. Future allocations may add other interference scenarios to the list.

BSs are located relatively close, but not too close, to each other. This is a tradeoff between the requirements of the two systems that is necessary in order to minimise the combined effect of interference from BSs and UEs respectively. The needed coupling loss between BSs deployed indoors and operating on the first adjacent channels here is again in the region of 70–80 dB.

Another consideration from the FDD network point of view is that an increase in interference in uplink for FDD MSs will force FDD MSs to increase their power. This will in turn cause more interference to FDD outdoor systems that may be using the same, or an adjacent, FDD channel. So, a complex interference environment within one building may in fact also eat capacity from outdoor systems.

In a building where an FDD system is already deployed, it could be possible for the operator to increase the capacity by deploying a TDD system in conjunction with the FDD system. If one operator is operating both systems, it would also be easier to avoid difficult interference scenarios since inter-system handover can be used. If networks are planned very carefully, the capacity of the combined systems may increase by 50–90% compared with the capacity where only a pico-FDD system is deployed.

Yet another scenario is when a building is covered not only by indoor systems, but also by outdoor systems – i.e., by a micro- or macro-BS. Assume, for example, that indoor coverage is provided both through a TDD pico-system (BSs located indoors) and by an FDD micro-/macro-system (BSs located outdoors). In this case, similar principles for locating the BSs as described above apply. For TDD system capacity it is advantageous if the FDD micro-/macro-BS is placed close to the building covered by the TDD pico-system, while for the FDD BS it is preferable if the considered building is located close to its cell edge. A tradeoff is therefore needed, and in [18] a separation between the pico-system and the macro-BS of approximately 300–500 m is suggested. Of course, this distance depends heavily on several parameters in the BSs as well as on the physical properties of the surrounding area and buildings.

One principle resembling the previous one, but which could make it substantially easier for an operator to control interference within his own network by means of good network planning, is to use FDD for outdoor coverage *only* and TDD for indoor coverage *only*. The combined network should then be planned so that the TDD system only covers the building and no outdoor area. Thereby, a very small amount of interference from the indoor system is experienced outdoors. Similarly, for the same reason, the FDD system should be planned so as to provide as little coverage as possible indoors. Inter-system handover is then used as soon as users move in or out of the building. This reduces the interference between the systems by, in essence, preventing any FDD connections from being maintained for users entering the building, and similarly by ending all TDD connections for exiting users.

If networks are deployed without taking these aspects into account, some serious QoS problems may arise. One particular problem worth mentioning, which becomes much more likely if network planning is poorly performed, is that a single UE connected to one mode could potentially block the uplink of a whole BS of the other mode if it is allowed to get too close to it. This event can occur in both directions – i.e., an FDD UE can block a TDD BS and vice versa – and is caused by lack of the

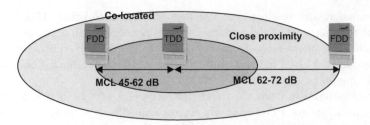

Figure 10.27   Co-location and close proximity placement of local area base stations.

'dying on downlink'[2] phenomenon that has been built into the FDD mode itself but is not functioning when both modes are involved. The current 3GPP standard [14] defines the Minimum Coupling Loss (MCL) between UE and TDD BS as 70 dB for the macro type of BS, 53 dB for the micro type of BS and 45 dB for the pico type of BS.

Lastly, interference from FDD UEs to TDD UEs cannot be controlled totally by means of network planning. However, the probability of a TDD UE losing coverage due to FDD UE interference can be kept at a predetermined level by carefully selecting nominal cell sizes, etc., but it can never be eliminated. Therefore, there is always a need to be able to perform intelligent inter-system handover as an escape mechanism for the victim UE.

10.2.8   Co-located and Close Proximity Local Area Base Station System (BSS)

To make a correct network plan, the planner is required to take into account possible inter-system interference as well as own inter- and intra-cell interference. There can be both co-located TDD systems and co-located TDD–FDD systems covering the same indoor area, and work on determining the required ACLR and ACS values for local area BSs was ongoing in spring 2005 in 3GPP ([12]–[14]).

In this text, BSs are classified as being co-located or operating in close proximity using different values for MCL. The MCL values which were suggested in [11] are given in Figure 10.27. Currently, MCL values are defined by 3GPP in [14] for the co-located case of two macro-BSs in TDD mode as 30 dB, and for co-location of two micro-BSs the MCL is 45 dB. In case of operation of BS TDD and BS FDD mode in the same geographical area the minimum MCL requirement between BSs is around 70 dB. Although there is a whole bunch of interference coupling scenarios defined in [14], some are not fully defined with MCL values yet.

10.2.9   Radio Performance

TDD radio performance is characterised by synchronisation needs, asymmetry and low cell breathing. Pole capacity and load factor methodology are not of particular interest

---

[2] Essentially meaning that a UE coming very close to a BS should be blocked by the BS long before it causes any harm in the other direction. Thereby, one link is dropped instead of all the links in a whole BS.

in the TDD planning procedure, since power-based load control is not needed to ensure system stability and coverage, etc. Code limitation in each slot will strike before coverage is badly affected. Also, pole capacity can be used only on a slot-by-slot basis in TDD, so its precision is poor, due, for example, to low spreading factors and few users per slot.

10.2.10   Time Division Duplex and Frequency Division Duplex Processing Gains

Processing Gain (*PG*) is calculated differently for FDD and TDD, since for TDD we need to take the slotted structure and the midamble into account. In FDD, calculations are straightforward according to the Erlang B Formula (10.1). In TDD, however, the frame is divided into 15 slots and the information is transmitted in one or several slots using one or more codes. If *PG* is calculated for data rates that can be supported by transmission in one or more slots, the following expression can be used:

$$PG = \frac{W}{R} \cdot \frac{k}{15} \cdot \frac{\text{Chips in slot} - \text{Midamble} - \text{Guard period}}{\text{Chips in slot}} \tag{10.8}$$

where k is the number of slots used for the considered service (here $k = 1$); W is the spreading bandwidth (3.84 MHz); and R is the bit rate which is given by:

$$R = \frac{\log_2(M) \cdot R_c \cdot (T_s \cdot W - GP)}{SF \cdot T_f} \tag{10.9}$$

where M is the size of symbol set (QPSK type – i.e., 4); T_s is the slot duration (666.7 μs); T_f is the frame duration (10 ms); R_c is the code rate (1/3); GP is the length of the guard period (96 chips); and SF is the spreading factor (from 1 to 16 in powers of 2). Assuming that burst type 1 is used for 12.2 kbps speech and burst type 2 for 128 kbps data, then the parameters are as follows: chips in slot = 2560, midamble (burst type 1/2) = 512/256, guard period = 96. Examples of *PG* for some services are given in Table 10.7.

10.2.11   Time Division Duplex Radio Link Budget Examples

The E_b/N_0 values considered in the radio link budgets for TDD are given in Table 10.7. They are obtained through link-level simulations assuming a so-called indoor A channel, which is a chip-spaced two-tap channel where the second tap has 9.6 dB lower average power than the first tap. In all cases, transmit power control and multi-user detection were employed.

In the example radio link budget given in Table 10.8, interference margins have been set to 8 dB for all services even though that figure does not correspond to the same load factor in all cases. However, as mentioned earlier, the load factor and its control are less important in TDD than in FDD due to the timeslot property – i.e., maintaining system stability is less of an issue. Therefore, the interference margin value above is used only for establishing a nominal cell range, which can be used in network planning. Within this margin, interference from other cells as well as interference from the own cell remaining after multi-user detection will be contained.

Table 10.7 Example downlink and uplink E_b/N_0 values for some time division duplex services (RxD = receive diversity). Values for both real time and non-real time high data rate services are given.

TDD	Downlink		Uplink – RxD		Uplink – no RxD	
Service [kbps]	12.2	128	12.2	128	12.2	128
E_b/N_0 [dB]	9.4	11.5 (RT) 6.7 (NRT)	1.7	1.0 (RT) 0.3 (NRT)	8.6	8.7 (RT) 6.4 (NRT)
Fast fading margin [dB]	5.5	3.5 (RT) 3.1 (NRT)	6.3	6.3 (RT) 3.4 (NRT)	6.3	6.3 (RT) 3.4 (NRT)
Processing gain [dB]	12	2.4	12	2.4	12	2.4

Table 10.8 Example time division duplex radio link budget for uplink (RxD = receive diversity). A pico-base station with internal antennas is assumed, thereby no cabling losses are included.

		Voice		NRT data	
		12.2 kbps RxD	12.2 kbps No RxD	128 kbps RxD	128 kbps No RxD
Transmitter (mobile station)					
Max. transmitter power	[dBm]	21	21	24	24
Mobile station antenna gain	[dBi]	2	2	2	2
Body loss	[dB]	3	3	0	0
Equivalent isotropic radiated power	[dBm]	20	20	26	26
Receiver (base station)					
Number of used slots in TDD		1	1	1	1
Thermal noise density	[dBm/Hz]	−174.0	−174.0	−174.0	−174.0
Base station receiver noise figure	[dB]	5	5	5	5
Desensitisation		0	0	0	0
Receiver noise density	[dBm/Hz]	−169.0	−169.0	−169.0	−169.0
Receiver noise power	[dBm]	−103.2	−103.2	−103.2	−103.2
Interference margin	[dB]	8	8	8	8
Receiver interference power	[dBm]	−95.9	−95.9	−95.9	−95.9
Total effective noise + interference	[dBm]	−95.2	−95.2	−95.2	−95.2
Processing gain	[dB]	12	12	2.4	2.4
Required E_b/N_0	[dB]	1.7	8.6	0.3	6.4
Receiver sensitivity	[dBm]	−105.5	−98.6	−97.3	−91.2
BS antenna gain	[dBi]	4	4	4	4
Cable loss in the base station	[dB]	0	0	0	0
Fast fading margin (= TPC headroom)	[dB]	6.3	6.3	3.4	3.4
Max. path loss	[dB]	123.2	116.3	123.9	117.8

Table 10.9   Example time division duplex radio link budget for downlink. A pico-base station with internal antennas is assumed, thereby no cabling losses are included. Transmit diversity is assumed not to be employed.

		Voice [12.2 kbps]	NRT data [128 kbps]
Transmitter (base station)			
Max. transmitter power	[dBm]	24	24
Base station antenna gain	[dBi]	4	4
Cable loss in base station	[dB]	0	0
Equivalent isotropic radiated power	[dBm]	28	28
Receiver (mobile station)			
Number of used slots in TDD		1	1
Thermal noise density	[dBm/Hz]	−174.0	−174.0
Mobile station receiver noise figure	[dB]	9	9
Receiver noise density	[dBm/Hz]	−165.0	−165.0
Receiver noise power	[dBm]	−99.1	−99.1
Interference margin	[dB]	8	8
Receiver interference power	[dBm]	−91.9	−91.9
Total effective noise + interference	[dBm]	−91.1	−91.1
Processing gain	[dB]	12	2.4
Required E_b/N_0	[dB]	9.4	6.7
Receiver sensitivity	[dBm]	−93.7	−86.8
Mobile station antenna gain	[dBi]	2	2
Body loss	[dB]	3	0
Fast fading margin (TPC headroom)	[dB]	5.5	3.1
Max. path loss	[dB]	115.2	113.7

Table 10.9 gives a corresponding link budget for downlink. For the choice of downlink interference margin values, the same approach as for uplink applies – i.e., that the value presented herein mainly sets the nominal cell range and contains both inter-cell interference and intra-cell interference not cancelled by the orthogonal codes used in downlink.

As can be seen, the required E_b/N_0 values are substantially higher for downlink than for uplink. This stems mainly from the fact that no diversity scheme is included in the downlink calculation. If that had been the case, link performance would improve and balancing of the two links would be easier. However, since transmit diversity is not as efficient as receive diversity, there will still be a higher requirement on downlink E_b/N_0 values. Due to the wish to balance the links and to get good capacity and coverage performance, it is anticipated that transmit diversity will be implemented in commercial BS products, at least in those aimed at outdoor deployment.

Note that the link imbalance will be even greater if one user is not allowed to use all available output power in the BS, a restriction that is quite common.

By using a simple propagation model without slow fading considerations, the resulting maximum propagation loss values in these example link budgets translate to a cell range of about 700 m in uplink (with receive diversity) and about 400 m in downlink.

10.2.12 Some Other Important Parameters and their Effect on the Radio Link Budget

10.2.12.1 Base Station Noise Figure

Desensitisation of the BS receiver in pico-cells has been proposed as one way of counteracting the possibly severe interference problems that may arise from time to time in TDD (and partly also in FDD). The idea, roughly described as 'increasing the BS receiver noise figure', is to make the receiver less sensitive to interfering signals coming from TDD and FDD MSs and other TDD BSs located close to the BS. However, this approach does not completely solve the problems and it also has the negative effect of increasing the overall interference level in the own system since, then, the wanted MSs must increase their respective output powers to compensate for the poorer uplink receiver performance.

10.2.12.2 Fast Fading Margin

This parameter is similar to transmit power control headroom in FDD. It ensures that the Transmit Power Control (TPC) scheme has enough room to vary the power to compensate for fading effects. If TPC is not used, this parameter will be set to zero.

10.2.13 Summary

10.2.13.1 Synchronisation and Interference

- Intra-operator cell synchronisation is necessary within each cluster of cells. In general, it is uplink that is most sensitive to interference.
- Inter-operator synchronisation is beneficial but not crucial. Without it, more attention has to be paid to how the other operator's network is deployed (BS locations, etc.).
- Some coordination of TDD and FDD network planning is needed to avoid the worst instances of inter-system interference. Still, not all interference may be avoided by planning, and where necessary other actions, such as intelligent handovers (escape mechanisms), are needed.

10.2.13.2 Coverage

- Absolute values of maximum allowed propagation loss are lower than for FDD due to the slotted property of TDD. However, another effect of this property is that coverage is also much more stable when loading is increased – i.e., 'cell breathing' is much lower.
- Without downlink transmit diversity, there is quite a large imbalance between uplink and downlink in terms of maximum allowed propagation loss.

10.2.13.3 Capacity

- Low rate services seem to be close to code limited in many cases, while high rate services are most often interference limited.

- A timeslot reuse factor of 1 may be possible for speech services, but probably not for data services with a speed of 56 kbps or more. However, the reuse factor may not have to be decided on a permanent basis, but could potentially result from an 'average utilisation effect' from channel allocation decisions made by a dynamic channel allocation scheme.

- Delay requirements impact heavily upon the throughput obtainable. Non-delay-sensitive services can reach a much higher throughput than delay-sensitive services.

- TDD capacity will be degraded by an FDD system deployed in the same area and operating on adjacent carriers, especially if the FDD system is highly loaded. Some coordination of TDD and FDD network planning is highly recommended. Also, some degree of coordination regarding parameter setting – e.g., for handover and load control purposes – is beneficial.

- The spectral efficiencies of TDD and FDD do not differ significantly from each other.

References

[1] Sipola, J., *Link Adaptation in EGPRS Radio Interface*, University of Oulu, Department of Electrical Engineering, 2001.

[2] Nikkarinen, S., *EGPRS Network Level Simulation Results Phase 1: Part B*, Nokia Research Centre, 2001.

[3] Meyer, M., TCP performance over GPRS. *IEEE Wireless Communications and Networking Conf. 1999*, Vol. 3.

[4] GSM 05.08, Digital Cellular Telecommunications System, Radio Subsystem Link Control.

[5] GSM 04.60, Digital Cellular Telecommunications System, Radio Link Control/Medium Access Control (RLC/MAC) Protocol (GPRS).

[6] GSM 03.64, Digital Cellular Telecommunications System, General Packet Radio Service (GPRS); Overall description of the GPRS radio interface, Stage 2.

[7] GSM 05.05, Digital Cellular Telecommunications System, Radio Transmission and Reception.

[8] GSM 03.60, Digital Cellular Telecommunications System, GPRS, Stage 2.

[9] 3GPP, TS 25.102, v5.6.0, UTRA (UE) TDD; Radio Transmission and Reception.

[10] 3GPP, TS 25.105, v5.6.0, UTRA (BS) TDD; Radio Transmission and Reception.

[11] Siemens, ACLR, ACS and Spectrum Emission Mask for Local Area BS in TDD mode. *TSGW4#16(01) 0270, Vienna, Austria, 19–23 February 2001.*

[12] Siemens, Receiver Blocking and Intermodulation Characteristics for Local Area BS in TDD Mode. *TSGW4#16(01) 0270, Vienna, Austria, 19–23 February 2001.*

[13] 3GPP, TR 101.112, v3.1.0, Universal Mobile Telecommunications System (UMTS); Selection Procedures for the Choice of Radio Transmission Technologies of the UMTS (1997-11), UMTS30.03.

[14] 3GPP, TSG RAN, 3G TS 25.952, v5.2.0, TDD Base Station Classification.

[15] Holma, H. and Toskala, A., *WCDMA for UMTS*, 3rd edn, John Wiley & Sons, 2004.

[16] Holma, H., Povey, G.J.R. and Toskala, A., Evaluation of interference between uplink and downlink in UTRA/TDD. *Proc. VTC'99, Fall, Amsterdam, The Netherlands, September 1999*, pp. 2616–2620.

[17] Holma, H., Heikkinen, S., Lehtinen, O.A. and Toskala, A., Interference considerations for the time division duplex mode of the UMTS terrestrial radio access. *IEEE Journal on*

Selected Areas in Communications, Special issue on Wideband CDMA, **18**(8), pp. 1386–1393, August 2000.

[18] Haas, H., McLaughlin, S. and Povey, G.J.R., The effect of inter-system interference in UMTS at 1920 MHz. *Proc. 3G, 2000*.

[19] Haardt, M., Klein, A., Koehn, R., Oesterreich, S., Purat, M., Sommer, V. and Ulrich, T., The TD-CDMA based UTRA TDD mode. *IEEE Journal on Selected Areas of Communication*, **18**(8), pp. 1375–1385, August 2000.

[20] GSM 05.02, Digital Cellular Telecommunications System, Multiplexing and Multiple Access on the Radio Path.

[21] 3GPP, TSG RAN, 3G TS 25.942, v5.3.0, Radio Frequency (RF) System Scenarios.

[22] Halonen, T., Romero, J. and Melero, J., *GSM, GPRS and EDGE Performance Evolution towards 3G/UMTS* (2nd edn), John Wiley & Sons, 2003.

Index
